RIVERSIDE COMMUNITY COLLEGE
1916

People and Environment: Behavioural Approaches in Human Geography

2nd edition

D.J. Walmsley and G.J. Lewis

Copublished in the United States with
John Wiley & Sons, Inc., New York

l Rachel
Natasha

Longman Scientific & Technical,
Longman Group UK Limited,
Longman House, Burnt Mill, Harlow,
Essex, CM20 2JE, England
and Associated Companies throughout the world.

Copublished in the United States with
John Wiley & Sons, Inc., 605 Third Avenue, New York, NY 10158

First published 1984
Second Edition 1993

ISBN 0–582–07866–0

British Library Cataloguing in Publication Data
A CIP record for this book is available from the British Library

Library of Congress Cataloging-in-Publication Data
Walmsley, D.J.
People and environment : behavioural approaches in human geography /
D.J. Walmsley and G.J. Lewis. — 2nd ed.
p. cm.
Rev. ed. of: Human geography. 1984.
Includes bibliographical references and index.
ISBN 0–470–22121–6 (USA only)
1. Human geography. 2. Human behavior. I. Lewis, G.J.
II. Walmsley, D.J. Human geography. III. Title.
GF41.W355 1993
304.2—dc20 93–985
CIP

Set by 8 in 9½/12pt Ehrhardt
Printed in Malaysia by PMS

Contents

List of Figures

List of Tables

Preface

This book comprises a second edition of *Human Geography: behavioural approaches*, first published in 1984. The first edition attempted to synthesize the massive volume of geographical literature to have appeared mainly since 1960 concerned with both how people come to know the environment in which they live and with the way in which such knowledge influences subsequent 'spatial behaviour' (a jargon expression denoting where people go and what they do). Our feeling at the time was that earlier attempts to review this research had been overly concerned with cognition and with images. They had therefore told us a great deal about how people build up 'mental maps' but had at the same time overlooked a number of important issues such as how people derive information from their environment, how they use public and private information channels, how they evaluate environmental information in making decisions, how preferred behaviour may be suppressed as a result of constraints of one sort or another, how environmental meaning can be socially constructed, and how people develop a sense of attachment and belonging to place. *Human Geography: behavioural approaches* sought to rectify this situation by addressing these issues and by taking a broad perspective that emphasized the variety and range of behavioural approaches adopted in the study of people–environment interaction. Particular attention was paid to the philosophical basis for behavioural research in human geography. This was because we believe that human geography in the decades before our book appeared had tended to be preoccupied with the methodological 'How?' at the expense of the philosophical 'Why?'.

In other words, in our view, insufficient attention had been given to the philosophical foundations of behavioural research. This is not to say that the first edition of the book propounded a single philosophical orientation or advocated a particular research paradigm. Rather an attempt was made throughout the first edition to illustrate the variety of approaches that had been adopted. Thus, although we have a leaning towards the transactional-constructivist position that highlights the way in which spatial behaviour is mediated by environmental knowledge, attention was also paid to other philosophical approaches, notably humanistic perspectives. Likewise, the book stressed that behavioural approaches should be seen as complementing rather than supplanting other approaches in human geography: for example, the study of everyday spatial behaviour within cities cannot ignore the findings of so-called 'structural' approaches (e.g. social area analysis), just as studies of travel behaviour can never overlook aggregate-level spatial interaction models and their attempts to isolate the general structure of behaviour irrespective of the identity of the individuals involved.

The same sentiments lie behind the second edition as lay behind the first. We believe that structural approaches fail to explain why people–environment interaction takes the form that it does, no matter whether those structural approaches emphasize socio-spatial structures, the structure of aggregate behaviour, or structural Marxism. These approaches therefore need to be complemented by the study of individual decision-making units. However, it is important that such micro-scale study adopt an authentic view of human behaviour that highlights distinctly human characteristics and activities, thereby dispensing with both the notion of the mind as a 'black box' and the idea that individuals are powerless in the face of reified concepts like class or culture. At the same time, of course, it is important to guard against a perspective that treats individuals as 'sovereign' decision-makers. Almost all people are constrained in their spatial behaviour and it is therefore important to appreciate how these constraints are generated and overcome.

As with the first edition, the rationale for, advantages of, and shortcomings with behavioural approaches are explored at length in both substantive chapters and in a number of detailed examinations of particular aspects of life in advanced Western society. The format of the book has, however, been changed for the second edition in order to take better stock of (1) the behavioural literature that has been published since the early 1980s, and (2) shifts of emphasis within the overall domain of geographical inquiry. For example, in the second edition humanistic work on experiential environments has been incorporated with so-called 'mainstream' work, thereby reflecting the fact that the importance of a humanistic perspective is now much more widely appreciated than it was some years ago. Similarly, the title of the book has been changed to emphasize that what is under consideration is the way in which behavioural approaches can complement other approaches in the overall study of people–environment interaction.

The two most obvious changes in format relate to our treatment of the overall approaches adopted by human geographers in the study of people–environment interaction, and to the replacement of 'case studies' by what we are choosing to label 'fields of study'. In terms of overall approaches to the study of people–environment interaction, we have done away with the distinction between micro-scale and macro-scale and instead focused our discussion on the distinction between 'structural' and 'behavioural' approaches, each of which is dealt with in two chapters in Part 2 of the book. This change has enabled us to look particularly at the interplay between patterns and structures on the one hand and processes and behaviour on the other hand. In this way we hope we have been able to demonstrate effectively the importance of choice and constraint in human use of the environment, notably the fact that many constraints are socially generated by the groups and cultures to which individual actors belong.

The 'case studies' in the first edition were very much literature-led. That is to say, we chose to focus on the issues that seemed to be exciting the greatest volume of geographical writing. In the second edition, we have abandoned this approach and focused instead on what we consider to be fundamentally important aspects of human existence in advanced Western society. Our aim has been to demonstrate the salience of each issue in contemporary life, to highlight the balance between choice and constraint and their respective influence on behaviour, and to point out gaps in knowledge and possibilities for further research.

Any book combining the work of authors separated by 12 000 miles is bound to have a protracted development. The idea for the first edition arose when we were both at the University of New England in 1979. However, it was 1981 before work on the project actually commenced. Since that time, we have been fortunate (to use the jargon of what is now regarded as 'time geography') that our respective 'life-paths' through 'time–space' have intersected on several occasions in both Britain and Australia. We are obviously grateful to the Universities of New England and Leicester for making this possible. On top of that we are very grateful to various individuals within those universities for helping enormously with the production of the second edition. Megan Wheeler, Bev Waters, and Judi Winwood-Smith have helped with typing, Terry Cooke has overcome the problems that inevitably arise even in 'user-friendly' computer systems, Wade Edmundson helped with the scanning of certain parts of the first edition, and Kate Moore and Ruth Pollington prepared the figures with great skill. Above all, though, we would like to thank our families, as in the first edition, for their continued love and support.

D.J. Walmsley
G.J. Lewis
Armidale and Leicester
July 1992

Acknowledgements

We are grateful to the following for permission to reproduce copyright material:

Academic Press and the author, Prof. J.D. Williams for Fig. 8.7 (Williams & McMillen, 1980); The American Geographical Society for Figs 2.2 (Morrill, 1965c) & 8.1 (Zelinsky, 1971); American Sociological Association for Fig. 2.1 (Isbell, 1944); Edward Arnold Publishers for Figs 1.3 (Downs, 1970a), 2.12 (Herbert, 1977), 7.2 (Sharpe, 1988), 8.2 (Cadwallader, 1989a) & Tables 7.1 (Stafford, 1972) & 7.2 (Britton, 1990); Edward Arnold Publishers for Fig. 4.6 by Golledge, R.G. from Learning about urban environments. In: Carlstein, T., Parker, D., Thrift, N.J. (eds) *Timing Space and Spacing Time*, Vol 1 *Making Sense of Time* (1978) & Table 2.4 by Cater, J., Jones, T. from *Social Geography: an introduction to contemporary issues* (1989); Association of American Geographers for Figs 5.8 (Desbarats, 1983), 5.10 (Pred, 1984), 11.2 (Rowe & Wolch, 1990), 11.3 (Ley, 1974) & 11.10 (Aitken, 1990); Butterworth Heinemann for Fig. 10.4 (Boniface & Cooper, 1987); Butterworth Heinemann and the author, Prof. M. Pacione for Fig. 2.9 (Pacione, 1990b); Carfax Publishing Company for Figs 3.1 (Harvey, 1974), 5.4 (Gould & White, 1968) & 7.3 (Taylor & Thrift, 1983b); Croom Helm Ltd. for Figs 1.2, 8.5 & part Table 1.1 (Golledge & Stimson, 1987); *Economic Geography* for Figs 5.5 (Horton & Reynolds, 1971) & 10.6 (Aldskǒgius, 1977); the author, Dr. H.W. Faulkner for Table 11.1 (Faulkner, 1978); Free Press for Fig. 4.10 (Wilson & Alexis, 1962); David Fulton Publishers for Figs 2.8 (Herbert & Thomas, 1982) & 7.1 (Thrift, 1979); the editor, *Geografiska Annaler Series B* for Fig. 9.3 (Potter, 1979); The Geographical Association for Fig. 4.4 (Kirk, 1963); Geographical Society of Ireland for Fig. 11.5 (Boal, 1969); Gower Publishing Co. Ltd. for Fig. 7.5 by Atkinson from The Changing Corporation In: Clutterbuck, D. (ed.) *New Patterns of Work* (1985); the author, Prof. D.T. Herbert for Fig. 11.9 (Herbert & Raine, 1976); Hutchinson Educational for Fig. 7.8 (Pred, 1977a) & Table 10.1 (Cosgrove & Jackson, 1972); Institute of Australian Geographers for Fig. 5.6 (Walmsley, 1979); The institute of British Geographers for Figs 2.11 (Giggs, 1973) and 3.2 (Gregory, 1981); Longman

Group UK Ltd. for Figs. 7.6 (Goddard, 1978), 11.7 (Herbert, 1975), 11.11 (Herbert, 1982) & 11.12 (Walmsley, 1988a); Lund Studies in Geography for Figs 4.9 (Pred, 1967) & 7.7 (Törnqvist, 1973); The MIT Press for Fig. 5.3 (from Figure 23.2, p437, by Appleyard In *Planning Urban Growth and Regional Development*, Rodwin (ed.) 1969) © 1969 by the Massachusetts Institute of Technology; Ohio State University Press for Fig. 8.3 by Mabogunje from Systems approach to a theory of rural-urban migration In: Geographical Analysis **2**, No.1 (January 1970) Copyright 1970 by the Ohio State University Press. All rights reserved; The Open University for Fig. 11.4 (Blowers, 1973); Pergamon Press Ltd. for Fig 4.7 (Walmsley & Jenkins, 1992a) Copyright 1992 Pergamon Press Ltd.; Population Association of America and the author, Prof. A. Speare Jr. for Fig. 8.6 (Speare, 1974); the author, Dr. J.D. Porteous for Fig. 11.1 (Porteous, 1977); Regional Science Association International for Fig. 9.2 (Huff, 1960); Routledge for Fig. 10.9 (Rapoport & Rapoport, 1975); Royal Dutch Geographical Society KNAG for Figs 2.5 (Lewis & Davies, 1974), 2.7 (Musterd, 1991) & 5.1 (Pocock, 1973); Sage Publications for Fig. 1.1 (Stea & Downs 1970); Tavistock Institute for Fig. 11.8 (Lee, 1968); The University of Chicago Press for Figs 2.3 (Park *et al.*, 1925) & 2.10 (Faris & Dunham, 1939) Copyright 1925 & 1939 The University of Chicago Press, Chicago 60637; Unwin Hyman Ltd. for Fig. 3.3 (Cooke, 1989a); Van Nostrand Reinhold for Fig. 4.2 (Moore, 1976) © 1976 by Dowden, Hutchinson and Ross, Inc.; John Wiley & Sons Ltd. for Fig. 7.4 (Rees, 1974) Copyright © 1974 John Wiley & Sons Ltd.

Whilst every effort has been made to trace the owners of copyright material, in a few cases this has proved impossible and we take this opportunity to offer our apologies to any copyright holders whose rights we may have unwittingly infringed.

Part 1 Introduction

Behavioural research in human geography really began in the 1960s. Its origins can be found in the dissatisfaction that was widely felt with the normative and mechanistic models of people–environment interaction that had existed before that time, based on such unreal notions as that of rational economic behaviour.

Basically behavioural approaches are non-normative and focus on the information-processing and acted-out behaviour of decision-making units (e.g. individual people, households, firms), special attention being paid to the constrained circumstances in which decisions are made. Behavioural approaches therefore focus on the 'how' and 'why' of behaviour and on the way in which individuals interpret and assign meaning to the environment. The scale and nature of such studies vary enormously. Early work tended to concentrate on overt behaviour (e.g. travel patterns) and on environmental perception (e.g. mental maps) but in recent years the scope of behavioural research in geography has widened considerably to take in the 'scientific' analysis of attitudes, decision-making, learning, and personality, as well as 'humanistic' examinations of the meaning that attaches to place and to landscape. Not surprisingly, there are parallels to be seen between what is happening in human geography and what is happening in other disciplines. Psychology, for example, has an increasing number of practitioners who argue that the discipline (1) should break away from its preoccupation with the behaviourist tradition (which essentially seeks to reduce complex behaviour to simple bonds between stimuli and responses); (2) should reduce its emphasis on artificial laboratory experiments; and (3) should look instead at behaviour as it occurs in natural situations. This viewpoint in clearly seen in ecological psychology and, in particular, in environmental psychology, and the links between these branches of the discipline and behavioural research in human geography are quite close.

One of the consequences of close links with psychology has been the adoption by geographers of psychological terminology. Very often this borrowing process has proceeded in a cavalier fashion. The term 'perception', for instance, is widely used and abused in the geographical literature. In many cases it describes a process that would be better labelled 'cognition'. Loose terminology, and a fascination for what goes on in the mind at the expense of acted-out or observable behaviour, has brought behavioural researchers in human geography a great deal of criticism. It has been pointed out, for example, that the links between attitudes and perceptions on the one hand and 'spatial behaviour' (choices as to where people will live, shop, work, play, etc.) on the other are neither clear nor direct. More fundamentally, behavioural studies in geography have been charged with viewing the environment as given, and hence with overlooking people's capacity to change the world. All too often, it is claimed, the focus of behavioural approaches has been on a value system that supports the status quo. In other words, behavioural approaches are often seen as being primarily concerned with the interpretation of behaviour as it is, with a result that little emphasis is given to the question of how behaviour might be changed so as to improve human well-being.

Despite these criticisms, behavioural research in geography has many practitioners. The distinction between objective and behavioural environments, that was postulated by early researchers, is a persistent and helpful one, as is the distinction between images and schemata. What seems to be needed in geography therefore is a greater attention to the philosophical and epistemological foundations of behavioural research so that a framework or paradigm can be devised that will bring together the varied efforts of all those researchers interested in people–environment interaction. Also needed is an explicit appreciation that much people–environment interaction is highly constrained, the important constraints in many cases being socially constructed (e.g. gender roles forced on people by social mores and by the mode of functioning of society as a whole).

Part One looks at the origins and growth of behavioural approaches to the study of people–environment interaction, particularly in relation to trends in other disciplines. It also outlines some of the objections that have been raised to behavioural research, and the reply that has been made to those objections. It is pointed out that the issue of how to relate form, structure, or pattern to behavioural process is one of the enduring problems in geography. Several philosophical approaches have been suggested as providing the key to overcoming this problem. Thus positivists argue that investigation of human behaviour by means of scientific method provides insights into the linkages between geographical form and behavioural process. Humanists, in contrast, are critical of scientific method because it concentrates only on what is measurable and therefore overlooks important issues like values and beliefs. Structuralists generally have no time for either positivists or humanists, preferring instead to view patterns of social life (e.g. human use of the environment) as resulting from the mechanisms (or 'structures') which underpin those patterns but which are not directly observable (e.g. structures in the brain, Marxist interpretations of the structure and functioning of political economies). The debate in geography between the positivists, humanists, and structuralists has at times been fairly vitriolic. Perhaps such differences of opinion belong to the past. Certainly, as Part One points out, there is evidence of a coming together of proponents of the various philosophical approaches. Both advocates of scientific method and humanists increasingly appreciate the importance of structures, whereas structuralists are coming to see that actual behaviour patterns may be one of the things that cause social life to take different forms in different places. The net effect of this discourse between the different positions is a revitalization of interest in behavioural approaches to the study of people–environment interaction.

1 People and Environment

Human geography has changed markedly since 1945 (Johnston 1979). The areal differentiation school of thought, which emphasized the delimitation of regions and the study of the characteristics that made them distinct, waned in popularity after having been the dominant geographical paradigm for much of this century. In its place came the quantitative revolution and the search for laws and theories. The application to geography of principles from both scientific method and the philosophy of science appeared to promise much (Harvey 1969a). A focus on the interrelationship between form and process, and in particular on the way in which behavioural processes brought about spatial patterns, encouraged geographers to turn their attention to model building. In human geography such models were probably best developed in what has become recognized as the locational or spatial analysis school of thought which sought to examine phenomena such as settlement patterns and the location of economic activity in terms of geometrical patterns and mathematical distributions (Haggett 1965). In many cases such model building derived its inspiration from micro-economics and was normative in approach in that it stipulated the spatial (i.e. geographical) patterns that should obtain given a number of assumptions about the processes that were supposedly operating. Commonly, the centre of attention was a set of omniscient, fully rational actors operating in a freely competitive manner on an isotropic plane. Indeed much of location theory was built around such ideas. Model building was seen by many as a way of making geography more 'scientific' and this was generally thought to be a good thing.

There have always been critics of normative model building in human geography and of attempts to develop the discipline along the lines followed by the natural sciences. Initially such criticisms came from two quarters: (1) an old guard of 'regional geographers' who defended areal differentiation as the core geographical method; and (2) those who believed that the goal of value-free, 'scientific' human geography was both undesirable and unrealistic (N. Smith 1979). Increasingly, however, criticism emerged from within the ranks of the practitioners of modern quantitative human geography. Disquiet focused on the fact that many normative models are grossly unrealistic in that they ignore the complexity of real-world situations and instead concentrate on idealized postulates such as rational economic behaviour. At the same time many models are overly deterministic in that there is only one logical behavioural outcome in any given situation. They are also reductionist in that they reduce everything to a question of economic returns and thereby overlook the fact that individuals have different attributes and motivations and respond in varying and very often non-economic ways to different environmental characteristics (Porteous 1977). Above all, the development of sometimes elegant models of spatial form and process, based around the notion of rational economic behaviour, has had little relevance for human geographers increasingly committed to using their skills in the solution of pressing societal problems.

It is not surprising, therefore, that dissatisfaction with the progress of the quantitative revolution in human geography, and qualms about the acceptability of normative models, have led to a number of developments within the discipline. For example, some researchers have turned away from the philosophy of science and sought guidance in moral philosophy. As a result, they have addressed themselves to questions of justice and equity in an attempt to appear 'relevant' (Harvey 1973). Others have abandoned the search for postulates of individual behaviour as a basis for understanding spatial form and have preferred to develop aggregate-level, descriptive approaches (exemplified by spatial interaction models) which conceptualize individual choice as a random process subject to certain overall constraints (Wilson 1971). Yet another approach, and one which is the focus of a good deal of attention in this book, has taken the view that a deeper understanding of people–environment interaction can be achieved by looking at the various psychological processes through which individuals come to know the environment in which they live, and by examining the ways in which these processes influence the nature of resultant behaviour. This does not, of course, mean that observed behaviour should be seen as the mere outworking of deep psychological processes. Such a view would be simplistic and wrong, given that the way in which people interact with the environment is often very highly constrained for one reason or another.

A behavioural emphasis in human geography is not of course entirely new: after all, the 'landscape school' in North American geography focused on humans as morphological agents and therefore tried to show how behavioural processes influenced landscape patterns. Similarly, advocates of human geography as a type of human ecology – a view that was very popular earlier this century – owed much to the possibilist philosophical position that stressed the significance of choice in human behaviour (Haggett 1965: 11–12). What distinguishes modern behavioural research in human geography from this earlier work is the primacy afforded to individual decision-making units, the importance given to both acted-out (or overt) behaviour and to the activities that go on within the mind (covert behaviour), a non-normative stance that emphasizes the world as it is rather than as it should be under certain theoretical assumptions, a concern to understand how socially generated constraints influence virtually all forms of people–environment interaction, and a willingness to consider those distinctly human characteristics that lead individuals to develop a sense of attachment to some places and a feeling of dislike for others. Not surprisingly, this new orientation has posed philosophical and methodological questions that have never before been tackled seriously by human geographers. Yet it would be wrong to view behavioural approaches to the study of people–environment interaction as a completely distinct branch of the discipline (possibly called 'behavioural geography', as advocated by some early researchers) because work in this field has tended to complement rather than supplant existing systematic branches of geography (e.g. urban geography, industrial geography). In other words, the subjects on which behavioural research has focused are often the same as those studied by both location theorists and relevance-orientated social geographers (e.g. the provision of urban services, travel patterns, migration). Even model building remains an important strategy with behavioural researchers, albeit with a positive rather than a normative research goal (i.e. a desire to understand the world as it is rather than as it should be under certain theoretical conditions). In order to appreciate the nature of behavioural approaches to the study of people–environment interaction, it is important to outline the character and significance of this work within human geography.

A behavioural paradigm?

To Gold (1980: 3), 'behavioural geography' (a term commonly used as a shorthand expression to signify the entire range of behavioural approaches used in the study of people–environment interaction) is a geographical expression of *behaviouralism*, which is itself a movement within social science that aims to replace the simple, mechanistic conceptions that previously characterized people–environment theory (e.g. the notion of rational economic behaviour) with new versions that explicitly recognize the enormous complexity of behaviour. To many, then, behavioural approaches in human geography constitute a point of view rather than a rigorous paradigm: the underlying rationale for the adoption of such approaches lies in the argument that an understanding of the geographical distribution of humanly made phenomena on the earth's surface rests upon knowledge of the decisions and behaviours which influence the arrangement of the phenomena rather than on knowledge just of the positional relations of the phenomena themselves. In other words, morphological laws that describe geometrical patterns are insufficient for understanding how those patterns come into being. Process can only be uncovered if attention is directed to the decision-making activities of the actors involved in creating a given pattern (Johnston 1979: 117). This inevitably involves consideration of the way in which behavioural intentions can be frustrated by constraints of one sort or another.

Early reviews of behavioural approaches in human geography pointed out that two types of study were dominant: analyses of overt behaviour patterns (often travel patterns) and investigations of perception of the environment (Gould 1969). The former tended to be based on an inductive approach that sought to observe reality as a prelude to arriving at generalizations which described the behaviour under study. In this way, the emphasis was on discovering the general in the particular. This *inductive* approach therefore differed from normative model building which usually began from the opposite perspective with simplified assumptions about the motivations underlying behaviour. These were then used as axioms from which *deductions* could be made (as, for instance, when the practice of distance minimization is deduced from the postulate of rational economic behaviour). The same inductive element was evident in perception studies. The aim was to study particular instances of environmental perception so as to identify general principles, the assumption being that comprehension of the way in which an individual perceives the environment would help in understanding that individual's behaviour. The rationale for this was set out clearly by Brookfield (1969: 53): 'decision-makers operating in an environment base their decisions on the environment as they perceive it, not as it is. The action resulting from their decision, on the other hand, is played out in a real environment'.

In these early behavioural approaches, no a priori assumptions were made about the perception process. Indeed, the interface between the environment and behaviour was thought of very much as a 'black box'

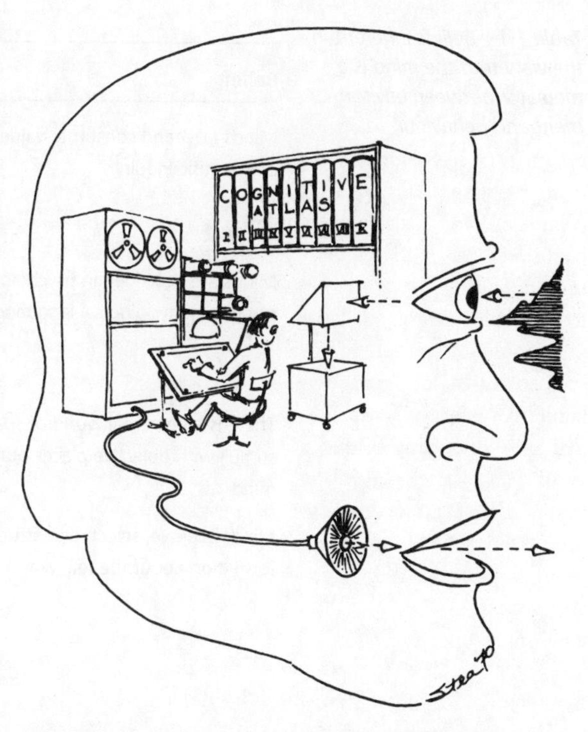

1.1 A primitive view of environmental perception (Source: Stea and Downs 1970: 5)

through which humans formed an image of their world (Fig. 1.1). Burnett (1976) has attempted to identify some of the beliefs about the mind as a mediator between the environment and overt behaviour that are common in such black box situations and has examined the implications of these beliefs for an understanding of people–environment interaction (Table 1.1). In principle, this focus on the mind should have encouraged geographers to explore the whole issue of human consciousness. However, according to Guelke (1989: 290), by focusing on the environment as perceived in a rather simplistic sense, 'geographers retained an essentially positivistic view of human geography'. In short, many researchers stuck with the tried and tested scientific method and focused only on what was measurable. They were therefore slow to realize the potential of those alternative approaches (often called 'humanistic') that emphasized human consciousness, values, subjectivity and intentionality.

Table 1.1 Beliefs inherent in the view that the mind is a mediator between environment and behaviour

Belief	Elaboration
Minds exist and constitute valued objects of scientific inquiry	We are more concerned with the description of preferences and perceptions than the description of conditions of neurons and nerve fibres
Minds are described in psychological and not neurophysiological language	Minds do not have peculiarly mental, non-material, or ghostly properties which would place them outside the realm of acceptable scientific discourse
There is an external world of spatial stimuli with objective places outside the mind	These include things such as industrial agglomerations, central places, residential sites, etc.
Minds observe, select, and structure information about the real world	Minds have processes corresponding to spatial learning and remembering and have streets somewhat corresponding to mental maps, perceived distances, awareness spaces, environmental cues, multidimensional image of shopping, residential and other locations, and more or less imperfect spatial knowledge
Mental events or processes occur that correspond to thinking about or evaluating spatial information	Minds have states describing action spaces and space preferences and utility functions
Minds are the seat of emotions and sensations, and are the seat of attitudes, needs, desires, and motives	Minds thus are the producers of satisfactions and dissatisfactions, environmental stress, and aspirations to optimize or satisfice in making location decisions
Spatial choices are made by thinking according to decision rules which are made in the mind and result from prior mental states, events, and processes	Choices are made among perceived alternatives. The decision rules may be viewed as methods of relating collated and evaluated information about alternatives on the one hand to motives on the other
Spatial choice decisions are the cause of an overt act, and over time sequences of spatial choices by individuals and groups cause overt behavioural processes; which in turn cause changes in spatial structures in the external world. Thus, ultimately location processes are explained (caused) by mental states, events, and processes	An overt act is such as a search for a new residence, purchase of a new industrial site, or a shopping trip. Sequences of choice over time are such as intra- and inter-urban migrations. Changes in spatial structures are like transitions in urban land use

Source: Golledge and Stimson (1987: 12)

The search for general principles or laws in people–environment interaction has led researchers in many directions. As a result there has been a plethora of specific studies but very little progress towards the formulation of an overall framework that ties together individual research efforts. Yet such a framework is important if research is to be other than piecemeal and if knowledge about people–environment interaction is to be cumulative. An early attempt at a framework, specifically for locational decision-making, was Pred's (1967) development of the behavioural matrix, the axes of which represented the quantity and quality of information available to a decision-maker and the decision-maker's ability to handle that information (see Fig. 4.9). Individuals could be schematically assigned to the matrix in any position from the rational economic individual to the rank incompetent. Unfortunately, however, the axes have proved difficult to define in a measurable way and the matrix is probably best remembered for the stimulus that it gave to the search for other frameworks.

After Pred, the main framework for research in what might be loosely termed 'behavioural geography', focused on environmental cognition. The rationale for looking at cognition is a simple one: 'if we can understand *how* human minds process information from external environments and if we can determine *what* they process and use, then we can investigate how and why choices concerning those environments are made' (Golledge and Rushton 1976: viii). To unravel the process of cognition is not, however, an easy task, partly because of the ever-present feeling that how we interpret the environment and how we behave is obvious and yet not easily articulated (Downs and Stea 1977: xiii). Indeed, it is very difficult to analyse what is often taken for granted and to come up with answers to questions such as how we know the world around us, how we learn about the environment, and how this learning influences our everyday lives.

One conceptual framework that describes the 'how' and the 'what' of environmental cognition has been provided by Golledge and Stimson (1987) (Fig. 1.2). This framework proposes that information from the environment is filtered as a result of personality, cultural, and cognitive variables to form two sorts of cognitive representations: *images* (which are pictures of an object that can be called to mind through the imagination) and *schemata* (which are the frameworks within which environmental information derived from experience is organized). Although important and useful, the distinction between images and schemata is one that has often been overlooked and many writers have preferred to talk about the general activity of *cognitive mapping* which is generally taken to encompass those 'processes which enable people to acquire, code, store, recall, and manipulate information about the nature of their spatial environment' (Downs and Stea 1973c: xiv). Of course, despite its overall attractiveness, an approach such as that of Golledge and Stimson does not fully and explicitly highlight the role of constraints in human spatial behaviour, nor does it give full recognition to feelings and values, and to the bonding that occurs between people and place.

The development of behavioural approaches in human geography

As the rationale for behavioural approaches within human geography became more acceptable during the 1960s and 1970s, the origins and coherence of this field of inquiry became the focus of a good deal of study. Several early reviews traced the development of behavioural approaches within the discipline, culminating in what was sometimes seen as a 'behavioural revolution' (Wood 1970; Brookfield 1969). Several later reviewers have been more sceptical of such attempts at 'finding continuity where none was experienced and progress where none was felt' (Downs and Meyer 1978: 59). Indeed Thrift (1981: 35) has argued that

> whereas with many other subject areas it is possible to establish a theoretical bloodline that stretches back beyond the 'quantitative revolution', it is hard to convincingly reconstitute the scattered set of papers that exist from before this point in time that can be labelled as 'behavioural' as in any way representing a coherent behavioural geography.

In many ways this criticism may be a little harsh since there is plenty of evidence to suggest that human geographers in early times had been concerned with the 'subjective' world, encompassing map imagery

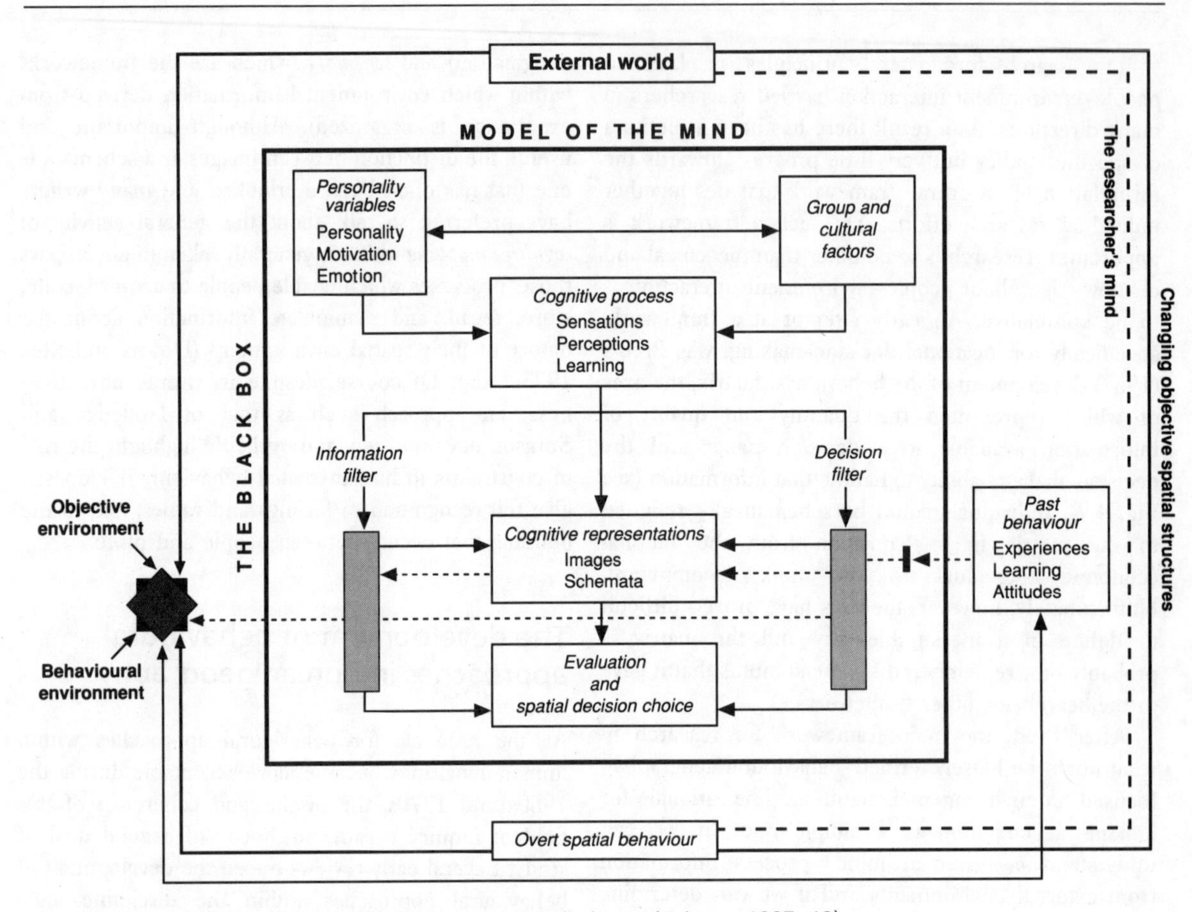

1.2 The process of environmental cognition (Source: Golledge and Stimson 1987: 13)

(Trowbridge 1913), the perceived world (Wright 1947; Kirk 1952), perceived spatial orientation (Gulliver 1908), and the perception of natural hazards (White 1945). However, it has to be conceded that it was not until the 1960s that behavioural approaches began to take some coherent shape within the discipline.

By this stage, Golledge *et al.* (1972) were able to identify five dominant areas of research that adopted a behavioural approach to the study of people–environment interaction: studies of decision-making and choice behaviour (especially locational choice, route selection, and patronage patterns); analyses of information flows (particularly in relation to innovation and diffusion); models of search and learning (often derived from theories in psychology); examination of voting behaviour; and perception research (focusing on hazard perception, image formation, and mental maps). During the 1970s this range of behavioural studies of people–environment interaction expanded to such an extent that some researchers began to use the term 'behavioural geography'. However, others continued to be hesitant about recognizing such a clear sub-field: 'the program of research endeavour envisaged by behavioural geography never involved the creation of a new branch of the discipline. Rather it was concerned with the elaboration of a distinctive approach to developing theory and solving problems in a wide variety of substantive areas in human geography' (Cox and Golledge 1981: xxvi). Some authorities even suggested that attempts to identify new sub-disciplines were rather futile. For example, according to Downs and Meyer (1978: 62), such attempts to trace the development of a 'new' field of study often degenerate into '*post hoc* structures which do not adequately reflect the context of ideas within which the field emerged' and, therefore, 'it is more important to try to capture and reconstruct the context, the intellectual framework and atmosphere within which perceptual geography emerged' than it is to trace the particular form that early studies took.

These differences of opinion as to the significance of the emergence of 'behaviouralism' in human geography may be due, in part, to the existence, by the 1970s, of two very different perspectives. According to Downs and Meyer (1978), by this stage a fundamental distinction could be made according to whether the studies were either positivistic or humanistic in orientation: the former strive for empirically verifiable, scientifically respectable measurement whereas the latter rely on description and empathy to reveal the meanings of different environments to the people who live within them. This dichotomy illustrates very clearly that there is no one behavioural approach to the study of people–environment interaction but rather a multiplicity of approaches.

Positivist research

One of the first topics to be studied by geographers from a scientifically oriented behavioural perspective that emphasized the decision-making process was human adjustment to natural hazards. Researchers at the University of Chicago, motivated by a desire to improve floodplain planning and management, conducted a number of studies into how floodplain dwellers evaluated the risks in their environment (White 1964). What emerged from these studies was a picture of people as boundedly rational animals: there were few signs of individuals consciously maximizing economic considerations but ample evidence of occasions when limited experience led to less than optimal behaviour (Kates 1962). In short, it became obvious from these studies that, in relation to hazard adjustment, individuals make choices between what they believe to be alternative courses of action but that the choice process is only boundedly rational in intent and is, in any event, based on limited and perhaps inaccurate information which is evaluated according to predetermined criteria that reflect both the experience and the whims of the individual concerned (Johnston 1979: 113–14).

The idea of bounded rationality has its origins in economics (Simon 1957). It refers to the tendency for individuals to simplify highly complex problems which they cannot fully comprehend and then to attempt to act rationally within this simplified model of reality. The idea was taken up by Wolpert (1964) who showed that, as a result of imperfect knowledge and a degree of aversion to uncertainty, farmers in central Sweden generally achieved only two-thirds of their potential output. To Wolpert, this suggested that the farmers were *satisficers* who, confronted with a problem, sought to find a satisfactory rather than an optimal course of action. No doubt encouraged by this result, Wolpert (1965) turned his attention to behavioural aspects of the decision to migrate and suggested that a full understanding of migration flows would demand consideration of the set of places of which an individual is aware (action space), the desirability and usefulness of each place to the individual (place utility), the motivation and goals of the decision-maker, and the stage the decision-maker had reached in the life cycle. This need to consider behavioural variables in migration had been noted earlier by Hägerstrand (1957) and was taken up again when that author made a plea for regional scientists to consider people and not just locations (Hägerstrand 1970). However, on this occasion, Hägerstrand argued that behavioural studies should consider time as well as space and should pay particular attention to the way in which movement is inhibited by transport constraints and other commitments that vary temporally (Pred 1977b). Hägerstrand, in other words, demonstrated a difference that emerged early on between North American and Scandinavian geographers interested in behavioural approaches to the study of people–environment interaction: the former tended to focus on sovereign decision-makers actively choosing from among environmental opportunities while the latter highlighted the way in which individuals react to constraints (Thrift 1981). Over time, of course, this distinction has weakened as all researchers have come to appreciate the importance of constraints in all behavioural research.

Work on bounded rationality and satisficing is typical of much behavioural research in human geography in that it tends to work 'backwards' from observable behaviour and to make inferences about the

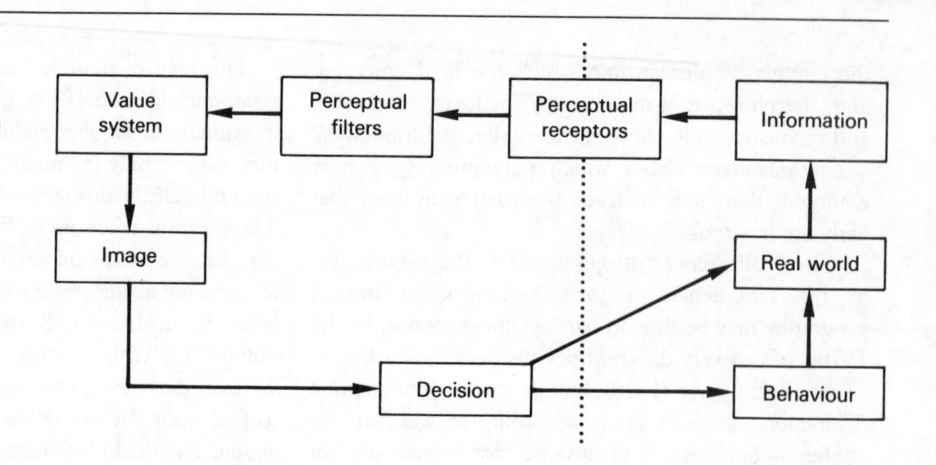

1.3 Environmental perception and behaviour (Source: Downs 1970a: 85)

mental processes that operate in the mind. Thus geographers have frequently used terms like 'attitude' and it has often been naïvely assumed that knowledge of attitudes will facilitate prediction of behaviour. In other words, there has been a tendency for geographers to adopt psychological terms without reflecting fully on what they might involve: for example, it is risky to talk of attitudes without recognizing that an attitude comprises a cognitive component (knowledge), an affective component (feeling), and a behavioural component (outcome or action) (Svart 1974). Perhaps the most commonly adopted psychological terms in geography have been 'perception', 'cognition', and 'learning'. Indeed, a great deal of the early work in 'behavioural geography' assumed that individuals derived information about the environment through perception and evaluated this information in terms of a value system (i.e. learning) to arrive at a cognitive image in respect of which decisions were made, much in the manner described in Fig. 1.3. Attempts to measure, in a verifiable way, the image that people hold in their minds generated a vast literature (Downs and Stea 1977).

Humanistic research

The concern for interpreting observable behaviour in terms of what goes on in the mind, that is evident among positivist researchers in the scientific tradition, is also apparent in those branches of geography which adopt a humanistic approach. However, with this type of inquiry, the scientific goals of replicable and verifiable measurement are generally discarded in favour of an emphasis on interpreting the world as seen through the eyes of the individuals whose behaviour is to be understood. The relevance of imagination (Lowenthal 1961), the importance of values and beliefs (Buttimer 1969, 1974), the nature of environmental meaning (Tuan 1971), and the significance of space and place were important among the early themes considered by humanistic geographers. Prominent among such work have been studies of landscapes (e.g. Lowenthal and Prince 1964, 1965). In this sense a landscape is something that is all around, something that is being continuously created and altered (often unconsciously as much as through conscious design), and something that is imbued with symbolic and cultural meaning (Meinig 1979a). The study of how landscape is interpreted is therefore integral to an understanding of a people's relationship to their environment because changes in the landscape give clues to changing evaluations of the environment (Prince 1971; Lowenthal 1975). This fact has of course been recognized by historical geographers for a long time and a good deal of effort has gone into contrasting myth and reality in the views of landscape held by early settlers in new lands (Watson and O'Riordan 1976).

Images of landscape, and of environment generally, are formed in the minds of all individuals. In fact so important are the conceptions of environment held by common people that Wright proposed a new field of study – *geosophy* – to take account of them. His proposal rested on the contention that 'geographical knowledge of one kind or another is universal among men [*sic*], and is in no sense a monopoly of geographers' (Wright 1947: 13). Wright also pointed out that there are no totally unexplored lands left on earth and suggested that the coming geographical frontiers are to be found instead in the *terrae incognitae* in the minds of individual human beings. The idea of subjective worlds, partly private and partly shared, emerged again in the work of Kirk (1952, 1963) who distinguished between phenomenal and behavioural environments: the former refers to 'real' situations and the latter to the 'psycho-physical field' in which facts are arranged into patterns or structures and evaluated in line with cultural influences. To Kirk (1963: 365), the behavioural environment has 'shape, cohesiveness and meaning added to it by the act of human perception'. What Kirk proposed, therefore, was that all individuals have personal worlds (behavioural environments) but that there also exist consensus views (Lowenthal 1961). In short, similarities exist between the behavioural environments of different people.

Acceptance of the fact that individuals have personal worlds encouraged geographers to look at the environment from an *experiential* perspective. For example, Tuan (1974a) introduced the term *topophilia* to describe the affective bond that develops between people and place. Place in this sense is a centre of meaning that varies in scale from a piece of furniture to a nation state (Tuan 1975a). Exactly how the meaning that a place holds for an individual is to be uncovered is of course a difficult problem. In one sense it demands a humanistic perspective that focuses on the earth as the home of humankind and geography as the mirror of humankind (in that knowledge of human nature may be deepened by studying people's relations with their environment) (Tuan 1971). In another sense it implies that geographers may have to turn to philosophical approaches to explanation that they have so far ignored. Tuan (1972) has suggested, for instance, that structuralism and existentialism may be studied with benefit by geographers: the former seeks to identify human thought patterns that transcend cultural differences (see Chapters 2 and 3) while the latter sees human beings as striving to define a self that is given neither by nature nor by culture (see Chapter 5). Other writers have argued for an apparently less rigorous, idealist position, derived from history (Collingwood 1956), that involves rethinking the thoughts of the people whose behaviour is under study in order that the researcher can explain human actions in a critical and analytical way by reconstructing the actions and the decisions underlying them (Guelke 1974).

During the 1960s and 1970s the sort of approach advocated by humanists did not attract large numbers of devotees. As a result, according to Saarinen (1976: 151), concepts such as 'geosophy did not catch on, nor did the term behavioural environment'. It is little wonder, therefore, that Kirk (1978: 388) himself has rather wryly alluded to

> that often quoted but, I suspect, less frequently read, paper of mine in 1952. The concept of the behavioural environment first formulated therein has had a chequered history in the past 25 years of geographical thought and in the process perhaps sight has been lost of the context in which it was first defined.

Instead of following early humanistic initiatives, most behavioural researchers in geography have preferred to remain, at least until relatively recently, within the positivist tradition and the scientific methodology in which they themselves had been schooled. To some extent this might now be changing because there is evidence of a coming together of positivistic and humanistic approaches, perhaps stimulated by criticism of the way in which behavioural research has developed in geography.

Behavioural approaches: a critique

In the 1960s the development of behavioural approaches in geography appeared to offer much. A substantial literature emerged, incorporating the early writings of Wright, Kirk, Lowenthal, Tuan, Golledge, and others, together with the work of the natural hazards group. This led Burton (1963) and Brookfield (1969) to claim that a behavioural emphasis rivalled the quantitative revolution in terms of the significance of the viewpoint that it provided. Such enthusiasm was not, however, universal. From early times Harvey (1969b) expressed doubts as to whether the unravelling of choice behaviour would be any more rewarding than the conceptualizing of human actions as the result of a probabilistic decision process. In his view, such statistical generalizations might be quite adequate for the purpose at hand in most geographical research. In a somewhat different vein, Hägerstrand (1975) and Cullen (1978) both criticized geographers for ignoring time as a behavioural dimension and for approaching most studies from the point of view of comparative statics (i.e. looking at what happened in a place at two points in time and inferring why change had occurred). Even harsher criticism came from those who believe that most behavioural research belies a commitment to the status quo in the sense that it is preoccupied with behaviour as it is rather than as it might be (Massey 1975b). This radical perspective, largely associated with Marxist thought, emphasizes the significance of societal and institutional constraints on behaviour, and thus argues that studies of decision-making, choice and preferences, and urban images are of little consequence. In the view of Jensen-Butler (1981), this criticism will be inevitable so long as behavioural research in geography is ontologically grounded in idealism rather than materialism. In other words, the sentiment behind this criticism of behavioural research contends that understanding of the world is less important than an ability to change it. Because of this dispute between the positivists and the Marxists, behaviouralism tended to make little headway. Furthermore, because of its perceived lack of social relevance at a time when social issues had become the major focus of human geography, behavioural research was often relegated to a minor role within the discipline (Smith 1977; Herbert and Smith 1979).

Despite the criticisms outlined above, the proponents of behavioural approaches to the study of people–environment interaction continued to argue that such approaches should be absolutely fundamental to studies in locational analysis, regional science, development, and planning. They reiterated the argument that it is impossible to plan for the future until present-day behaviour is properly understood. However, in order to assert its relevance, the nature of behavioural research needed to change. According to Cullen (1976), 'behaviouralism' had become something of an intellectual backwater in geography because researchers rather blindly borrowed the scientific paradigm and then let that paradigm determine the nature of the problems that were to be investigated. Thus the independent–dependent variable format of much scientific endeavour was overused. For example, it was common practice to compile aggregate images of an area in the manner devised by Lynch (1960) and then argue that such aggregate images influence behaviour. Image was the independent variable and behaviour the dependent variable. In Cullen's view, the ubiquity of this framework is also its principal failing: it can be applied in a wide variety of cases because it is almost wholly lacking in content, explains nothing, and generates no testable hypotheses. His solution is to focus on the individual *qua* individual and to try to understand mental processes at that level.

Although Cullen's view is plausible, there are other interpretations of the lack of interest in behavioural research in geography in the 1980s. Brookfield (1989), for instance, argued that the principal reason for the lack of popularity of behaviouralism in human geography lay in its rather close links with cognition and in associated problems of measurement and generalization. His solution is to revert to the original purpose of what became known as behaviouralism – the study of decision-making in its broadest sense, focusing on decision-makers subjected to a wide variety of constraints both in time and space. Bunting and Guelke (1979) made similar points over 10 years ago in suggesting that the blame for the weaknesses in behavioural approaches in geography lay in the theoretical framework that had been adopted in most studies. Bunting and Guelke argued that, by assuming that individuals react to their environment as they perceive and interpret it through previous experience and knowledge, much behavioural research in geography has all too often put too much emphasis on ego-centred interpretations of the environment. They were also critical of the assumption that there exist strong relationships between revealed images or preferences and actual, real world behaviour. As far as Bunting and Guelke were concerned, this was an unfounded assumption because little research had looked at the congruence between image and behaviour.

Bunting and Guelke's comments provoked a counter-attack by several very prominent researchers who had used behavioural approaches. For example, Rushton (1979) accused Bunting and Guelke of drawing a caricature of 'behavioural geography', while

Saarinen (1979) responded by pointing out that the problems of measuring images were overdrawn and by suggesting that the subject matter with which Bunting and Guelke were concerned was but one small part of a newly developing interdisciplinary field which transcends geography to include many of the other social sciences. In a similar vein, Downs (1979) made the point that behavioural research in geography has come a long way in a short time. Of course it would be wrong to see Bunting and Guelke's argument as a rejection of all that behavioural research in geography stands for, since their main concern was to assert the primacy of overt behaviour over mental imagery as a focus of geographical attention. Briefly, they argued that geographers should begin with what is known, with what one can measure with confidence, and with what can be directly sensed, and only then move on to a consideration of vague and difficult to measure images and thoughts which may provide the key to understanding overt behaviour (Bunting and Guelke 1979).

Where Bunting and Guelke differ from mainstream behavioural researchers in human geography is in their assertion that it is better to observe behaviour and then infer what goes on in the mind than to postulate mental processes and phenomena (e.g. cognitive images) and then look for manifestations of those processes and phenomena. This difference of opinion harks back to one of the most basic problems in modern geography, that is, how to relate form and process. Olsson (1969) introduced the term 'geographic inference problem' to refer to the problem of relating form, structure, and pattern (which is usually a macro-scale phenomenon such as land use) to process (which is usually specified at the micro-scale, as in assumptions about how individual human beings behave) at a satisfactory level of aggregation. His suggested solution of using many-valued logics has found few adherents (Olsson 1970). More prosaic attempts to use mental models derived at the individual scale as input for models of aggregate behaviour patterns at the macro-scale have met with a similar lack of success (Golledge and Stimson 1987).

This failure to find a satisfactory way of relating form and process means that behavioural researchers in geography must be on their guard against the dangers of *psychologism* (Mills 1970), that is the fallacy whereby social phenomena are explained purely in terms of the mental characteristics of individuals. This is particularly important because one of the most telling criticisms yet made of behavioural approaches in geography is that they frequently view an individual as *homo psychologicus* and tend to treat environmental behaviour as a one-dimensional phenomenon to the extent that the economic, social, and political considerations that act concomitantly with environmental influences are frequently overlooked (Thrift 1981). This is a weakness because, quite obviously, spatial behaviour (e.g. travel patterns) is not simply a function of environmental knowledge. Only when the whole range of influences and constraints is assessed and analysed will a full understanding of real world behaviour be possible. In this context, Mercer (1972) has suggested that the focus of attention for geographers adopting behavioural approaches should be the multi-faceted *situation* in which behaviour occurs and the way in which meaning is ascribed to both the situation and resultant behaviour. The *taken-for-granted world* might be particularly important in this context but it is invariably overlooked by many researchers (Ley 1977). What geographers need is perhaps a 'geographical imagination', paralleling the sociological imagination (Mills 1970), to enable them to take account of the individual's recognition of the role of space and place in her or his own biography, and of the way in which transactions between individuals and organizations are affected by the space that separates them (Harvey 1990).

Revitalization and convergence

Despite the conflicting views of advocates of strict scientific method (positivists), humanists, and radicals on where geography should be heading, Aitken (1991a) has recently claimed that during the 1980s behavioural research in geography has at last come of age as a major area of study within the discipline. This is partly because, over the last decade, researchers have broadened their perspective and developed a wider range of methodologies than previously existed (Couclelis and Golledge 1983; Saarinen *et al.* 1984; Walmsley and Lewis 1984). This revitalization of interest in behavioural approaches within human geography stems from two sources: first, the recognition that if behavioural research in geography was to make a contribution to current local and world issues then it had to shift its attention away from themes that were readily amenable to behavioural analysis and to the study of problems; and second, growing dissatisfaction with both the deterministic overtones of much structuralist thinking and with the idiographic nature of much pure humanistic geography has led to a re-examination of the potential of behaviouralism as a means of overcoming these deficiencies (Aitken and Sell 1989). Therefore, by the late 1980s, the vehemence of the exchange between the humanists, structuralists, and positivists began to ebb and some authorities, albeit to varying degrees, began to conclude that the long-awaited convergence of these apparently opposing discourses was under way (Seamon 1987; Golledge 1988; Cadwallader 1988). This is clearly illustrated in a series of essays which emphasize both accommodation and tension in the shift towards convergence (Kobyashi and MacKenzie 1989).

In the revitalization of interest in behavioural approaches in geography and the convergence of positivism, humanism, and structuralism, five major themes of behavioural research can be identified (cf. Aitken *et al.* 1989). In many respect these major research foci represent outgrowths of the work done by early behavioural researchers. They are considered in detail here because, in some respects, they represent the 'state of the art' in behavioural approaches to the study of people–environment interaction and indicate the ways in which the various approaches complement each other.

Analytical behaviouralism

The positivistic perspective which characterized much of the behavioural research in geography in the 1960s has, despite the criticisms of the humanists and the Marxists, continued to play a significant role within the discipline. Even though reviews of the literature nowadays commonly emphasize the significance of the non-scientific perspective, it is important also to acknowledge the fact that much empirical research is still characterized by positivistic overtones (i.e. a concern for scientific method). An inspection of current geographical and interdisciplinary journals bears witness to this claim. Therefore it is little wonder that Golledge and Stimson (1987: i) were able to state unashamedly that 'we believe that research requiring an analytical mode has produced the greatest . . . contribution to the entire area of behavioural research in geography'.

In the continued significance of analytical behavioural geography, three major trends are apparent. First, following the suggestion of Bunting and Guelke (1979), there has been a shift away from perception studies and to a broader focus on spatial behaviour at the micro-scale, particularly overt behaviour. This is evident in studies of migration (particularly those concerned with choosing a home, retirement destination, and mode of transportation), consumer behaviour, and recreation and leisure (Golledge and Timmermans 1990; Timmermans and Golledge 1990). Second, explanations of these behaviours have frequently been conceived of within the context of mathematical modelling. Often these models (such as discrete choice models, decompositional multi-attribute preference models, and dynamic choice models) have been derived from cognate disciplines (such as marketing, economics, and psychology) and applied within a spatial context (Louviere and Timmermans 1990a). The proponents of these models claim that they get closer to understanding behaviour than any other perspective because they provide a measure of both the input into, and output from, the so-called 'black box' of the mind. Although what is produced from such studies is a simplified view of reality (in that it assumes that people make choices according to relatively simple rules), it nevertheless often provides a reasonably meaningful basis for planning strategies and for the development of policy (Golledge and Stimson 1987). The third trend within analytical behaviouralism has been a shift towards more socially relevant issues and a greater policy orientation (Golledge *et al.* 1983; Golledge and Timmermans 1988; Clark and White 1990). Clearly these innovations illustrate how behavioural research in geography has shifted its focus beyond that of strict positivism. A greater emphasis is now being placed on the 'ongoing transactions that provide the essential structure and meaning of the relationship between people and their milieux' (Aitken 1991a: 180).

Sense of place

Following the pioneering papers by Wright (1947), Lowenthal (1961), and Kirk (1952, 1963), the humanistic perspective provided a major critique of positivism within human geography. The initial preoccupation with the relevance of imagination (Lowenthal 1975), values and beliefs (Buttimer 1969, 1974), and environmental meanings (Tuan 1971) rapidly evolved into a full-scale attempt to restore human subjectivity and consciousness to a position of prominence within geography (Ley and Samuels 1978a). This took several directions, including a greater emphasis on human values and attitudes (Bowden and Lowenthal 1975; Pred 1990), cultural patrimony (Rowntree 1988), the aesthetics of landscape and architecture (Cosgrove 1985a; Cosgrove and Daniels 1988), and the emotional significance of place in human identity (Pocock 1981c; Entrikin 1990; Eyles 1985; Eyles and Peace 1990). Some protagonists of the humanist perspective have even advocated the need for human compassion and engagement in the resolution of social and economic problems (Guelke 1985; White 1985) and Tuan (1990) has gone so far as to claim that, for all its realism, geography would be a sterile subject without the occasional tug of fantasy. The underlying basis of much of this perspective is well illustrated by Pred's (1991: 4) introduction to a journal issue on the theme of meaning and modernity:

> for me, what goes on inside people's minds, the invisible geography of human consciousness, or the temporal and spatial variation of human consciousness, cannot be separated from the human geographies women and men individually construct. For me, what goes on in people's minds cannot be separated from the construction of unevenly developed built environment, from the shaping of landscape and land-use patterns, from the appropriation and transformation of nature, from the organization and use of specialized locations for the conduct of economic and social practices, from the pattern of movement and interdependence between such localized activities, from the formation of symbolically laden ideology-projecting sites and areas. Human consciousness and human geography produce one another.

Despite the swing away from positivism during the 1980s, in particular with the emergence of a 'sense of place' as a focus for behavioural research, many have looked upon the sentiments expressed by Pred with alarm. Some have viewed this turn to humanism as a kind of amnesia, a turning away from problems, and a retreat into esoterica (Entrikin 1976). Brookfield (1989: 313), for instance, has pointed out that he finds 'much of the humanistic writing, especially that focussing on the "sense of place", somewhat precious. No doubt it was a healthy reaction to both the excesses of the positivists and the arrogant social determinism of the radicals, but at least these latter people are dealing with problems in the real world'. In a powerful rebuttal of these criticisms, Buttimer (1990) argues the need for a *greater* degree of humanism, not only at the micro-level but also as an aid to understanding of the destruction of the global environment and the radical transformation of culture and politics. Yet, to fully employ the humanistic perspective in research remains problematical, although the long-established anthropological tradition of participant observation does provide a way forward (Lewis 1986; Jackson and Smith 1984). Unless such a procedure is adopted, there is a great danger of the humanists in geography becoming preoccupied with esoteric case studies thus meriting the same criticism as that made by Leach (1961: 41) of much of anthropology:

> case history material . . . seldom reflects objective description. What commonly happens is that the anthropologist propounds some rather preposterous hypothesis of a very general kind and then puts forward his [*sic*] cases to illustrate the argument. . . . Insight comes from the anthropologist's private intuition; the evidence is only put forward by way of illustration.

Fortunately, recent trends in humanistic geography suggest that this is not likely to occur since there is now a recognition that some positivism may be of value in humanistic research and that, conversely, a humanistic perspective can help to frame scientifically orientated investigations (Seamon 1987; Ley 1989).

Knowing the environment

Much of the early behavioural work in geography comprised what has become known as 'perception studies'. These examined such themes as attitudes and expectations (Saarinen 1966), risk and uncertainty (Wolpert 1964), learning and habit (Golledge and Brown 1967), decisions and choice (Pred 1967; Burnett 1973), preferences for places (Gould and White 1974; Rushton 1969a), cognitive maps (Stea 1969; Downs 1970b), and the general process of acquiring spatial knowledge (Golledge and Zannaras 1973). Much of this work was set within the context of attempts to understand the decision-making process (see Fig. 1.3). By focusing on the components of this process, and by interrelating each within a system-like framework, it was hoped that understanding would be achieved. However, following the powerful critique by Bunting and Guelke (1979) of the supposed linkage between the 'environment-as-perceived' and actual behaviour, it rapidly became apparent that studies of cognitive images are not an end in themselves and are only of value in geographical inquiry if they are conceived of as a means to an end (i.e. a deeper appreciation of why a course of action was taken). As a result there has been a shift in geographical research on environmental cognition away from simplistic explanatory linkages (the notion that knowledge about an image is the key to understanding subsequent behaviour) and towards attempts to gain understanding of what might be termed the process of 'knowing the environment' (a focus on how the environment becomes known and acquires meaning). According to Walmsley (1988a: 11), three questions are particularly important: How does environmental knowledge come about? Does knowledge involve the gradual accumulation of bits of information or does it entail the awakening of a priori structures in the brain that cause individuals to think of the environment in terms of a limited number of predetermined ways? And what is the nature of the relationship between knowledge and behaviour?

Both scientific and non-scientific traditions pervade the literature that addresses these questions and the links with cognate disciplines have been strengthened as a result of a focus on these issues (Spencer and Blades 1986; Saarinen *et al.* 1984; Aitken and Sell 1989). As a result, it is becoming increasingly evident that environmental knowledge is not something that explains 'spatial behaviour'. Rather, a study of the acquisition of environmental knowledge is important to the extent that it provides an appreciation of the 'geography' inherent within issues. In other words, understanding how an individual becomes aware of the environment and how the environment takes on meaning can provide insights into why that individual is a particular sort of person and behaves in a particular way. This point of view can be illustrated by two examples. First, Ekinsmyth (1988) has traced the learning by new residents of an urban environment over a period of 2 years by means of a series of longitudinal surveys. By emphasizing the 'how' of the process of learning, it was possible to identify not only social and cultural differences in the learning process but also significant psychological factors such as the way in which individuals oriented themselves spatially. In the second example, Matthews's (1992) study of children's understanding of large-scale environments demonstrates that, either at birth or shortly after, all children are 'natural' environmental mappers and 'proto-geographers'. This implies that the environmental capability of children has been underestimated within educational psychology, partly because of an uncritical acceptance of the Piagetian theory of human development and partly because of an over-reliance on novel, small-scale laboratory tests that are divorced from the children's world of experience.

Place and behaviour

A significant theme in the early development of behavioural approaches in geography, particularly at the interface with psychology, was the relationship between place and behaviour, most vividly illustrated in geography in early hazard studies (Burton *et al.* 1978). Within psychology itself, the concern with place and behaviour was most evident in work on crowding and architectural design (Mercer 1975). In geography the deterministic overtones of much of this research consigned such place–behaviour studies to a minor role for a long time. Indeed, it is only recently, as part of a growing desire to understand behaviour at the micro-scale, that studies of the link between place and behaviour have acquired a more central position within geography. According to Aitken (1990), this revival of place–behaviour studies in geography can be related to two factors. First, there is a growing dissatisfaction with structuralist approaches and in particular with the tendency of the neo-Marxists to overlook the importance of individual actions and the dynamics of people's daily lives in the restructuring of society (Cadwallader 1988). Second, the resurgence of interest in the study of place and behaviour owes something to the growing realization among planners and designers that the production of the built environment is inextricably tied to behavioural and social processes. According to several authorities, micro-level research is therefore in growing demand among practitioners in the field of planning (Zube and Sell 1986; Huttman and van Vliet 1987).

Decision-making

When behavioural research in geography began to emerge as a distinct area of study, its prime focus was upon locational decision-making, initially conceived of in terms of risk and uncertainty and the notion of imperfect rationality, and then eventually thought of within the context of cognition. The way that this approach developed during the 1970s was criticized severely from several quarters. For example, a recent review of early attempts to analyse the decision-making process concluded that 'all behaviour location studies risk trivializing locational decisions precisely because they treat these decisions as strongly isolated from any thoughts, ideas, opinions, or speculations that decision-makers might possess' (Philo 1989: 211). In other words, many studies of decision-making adopted a very narrow perspective that looked at 'spatial' variables to the exclusion of virtually all others with the result that research showed little awareness of the overall context within which decisions were made. At the end of the day, therefore, geographers adopting a behavioural approach to the study of decision-making could 'say little about geographical actions other than routinized and trivial shopping trips' (Philo 1989: 215). It is little wonder, then, that research attention shifted away from decision-making as such and to the context within which decisions are made.

Brookfield (1989) has argued most cogently for a more realistic and relevant conception of the decision-making process than that employed by analytical behavioural geographers. He argued that, on the one hand, every locational decision involves several decision-takers at different levels and, on the other, the behavioural environment of each decision-maker includes important economic and structural components as well as a constantly changing 'information content'. Because of this, decisions need to be considered in a context of available resources and constraints, that is the political economy – both micro and macro – which surrounds the decision-makers. Specifically, Brookfield (1989: 315) highlighted the complexity of the decision-making process:

> A large subset of all decisions of consequence have 'downstream' effects: they impact on others. These impacts may or may not have an effect on the decision; they constantly have consequences, which include subsequent decisions which have to be taken by widening chains of people. By the same token, a further large subset of decisions has 'upstream' preconditions, the consequences of former decisions. In this direction the chains that could be traced are endless, but once again some past decisions have much greater 'downstream' effect than others.

Such a conception of decision-making has been largely ignored by geographers, the notable exception being those who have adopted a 'business organization' approach to location studies (McDermott and Taylor 1982). Recently, however, the infusion of a more structurally aware behaviouralism into geography has become apparent. For example, Blaikie and Brookfield (1987), in their study of land degradation, focused upon the conditions under which decision-making was undertaken and, by so doing, pointed the way towards a focus on both structure and behaviour in geographical inquiry. What this revival of interest in behaviouralism is arguing is that the dogmatic claims of the structuralists need exploding because the conditions for decision-making are governed not only by structures but also by values and cognition, and by information and the ability to handle information.

A parallel line of questioning of the effectiveness of structuralism as an explanation of spatial behaviour, and advocacy of the need to incorporate human agency in some form or another in any attempt to understand spatial behaviour, is to be found in the work of Gregory (1978). The means by which the integration of structure and behaviour may be achieved has also been explored in Giddens's (1984) structuration theory, which links structure and human agency, and Parkin's (1979) closure theory, which links voluntaristic social action and the system of constraints by which such actions are limited (see Chapters 2 and 6). Both these contributions are, however, highly theoretical and still need to be made operational in the realm of applied research. Significantly, however, even many structuralists now concede that a focus on decision-making is essential and that behaviouralism has an important role to play in the study of human affairs providing it is conceived of within a meaningful social context (Pratt 1989). Despite this, theorizing about the relationship between individual behaviour and macro-structures remains a thorny issue (Desbarats 1983; Couclelis 1986; Cadwallader 1988).

The rationale for a renewed interest in behavioural approaches to the study of people–environment interaction stresses the importance of such approaches as a vehicle for linking structure and agency in attempts to provide a more meaningful understanding of contemporary issues, no matter whether they be personal concerns (e.g. loneliness and homelessness) or global concerns (e.g. world famine and environmental change). As society shifts to a postmodern world, the need to understand how such issues are manifest at the individual and local level becomes increasingly urgent. To ignore such local manifestations is to render the issues as little more than academic abstractions (Moulaert *et al.* 1988; Entrikin 1990; Harvey 1989). In short, geography is becoming increasingly confronted by new issues. To tackle these issues effectively might necessitate a 'new geography', or at least a geography that encompasses new approaches. Certainly, claims have been made that a new form of regional geography might offer the key to understanding how and why human well-being varies from place to place. However, as Curry (1991) has succinctly argued, much of what is claimed to be the 'new geography' of the 1990s has its roots in earlier times. Behavioural approaches in geography are a case in point because new perspectives arising from the fusion of the ideas contained in structuralism, humanism, and behaviouralism hark back in many instances to issues first confronted in the 1960s by both geographers and other social scientists.

Behavioural studies in geography and cognate disciplines

A major factor in the emergence of behavioural approaches in human geography, and in their recent revitalization, has been the interdisciplinary character of the perspective involved, notably the way in which it offers links between geography and environmental psychology (Spencer and Blades 1986), cognitive psychology (Golledge 1982), gerontology (Warnes 1987, 1989; Smith 1991), hazard studies (Mitchell 1984; Rochford and Blocher 1991), environmental design studies (Aitken and Sell 1989), and planning (Golledge and Timmermans 1988). A significant development in this trend from the mid-1970s onwards has been the forging of stronger links with psychology. At first this link was largely focused on conceptual matters but, of late, it has been strengthened by the common use of measurement techniques and experimental designs.

Until recently psychologists had shown little interest in people–environment interaction. There are a number of reasons for this. For example, there has been a tendency for psychologists to be more interested in mental processes in an abstract sense than in environmental cognition *per se*, and in laboratory-based experimental designs rather than the real world. Above all, much of psychology has adopted a *behaviourist* (cf. behaviouralist) perspective. As a result, there has been a tendency for it to be reductionist in that complex situations are reduced to simple bonds between stimuli and responses (S–R bonds). The environment, from this perspective, is thought of only as a set of stimuli to which individuals respond in a more or less deterministic manner. Not surprisingly, given this view, much of the experimentation on which S–R theory in particular, and behaviourist research generally, is based has been conducted with animals. As a result, its extrapolation into the realm of human behaviour necessitates what Koestler (1967) has trenchantly described as a 'ratomorphic' view of human beings. Obviously human beings are not rats and do not behave as rats do. In short, behaviourist theories are elegant but unhelpful when it comes to understanding real world people–environment interaction (Getis and Boots 1971). The difference between behaviourism and behaviouralism may appear to be both pedantic and semantic but it is in fact fundamental. The former attempts to reduce behaviour to S–R bonds whereas the latter emphasizes the significance of predisposing factors (attitudes, beliefs, values) as well as the manner in which individuals make decisions about where to go and what to do within a set of socially generated constraints and intersubjectively shared environmental meanings.

It would be wrong, of course, to characterize all psychology as having been conducted within the behaviourist tradition. Indeed, it is to work in non-behaviourist psychology that many behavioural researchers in geography have turned for inspiration. One fruitful source of inspiration has been *Gestalt* psychology (Koroscil 1971; Boal and Livingstone 1989). Part of the appeal to geographers of the *Gestalt* approach lies in the importance it attaches to perception. Gestaltists argue that perception proceeds according to innate abilities which organize environmental stimuli into coherently structured forms or patterns (Kirk 1963). Thus each person has a behavioural environment which is the environment as perceived. Although each individual perceives uniquely, the resultant behavioural environments have much in common because they are derived from both common neurological mechanisms that are innate in people and from common superimposed socializing experiences (Ittelson *et al.* 1974: 68). One of the most prominent *Gestalt* psychologists was Lewin (1936, 1951) and his work is frequently cited by geographers because he was acutely aware of the significance of the environment and of the way in which the nature of the behavioural environment influenced an individual's actions. Lewin in fact thought of human behaviour as a stream of activity that resulted from the interaction of factors within the person (needs, values, feelings, predispositions) with external factors as perceived in a given behavioural setting (Ittelson *et al.* 1974: 69). The unit of study was therefore the individual together with her or his environment and Lewin devised the notation $B = f(P, E)$ to demonstrate the way in which behaviour is a function of both the person and the environment. This interaction of the individual with the behaviour setting is sometimes known as the *life space*: it is a cognitively structured psychological field that extends from the past, through the present, and into the future (Pocock and Hudson 1978).

Gestaltists are not the only psychologists to focus on people and their behavioural setting, and life space is not the only concept that has been used to describe that focus. For example, over three decades ago Sprout and Sprout (1957) introduced the term *milieu* to encompass the human and non-human tangible objects, the social and cultural phenomena, and the images that impinge on behaviour. Likewise, a number of researchers have argued for a 'psychology of situations' that pays explicit attention to the physical–geographical, biological and socio-cultural environments that mediate actual behaviour (Magnusson 1981). More significantly, Lee (1968, 1976) borrowed the notion of a *schema* from neurology and used it to characterize the inner representations that individuals have of external environments. In this view, individuals organize knowledge in their minds according to the places to which that knowledge relates. Lee's (1976: 36) argument was that 'just as everything must be some*thing*, so it must also be some*where*'. Thus individuals build up *socio-spatial schemata*, which they can add to or modify, for coming to terms with their everyday environment. These are the frameworks in the mind on which individuals 'hang' information about the environment. Precisely how people do this is of course a critical consideration in the design professions, and architects in particular have put a great deal of effort into studying the problem of how individuals use schemata to interpret the environment (Altman 1975). Generally speaking, such work has concentrated on small-scale structures and therefore on *place* rather than space (Canter 1977). As a result it has been prone to emphasize the idea of *images*. As introduced by Boulding (1956: 17), an image is built up by an individual from information derived from the enveloping social and physical milieu of the individual over her or his entire life history.

Another approach that emphasizes the influence of environment on behaviour is to be found in *ecological psychology*. This branch of psychology asserts that real world conditions cannot be simulated in contrived situations and that psychologists must therefore break away from their traditional role as 'operators' (who create situations to which subjects have to react) and take on the role of 'transducers' or, in other words, interpreters of behaviour as it occurs (Barker 1968). The focus of attention in ecological psychology is the *behaviour setting*. This setting is bounded in space and time and has a structure that relates physical, social, and cultural phenomena in such a way as to elicit common or regularized forms of behaviour (e.g. studies of the way in which classroom layout can influence student behaviour) (Ittelson *et al.* 1974).

Of greater interest to the geographer than ecological psychology has been *environmental psychology* (Altman and Christensen 1990; Spencer and Blades 1986). This is a relatively new branch of psychology that shares with the ecological perspective the view that the object of study should be the real world rather than laboratory behaviour (Craik 1968, 1970). Environmental psychology has in fact a dual origin: it began partly as a result of a realization that psychology generally has had little to say, in the face of increasing environmental problems, about human impact through planning and design on the environment (Proshansky *et al.* 1970b; Wohlwill 1970), and partly as a result of the discovery of engaging intellectual puzzles in the fields of environmental perception, cognition, and assessment, environmental personality, and environmental adaptation (Craik 1977). According to Ittelson *et al.* (1974: 12–14), the sub-discipline of environmental psychology is now at the stage where it can be characterized by eight assumptions in respect of the environment:

1. The environment is experienced as a unitary field.
2. People are an integral part of the environment rather than objects within it.
3. All physical environments are inescapably linked to social systems.
4. The influence of the environment on individuals varies with the behaviour in question.
5. The environment often operates below the level of awareness.
6. There may be significant differences between 'observed' and 'real' environments.
7. Environments can be cognized as a set of mental images.
8. Environments have symbolic value.

In simple terms, environmental psychology seeks to be of applied value by aiming to tell designers something of how people experience the environment (Craik 1973). It is, however, far from deterministic in its approach in that it explicitly recognizes that humans are goal-directed animals that influence the environment and are influenced by it (Stokols 1977).

As with behavioural research in human geography, the range of issues with which environmental psychologists have been concerned is very broad and a variety of behaviours have been examined at a number of scales and with a fair degree of methodological eclecticism (see, for example, Altman and Werner 1985; Altman and Wandersman 1987; Altman and Zube 1989). One means of describing this work has been provided by Stokols (1978) who argued that people–environment relations can be differentiated in terms of whether they involve a cognitive or behavioural transaction and whether the focus is on people actively interpreting the environment or on people reacting to environmental attributes. In this way it is possible to develop a fourfold typology of interpretive, evaluative, operative, and responsive studies (Table 1.2). Clearly, on this basis, a great deal of behavioural research in human geography can be designated as interpretive or evaluative in emphasis, as is shown by attempts to describe images, schemata, mental maps, and preference surfaces.

In view of this similarity between the disciplines, it is worth noting the claim that environmental psychology has a chance of both promoting the welfare of humankind and of jerking establishment research out of its deepening rut of statistical trivialization (Mercer 1975). By the 1980s this became evident in the increased interest that was shown in the problems of groups such as children, the elderly, the poor, and the disadvantaged (Spencer and Blades 1986; Stokols and Altman 1987). Indirectly, this shift of interest in environmental psychology has played a part in the revitalization of behavioural approaches in human geography. In fact, it is with the very same issues that much behavioural research in geography, and thus the content of this book, is concerned.

Table 1.2 *A typology of people–environment interaction*

	Form of transaction	
	Cognitive	Behavioural
Phase of transaction	**Interpretive**	**Operative**
Active	Cognitive representation of the spatial environment	Behaviour modification to adapt to the environment
	Personality and the environment	Human spatial behaviour (personal space, territoriality, crowding)
Reactive	*Evaluative*	*Responsive*
	Environmental attitudes	Human responses to the physical environment
	Environmental assessment	Ecological psychology

Source: Adapted from Stokols (1978: 264)

The scope of the book

The book is divided into three parts, each of which has an introduction designed to set the scene and introduce the reader to the issues discussed in that part. Part 1, as has been seen, looks at the origins, growth, and present state of development of behavioural research in geography, particularly in relation to trends in other disciplines. It also outlines some of the objections that have been raised to behavioural research, the reply that has been made to these objections, and the revitalization of interest in behavioural approaches to the study of people–environment interaction. Part 2 examines the two perspectives widely adopted in studies of spatial behaviour: 'structural' approaches and 'behavioural' approaches. The former is the centre of attention in Chapters 2 and 3; the latter is the centre of attention in Chapters 4 and 5.

In this context it is important to be clear about the interpretation that the book places on the terms 'structural' and 'behavioural'. In particular, it needs to be appreciated that the book adopts a very broad interpretation of the term 'structural approaches'. This term is used in an almost colloquial sense to cover both the sort of studies of 'spatial structures' that geographers have engaged in for many years (e.g. social area analyses within cities) *and* rather more recent attempts to apply the philosophical position of 'structuralism' in the geographical domain. The former approach (including attempts to study the 'structure' of behaviour through things like spatial interaction models) is described in Chapter 2 as 'empirical structuralism'. The latter approach, adopting a more philosophically pure form of structuralism often associated with political economy approaches in general and structural Marxism in particular, is labelled 'transformational structuralism' and is discussed in Chapter 3. Some authorities might view this broad use of the term 'structuralism' as somewhat unconventional and unusual. It is nevertheless useful in highlighting the rich variety of work in human geography and in focusing attention on the changing view of 'structure' within the discipline. In short, the distinction between 'empirical structuralism' and 'transformational structuralism' provides a convenient framework for organizing the work of geographers.

The definition of 'behavioural approaches' is less problematical. At issue are the assumptions that have been made from time to time about how people behave when they interact with the environment. Understandably, the volume of research on this issue is enormous. For convenience it can be divided according to two themes. The first concerns how people get information about the environment and how they use that information to make decisions about where to go and what to do. The second theme concerns the way in which individuals build up simplified images of environments that are too big and too complex to be known in their entirety, how these cognitive representations can serve as bases for decisions as to behaviour (subject to a variety of constraints), and how the environment acquires meaning. The issues of information and choice are dealt with in Chapter 4 and those of imagery, behaviour, constraints, and meaning are dealt with in Chapter 5. Part 2 concludes with a critique of both structural and behavioural approaches (Chapter 6) that explores the interplay between choice and constraint.

Part 3 illustrates the potential application of behavioural approaches in five areas of everyday life within the context of rapid social, economic, and technological change. The focus of attention in each of the five chapters is on the main actors and how they reach decisions, albeit often very highly constrained decisions. Chapter 7 is concerned with jobs and work, including decision-making in relation to the location of economic activity and the organization of work. Chapter 8 focuses on housing and migration, in particular on why people migrate, how they choose where to go (or why they are highly constrained in their mobility), and the role of the housing market. Shops and shopping are the subject of Chapter 9 with attention being paid to the provision of retail facilities and the changing nature of consumer behaviour. Alongside work, housing, and shopping, leisure and recreation are important components of everyday life in advanced Western society. Accordingly, Chapter 10 discusses various approaches to the study of leisure time activities. Finally, Chapter 11 is concerned with a sense of belonging and its influence on community development and human well-being. Throughout Part 3, the chapters openly acknowledge that there are many things we do not know about behaviour in everyday life and that, as a result, there is considerable potential for continued research in the field.

Part 2 Approaches to the Study of People and Environment

For much of this century, human geographers have been interested in aggregate patterns of behaviour rather than in people as individuals. At the centre of attention have been the features that can be observed when a group of individuals interact with their surroundings. This approach is clearly seen in a variety of spatial interaction models (such as the so-called 'gravity model') that investigate how the flows of humans and of human artefacts (e.g. phone calls) between places diminish as the distance between places increases. The development of such models rests on the rationale that there are characteristics of aggregate spatial behaviour that are observable irrespective of the identity of the individuals under study. This is possible in so far as the individuals in question are assumed to adopt a common, largely economic frame of reference in deciding where to go and what to do. In short, in undertaking such studies, human geographers attempt to identify the *structure* of behaviour.

A similar concern with structures is evident in the study of settlement patterns within cities. In this case the focus of attention is on describing the way in which a myriad of individual decisions concerning residential location result in a pattern of social areas that reflects the overall *social structure* of the society within which the cities are located. With its origins in the Chicago School of Human Ecology, and manifestations in social area and factorial ecologies of city structure, this approach highlights the relationship between socio-spatial structures on the one hand and individual behaviour and levels of well-being on the other (e.g. the possible link between social milieux and social pathology).

Both the spatial interaction and the socio-spatial approaches in human geography are concerned with structures and can be thought of, in philosophical terms, as examples of *empirical structuralism*. The aim of researchers working in both areas is to identify the *structure of behaviour* through empirical measurement of aggregate patterns. Of course, researchers working in these areas very rarely describe themselves as 'structuralists', although a concern for what is described as 'spatial structures' often pervades their writings. This lack of explicit concern for 'structuralism' reflects the fact that much of the work described here as 'empirical structuralism' had its intellectual antecedents in a time before the word 'structuralism' entered the geographer's vocabulary. In short, much work was undertaken in the past with relatively little consideration of the appropriateness (or otherwise) of the underlying philosophical position. Indeed, it is only really in the last two decades that human geographers have become explicitly concerned in a major way with the philosophy underlying their work. With this deeper concern with philosophy has come an interest in what might be termed *transformational* (cf. empirical) *structuralism*. This approach switches attention away from patterns of aggregate behaviour. In fact the term implies a much purer and philosophically more specific use of the word 'structuralism'. It talks of 'structuralism' in a sense that would be widely understood throughout contemporary social science.

The argument underlying the adoption of transformational structuralism is that a concern for monitoring individual actions, even in the aggregate, is unwarranted because what is important to understanding people–environment interaction is an appreciation of how *unobservable structures* influence behaviour. In other words, real-world behaviour is viewed as merely the outworking of deep-seated structures whose operation cannot be observed directly but rather must be inferred from an examination of the manifestations that are observable. There are of course different sorts of transformational structuralism. To some, the deep-seated structures that are of concern are to be found in the study of the growth, structure, and functioning of the human brain. Piaget's work is a case in point. More commonly in the geographical literature, a concern with structure has led to an interest in political economy generally and structural Marxism in particular. Of special attention has been the argument that social phenomena (which include much of the people–environment interaction studied by human geographers) can only really be understood if they are seen as part of a superstructure that is related to the material needs (or 'mode of production') on which society is based. Geographical studies incorporating empirical structuralism are discussed in Chapter 2 and transformational structuralism is discussed in Chapter 3, together with an approach that has become known as 'structuration'. In short, the distinction between 'empirical structuralism' and 'transformational structuralism', although somewhat unusual and unconventional, is a useful way of classifying a good deal of geographical work that has had as its focus a concern for understanding overall patterns of people–environment interaction.

One of the criticisms levelled against the structuralist approaches adopted in human geography is that such approaches reveal next to nothing about why individual behaviour takes the form that it does. In other words, a knowledge of structure (e.g. in the sense of social areas within a city) says nothing about the processes that bring those structures into being (e.g. migration decisions by residents of the city). Likewise, a concern for structures in a Marxist sense (e.g. modes of production) leaves unclear the extent to which individuals can act autonomously. It is not surprising, then, that a good deal of effort in human geography in recent years has gone into studying people–environment interaction from the perspectives of the 'actors' involved. These behavioural approaches are reviewed in Chapters 4 and 5.

There are two main issues in Chapter 4: how individuals come to know the environment; and how they differentiate between places and make decisions as to which places best suit their needs for housing, recreation, shopping and many other forms of behaviour. Essentially, the position adopted in Chapter 4 is one that derives from transactional constructivism. According to this view, individuals 'construct' their own notion of 'reality' in their minds while they are engaged in interaction with the environment. This is very different from seeing the environment as a set of stimuli to which humans respond. One corollary to the adoption of a transactional-constructivist position is that the behaviour of individuals can only be understood if due attention is paid to the way in which they experience the environment. This topic is taken up in Chapter 5 where attention shifts from information, choice, and decision-making to images, to constraints on behaviour, and to the way in which people interpret their surroundings. Three particular issues are addressed: how people develop simplified images of places that are too big and too complex to be known in their entirety; how environmental knowledge and environmental images influence the way in which individuals make decisions about where to go and what to do; and how parts of the environment take on meaning for the people who interact with that environment. Although there have been attempts to measure environmental experience through the application of quantitative, empirical techniques from mainstream social science (what might be termed the 'positivist' approach), the topic of experiential environments has also stimulated a great deal of interest in what might be termed 'humanistic' approaches. Such approaches focus on the way in which people ascribe meaning to their surroundings and on the extent to which this meaning is intersubjective, that is to say, shared by a group of people. Therefore of considerable concern to human geographers is the way in which people develop a sense of belonging to place and the way in which much human interpretation of the environment is taken for granted. These issues are addressed in Chapter 5.

Three considerations underlie both 'structural' and 'behavioural' approaches to the study of people–environment interaction:

1. How to relate behaviour to structure;
2. How to demonstrate that structure, and in particular the spatial organization of society, is both a creation of human beings and a constraint on their activity and behaviour;
3. How to describe people–environment interaction in such a way as to be useful to planners and policy-makers without losing sight of the rich variety of human behaviour.

The position adopted in this book, and set out at some length in Chapter 6 in a critique of both the 'structural' and 'behavioural' approaches, is that a multiplicity of viewpoints offers different and often complementary perspectives on why the world is as it is. In a postmodern world, where the conditions of human existence vary markedly from place to place, there is as yet no single overarching theory that satisfactorily explains the variability in people–environment interaction. In the absence of such a theory, what is needed is an appreciation that a balance exists between choice and constraint, an understanding of the reciprocal relationship between structure and behaviour, and an awareness of the complex ways in which micro-scale activity (e.g. individual decisions) can lead to macro-scale patterns (e.g. social areas within cities).

2 Spatial Interaction and Spatial Structures

Human geography is concerned with the way in which people interact with the environment. This interaction often results in spatial patterns of one sort or another (e.g. patterns of land use within the city). Human geography is therefore very much concerned with the use of 'space'. However, as Knox (1982: 2) argues,

> space itself should not simply be regarded as the medium in which social, economic and political processes are expressed. It is of importance in its own right in contributing both to the pattern of urban development and to the nature of the relationships between different social groups within the city. . . . [Space] . . . also emerges as a significant determinant of the quality of life in different parts of the city because of variations in physical accessibility to opportunities and amenities.

Originally, geographers adopted an *idiographic* approach, focusing on an individual place and the human activities associated with that place; nowadays this approach has given way to a more *nomothetic* emphasis that attempts to find the general in the particular in order that statements can be made about people–environment interaction such that these statements will obtain in a wide variety of cases. This search for generality has invariably involved attempts to explore the structures of space and society. In short, human geographers have become interested in what might be loosely termed 'structuralist' approaches. According to Rossi (1981), a structuralist approach is one which seeks explanation for observed phenomena in the general structures that underpin all phenomena. Structuralism, in this sense, comes in two forms: empirical structuralism (or structural functionalism) and transformational structuralism. The former is characterized by a system-maintaining perspective which allows widely differing societies to be analysed by emphasizing their basic functional characteristics. These, in turn, it is argued, reflect deep-seated, relatively permanent, structural characteristics (Morgan 1975: 291).

The study of spatial structures is obviously a key part of such an approach, space itself being conceptualized in terms of points, lines, and areas at several different scales: world, nation, region, neighbourhood, domestic (Haggett 1965). The assumption is often made, for example, that certain spatial structures are associated with certain spatial behaviours. This is in fact a point of view that underlies much of location theory where rational human beings are presumed to follow economic motives that lead them to act in a certain way (e.g. minimizing costs, maximizing profit or utility) in a given environment. It is also a point of view that underlies what can be thought of as deterministic gravity models. In simple terms such gravity models seek a functional or statistical relationship between external stimulus variables (e.g. city size) and responses (e.g. migration) in a repertoire of human spatial behaviour. The goal in such studies is not, however, to understand the behaviour of any given individual so much as to understand the behaviour of 'statistically average people', that is to say, the overall features that emerge in an aggregate pattern of behaviour (Hudson 1976b).

Such spatial interaction models are, then,

macro-scale 'structural' models in the sense that they attempt to identify order in the spatial behaviour of collections of individuals. They are not explicitly concerned with what motivates each individual to behave in a certain way. They are more interested in the overall 'structural' properties of behaviour *in toto*. The same can be said of studies concerned with the identification of social areas, for example the proliferation of attempts to identify and explain the organization of social space. With its origins in the Chicago School of Human Ecology, this work highlights the unequal competition between different populations (and the way this competition is manifest in different land uses) as well as the relationship between social milieux and social pathologies. All these sorts of 'empirical structuralism' are examined in this chapter. 'Transactional structuralism', which involves a much more precise and philosophical purer use of the term 'structuralism', is examined in Chapter 3.

Spatial interaction

Over a century ago Carey (1858) suggested that the laws of physics could be applied to the study of human behaviour since the flow of people or of human artefacts (e.g. trade, telephone calls) between locations can be explained by the laws of gravitation. The application of the *gravity model* was fundamental to early studies of spatial interaction and subsequently in an approach that became known as 'social physics' (Ajo 1953). Although most social scientists would now frown on the idea that the complexities and subtleties of human behaviour can be reduced to simple physical laws, the gravity model has continued to attract interest. In its simplest form the model can be expressed as follows:

$$I_{ij} = \frac{k\,(P_i P_j)}{D_{ij}^{\,b}}$$

where I_{ij} is the interaction between town i and town j, P_i and P_j the population of towns i and j respectively, D_{ij} the distance between towns i and j, k a constant and b an exponent.

In short, according to the gravity model, the interaction between two places is determined by the population of those places (a surrogate measure for mass) and the distance between places raised to a power (most commonly squared). In other words, knowledge about the key elements of a structure (settlement size and distance) leads to the prediction of behaviour. The enumerator in the equation measures the potential for interaction and the denominator the impedance function. For example, if town i had a population of 7, and town j population of 8, then there would be 56 potential interactions. The extent to which these potentials are realized is determined by the impedance function. The model can be varied to fit different circumstances by defining distance in terms of geography, time, or money. In this way the model has been applied with success to predict such interactions as inter-city migration, traffic flows, telephone calls, and trade (Olsson 1965). It has even been reformulated in probabilistic rather than deterministic terms (Huff 1963).

Closely related to the gravity model proper are potential models. If the gravity model is applied in relation to an individual person rather than a town, then one of the two population figures in the enumerator of the gravity model can be set at unity. Moreover, if the impedance function is set at the square of the distance (as is common in many gravity models), then the potential interaction of an individual with a place can be calculated from

$$I = \frac{P}{D^2}$$

where I is the interaction, P the population of the place in question and D the distance of the individual from the place in question. This means that, if two places are equidistant, an individual will have more potential interaction with the bigger place. Perhaps a more significant use for this potential model is in describing the total potential interaction of an individual in a given place with a broad area in the surrounding environment. When this information is displayed cartographically for a number of places, a ***population potential surface*** results.

Many variations on the gravity model have been suggested in the study of spatial interaction (Batty 1976). For example, in the field of consumer behaviour, Cadwallader (1981) has suggested a 'cognitive gravity model' whereby interaction is directly

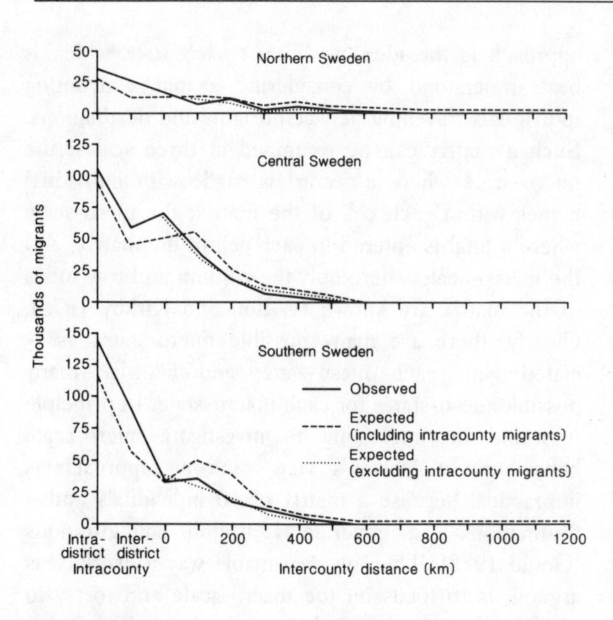

2.1 Migration and intervening opportunities in Sweden, 1921–30 (Source: Isbell 1944: 630)

proportional to subjective evaluations of the attributes of a shopping centre and inversely proportional to cognitive distance. The idea that individuals evaluate what is important, and the notion that such evaluation is reflected in aggregate spatial behaviour, have antecedents in Stouffer's (1940) *intervening opportunities model*. According to this model, the number of people going a given distance is directly proportional to the number of opportunities at that distance and inversely proportional to the number of intervening opportunities. In other words, the number of people travelling a given distance is directly proportional to the percentage increase in opportunities at that distance. Isbell (1944) successfully tested this model in relation to inter-county migration in Sweden between 1921 and 1930 (Fig. 2.1). Stouffer (1960) also suggested that the model was useful in predicting the selection of colleges by undergraduates, the selection of marriage mates, and vacation travel.

One of the main problems with Stouffer's model is that of measuring 'opportunities'. Adequate data are seldom available. As a result, the intervening opportunities model has not attracted the attention that has focused on the gravity model with its more easily measured elements of distance and population. The gravity model, however, is not without its problems. Although it is lawlike (in that it is based on an invariable and general relationship between the concepts of mass and distance), the gravity model is not embedded in any wider scientific theory (Amedeo and Golledge 1975: 29). It does not, for instance, allow for the application in social science of any general theory from physics. It remains simply an empirical regularity that satisfactorily describes human spatial behaviour in a number of contexts. The appellation 'gravity model' is therefore a misnomer because there is no way in which energy force fields can be invoked to explain why the observed pattern came about, as in the case of physics (Schneider 1959). Moreover, because the model is calibrated to fit different circumstances, both the constant and the exponent in the equation have been found to vary considerably. As a result, like all models that are derived from observations in a given study area, the gravity model provides mere quantitative description of the history of the recent past for that study area and reveals little of a generalizable nature (Louviere 1981). It is therefore dangerous to use the gravity model as a basis for planning since, in a very fundamental sense, it describes only the status quo; changes in behaviour are assumed to reflect what people want to do and the environment is assumed to be arranged or planned to accommodate desired behaviour (Burnett 1981). Moreover, description is only in aggregate terms. There is no behavioural base to the gravity model. In fact what the model says about behaviour is essentially a circular argument: if many people go to a place, it must be attractive; and because it is attractive, many people go there (King and Golledge 1978: 290).

Statistical mechanics

In overall terms, the sort of spatial interaction model characterized by the gravity model and the intervening opportunities model asserts that interaction (I_{ij}) between an origin zone i and a destination zone j for a particular flow is directly proportional to the demand requirements generated in origin zone i (D_i), directly proportional to the number or size of the opportunities in zone j (O_j), and proportional to some function of the distance or cost involved in travelling from i to j. In formal terms this can be expressed as follows:

$$I = k\, D_i\, O_j\, f(C_{ij})$$

It should be noted, however, that traditional gravity models are only one part of a family of spatial interaction models (Wilson 1971). Classical Newtonian physics, on which the traditional gravity model is based, works well where the basic parameters of the system can be specified, as in the prediction of flows between two distinct and discrete towns. It is less than satisfactory, however, when it comes to predicting interaction between a large number of elements (as, for example, with interaction within a system of cities). In such cases, the branch of physics known as *statistical mechanics* may provide a better basis for spatial interaction models than that afforded by Newtonian mechanics.

The field of statistical mechanics has been explored by Wilson (1970) who has shown that many spatial interaction models have a common foundation in information theory and can be constructed by an *entropy maximization procedure.* In other words, in Wilson's view, there is a common theoretical base to all models of spatial interaction and the difference between models appears to be essentially one of emphasis and approach. The central notion of entropy maximization is one that is quite difficult to come to terms with. What it involves is a method for making the best estimate of a probability distribution from the limited information that is available, given a situation where everything that is known about a system can be expressed as a series of constraints. Entropy maximization therefore provides a way of saying something about the overall state of a system while not overlooking the fact that uncertainty surrounds the individual elements within the system (Wilson 1970). In other words, it is the macro-properties of the system (i.e. its overall structural characteristics) that are the focus of attention in statistical mechanics and a state of maximum entropy can be thought of as the most probable state within a system.

Although the notion of entropy maximization sounds simple, the mathematics on which it is based is extremely difficult. Gould (1972), for example, described Wilson's (1970) book as the most difficult geography book he had ever read. The key to Wilson's approach is the idea of *the most likely state*, which is best understood by considering a matrix denoting individuals travelling between origins and destinations. Such a matrix can be examined at three scales: the micro-scale where a record is made with individual names within each cell of the matrix; the meso-scale where a total is entered in each cell of the matrix; and the macro-scale where only the column and row totals of the matrix are known (Wilson and Kirkby 1975). Clearly, there are many possible micro-states associated with each meso-state, and likewise many possible meso-states for each macro-state. In principle it would be interesting to investigate micro-scale behaviour. In Wilson's view such an approach is impractical because a matrix of 10 individuals and 4 destinations can generate 1 trillion configurations (Gould 1972). The only reasonable way to work, it is argued, is to focus on the macro-scale and to try to identify the most probable macro-state. The most likely macro-state is of course that which accommodates the greatest number of meso-states (just as the most probable meso-state is that with the greatest number of micro-states associated with it). Entropy maximization provides a method of calculating which states are most probable. For example, for any distribution, the number of possible configurations is given by:

$$W(T_{ij}) = \frac{T!}{\underset{\text{ij}}{\pi}\, T_{ij}!}$$

where $W(T_{ij})$ is the number of configurations that trips can take, T the total number of trips and T_{ij} the trips in the cells of the matrix. This equation is usually subjected to constraints, notably in terms of the fact that the number of trips emanating from each origin zone is known, as is the number of trips terminating in each destination zone.

The entropy maximization model of spatial interaction has been applied with success by Batty (1970a, b) in a study of both the English Midlands and north-west England. It has also been suggested that the model could be used as a basis for a reformulation of Reilly's law of retail gravitation so as to take account of simultaneous competition between a multiplicity of shopping centres (Batty 1976). In Wilson's (1970) opinion, the entropy-maximizing methodology holds a number of advantages over the traditional gravity model: in the first place it breaks away from the deterministic analogy with gravity and therefore provides a broader statistical foundation for model building; secondly, it provides a means of both improving spatial interaction models in particular situations and extending the realm of such models to include complex situations that have been beyond the realm of traditional gravity formulations; and finally it provides a stimulus which will lead to the development of more dynamic models to replace the essentially static equilibrium models available at present.

Of course not all commentators share Wilson's enthusiasm. Van Lierop and Nijkamp (1980), for instance, evaluated gravity models and entropy models on methodological, theoretical, logical, and practical criteria and compared them with logit models (where the chance of a certain alternative being chosen by a certain trip-maker is determined by the utility of that alternative compared to the utility of all other alternatives) and probit models (where the probability of a trip being made is related to the attitude of the trip-maker which is in turn related to a series of variables that measure personal characteristics). Their conclusion was that there was little to choose between gravity and entropy models and that both were inferior to logit models. This suggests that an approach that pays some attention to individual evaluation is to be preferred to the approach of statistical mechanics. A similar point of view has been expressed by Hudson (1976b) who has drawn attention to the fact that, in entropy maximization, individual decisions are treated as random phenomena. Macro-scale description, in other words, fails to provide any understanding of why individuals behave in the way that they do. Wilson (1970) anticipated this criticism by suggesting that disaggregation of entropy models will improve their level of 'explanation' but the fact remains that, with the approach of statistical mechanics, individual decisions are interpreted as random events that are not deserving of attention in their own right.

Despite fundamental criticisms of spatial interaction models and statistical mechanics, alternative modes of analysis have not completely overshadowed other macro-level models in human geography during the 1980s (Boyce 1988). Such macro-scale models have continued to be of significance not only because of the increased ability to handle large data sets brought about by the widespread availability of computers but also because of several attempts to apply the models in more politically 'relevant' directions. For example, *regional analysis* has been a major focus of attention, largely as a result of its significance for assessing the impact of technological change and for forecasting future development (Anselin and Madden 1990). The majority of these regional models have been couched in input–output terms (Batey and Rose 1990; Hewings *et al.* 1988). Paralleling these developments has been a revival of interest in what is called 'marketing geography'. This work emphasizes the opimization and delimitation of market areas (Benson and Faminow 1990; Lentiner *et al.* 1988). However, probably the most interesting change to have occurred in the use of macro-scale approaches to the study of structure in people–environment interaction has been in relation to modelling social issues such as gender (Madden and Chen Chui 1990), disease (Thomas 1990), housing (Clarke *et al.* 1989), and externality effects (White and Ratick 1989; Barnett *et al.* 1990). The relevance and significance of these analyses has still to be fully assessed.

Simulation

From a geographical point of view, a major deficiency in all spatial interaction and regional models is their failure to consider the *time* perspective. Therefore, as an alternative, some geographers have developed simulation techniques that explicitly set distance specifically, and space generally, within a temporal framework. A major contribution to this approach has been Morrill's (1965a, b, c) simulation of the development of towns in Sweden, the growth of the Negro ghetto, and the extension of the urban fringe in American cities. The nature of such simulation models can be illustrated by reference to Morrill's (1965b) study of the expansion of the Negro ghetto in Seattle, Washington (Fig. 2.2). The model adopted in this case was of the Monte Carlo variety in that locational decisions were conceived of as being subject to both errors and uncertainties that could not be specified or ignored. The migration itself was viewed as being a

2.2 (a) The migration probability (mean information) field; (b) simulated moves from three sample blocks; (c) simulated migration patterns, Seattle 1948–50 (Source: After Morrill 1965c: from Expansion of the urban fringe: a simulation experiment, *Papers of the Regional Science Association* **15**, 185–202 by permission of the American Geographical Society)

1	2	3	4	5	6	7	8	9
10	11	12	13	14-15	16	17	18	19
20	21	22	23	24-25	26	27	28	29
30	31	32	33-34	35-37	38-39	40	41	42
43	44-45	46-47	48-50	x	51-53	54-55	56-57	58
59	60	61	62-63	64-66	67-68	69	70	71
72	73	74	75	76-77	78	79	80	81
82	83	84	85	86-87	88	89	90	91
92	93	94	95	96	97	98	99	00

a

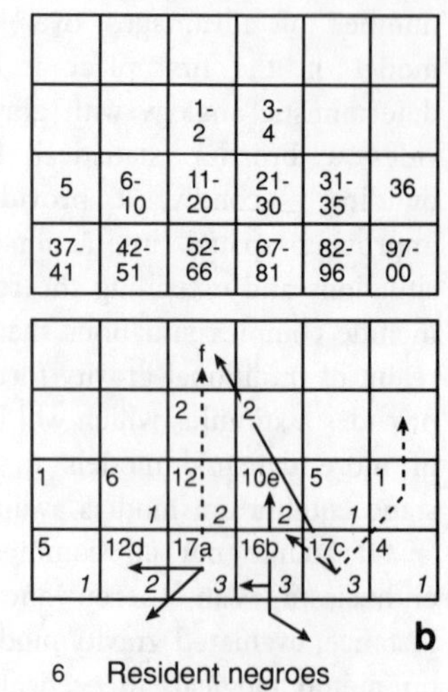

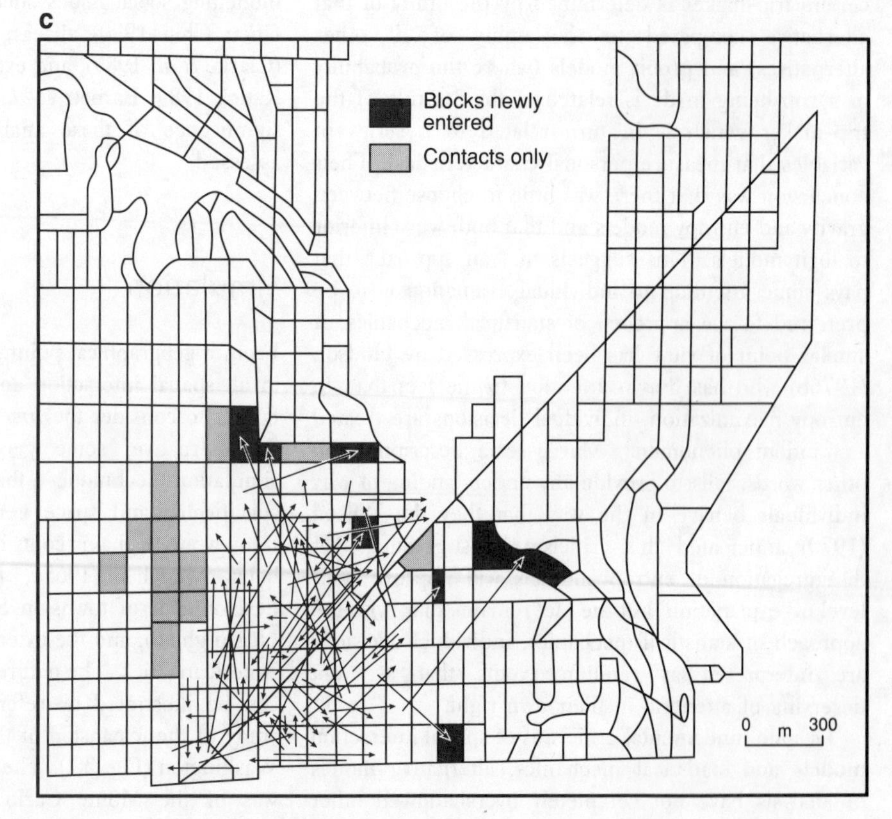

function of:

1. The distance between an origin and all possible destinations;
2. The relative attractiveness of an origin and potential destinations;
3. The availability of information about likely destinations.

The model was made operational by initially assigning a probability-based index of attraction to each of the sub-areas of Seattle. This index reflected rates of population growth, the initial population size of the sub-area, and the size of the Negro population. A contiguity constraint was introduced to simulate the wave-like form of ghetto expansion, as well as the tendency for black households to avoid moving into non-contiguous all-white blocks. Since the probability of moving to all possible destinations varies inversely with distance, a mean information field was constructed wherein probabilities of migration decline with distance from the point of origin (Fig. 2.2(a)). By allocating random numbers to potential migrants in turn, the number of movers and non-movers, and the direction and distance of their migration, were statistically simulated (Fig. 2.2(b)).

Despite the valuable insights they provide into spatial processes, simulation models such as that developed by Morrill suffer from a number of weaknesses. One of the major difficulties concerns the weighting given to the index of attraction, which unfortunately has to be empirically determined and statistically tested. The predictive value of these models is also limited on account of the fact that they are ill equipped to probe the underlying causes of the behaviour in question (e.g. Ossowicz and Slawski 1989; Fotheringham *et al.* 1989). Thus, although simulation models have the potential to describe the structural properties of behaviour in people–environment interaction, and the potential to incorporate time more fully into the study of spatial behaviour, they suffer – like all structural approaches – from the fact that they provide little insight into *why* behaviour takes the form that it does. The underlying proposition in simulation approaches is that people–environment interaction is structured in predictable ways. Just why this is so is rarely questioned. A similar viewpoint underlies the large volume of work that deals with the way in which such structuring manifests itself in spatial patterns of one sort or another.

Socio-spatial structures

A fundamental and long-standing deficiency of spatial interaction and simulation models is their failure to distinguish the 'social' component in human behaviour. 'Space' has all too often been regarded as unproblematic and an essentially 'physical' phenomenon. Geographers have long recognized this deficiency though attempts to incorporate the 'social' into interaction models have, as yet, proved extremely difficult. It is not surprising, then, that the focus of much geographical attention has been on social space in an areal as opposed to an interactionist sense. In this regard, geographers have not been alone. Indeed, much of their research was inspired by the work of urban ecologists and sociologists.

Traditionally, social scientists have viewed the artificiality, strangeness, and diversity of the urban environment as fundamental conditioners of individual behaviour and social organization within the city (Bulmer 1984). Much of this tradition stems from the attempts by philosophers such as Tönnies, Simmel, Weber, and Durkheim to interpret the effect that the Industrial Revolution had on urban living. In simple terms, these writers argued that pre-industrial society was characterized by small homogeneous groups of people who performed similar tasks and had similar interests that led them to think and behave alike, thereby giving rise to a uniform way of life. The coming of the Industrial Revolution resulted in new forms of production and new ways of living centred around economic specialization and advancements in modes of communication and transportation. It therefore involved the agglomeration of large numbers of people from diverse backgrounds, a concomitant increase in the number of people and the variety of situations with which individuals were in contact, and the emergence of a greater division of labour. In such a situation the sustaining of close primary relationships with friends and relatives became more difficult and, at the same time, social differentiation resulted in a divergence of life styles which weakened group consensus and cohesion and threatened to disrupt the existing social order. Formal controls commonly emerged to counter these disruptive trends. Often these controls worked but where they failed there was, according to theorists, an increase in what has become known as social disorganization and deviant behaviour.

In the inter-war period these ideas were modified and developed in two schools of thought that had a significant impact on human geography (Matthews 1977; Kurtz 1984). The first comprised the researchers of the Chicago School of Human Ecology (sometimes also termed 'Urban Ecology' or 'Social Ecology') under the leadership of Park (1936). Like the earlier theorists, these sociologists believed that urbanization created new societies, new ways of living, and different types of people in 'a mosaic of little worlds which touch but do not interpenetrate' (Park 1936: 608). The distinctiveness of the contribution of the Chicago School lies in its conception of the urban environment as a form of social organism in which individuals struggle for survival in a way similar to the struggles that go on in the plant and animal kingdoms (D. Smith 1989). The basic concept of this biological analogy is that of *impersonal competition* between individuals and groups for desired locations within the city. This competition operates through the market mechanism. It results in variations in land rents and leads to a segregation of different groups according to their ability to pay the rents of different sites. Therefore, the predominance of a particular group within a particular part of the city results from its relative ability to compete for space. According to the Chicago School, different socio-economic groups come to occupy different concentric zones in the city, based on their ability to pay for the land in question. Thus lower status groups focus on cheaper land in the inner suburbs and higher status groups occupy more sought after suburban locations.

The human ecologists also observed *symbiotic* functional relationships between individuals. Where these relationships were localized, there emerged 'territorial' units whose distinctive characteristics – physical, economic, and cultural – are the result of the unplanned operation of ecological and social processes (Burgess 1920: 458). Such territorial units were termed *natural areas*, or *communities*. Such communities tended to develop as areas within the concentric, socio-economically based zones. Thus there emerged, in many cities, a 'Little Italy' and a 'Deutschland'. With the passage of time, the relative attractiveness of location changes, as does the competitive power of different groups. As a result natural areas and communities disappear in some parts of the city only to reappear in other parts. Again biological concepts, such as invasion and succession, were employed to describe the manner in which these changes took place.

These concepts were summarized by Burgess (1920) in his model of residential differentiation and neighbourhood change (Fig. 2.3). As noted, the resultant zonation of the city was viewed by Burgess as arising from the differential economic competitiveness of different functions and different social groups, while the segregation of small groups within each zone (e.g. the ghetto, Chinatown) arose as a result of a symbiotic relationship based on language, culture, or race. In short, this model describes both the dynamic process whereby cities grew and changed, and the spatial arrangement of land uses and social groups.

The second school of thought to look at the impact of urbanization on society was that associated with Wirth, and particularly with his influential paper on urbanism as a way of life (Wirth 1938). According to Wirth, the social and psychological consequences of increasing urbanization stem from three factors: increased population size; increased population density; and the increased heterogeneity (or differentiation) of populations. Put simply, this means that, at the personal level, the individual has to face – and cope with – the abundant and varied physical and social stimuli that derive from a large, dense, and highly diverse environment. Although coping strategies can take many forms, certain types of behaviour become 'normal'. For example, many urban residents tend to become aloof, impersonal, and indifferent in their relationship with others. Likewise, many suffer from anxiety and nervous strain that is attributable to stimuli in the urban environment. Furthermore, the resultant loosening of personal bonds often leaves individuals lacking support in times of difficulty and free to pursue ego-centred behaviour. The effect, on the one hand, is an increase in the incidence of loneliness and mental illness and, on the other hand, an increase in the level and diversity of what might be called 'deviant

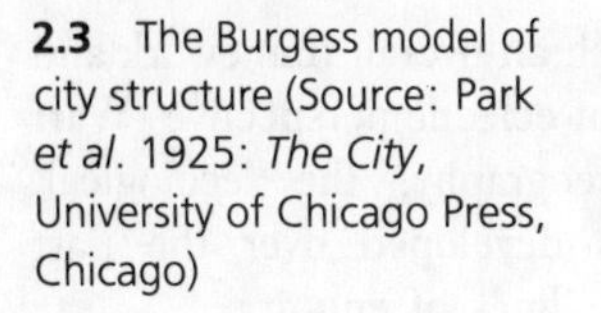

2.3 The Burgess model of city structure (Source: Park *et al.* 1925: *The City*, University of Chicago Press, Chicago)

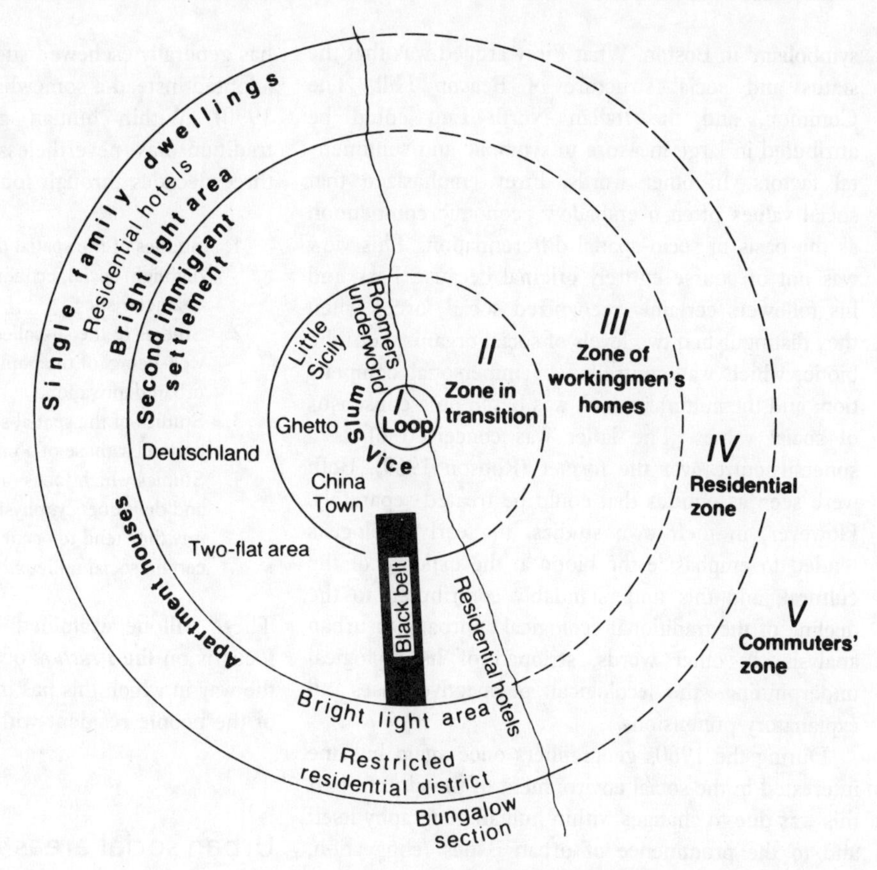

behaviour'. To some commentators, this characterization of large cities as hostile environments serves to explain the tendency for the walking speed of pedestrians to increase with city size (see Walmsley and Lewis 1989).

In addition, Wirth (1938) suggested that the city's size, density, and heterogeneity provide the basis of social change. The division of labour and the residential differentiation of the urban population leads to a separation of activities between home, school, workplace, friends, and relatives, thus weakening the support and control of the family, friends, and neighbours, leading to an absence of social control and an increase in social disorganization. As a result of this weakening of social norms, society has developed 'rational' procedures (such as criminal codes) and institutions (like welfare agencies) to provide the necessary support and control. Yet, according to Wirth, this new order cannot replace a communal order entirely and, as a result, a condition known as *anomie* develops: some people are unclear about what the new order involves, while others disregard it entirely, thereby creating another basis for deviant behaviour.

Despite the obvious relevance of the human ecology school of thought to the study of social space, and its significance in the development of human geography, it was heavily criticized during the 1940s and 1950s and, as a result, almost slipped into oblivion. Much of the criticism focused on the biological analogy of the ecologists, with many arguing that it involved an excessive reliance on competition as a basis of social organization. In these terms, the failure of studies to identify the concentric zones and natural areas predicted by the theorists was attributed to the unreality of the underlying propositions (Cater and Jones 1989). A further serious weakness of the ecological approach has been its failure to consider cultural and motivational factors, as was illustrated by Firey (1945) in his classic study of 'sentiment and symbolism' in Boston. What Firey argued was that the status and social structure of Beacon Hill, The Common, and the Italian North End could be attributed in large measure to symbolic and sentimental factors. In other words, Firey emphasized that social values often overshadow economic competition as the basis of socio-spatial differentiation. This view was not of course entirely original because Park and his followers certainly recognized social forces when they distinguished two levels of social organization: the biotic, which was controlled by impersonal competition; and the cultural, which was based on a consensus of social values. The latter was conceived of as a superstructure over the former (Robson 1969). Both were seen as entities that could be treated separately. However, in their own studies, the early ecologists tended to emphasize the biotic at the expense of the cultural, and this understandably contributed to the decline of the traditional ecological approach in urban analysis. In other words, stripped of its biological underpinning, the ecological perspective loses all explanatory pretensions.

During the 1960s geographers once again became interested in the social environment of the city. In part this was due to changes within human geography itself and to the prominence of urban issues (congestion, renewal, ghettos) in the world of politics. In part also it arose from reformulations of the original concepts of the urban ecologists. For example, Hawley (1950) revised the ecological perspective by conceiving of it as the study of the development and structure of communities, thereby emphasizing the functional interdependence within communities that results from collective adaptation to competition.

Somewhat earlier, in the 1950s, a school of economists had produced a justification of concentric rings based on neo-classical assumptions and couched in the language of land values (Evans 1973). In a somewhat different vein, Schnore (1965) elaborated upon the underlying basis of the work of the urban ecologists, and suggested, like others, that the ecological approach should be conceived of as a framework for analysing the *spatial organization* of urban residents. Such a framework can offer a great deal to the understanding of a number of urban issues (such as neighbourhood change, social inequality, ethnicity, and social pathology). However, examination of such issues has generally eschewed such an overall framework and adopted instead a somewhat eclectic perspective (Warf 1990). Within human geography, the 'ecological' tradition has nevertheless developed over the past three decades through four lines of enquiry:

1. Studies of the spatial distribution of population that highlight social, economic, and demographic characteristics.
2. Studies of the so-called 'quality of life', or 'social well-being' of the population, with an emphasis on urban deprivation.
3. Studies of the spatial segregation of social groups, in particular those of a minority status.
4. Studies which focus on social problems (e.g. crime and delinquency, physical and mental health) and the way they tend to occur in certain parts of the city or certain social milieux.

These will be examined in turn. In each case, the focus is on the *structure* of the social environment and the way in which this has implications for the *behaviour* of the people resident within that environment.

Urban social areas

Over the past 30 years, attempts to identify the nature of social areas within cities have involved increasing levels of sophistication, often based on multivariate statistical techniques (Knox 1987). Of crucial significance in the study of urban social areas was the development by Shevky, Bell, and Williams, in their attempt to determine 'community areas' in Los Angeles and San Francisco, of what has become known as *social area analysis* (Shevky and Williams 1949; Shevky and Bell 1955). Shevky and his followers realized that a single variable (such as occupation or education) was not truly diagnostic and so they developed a classificatory criterion based on a series of postulates about social differences in post-war United States. Using many of Wirth's ideas, it was argued that urban social differentiation was the product of the increasing 'scale' of society. This involved three major trends: (1) changes in the range and intensity of relations; (2) increasing differentiation of functions; and (3) increasing complexity of social organization.

Table 2.1 Constructs in social area analysis

Statements relating to the character of industrial society (aspects of increasing scale)	Concomitant trends	Changes in the structure of the social system	Constructs	Census measures designed to reflect constructs in previous column	
Change in the range and intensity of relations	Changing distribution of skills (lessening importance of manual productive employment, growing importance of clerical, supervisory, managerial employment)	Change in the arrangement of occupation based on function	Social rank (economic status)	Occupation Education Rent	Index 1
Differentiation of function	Changing structure of productive activity (decline of primary production, growing importance of city-central relations, declining importance of household as an economic unit)	Changes in life style Movement of women into urban employ ment Development of alternative family patterns	Urbanization (family status)	Fertility Women in the labour force Single-family dwelling units	Index 2
Complexity of organization	Changing composition (increased spatial mobility, changes in age and sex structure, increasing heterogeneity of population)	Redistribution in space Changes in dependent population ratio Isolation and segregation of groups in the population	Segregation	Racial and national groups in relative isolation	Index 3

Source: Adapted from Shevky and Bell (1955)

These trends, according to Shevky *et al.*, resulted, respectively, in changes in the division of labour (with a growth in clerical and managerial employment), changes in family life style (with an increase in paid female employment), and changes in the composition of local populations (as a result of greater mobility and, consequently, greater spatial segregation). Evidence of these trends was sought by translating each into a construct – social rank, urbanization, and segregation – which was then measured by a series of census-derived variables (Table 2.1). For each construct, an index was derived based on an unweighted average of standardized scores for each variable. The classification of each census tract was derived by the plotting orthogonally of the social rank and urbanization indices, divided into four equal parts so as to provide sixteen different types. A further division was provided by above- and below-average scores on the segregation index. Figure 2.4 illustrates such social space diagrams for four towns in Britain and provides a comparative

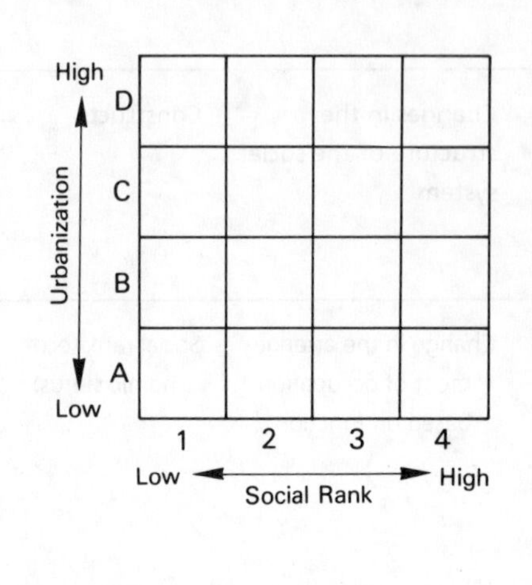

Newcastle-under-Lyme

7	4		
26	17	11	5
13	10	3	2
1	2		

Sheffield

2	7	8	11
93	43	6	11
31	4		

Bournemouth & Poole

	1	4	4
3	13	23	5
19	33	4	1
1	5		

Kingston-upon-Thames

		4	
2	5	32	8
5	31	1	2
	6	1	

2.4 Social space in four British towns, 1971. The numerals refer to the distribution of enumeration districts classified according to the constructs of social rank and urbanization

summary of their social composition. By mapping these data, it is possible to identify social areas that 'generally contain persons having the same level of living, the same way of life, and the same ethnic background', which, in turn, leads to the conclusion that 'persons living in a particular type of social area systematically differ with respect to characteristic attitudes and behaviours from persons living in another type of social area' (Shevky and Bell 1955: 20).

In a detailed review of the social space schema, Timms (1971) outlined many of its shortcomings, particularly emphasizing its tendency to conceive of social change largely in economic terms and to ignore differences in people's social values and in access to power. Partly as a response to these difficulties and the rapid growth in computer technology during the late 1960s, social area analysis was superseded by *factor analysis* as a means of identifying the major dimensions of social differentiation in cities. In contrast to the essentially deductive theory of the social area schema, factor analysis is an inductive procedure by which to analyse the relationships between a wide range of social, economic, demographic, and housing characteristics, with the object of establishing what common patterns, if any, exist in the data. In essence, factor analysis (like the closely related principal components analysis) begins with a matrix of variables against areas. It then seeks to reduce the original variables to a smaller number of synthetic variables ('factors', 'components', or 'dimensions') that take stock of the interrelatedness of the original data.

The nature of factor analysis can best be illustrated with reference to one example, the case of Leicester in 1966 (Lewis and Davies 1974). The study employed fifty-three census-derived variables, on an enumeration district basis, to represent the broad spectrum of the city's demographic, social, economic, and housing structure. Using a varimax rotation of a principal axes solution, the input variables collapsed into eight major dimensions, which together accounted for 72 per cent of the variance in the initial data set (Table 2.2). Of these components, the first three were particularly interesting. Component I revealed high loadings on variables reflecting socio-economic status (such as income, education, occupation, and material possession). In contrast, Component II represented a life-cycle dimension (contrasting the distribution of older persons with those of large households, young children and economically active males) and Component III suggested an association of variables measuring levels of mobility (including migration and journey to work). Figure 2.5(a) reveals that Component I had an identifiable sectoral distribution with high-status residents concentrated to the south-east of the city centre and in a number of new peripheral suburbs to the north and west, while low-status areas were concentrated in a zone stretching from the south-west and west across the inner city towards the east and north-east, thereby linking the late nineteenth-century terraced housing areas with the inter-war and post-war public

Table 2.2 Factor loadings in Leicester, 1966

Component I. Socio-economic status
+0.85 Two-car households
+0.85 High social class
+0.84 Employers–managers
+0.78 High car ratio
+0.75 Professional workers
+0.71 Intermediate non-manual workers
+0.65 Distribution–service workers
+0.63 Car commuters
+0.61 Owner occupiers
+0.32 Government workers
−0.40 Irish
−0.56 Foremen–skilled workers
−0.58 Pedestrian commuters
−0.59 Tenants unfurnished property
−0.66 Non-car households
−0.69 Industrial workers
−0.69 Unskilled manual workers
−0.83 Low social class
−0.82 Personal service–agricultural workers
Per cent variance explanation = 17.6%

Component II. Stage in life cycle
+0.77 Old aged
+0.71 Middle aged
+0.59 Rooms per person
+0.48 Households with pensioners
+0.42 Tenants unfurnished property
+0.37 Sparse occupancy households
+0.35 Owner occupiers
+0.33 Pedestrian commuters

−0.55 Mature adults
−0.58 Council tenants
−0.81 Large households
−0.85 Children
−0.93 Single ratio
Per cent variance explanation = 9.9%

Component II. Mobility
Positive – None

−0.31 One-person households
−0.41 Mature adults
−0.53 New residents
−0.41 Households sharing dwellings
−0.65 Recent movers
−0.65 Local movers
−0.82 Single-person movers
−0.93 Movers within five years
−0.96 Female movers
Per cent variance explanation = 8.2%

Source: Adapted from Lewis and Davies (1974: 198)

housing estates. On the other hand, Component II was much more concentric in its distribution, often cutting across the socio-economic divisions, with young families predominating on the city's periphery, giving way towards the centre to an older population, and thence to a zone of younger families again close to the city centre (Fig. 2.5b). The latter is particularly interesting since it reflects the 1950s and 1960s influx of immigrants, particularly from the Indian subcontinent. The mobility dimension (Component III) reveals a much more complex pattern: it identifies not only the recency of the immigrant settlement and the development of new suburban private and public estates but also the high turnover in many middle-class neighbourhoods. The immobility of the older population of the intermediate concentric zone is particularly striking (Fig. 2.5c).

In line with tests of the social area schema, the majority of factorial studies of North American cities have confirmed the significance of social rank, urbanization, and segregation as basic underlying dimensions of socio-spatial differentiation. Salins (1971) has even shown some consistency in these dimensions through time for four United States cities: social status remained sectoral in 1940, 1950, and 1960, family status exhibited a zonal gradient, but ethnicity was subject to greater change. From such evidence many have claimed to derive a model of the Western city, yet evidence from Western Europe is nowhere as consistent as from North American cities (Table 2.3). Many have argued that the existence of extensive public housing is a major difference in the British context, while the absence of large-scale immigrant populations in Scandinavian cities needs

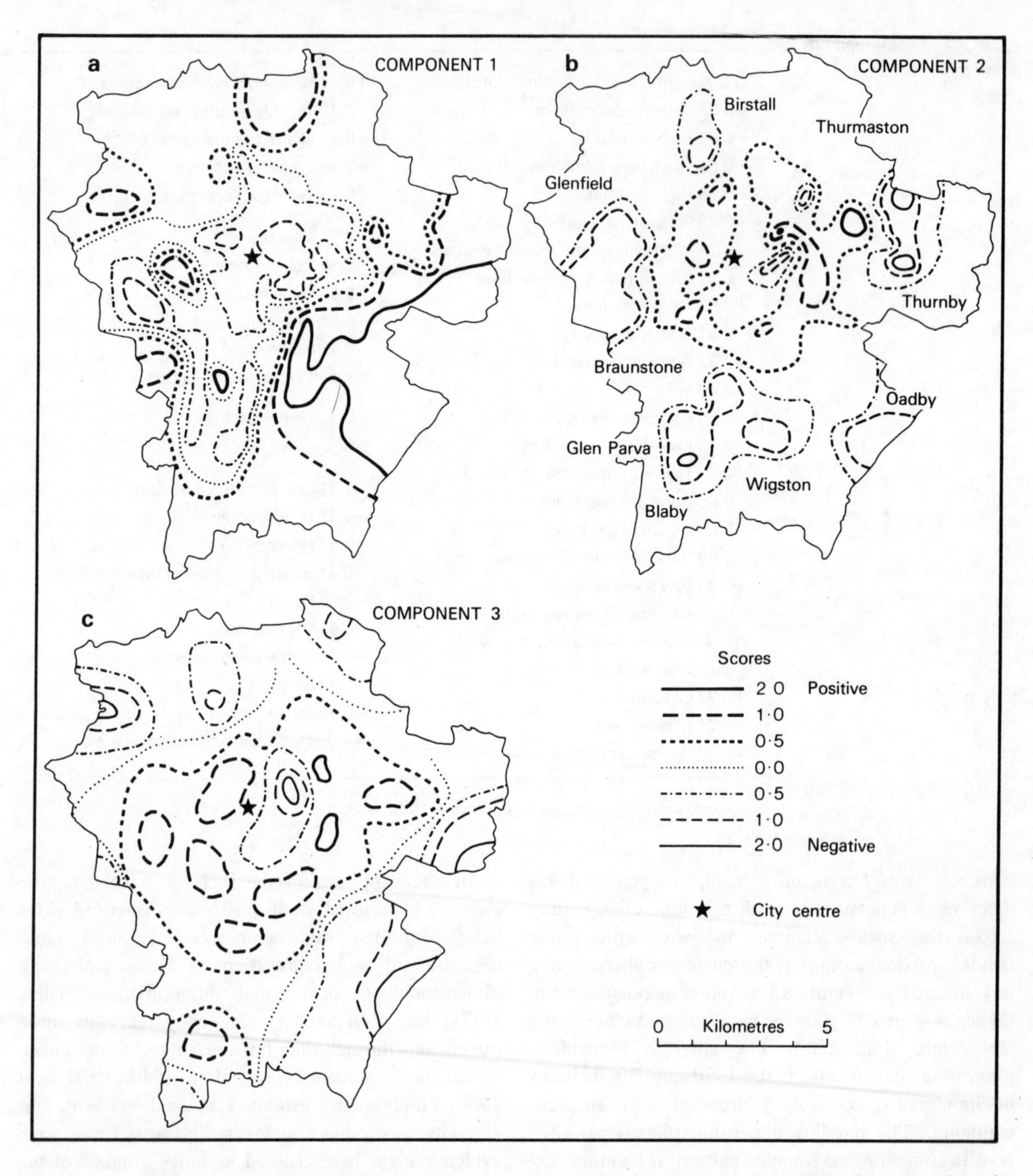

2.5 Factorial ecology of Leicester, 1966: (a) socio-economic status; (b) stage in life cycle; (c) mobility (Source: Adapted from Lewis and Davies 1974: The social patterning of a British city: the case of Leicester, *Tijdschrift voor Economische en Sociale Geographie* **65**, 199–203)

Table 2.3 Selected empirical findings from factorial ecologies

Author	City	Observed pattern of variance		
		Social status	Familism	Ethnicity
Rees (1970)	Chicago	Sectoral	Zonal	Clustered
Murdie (1969)	Toronto	Sectoral	Zonal	Sectoral
Timms (1971)	Brisbane Auckland	Sectoral	Zonal	Clustered
Robson (1969)	Sunderland	Zonal and sectoral	Weak	—
Herbert (1972)	Winnipeg	Sectoral	Zonal	Clustered
Davies and Lewis (1973)	Leicester	Sectoral	Zonal	Clustered
Johnston (1973b)	Auckland Wellington Dunedin Christchurch	Sectoral	Zonal	Clustered

also to be taken into account (Knox 1987). However, as illustrated in the case of Leicester, these two likely distinguishing features are more apparent than real with a result that they do little to detract from the universality of social status and family status as key elements in urban structure. Such a view fits in with Abu Lughod's (1969) long-established claim that:

1. Residential segregation according to socio-economic criteria will occur if the rank ordering of society is matched by a corresponding ordering of the housing market;
2. Family status will be a determinant of urban structure where families at different stages of the life cycle reveal different residential needs and where the available housing is able to fulfil these needs.

Despite the widespread adoption of factorial methods, considerable difficulties are involved in their use and in the interpretation of their results. For example, inevitable differences in input data make generalizations exceedingly hazardous, particularly where census sub-areas do not match actual patterns of residential variation. Additionally there is the problem of spatial autocorrelation (Cliff and Ord 1973) and the tendency in many studies to interpret factors (in other words, give them labels) using only a few major loadings, thereby obscuring or distorting more complex relationships. Researchers interested in overall spatial patterns rather than underlying social dimensions have attempted to avoid some of these difficulties by adopting grouping procedures, such as cluster analysis and multiple discriminant analysis, which provide a classification of census sub-areas. Figure 2.6 shows the result, again for Leicester, of a cluster analysis of factor scores for each area on the leading components (socio-economic status, mobility, life cycle, ethnicity, and housing quality) of an analysis, similar to that in Fig. 2.5 but this time based on 1971 data. Visual inspection of the dendogram resulting from the cluster analysis suggested an eight-group classification. Groups A and B form the most disadvantaged parts of the city, incorporating high density, low income, and generally poor housing. The difference between the two stems from the fact that A is characterized by high proportions of New Commonwealth migrants while B includes more elderly native residents. In contrast, Group C includes areas of more diverse population and may best be described as the city's 'rooming house district'. Those parts of the city developed during the inter-war period are picked out by Groups D (mainly low status public housing) and E (middle to high status private housing). Both have low fertility ratios and an ageing population. The post-war local authority housing estates, with low status and high fertility ratios, are identified in Group F. Group G comprises the most prestigious residential areas,

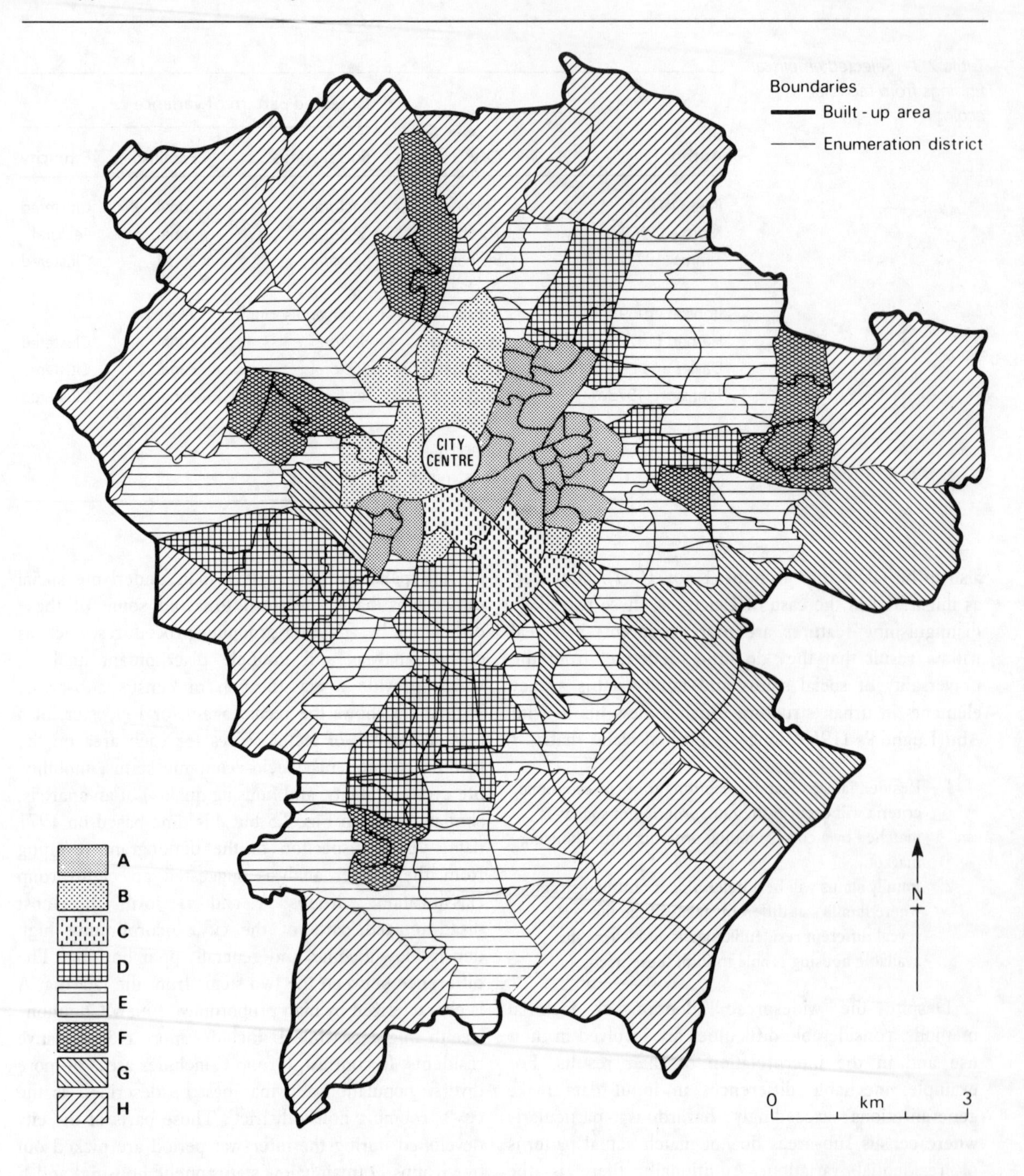

2.6 Cluster analysis for Leicester, 1971 (Source: Lewis 1981: 42)

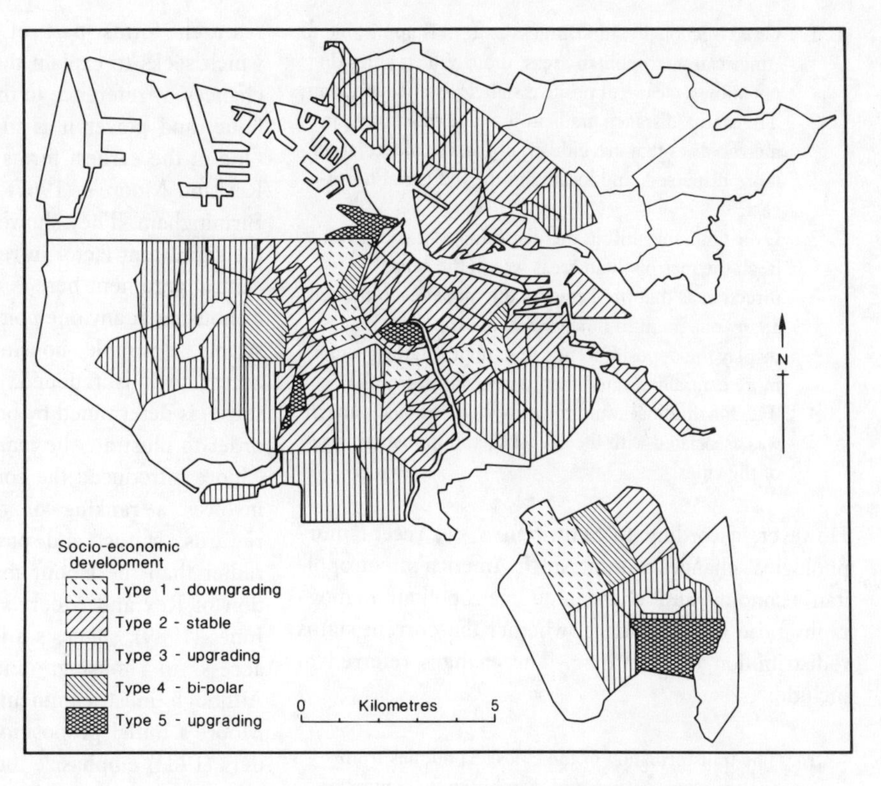

2.7 Socio-economic upgrading and downgrading in Amsterdam (Source: Musterd 1991: Neighbourhood change in Amsterdam In: *Tijdschrift voor Economische en Sociale Geografie* **82**, 30–9)

with high social status and low density. Slightly lower status and higher fertility ratios are to be found in Group H.

Of late, the focus of attention in geographical studies of social space has shifted away from attempts to identify the basic structure of residential differentiation and to the exploration of neighbourhood change (Cater and Jones 1989). Two types of studies are apparent. The first is concerned with the social and physical processes involved in the changing status of urban sub-areas (Goetze and Colton 1980; Herlyn 1989). For example, in a study of neighbourhood change in Amsterdam during the 1980s, Musterd (1991) used two indicators – the change 1986–87 in the percentage of households with relatively high income (above 2000 guilders after tax per month) and the change 1982–86 in the percentage of households on unemployment benefits – to classify neighbourhoods into four types according to whether they were socially stable, socially upgrading, socially downgrading, or characterized by bipolar development. Figure 2.7 reveals that the inner city was predominantly stable though some parts, particularly those neighbourhoods on its outer fringes, were in the process of downgrading while other parts had a mixture of development and decline. The latter were neighbourhoods in the throes of gentrification (Hamnett 1984). In general, the peripheral parts of the city were being upgraded though the experience of the neighbourhoods in the south-east of the city was diverse. In particular, the recently built Bijlemermeer district was unmistakably undergoing socio-economic decline.

The second group of studies concerned with the structure of neighbourhoods has focused on the broader question of the changing location of social groups within cities. For instance, Schwirian *et al.* (1990: 1143) posed the question 'how generalizable is the Burgess growth model in describing the residential redistribution of social status groups in American metropolitan areas?' In answering this question, using data for 318 American metropolitan areas from 1950 and 1980, their study showed that, over the 35-year period:

1. Cross-sectionally, the Burgess pattern is applicable to American metropolitan areas, though most cities do not show a clear-cut positive status/distance gradient;
2. The status/distance gradient is found most often in metropolises that are industrially based, older, larger, more dispersed, and located in the industrial north-east;
3. Over time, the shift in the distribution of social status in all the metropolitan areas was in the same direction as that predicted by the model even though at any one point in time several metropolises may display the opposite trends (e.g. upper status groups more centrally located than lower-status groups);
4. The longitudinal trend towards the predicted pattern was associated with the age, size, and industrial base of the cities.

However, according to Schwirian *et al.*, recent morphological changes within North America's metropolitan economy and changes to metropolitan demography raise the question of whether the current status redistribution will continue. The changes referred to include:

1. The transformation of the cities' economies from centres of production and distribution to centres of information exchange, service provision, and corporate and governmental administration;
2. An increased degree of decentralization of employment from the city to the periphery;
3. A shift in central city population composition from predominantly white to predominantly black, Hispanic, or Asian;
4. An increased surburbanization of the black population, albeit to segregated black suburbs.

Clearly, these four changes can affect the pattern of status redistribution though the nature and form of that redistribution are far from clear.

Paralleling the growth of interest in neighbourhood change has been a shift away from an ecological interpretation of city structure and towards a much broader analysis focusing upon the role of housing. Following Abu Lughod's (1969) claim that the prime determinant of residential differentiation is the form and workings of the housing market, considerable efforts have been made to discover the link between the location of housing and the location of social groups (Cater and Jones 1989). Essentially, this research forms part of the *managerialist perspective* which seeks to explain the unequal distribution of life chances by reference to those agencies whose responsibility and power it is to undertake that distribution. One of the earliest forms of urban managerialism was Rex and Moore's (1967) classic study of Sparkbrook, Birmingham. They showed that access to housing was the significant factor in residential differentiation. The central argument here is that, although the total stock of housing at any one point in time may not be in short supply, desirable housing usually is and therefore access to it is rationed. Housing desirability, in this sense, is determined by both tenure and quality, and in order to illustrate the managerial perspective, Rex and Moore introduced the concept of *housing classes* which involves a ranking of the population in terms of rewards, status, and power in the housing market rather than the labour market. Table 2.4, an adaptation of Rex and Moore's original ideas by Cater and Jones (1989), suggests a link between housing classes, access to housing, and neighbourhood location. Although most commentators agreed with Rex and Moore's initial proposition, recent writers like Saunders (1984) emphasize the need to consider to a much greater extent government housing strategies. For example, the British Government's intensified financial support for owner occupation, coupled with the impoverishment and progressive privatization of the public sector in the 1980s, is converting public housing into a residual and ever more undesirable form of tenure. Thus policy is impacting on the social structure of cities to the point where some areas are possibly deprived.

In this context, it is important to note that some groups may be becoming excluded from the allocation process. Parkin (1979) has theorized about this, focusing on Weber's concept of 'closure'. What Parkin did was attempt to link voluntary action and the system of constraints by which such actions are limited. 'Social closure' was defined as 'the process by which collectivities seek to maximize rewards by restricting access to resources and opportunities to a limited circle of eligibles' (Parkin 1979: 44). From this, a conceptual framework was built which illustrates how the activities of one group constrain the actions of others. In other words, closure is a means of mobilizing power for the purpose of engaging in the

Table 2.4 A modified version of Rex and Moore's housing class scheme

Class	Tenure, occupants, and typical location	Entry requirements	Utility derived by occupants			
			Material	Status	Power	Neighbourhood
1A	Owner occupiers (outright or mortgaged) of legitimate housing; suburban (or sometimes conserved or gentrified)	Wealth or creditworthiness; meet building society criteria	Acceptable shelter at a price subsidized by tax relief; appreciation of assets	Property ownership held in high esteem	Exclusivity; autonomy; mobility and choice	Local access to superior environment, schools and services
1B	Renters of legitimate council housing, usually suburban	Meet local authority 'points' and other criteria – prove need and worthiness. 'Respectable' working class	Acceptable shelter; some rent subsidy	Stigma attached to council renting	Dependent on local authority provision and rules; limited choice and movement	Local services may be inferior; mundane and uniform design; poor location
2A	Owner occupiers of low-grade property; usually old and inner city	Failure to qualify for 1A and 1B – low income, unemployment, perceived deviancy (e.g. non-white, immigrant, single parent, 'problem' family, mental illness, criminal record – all may act as disqualifier)	Frequently unmodernized, overcrowded and lacking amenities; capital appreciation minimal and little tax relief	Pale imitation of 1A	Independent, but often lack capital to maintain or improve	Old 'worn out' neighbourhood; high population density, mixed land uses; hazards and dangers include noise, dereliction, traffic, pollution, petty crime, vandalism, filth; poor service provision; negative image etc.
2B	Tenants of undesirable council property; inner city or overspill		Overcrowding, lack of amenities, disrepair	Second class version of 1B; highly stigmatized	Dependency on often substandard local authority provision	
3	Private renters; mostly inner city	As 2A and 2B	Beset with similar problems to 2B, but higher rents, multi-occupation and shared facilities in many cases	Slum dweller	Dependent and vulnerable, though some legal protection	

Source: Cater and Jones (1989: 61)

struggle over the distibution of scarce resources. It can therefore take at least two forms: *exclusionary closure* (the attempt by one group to subordinate another in order to gain a privileged position) and *usurpationary closure* (the attempt by the subordinated group to use power in an 'upward' direction by threatening the privileges of the legally defined superior group). Of course, confrontation can also occur within the excluded group (the notion of *dual closure*) as when social closures are directed against ethnic minorities within the already excluded group. In short, Parkin conceives of social groups as being neither a set of consensual individuals nor individuals forced by their designated structural positions into conflict. Rather individuals attempt to secure economic and social advantage in a world which fails to distribute resources equitably. Jackson and Smith (1984: 64) claim that 'in suggesting that classes may be defined with reference to their mode of collective action and not simply their place in the productive process or the division of labour, Parkin provides an appealing conceptualization of the articulation of voluntarism and determinism in social life'. In many senses, then, Parkin's views not only throw new light on ideas such as 'housing classes' but also anticipate the debate about structure and agency that is at the heart of the writings on 'structuration' that are discussed later in Chapters 3 and 6.

Urban deprivation

The current concern among geographers with urban deprivation has its intellectual antecedents in studies by urban ecologists of issues such as mental disorders, delinquency, and prostitution (Kurtz 1984). Urban deprivation finds expression both as problems *in* the city (the notion that the problems in question are found generally throughout society, but are particularly concentrated within urban areas) and as problems *of* the city (the notion that the problems are generated by the urban environment itself). Poverty may be an example of the former and noise pollution an example of the latter. A variety of terms such as 'quality of life', 'social well-being', and 'life chances', are often used interchangeably with that of 'deprivation' to denote a concern for assessing the 'social' and 'psychological' aspects of urban living, as well as the 'economic' (D.M. Smith 1977, 1979b). In short, the aim of work in this field is to answer the question 'who gets what where?' in order to identify where there are high and low levels of well-being. However, considerable problems are encountered when attempting to measure concepts like 'well-being'. Probably one of the earliest attempts to enter this field was the Committee on Social Trends set up by President Hoover in the United States in 1929. Social indicators (social statistics that monitor things such as unemployment rates, educational attainment, and housing standards) were used to measure the condition of the nation. In order to identify regional and local variations, these indicators were disaggregated into smaller geographical units, a procedure which results in what are nowadays referred to as *territorial social indicators*. Hoover's initiative was shortlived, largely because of

Table 2.5 Criteria for defining human well-being

1. *UN components of level of living*
   Health, including demographic conditions
   Food and nutrition
   Education, including literacy and skills
   Conditions of work
   Employment situation
   Aggregate consumption and savings
   Transportation
   Housing, including household facilities
   Clothing
   Recreation and entertainment
   Social security
   Human freedom

2. *Criteria of social well-being in the United States*
   Income, wealth, and employment
   The living environment
   Health
   Education
   Social order
   Social belonging
   Recreation and leisure

Source: Adapted from Smith (1979c: 21)

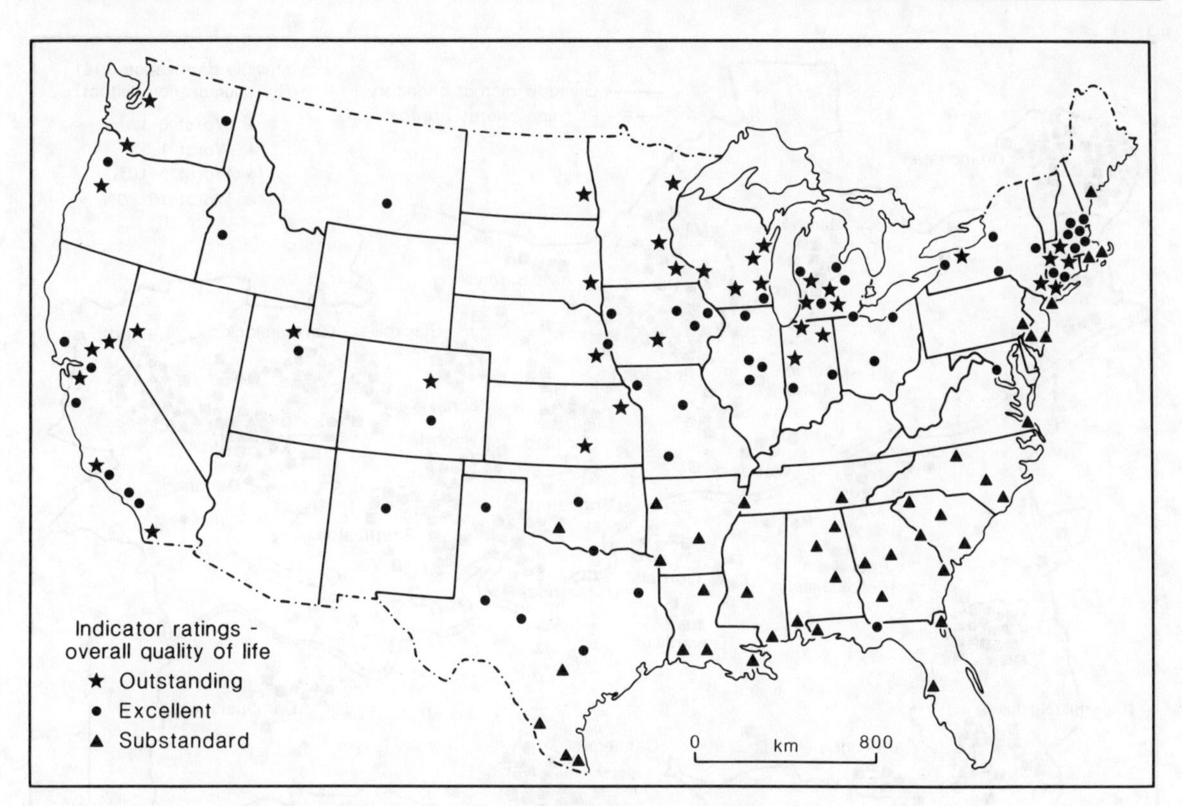

2.8 Overall quality of life in American cities (Source: Herbert and Thomas 1982: 412)

more pressing government initiatives during the Depression of the 1930s. Only in the 1960s did *geographers* begin to focus on social indicators and thereby contribute to the understanding of spatial variations in the quality of life. This was partly as a response to pleas for greater relevance in geography and also as a result of the greater availability of suitable data. Table 2.5 shows, for example, two sets of criteria used in major studies of the quality of life.

At an *inter*-urban scale, there have been several attempts to classify cities according to quality of life based on social indicators. For example, Liu (1975) used five broad classes of social indicator – economic, political, environmental, health/education, and social – to measure variations in the quality of life among US cities (Fig. 2.8). In this study, large metropolitan areas generally tended to score highly on the economic component (notably Dallas, Houston, and Portland in the south and west, a cluster of metropolises in the Midwest, and emerging clusters of smaller cities in 'sunbelt' states such as Texas). On the social component, the high-scoring metropolitan areas were found in the Rocky Mountain states and on the west coast, together with the 'newer' urban areas of the plains. In contrast, the low-scoring metropolitan areas were to be found east of the Mississippi and in older urban America.

In contrast to these inter-urban analyses, many more studies have focused on *intra*-urban variations in the quality of life (Herbert and Smith 1979). Overwhelmingly, these studies have emphasized that most urban problems are to be found in the inner city. The inner city is characterized, typically, by a falling population, high unemployment, poor housing, overcrowding, and a preponderance of the disadvantaged, vulnerable, and victimized (Robson 1988). Many of these residents form an 'underclass' of 'persons who lack training and skill and either experience long-term unemployment or have dropped out of the labour force altogether' (Wilson 1985: 174). An interesting recent

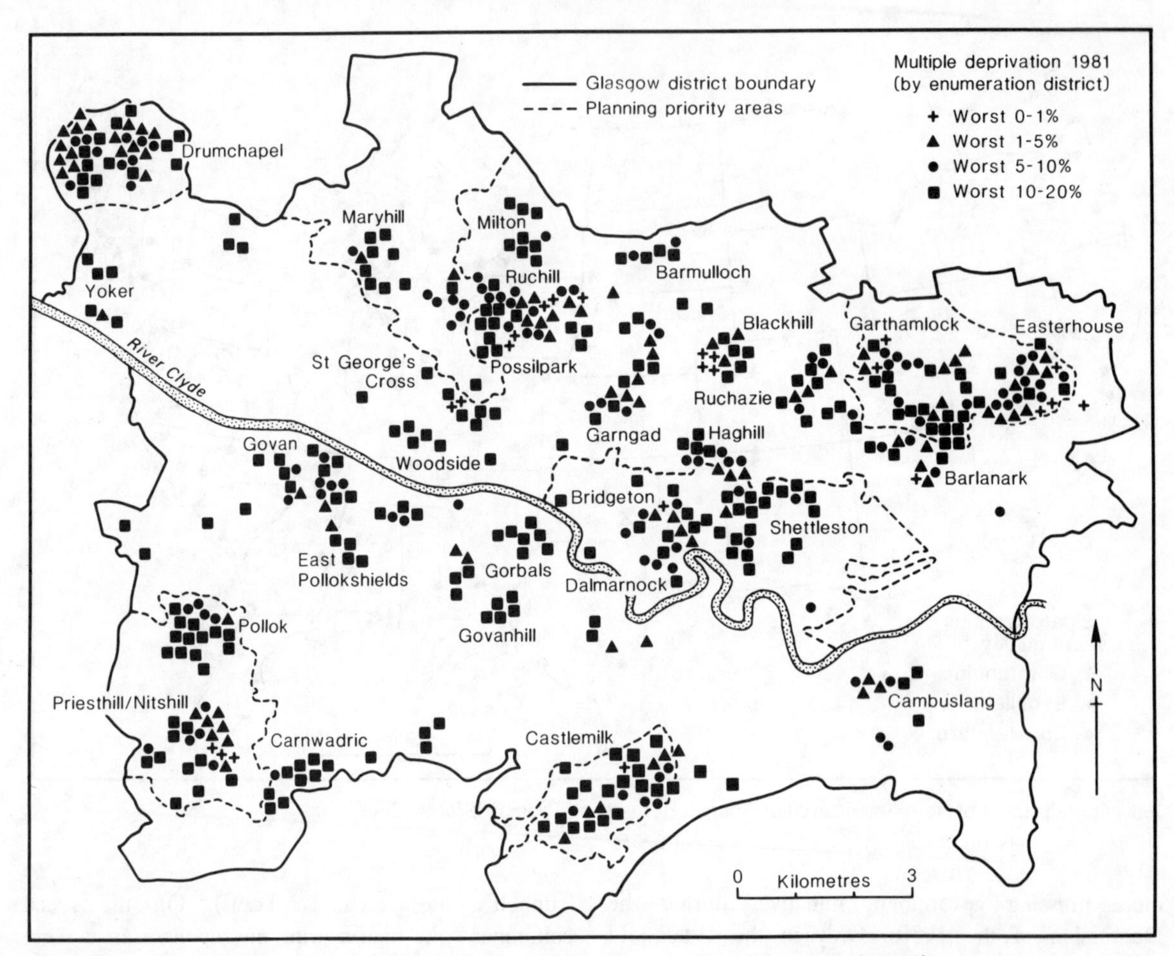

2.9 Distribution and incidence of multiple deprivation in Glasgow (Source: Pacione 1990b: 308)

contribution to the understanding of urban deprivation was Pacione's (1990b) study of Glasgow in 1981 (Fig. 2.9). By comparing the present situation with that in 1971, Pacione concluded that inner-city deprivation had attenuated, though it was still evident in Gorvan, Possilpark-Springburn and parts of East End. In contrast, in four large peripheral public housing estates – Castlemilk, Easterhouse, Pollok, and Drumchapel – deprivation had increased significantly. Of course, over the 10-year period in question, the inner city had experienced an extensive clearance and redevelopment programme, including the large-scale relocation of residents to the periphery. As a result, the inner city now contains a much reduced and ageing population while the outer public housing estates have a much younger population, residing in adequate housing, but experiencing serious social problems including overcrowding, unemployment, and single parenthood.

Within the literature on the quality of life, the controversy about the most suitable means of measurement is overshadowed by controversy as to the cause of deprivation (Townsend 1979). Table 2.6 summarizes the four main theories. Significantly, the first – the *cycle of deprivation* – has an affinity with the ecological perspective. It argues that children born into deprived households and into deprived neighbourhoods face fewer opportunities because of their limited access to alternative paths and possibilities. So a 'cycle' is perpetuated as these children go to deprived schools, thus acquiring few usable skills, few quali-

Table 2.6 Theories of poverty and deprivation

Theory	Source of problem	Characteristics of problem	Perpetuating features	Main outcome
Structural class conflict	Social formation	Unequal distribution of power Maintenance of disparities	Organization of labour Class distinctions	Inequality
Institutional management	Allocative system	Uneven distribution of resources Inefficient bureaucracies Weak communications/awareness	Maintenance of élitism Low welfare inputs Non-sharing of opportunities	Disadvantage Underprivilege
Cycle of deprivation	Social group Residential area	Few opportunities Limited access to social mobility Transmission of poor attitudes Disadvantaged environment	Sub-cultural norms Lack of positive interventions	Deprivations Low aspirations Low achievement
Culture of poverty	Individual inadequacies Family background	Group apathy Inherited deficiencies	Fatalism Failure of welfare services	Retardation Poverty

Source: Adapted from Herbert and Thomas (1982: 414)

fications, and sets of values that are no different from those of their parents. Thus, in adulthood, like their parents, they find themselves failing to compete successfully in the labour and housing markets. In policy terms, then, this 'theory' emphasizes the need for positive intervention in the home, school, and neighbourhood in order to compensate for inadequate aspirations and unacceptable behavioural norms. The *culture of poverty* argument also has its roots in urban ecology. This 'theory' suggests the existence of a general group attitude in which low-key aspirations and fatalistic assessments of low achievement are typical. However, this perspective also emphasizes some individual genetic connotations. Again, this theory calls for positive intervention in the education and welfare systems, but with a greater emphasis on the individual than is the case with the cycle of deprivation perspective.

In contrast, the remaining two theories take a much broader interpretation. Following the managerialist perspective discussed earlier, the *institutional management* approach to the study of urban deprivation suggests that the allocative systems of government and of private institutions fail to channel goods and services in ways sufficient to reduce deprivation. In other words, this 'theory' focuses on the need to reform the allocative system. *Structural class conflict* theories, on the other hand, suggest that all policies aimed at tackling deprivation are only palliative since they fail to tackle the root cause – the particular social formation that causes the problem. This group of 'theories' argues that capitalism invariably produces inequality because it allocates differential rewards in a competitive society with the result that some people win and others lose. From this perspective, the remedy to deprivation lies in considerable change to both societal structure and attitudes. Of course, all four perspectives set out in Table 2.6 are lacking in one fundamental sense: they fail to acknowledge that the concept of quality of life has different meanings to different people. Research therefore needs to develop indicators based on the feelings and values of the people under study, just as much as it needs 'objective' indicators of their housing conditions and educational attainment.

Spatial segregation of ethnic groups

Another significant focus in the work of urban ecologists has been upon the tendency for minority groups to form distinctive *de facto* territories. According to Boal (1976), such spatial segregation is inversely related to the process of assimilation within the host society, a process which itself is governed by different forms of group behaviour designed to minimize real or perceived threats to the group from outsiders. The process of assimilation can take place at different speeds for different groups. Normally, behavioural assimilation – the acquisition by the minority groups of a cultural life in common with the host society – will take place faster than structural assimilation – the diffusion of members of the minority group through the social and occupational strata of the host society. Most of the research on minority groups has been concerned with those ethnic minorities for whom the constraints upon assimilation are high and for whom levels of discrimination are greatest. These groups are often racially distinct and it is for such groups that membership of a minority is most likely to lead to residential segregation. In other words, residential segregation is an expression of subordinate status.

Despite this, Boal (1976) has claimed that such segregation should not be viewed entirely in a negative fashion because ethnically segregated areas fulfil a number of functions. For example, they fulfil a 'defensive' function, thereby enabling the isolation of individual members of an ethnic minority to be reduced. At the same time, residential segregation permits an organized defence of the group within a defined area to be developed. In their respective enclaves, members of minority groups can thus feel protected. A part of this defence involves what might be thought of as a 'preservation' function, the aim of which is to maintain the group's identity and cultural heritage through such things as language, religion, and marriage customs. Ethnic residential segregation also has a self-supporting role, or 'avoidance' function, whereby recent migrants are initiated into, and familiarized with, city life within the confines of the group. Finally, for some groups, territory is used to fulfil a 'resistance' function. In this sense, group affiliation with an area serves as a foundation for a power base for action against wider society.

Currently, two methods are widely used to compute the degree to which an ethnic group is residentially segregated (Duncan and Duncan 1955). The first, the *dissimilarity index*, measures the proportional difference between the members of two groups within a sub-area (usually a census tract) in a city and indicates the minimum proportion of either group who would have to change their sub-area residence to obtain an even distribution across all sub-areas of the city. The second index, the *exposure index*, is a measure of one group's residential isolation from another group. Under complete separation, members of a group would tend towards contact only with members of the same group; in contrast, if there was balance, they would meet members of other groups at a rate equal to the city-wide proportions of those groups.

In North America a number of studies have revealed that, of all the ethnic minorities in the metropolitan cities, the most segregated are the blacks. For example, a study by Taeuber (1965) found that the median dissimilarity index was 80, while Van Valey *et al.* (1977) found that over half of the 237 cities they analysed recorded dissimilarity values in excess of 70. In comparison, ethnic residential segregation in European cities is much lower. In Britain the index values of migrant groups (such as the Afro-Caribbeans, Indians, Pakistanis, and Bangladeshis) range from 40 to 70, although at the local level in small areas it could be much higher (Cater and Jones 1989). Some changes in segregation patterns have of course occurred during the past three decades, though significant levels of segregation still exist between blacks and whites in both American and British cities (Lighter 1985; D.J. Smith 1986). For example, in a study of 38 metropolitan areas, Clark (1986b) found that, between 1960 and 70, the *majority* of metropolitan areas experienced an increase in black segregation though some did have lower levels of segregation.

However, during the next decade there was a decline in segregation in all cities. Indeed twenty-five metropolitan areas had a decline of 5 percentage points or more in their dissimilarity indices. In a detailed study of the San Francisco Bay area between 1970 and 1980, Miller and Quigley (1990) confirmed these trends. At least part of the explanation for this decreasing level of segregation between 1970 and 1980 was the increase in black suburbanization.

Although such suburbanization was not new, its increase in the 1970s was dramatic even though its pattern was far from uniform across all US cities (Rose 1976). Certainly the suburbanization was very evident in the older and larger metropolitan areas (Clark 1986b). This is not surprising because it was in these cities that there was a large black population in the central city and it was from these concentrations that nearby suburbs drew large numbers of blacks (Johnson and Roseman 1990). However, the significant feature of the 1970s and 1980s was that black suburbanization was more than simply a spill-over phenomenon because it also included the relocation of comparatively young, affluent, and well-educated blacks into predominantly white areas (Frey 1985).

Almost all studies recognize that any explanation of residential segregation must adopt a multi-causal framework which involves affordability, social preferences, urban structure, and discrimination. Despite this, some researchers strenuously hold to the view that discrimination, largely that of local and national government, is the principal cause of segregation (Jackson 1987). Of course, not all agree and, of late, the relative importance of the various factors has become a source of considerable academic debate as researchers wrestle with the question of whether segregation results from a dominant factor – discrimination – or from of a complex reinforcing set of factors that in concert generate the observed geographical distribution of racial groups (W.A.V. Clark 1988; Galster 1988).

Social issues and social milieux

At the heart of studies of racial segregation is a concern for the well-being of different ethnic groups. Inevitably the study of social well-being involves examination of a wide range of conditions under which social problems occur. Once again, the study of these social conditions, and of the links between them and social problems, has its intellectual origins in studies by urban ecologists of issues such as mental disorders, delinquency, and prostitution. The prime concern in such work was the identification of the incidence of social issues and description of the physical and social environment which generated their occurrence. The implication was that an understanding of social milieux would lead to an understanding of social problems.

An example of this type of research was Faris and Dunham's (1939) study of several types of mental disease and associated pathologies including schizophrenia, manic depression, alcoholism, drug addiction, senility, and psychoses in Chicago and Providence, Rhode Island. Apart from the case of manic depression, the highest incidence of mental disorder was found in the heart of the city, and, by combining the rates for specific disease categories into a general insanity rate, a distance decay effect was demonstrated whereby the lowest rates were at the greatest distance from the city centre (Fig. 2.10). These results confirm the earlier evidence of Shaw and McKay's (1929) classic study of delinquency areas. They have also been supported by a large number of subsequent studies (Dunham 1937). Central city dominance is most vividly illustrated by schizophrenia which, according to Faris and Durham (1939), was found in high rental apartment districts, skid row areas, and predominantly black areas. More recently this pattern has been confirmed by Giggs (1973) in Nottingham, where 68 per cent of schizophrenia patients lived within 4 km of the city centre (Fig. 2.11). However, Giggs took care to point out that many of these patients were psychologically vulnerable persons who had moved into the inner-city locations. In other words, central city areas do not cause schizophrenia so much as attract schizophrenics. There is, then, a degree of confusion as to whether residence in a given area is a cause or a symptom of mental illness. The pattern is a little less confused for crime and delinquency.

Crime and delinquency are distributed within cities in a similar fashion to mental disorders; in fact indices of mental illness have frequently been combined with delinquency scores to form composite measures of social malaise. Since the early study by Reckless (1934), the majority of subsequent research has confirmed the tendency for crime and vice to be concentrated in and around the city centre. For example, Shaw and McKay's (1942) classic study in Chicago found that crime and delinquency occurred around the 'Loop' (essentially the central business district) and thence declined in a gradient towards the periphery. What they noted also were the recurrent

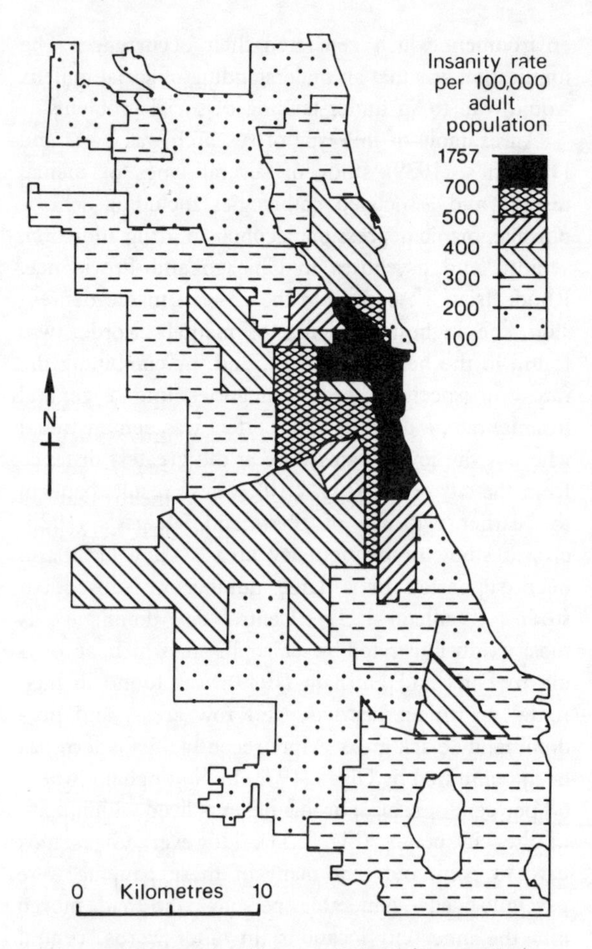

2.10 Insanity rates in Chicago, 1930–31 (Source: Faris and Dunham 1939: *Mental disorders in Urban Areas*, University of Chicago Press, Chicago)

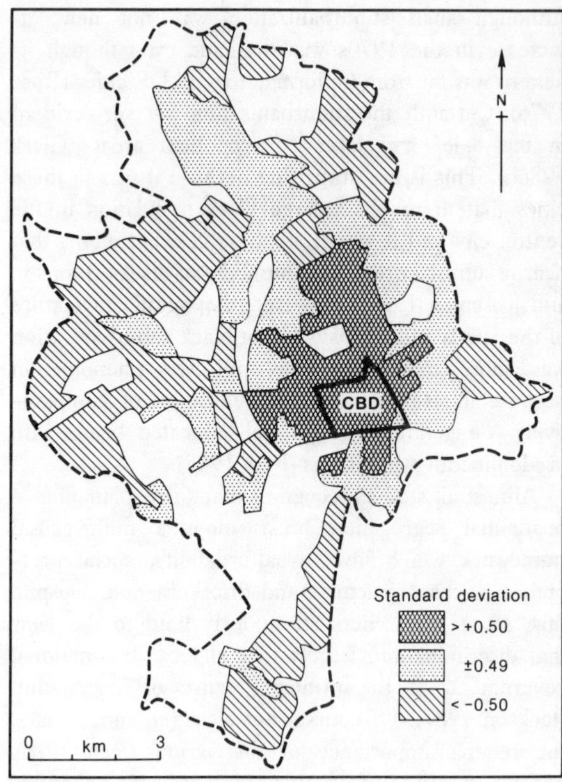

2.11 Schizophrenia in Nottingham 1963–69 (Source: Giggs 1973: 62)

correlates of crime, particularly substandard housing, poverty, foreign-born population, and high mortality. This led Shaw and McKay to draw a distinction between the central districts of poverty and physical deterioration, whose inhabitants were mobile and possessed confused cultural standards, and the stable family suburbs where delinquency was invariably low. Over the next 40 years, several major studies confirmed these findings though it has become increasingly evident that, in order to understand delinquency and crime, attention must be given not only to where offenders live, but also to where the crimes are committed (Harries 1980; Herbert 1982; Davidson 1981).

Statistics on the occurrence of crime reveal variations with the type of offence. Pyle (1974), for instance, demonstrated that, in the United States, shoplifting and cheque fraud offences are concentrated in the city centre, burglary and larceny in the suburbs, and robbery and drunkenness in the 'skid-row' parts of the city. Later Pyle (1976) also found a distinct association between low-income neighbourhoods and crimes of violence (including murder, rape, and assault) and between the 'transitional' areas of the city (characterized by a decaying physical environment and a concentration of manufacturing and wholesaling) and offences such as robbery, larceny, and car theft, as well as assault and murder. In contrast, middle to upper-income suburban neighbourhoods were associated with property crimes such as burglary, larceny and car theft. This pattern suggests that there might be marked spatial variations in opportunities for crime, particularly in large cities.

The centre–periphery dichotomy in the incidence of several urban pathologies has traditionally been explained in terms of the physical and human milieux of the city. Initially much emphasis was placed on a physical environmental explanation generally involving noise, pollution, and inadequate housing conditions. Admittedly noise is particularly high in and around the city centre but as yet its relationship with various forms of pathology has not been clearly demonstrated. On the other hand, McHarg (1969) and Esser (1971) have revealed links between levels of physical and chemical pollutants, which are highest over city centres, and both physical and mental abnormalities. However, with greater pollution control, this causal link may be of declining significance. The question of housing has been reviewed by Schorr (1963) who concluded that housing conditions clearly influence physical and mental health, especially where the housing quality is extremely inadequate. Wilner and Walkley (1963) confirmed this view when they found that, of forty studies, twenty-six showed positive associations between housing conditions and health. In other words, as housing quality improves, so too does health.

Of course, an obvious explanation of the high incidence of pathology in certain areas relates deviant behaviour to poverty and overcrowding. It is little wonder, therefore, that Martin (1967) concluded that if overcrowding, air pollution, and poor socio-economic conditions are eliminated, other factors are of little influence in determining the incidence of ill health. The vicious circle of poverty, physical blight, and deprivation is often encompassed by the term 'social disorganization' and seen, in that context, as a cause of mental illness. For example, McNeil (1970) has suggested that having employment, education, and a stable marriage promotes security while a lack of these 'stabilizers' diminishes an individual's capacity to adjust and may, in severe cases, lead to psychoses. Controversy still remains, however, over whether the physical environment of the central city itself creates a pathology, or whether those prone to exhibit pathological tendencies drift into apartments and rooming houses close to the urban core. There is ample evidence for both hypotheses but nothing conclusive in either direction. As a result, an early observation by Schmid (1960: 672) about the analysis of social space still holds true today despite the use of more data sets and more sophisticated statistical manipulations: 'any conclusions that might be derived directly from the analysis logically pertain to areas and not to individuals – although one may be strongly tempted to infer causality, the results of factor analysis merely measure the degree of concomitance of community structures and characteristics'. The fact that an area has a high level of social pathology does not mean that all residents of that area exhibit pathological behaviour. And the fact that pathology is not universal in an area weakens the case that its cause is to be found in the local physical and social milieux.

A more sensible approach would be to look at multi-factor explanations. Such an approach was developed by Herbert (1977) with reference to crime and delinquency (Fig. 2.12). The focus of this model is on poverty, which is seen as the product of structural factors such as differential access to employment and educational opportunities. The net effect of these factors is to produce a number of 'losers', notably the unemployed, the aged, misfits, and members of minority groups. It is not surprising, then that some researchers have suggested that research attention should shift from looking at structures to examining individuals and how they adapt and adjust to the environment in which they live (S.J. Smith 1986). In short, structural approaches need to be complemented by behavioural approaches. However, before exploring behavioural approaches (Chapters 4 and 5), it is essential to look at the second type of structural approach highlighted at the beginning of this chapter, namely transformational structuralism.

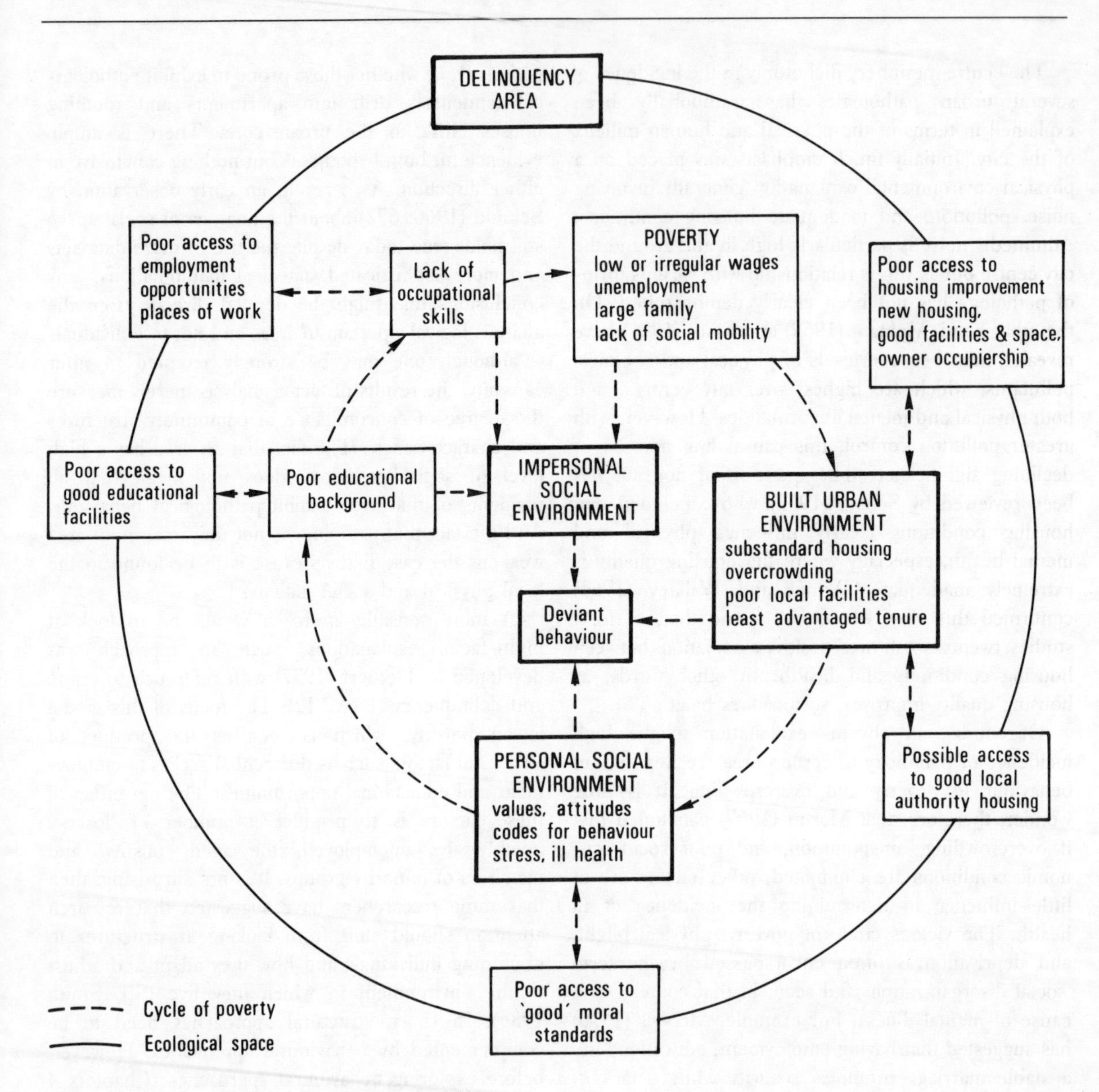

2.12 Delinquency: the cycle of disadvantage and its spatial consequences (Source: Herbert 1977: 227)

3 Structures in the Mind and Social Structures

Chapter 2 is concerned with structuralism conceived of in spatial terms. From that perspective, explanation of behaviour is sought in the interrelationship of the elements which make up the spatial structure of society. Many researchers argue, however, that such empirical analyses are too limited: 'we have to go beyond surface structures to find the deep and real structures that account for the variety of observable phenomena or conscious explanations and their apparent contradictions' (Rossi 1981: 63). This process of 'going beyond surface features' has taken human geographers into the realm of what is known as 'transformational structuralism'.

Transformational structuralism strives to understand the way in which society organizes itself without necessarily concerning itself with the individuals that make up society, even in aggregate terms. Rossi (1981: 63) has emphasized, for instance, that transformational structuralists 'find not only naive but also false assumptions that people's conscious explanations and overt behaviour are to be taken at face value as objects of scientific analysis'. In other words, transformational structuralists believe not only in the need to delve beneath surface appearances to discover explanations but they also argue that such analysis will identify the universal structures that provide the forces at work in societies. Much of this work focuses on a socio-structural level, best seen in the 'political economy' approach in general and structural Marxism in particular. This viewpoint argues that social and economic phenomena can only really be understood if they are seen as part of a *superstructure* that is related to the material needs, or *mode of production*, on which society is based. Many human geographers have pursued structural Marxism with great enthusiasm. Marxist and neo-Marxist approaches have therefore been very much in vogue and the focus of much geographical attention in the 1980s has been on 'structure' (see Peet and Thrift 1989).

Surprisingly, given this overall concern with structures, geographers have tended to overlook the existence of a unifying structure formed through the way individuals perceive and interpret the world around them (what Gregory (1978: 99–101) has referred to as a 'neuro-structural' approach). For example, cultural phenomena (such as language, religion, taboos, and kinship ties) which appear diverse can in fact be viewed as a series of basic structures which are universal in the sense of reflecting common elements in the human mind. Such deep seated structures are impossible to analyse directly and, therefore, can only be identified by observation of behaviour in much the same way that the workings of a political economy are impossible to observe directly and can only be inferred through their various manifestations. Neuro-structural and socio-structural approaches are both examined in this chapter, beginning with structures in the mind.

Structures in the mind

Over the years, geographers have come to realize that there exist unifying elements within the human mind which form 'deep structures' that influence behaviour.

This view had its origins in linguistics. It has since been adopted in several fields of social science, notably under the leadership of Lévi-Strauss (1963) in anthropology and Piaget (1970) in the study of children's learning processes. The basis of this type of work is the belief that the 'diverse structured patterns discernible in directly observable cultural phenomena are all transformations of a narrow range of mentalist structures which are common to all humanity' (Leach 1981: 29). However, these 'deep' structures are not easily analysed by conventional social scientific methodology. Indeed they can only be uncovered through the study of their 'transformations' into observable phenomena and behaviour.

The beginnings of linguistic structuralism can be traced back to the early part of this century, particularly to the work of de Saussure (1966). The fundamental aim of his approach was to distinguish the underlying concept from its transformation in everyday language. For example, the focus of attention was on the difference between the term used in a particular language (the *signifier*) and the concept which the term represents (the *signified*). The focus was also on the difference between the speech act of everyday life (*parole*) and the language system whose rules are understood so that the speech acts are interpretable (*langue*). The development of transformational structuralism in linguistics is associated primarily with the work of Chomsky (Smith and Wilson 1979). Particular attention has been paid to the way that children learn language from listening to their parents, thus leading to the claim that language systems must be 'biologically determined and genetically transmitted from parents to their children' (Lyons 1977: 7). It is argued, from this perspective, that language systems are based on a set of principles which 'form deep structures whose identification is the purpose of structuralist inquiry' (Smith and Wilson 1979: 26).

A similar viewpoint underlies the notion that the *general* learning of children, not just their linguistic ability, is influenced by the structure of the brain. For example, in studies of how children learn, Piaget (1970) adopted a structuralist approach to highlight the way in which the basic structure of the brain controls the way human intelligence develops. According to Piaget, all humans have the same innate modes of intellectual functioning, in particular *assimilation* (the incorporation of new information into existing schemata) and *accommodation* (the readjustment of schemata in the light of assimilation). By means of these two abilities, learning takes place and can be thought of as a continuous process of transformation that leads to an increase in intelligence and in the ability of an individual to cope in the real world. So, 'whereas other animals cannot alter themselves except by changing their species, man [*sic*] . . . can structure himself by constructing structures; and these structures are his own, for they are not externally predestined either from within or without' (Piaget 1970: 118–19). In short, the development of the brain determines the sorts of 'constructions' of which humans are capable.

The impact of Piaget's work in human geography has largely been in studies of the conception of space, particularly that of children. In simple terms, he argued that there are four levels to the conception of space: the *sensorimotor* level based on representation through action (e.g. the manipulation of objects); the *pre-operational* level based on simple mental constructs and intuitive, uncoordinated images of the world derived from the memory of previously perceived objects; the *concrete operational* level where simple logical thought permits mental representations of the environment that are symbolized and schematic; and the *formal operational* level at which the individual is capable of hypothetico-deductive reasoning and can handle abstract spatial notions that are not dependent upon real actions, objects, and spaces (Moore 1976: 148) (see Chapter 4). However, interesting as such work may be, Gould (1973a: 185) has argued that study focusing on children's conception of space 'does not deal with the essentially geographic images that children hold or the way that they learn about them'. In short, there is more to knowledge of the environment than mere conceptions of space. As a result, it is perhaps not surprising that, despite the growth in interest in cognitive mapping and the potential of the Piagetian approach, only a few studies have been conducted by geographers in a structuralist mould (see Golledge and Stimson 1987; Boal and Livingstone 1989; Matthews 1992). Nevertheless, in many areas of human geography, Piagetian structuralism is an approach with much latent potential.

Another form of transformational structuralism

largely ignored by geographers has been that associated with Lévi-Strauss (1963) in the field of social anthropology. Once again, the concern in this approach is to uncover common features of the human mind, irrespective of individual variations or particular cultural settings. Using the ideas of structural linguistics, Lévi-Strauss developed the structuralist method of binary opposition with reference to kinship and mythology. The methodology adopted involved three principal stages:

1. The definition of the phenomena under study in terms of a relation between two or more terms.
2. The construction of a table of all possible permutations between these two or more terms.
3. These permutations become the object of study. Such study yields a variety of connections, the empirical phenomenon considered at the beginning being only one possible combination among many. Thus understanding of the empirical phenomenon rests on the complete system of connections which must be reconstructed before understanding can be achieved (Lévi-Strauss 1973: 84).

An illustration of this methodology is Lévi-Strauss's study of kinship systems. The 'elementary structures' of prescriptive alliance were conceived of as comprising the most basic differentiation between nature and culture, determined by an elaboration of an incest taboo. 'Elementary structures' of kinship were defined as 'those systems . . . which proscribe marriage with a certain type of relative, or, alternatively, those which, while defining all members of the society as relatives, divide them into two categories, viz, possible spouses and prohibited spouses' (Lévi-Strauss 1969: xxiii). Another example of Lévi-Strauss's methodology is illustrated in his reinterpretation of Evans-Pritchard's (1940) classic study of the Nuer, in particular his charge that Evans-Pritchard failed to identify the general significance of his findings. For example, Evans-Pritchard interpreted the tendency in Nuer thought to equate twins with birds, in terms of their theology; on the other hand, Lévi-Strauss (1973) argued that this equation was a particular instance of a general symbolic relation between human groups and animal species. In other words, in Nuer cosmology, twin births are viewed as being supernatural and are equated with the sacred order of the 'above'. Because of their ability to fly, birds are also viewed by the Nuer as 'of above'; thus birds are appropriate symbols in the instance of twin births. In short, understanding of the world view of the Nuer can only be achieved, according to Lévi-Strauss, through an appreciation of the wider relations that structure their world view. Lévi-Strauss also claimed that the resolution of binary opposition (e.g. kin–not kin) allows the identification of the basic elements which can then be projected on to other phenomena. Study of the transformation necessary to achieve successful projection leads to the identification of 'deep structures'.

A good deal of controversy surrounds the work of Lévi-Strauss and the structural anthropologists, not only because of its complexity but also, according to many, because analysis of the concept of 'deep structures' has still to be transformed into worthwhile scientific procedures. Kurtzweil (1980: 244–5) summed up the mixed feelings that many have about the work of Lévi-Strauss by claiming that structuralists 'have not answered the question of human origins and of existence as they originally promised. Perhaps the aims of the structuralists were unattainable but they did introduce some intriguing and suggestive modes of analysis'. In response to these criticisms, it must be conceded that Lévi-Strauss (1973: 313) has emphasized that a structuralist approach can aid researchers to 'distinguish between what is primordial and what is artifical in man's [*sic*] present nature'.

Generally, little attention has been paid by geographers to the ideas of Lévi-Strauss, though some attention has been given to the implications of the concept of transformation for the study of spatial structure (Harrison and Livingstone 1982; Gregory 1978). Probably two reasons can account for this lack of geographical interest: first, the structuralists' apparent disregard for empirical detail and their disdain for the uniqueness of particular instances has clashed with the geographer's long-standing interest in what happens where; and second, geographers have been put off by both the tendency for structuralists to conceive of phenomena as static and by their tendency to adopt a cross-sectional framework that emphasizes signs and symbols. Moreover, by the 1970s, any attempt to develop Lévi-Strauss's ideas as a major methodology within geography would have been quickly overtaken by the emergence of a 'political economy' and neo-Marxian perspective within the discipline. A focus on a material interpretation of history, and an insistence on a dynamic view of social change within a critical 'political economy' framework, was more in line with the times. Indeed, by the early 1990s, structural Marxism had become the dominant form of 'structuralism' within geography.

Structural Marxism

A complex and confusing terminology has developed among political economists as different writers have either displayed a fair degree of eclecticism in their choice of source material or have highlighted subtle nuances in interpretation. In simple terms, the political economy perspective adopts a holistic approach to the study of society and argues that the institutions and ideas that are characteristic of society cannot be understood as separate from the underlying material needs of society. Within this perspective, political economy can be viewed at a variety of scales: the world economy can be seen as the scale of reality, the state and the nation as the scale of ideology, and the city as the scale of experience (Taylor 1982). In all cases, however, the superstructure of society is seen as founded on, and intricately related to, material needs. It is therefore reasonably legitimate to label the political economy perspective as one that advocates structural Marxism, although there are obviously some individuals who would describe themselves as political economists but not Marxists.

The term 'Marxist' is of course a vague one. Thompson (1978), for example, has noted no fewer than four usages for the term: as a set of doctrinal beliefs applicable to any and all situations; as a method of analysis, based on Marx's dialetical notion of thesis–antithesis–synthesis, that provides insights into the functioning of society; as one of the 'great ideas' developed in the course of human history; and as an ongoing and evolving intellectual tradition. Perhaps a more telling distinction is Gouldner's (1980) differentiation of *scientific* and *critical Marxism*. The former involves attempts to develop a general body of theory that can be applied universally while the latter uses historiographic study to gain understanding of specific events. It is the former – scientific Marxism – that has been most commonly adopted by geographers in the study of political economy. Although interpretations vary, this form of structural Marxism can be summarized in simple terms. All social phenomena (e.g. institutions, ideology, government) are part of a 'social formation' that is a superstructure founded on the prevailing 'mode of production' which is the term given to the material economic base for life. Everything in the superstructure derives from the mode of production. Moreover, each mode of production builds up internal contradictions and dialectical accommodation of these contradictions leads to the replacement of one mode of production by another. Thus, according to Marx, there exists a historical sequence that proceeds from tribalism to feudalism, to capitalism, and ultimately to socialism.

'For Marx, the scientific understanding of the capitalist system consists in the discovery of the internal structure hidden behind its visible functioning' (Godelier 1972: 335–6). Since capitalism is a social relationship, rather than a theory, its hidden structure cannot be apprehended by empirical means. So, 'Marx's historical materialism involves the development and refinement of plausible theories about the dominant forces in society, plausible both as explanations and as guides to action' (Johnston 1983: 99). Although this view is widely accepted by many, Saunders (1981: 18) suggests a need for some caution since 'it follows that there is no necessary and compelling reason to accept such theories other than one's own political values and purposes. Marxism, in other words, is as much a guide to political practice as a method of scientific analysis'.

The synopsis provided in the previous paragraph is of course a very crude summary of a very complex theory (Harvey 1973; Peet and Lyons 1981). Part of the appeal of structural Marxism in geography seems to lie in its usefulness in making sense of many of the ills that currently bedevil advanced, industrial economies (Peet 1977). Certainly the approach has provided useful insights into the structuring of the housing market (Harvey and Chatterjee 1974), residential segregation (Harris 1984), the role of capital (Harvey 1982), inequality and poverty (Peet 1975), and the role of the state (Taylor 1985).

The importance of the structural Marxist perspec-

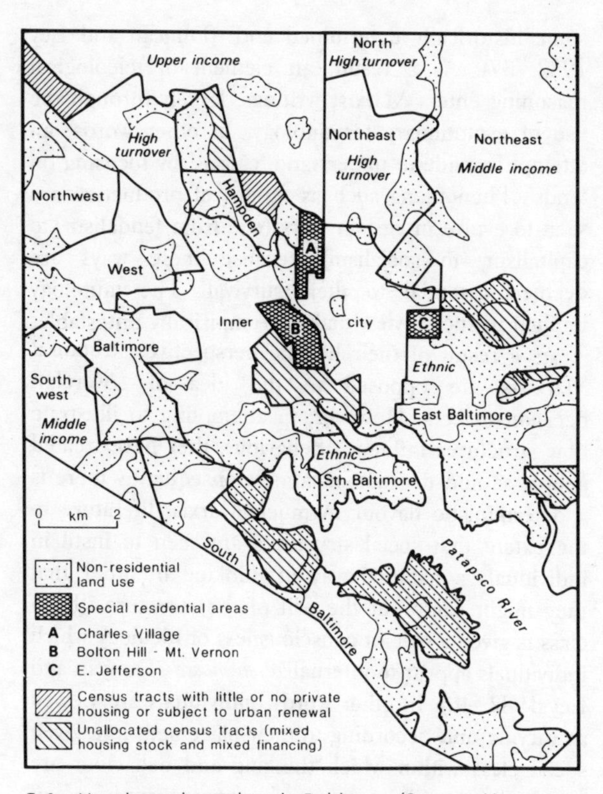

3.1 Housing sub-markets in Baltimore (Source: Harvey 1974: 246)

tive can be illustrated by looking briefly at recent studies of residential segregation in cities. Unlike the urban ecologists, the structuralists argue that, as the relations of production are continually changing, so society is changing. Moreover the process of change is characterized by conflict rather than consensus, within a highly constrained situation. In other words, from this perspective, it is the class relations of production, rather than decision-making and choice, which are the major determinants of residential differentiation in so far as 'the rich group can always enforce its preferences over a poor group because it has more resources to apply either to transport costs or to obtaining land in whatever location it chooses' (Harvey 1973: 135). For example, Harvey (1975) suggested that four determining processes were at work in influencing the pattern of residential segregation in inner Baltimore (Fig. 3.1). Residential segregation, in Harvey's view, (1) reflects the reproduction of social relations under capitalism; (2) mirrors the contradictions that exist within a class-bound society; (3) results in distinct communities which provide the milieux within which residents are socialized; and (4) brings about the fragmentation of the class consciousness of the proletariat, thus hampering the transformation to socialism.

It is frequently argued by structuralists that the sort of processes discussed by Harvey are controlled by a variety of institutions and, therefore, any explanation of segregation must be concerned with the nature and impact of these institutions. There are obvious parallels here with the urban managerialist perspective. However, the growing interest during the 1970s in 'urban managerialism' has been severely criticized by structuralists for its tendency to focus on the decision-making of institutions rather than the context within which they operate. This viewpoint is most clearly expressed by Castells (1977) as well as in a series of later studies (Castells 1978; Walker 1978; Bassett and Short 1980; Badcock 1984).

Of course, the political economy approach itself has come in for stern criticism. Ley (1978: 46) has pointed out, for example, that structural Marxism depends on a materialist epistemology that emphasizes functional economic relations (e.g. the relationship between capital and labour in a given mode of production), and it therefore reduces the position of values, ideas, and individual human action to the status of epiphenomena that are merely derivative of the mode of production and its consequent power relations. As a result of this 'inauthentic' model of human behaviour, Marxism deals in categories that are only imperfect reflections of the world of everyday life and consequently there is often a low degree of correspondence between the world of the Marxist commentator and the world of everyday experience (Ley 1978: 50). The human consciousness that is so very much a part of everyday life is of little significance in structural Marxism. Many structural Marxists in geography have in fact chosen to work at developing the general theory of political economy rather than at examining real events in specific times (Johnston 1980).

The most thorough and telling criticism of structural Marxism in geography has come from Duncan and Ley (1982). They begin by drawing a distinction in social science research between individualists and holists. The former argue that large scale social phenomena (e.g. culture, capitalism) are the products of the actions of the individuals who participate in these phenomena and hence explanation of these phenomena can theoretically be reduced to statements about individual behaviour. Holists, in contrast, believe that such reduction is theoretically impossible and that large-scale social phenomena have an existence in their own right that is unrelated to the conscious actions of the participating individuals. In terms of this dichotomy, structural Marxism, particularly the work of Althusser (1969) and Castells (1977, 1978) that has been so popular among geographers, is firmly in the holist school of thought. Duncan and Ley challenge the autonomy of macrostructures that is inherent in the holist view. They argue that such macrostructures are ultimately reducible to cumulative human actions: 'though individuals may not be free to transcend their social context, neither are they passive agents of a larger force such as "culture", the "logic of capitalism", or the "mode of production"' (Duncan and Ley 1982: 32). This is not to say that macrostructures can be ignored. Rather the message is that these structures form the context not the cause of behaviour and are, in any case, ultimately traceable to individual actions.

Duncan and Ley identify a number of specific weaknesses in the Marxist position. To begin with, structural Marxism is overly influenced by an organicist mode of thinking. In other words, the totality of existence is looked at very much as an organism: the whole is greater than the sum of the parts. For example, Marxists hold that the analytical approach characteristic of positivistic science is inadequate, that the whole determines the nature of the parts, and that the parts of a political economic system cannot be understood in isolation from the whole because the parts are dynamically interrelated and interdependent. This holistic view usually involves *reification*, whereby the whole and the parts (e.g. mode of production, class, capitalism) are seen, not as mental concepts or abstractions, but rather as things that have substance and causal efficiency. In other words, power, activity, and sometimes intentionality are ascribed to mental constructs. They are treated as 'things' rather than 'ideas'. Moreover, 'the reified categories of the Marxists are not only granted a life of their own, but they also have a purpose, or a telos: they move towards some historically determined end' (Duncan and Ley 1982: 37). As a result, an element of teleological reasoning enters Marxist writings, largely through the use of evolutionary terminology. In other words, an attempt is made to understand 'causes' by focusing on 'ends'. Phenomena such as modes of production are seen to evolve in certain ways (e.g. from feudalism, to capitalism, to socialism) because these ways are deemed essential to their survival. The tautology implicit in such a view makes it empirically untestable.

As a result of their holistic perspective, structural Marxists are opposed to the idea of sovereign decision-makers. However, in attempting to illustrate how structures influence behaviour, they have treated people as non-decision-makers. Consequently there is a deterministic flavour in much Marxist literature to the extent that social structures are seen to instil in individuals a habitual way of thinking in order that they might carry out the will of the structure. 'Each class is given a type of consciousness or ideology which individuals appear to internalize *en masse*' (Duncan and Ley 1982: 40). In other words, individuals are viewed as varying only according to their placement in a given social class within which thinking and behaviour are standardized. This sort of conceptualization of individuals emphasizes economic rationality and therefore has closer parallels with neo-classical economics than most Marxists care to acknowledge. In view of this, it is hardly surprising that those researchers in the vanguard of Marxism in geography were often the same ones that had been in the vanguard of the movement towards positivism and quantification. After all, much early positivistic geography borrowed heavily from neo-classical economics. A preoccupation with economic matters also results in structural Marxists taking a rather distorted perspective on reality. In the words of Duncan and Ley (1982: 45), 'to collapse the range of social experience to the outworking of deep economic structures is to present an impoverished view of the social, cultural, and political realms of life'. Given this blinkered view, it is not surprising to find that much of the Marxist literature in geography uses fragments of reality to illustrate theory rather than using theory to provide insights into the real world (Gregory and Urry 1985).

3.2 The nature of structuration (Source: Gregory 1981: 9)

a: REIFICATION: typified in social theory by Emile Durkheim and by some neo-Marxist formulations

SOCIETY

INDIVIDUAL

society is a reality which is external to and constraining upon human agency

b: VOLUNTARISM: typified in social theory by Max Weber

SOCIETY

INDIVIDUAL

society is constructed by intentional action

c: DIALECTICAL REPRODUCTION: typified in social theory by Peter Berger

SOCIETY **SOCIETY**

INDIVIDUAL **INDIVIDUAL**

society forms the individuals who create society in a continuous dialect: society is an externalisation of man, and man a conscious appropriation of society

d: STRUCTURATION: typified in social theory by Jürgen Habermas and Anthony Giddens

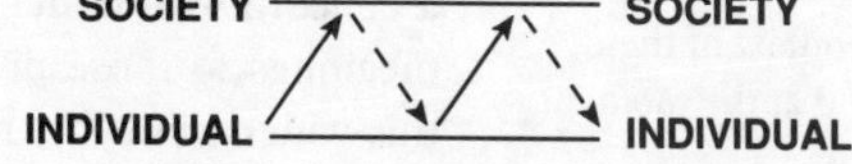

social systems are both the medium and the outcome of the practices that constitute them: the two are recursively separated and recombined

Structuration

The Marxist response to such criticisms has frequently involved a claim that what is under attack, in the guise of structural Marxism, is really a caricature of some of the main features of structural Marxism. The subtlety and complexity of Marxism, it is argued, are ignored. There is some substance to this claim. However, there is another way of overcoming some of the criticisms levelled at structural Marxism and this is contained in the theory of *structuration* devised by the sociologist Giddens (1979: 2) as a response to what he saw as 'the lack of a theory of action in the social sciences'. According to Giddens (1979: 2), the Marxists and their supporters 'in their eagerness to "get behind the backs" of the social actors whose conduct they seek to understand ... largely ignore just those phenomena that action and philosophy make central to human conduct'. What is being argued from this perspective is that the role of human agency needs to be emphasized since 'the production of society...is always and everywhere a skilled accomplishment of its members', although it is also necessary 'to reconcile such an emphasis with the equally essential thesis ... that if men [*sic*] make society, they do not do so merely under conditions of their own choosing' (Giddens 1976: 126).

The principal aim of structuration is to remove what has typified social theory for so long, namely the dichotomy between individual behaviour and the structure of society (Fig. 3.2(a)–(c)). What is proposed is a contextual theory of action in which social structures are held to be continually reformed through the rhythms of daily life: thus, human behaviour creates the structure of society; those structures provide the context for the socialization of humans;

and, in turn, human behaviour reflects and re-creates the structures (Fig. 3.2(d)). In other words, Giddens's approach envisages the fusion of society and the individual in the 'duality of structure', a term which 'relates to the fundamentally recursive character of social life, and expresses the mutual dependence of structure and agency . . . [such that] . . . the structural properties of social systems are both the medium and the outcome of the practices that constitute those systems' (Giddens 1979: 69). In this sense, structuration emphasizes that social relations cannot be simply interpreted in class terms. It therefore argues for a contingent theory of social action which is sensitive to the temporal and spatial specificity that underlies the unfolding of wider processes (Gregson 1987; Bryant and Jary 1991).

Essentially, the structuration of society involves three concepts:

1. The *system*, or the activities in time and space of the agents that comprise society. These activities are reproduced relations which stretch across time and space and are connected with past activities and ones yet to be realized.
2. The *structure*, or the recursive rules and resources which guide human agents in the context of their daily lives and which are reproduced at the moment they are drawn upon. In this way

    structure is the medium whereby the social system affects individual action and the medium whereby individual action affects the social system. The outcome of these individual–system interactions always (in varying degrees) affects the structural rules governing the next interaction. Thus, the theoretical separation of structure and system enables Giddens to capture both agency and structure in the production and reproduction of social life without according primacy to either (Moos and Dear 1986: 250).

3. *Structuration*, or the structuring of social relations in time and space within the context of the dual relationship between agency and structure. In other words, structuration refers to the means by which human agents govern the continuity and/or transformation of structures, and thereby influence the reproduction of the system obtaining in any given case.

In other words, Giddens (1984) conceives of structures as social systems which enable humans to live their daily lives. For this to occur, these systems have rules so that everyday living becomes routinized. In this way individuals can integrate into the local system. In advanced societies the rules are numerous and varied and so institutions (e.g. state, local government, trade unions, family) are formed in order to enforce the rules. These rules apply in particular places and their integration at these places requires that individuals be seen to exist in particular times and spaces. Thus, Giddens argues that the organizing units of societies are local systems, or *locales*. In this sense, structuration theory has much affinity with Hägerstrand's (1970) time–space perspective. Initially, in geography, time was viewed as a constraint on behaviour. Thus, the emphasis in many studies was on tracing individual paths through time and space (see Parkes and Thrift 1980). However, there is plenty of evidence to suggest that time–space was conceptualized by Hägerstrand and his associates in a much wider sense that involved a concern with understanding both situations and their meanings to those present (Hägerstrand 1984). Given this wider view, it is not surprising that there has been an increasing acceptance of the idea of 'the spatial basis' in the study of social systems, and an increasing awareness of the importance of time–space constraints in the study of human behaviour. Both Hägerstrand and Giddens emphasize this point.

Within the context of research in geography, the introduction of structuration theory raises two significant issues: from the point of view of study, how do researchers break into a continually reflexive process? And how can the concept of a locale be made operative? The first of these two issues has been a focus of some debate in the literature (Gregson 1986). Since all structures may be conceived of as being socially reproduced, Gregory (1980) was adamant that agency must be the starting point of any research. If that is the case, a deep understanding of structuration can only be achieved by means of a historical perspective. The trouble with this argument is that, according to Harvey (1987b), any such understanding needs to adopt a valid theoretical framework. In

Harvey's view, this is to be found in Marx's writings on historical materialism since these provide a clear means of understanding the origins of capitalism and its uneven spread in time and space. However, Marxism tends, as has been shown, to ignore human agency, that is, one-half of what the theory of structuration is concerned with.

The second issue of concern is the nature and form of space and place in the structuration process. After decrying the significance of space, contemporary structuralists, particularly those influenced by Giddens, have become aware of its potential in providing a fuller understanding of economic restructuring and social recomposition. Giddens (1984), for instance, has clearly argued that the structuration process can only be understood within a time–space framework and, in particular, in the context of a locale. 'Locales refer to the use of space to provide the settings for interaction, the settings of interaction in turn being essential to specifying its contextuality' (Giddens 1984: 118). Essentially, locales refer to the way in which space is part of the interaction between humans and structures. In this sense, they may be interpreted at all geographical scales. In turn, a locale may itself be internally regionalized in such a way as to reflect different social interaction in time and space. Thus, what is being emphasized in the notion of 'locale' is that localizing in time and space is an important structuring element of everyday life. In other words, locales provide various opportunities for, and constraints upon, behaviour. Thrift (1983) has added to this view of the constraining role of locales by conceptualizing them as the site of the determinate working of an objective social structure. According to Thrift (1983: 40), locales structure individual life paths in time and space, influence life paths, provide a means for interaction, provide the activity structure of daily life, and are the principal sites in the process of socialization. Harvey (1985a) also emphasized the significance of locales through the concept of *structured coherence*. In his view, the emergence of a coherence in human affairs may be structured not only by the prevailing form of production, but also by 'standard of living, the qualities and styles of life, work satisfaction (or lack thereof), social hierarchies, and a whole set of sociological and psychological attitudes towards working, living, enjoying, entertaining and the like' (Harvey 1985a: 140). Such coherence may vary in time and space, sometimes in such a way as to render the coherence temporary and often partial.

Emerging out of this concern by structuralists with time and space has been a growing interest in the concept of *locality* and its potential for interpreting economic restructuring and social recomposition (Swyngedouw 1989). According to Cooke (1989b: 12):

> locality is the space within which the large part of most citizens' daily working and consuming lives is lived. It is the base for a large measure of individual and social mobilization to activate, extend or defend those rights, not simply in the political sphere but more generally in the area of culture, economic and social life. Locality is thus a base from which subjects can exercise their capacity for pro-activity by making effective individual and collective interactions within and beyond that base. A significant measure of the context for exercising pro-activity is provided by the existence of structural factors which help define the social, political and economic composition of locality.

In Britain, for example, a series of what have become known as *locality studies* have been undertaken in places with different socio-economic, cultural, and political structures (see Fig. 3.3) (Cooke 1989a). Essentially, the purpose of these studies was to identify in detail how wider social and economic processes worked themselves out at the local level, thus enabling 'the finer-grained impact of local economic restructuring to be clearly understood' (Cooke 1986: 2). In a similar vein, Urry (1986: 2) has claimed that 'localities are distinctive sociological entities' but warned that 'although they are increasingly subject to structural change, this does not mean that their social and cultural characteristics can be reduced to the causes of such change'.

The majority of locality studies have adopted a historical perspective, viewing economic restructuring in terms of successive rounds of investment. For example, neighbourhood change in Brooklyn, New York, was interpreted by Warf (1990) in terms of the changing types of jobs available at successive time periods, although it was emphasized that the role of the local division of labour at different times was more significant than the generation of jobs. In turn, such

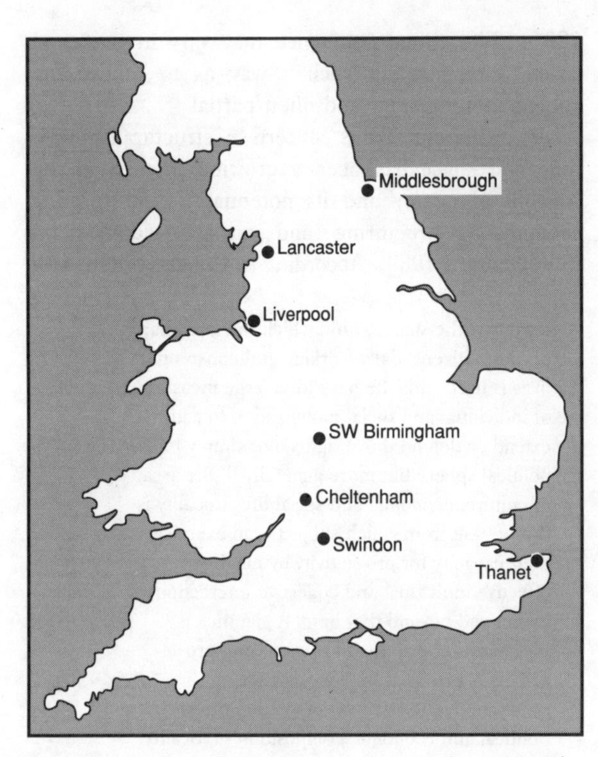

3.3 Locality studies in England (Source: Cooke 1989a: 2)

changes could only be understood against a background of linkages with the global economy, from the German revolution and the Irish Potato Famine of the 1840s to the more recent internationalization of capital and the way in which this has encouraged the recent wave of gentrification. At different time periods, then, neighbourhoods were created by different immigrant groups. These neighbourhoods derived their character from the interplay between job opportunities, cultural practices, and the changing housing market. Each round of investment was overlain upon, responded to, and modified by what happened in earlier periods. An important resource and constraint in shaping interaction within the local labour and housing market was that of ethnicity. Warf's study of Brooklyn, along with several others, also emphasizes that the results of restructuring in different localities tend to differ quite markedly, reflecting the historical and specific characteristics of each locality. 'Whereas national and international restructuring processes have generated new rounds of investment in localities, a related process has also occurred whereby investment is channelled towards a type of locality possessing historically based characteristics not simply economic in origin' (Cowen *et al.* 1989: 127).

A good deal of criticism has been levelled at locality studies, in particular for their reliance on narrative, for an overemphasis on uniqueness, and for the tendency of some writers to interpret all relations in terms of class, often at the expense of an appreciation of the composition of the locality itself (Sayer 1991). In particular, locality studies have been criticized for their overemphasis on employment and on the division of labour at the expense of local cultures. In the eyes of some, this has led to a limited interpretation of urban and regional change (Rose 1988; Jackson 1991). Lee (1990: 111), for example, has argued that 'not only do the localities become dependent but also much of their richness, significance and meanings as localities is lost' if attention is focused overwhelmingly on the division of labour. Given these criticisms, it is understandable that several geographers have claimed that geography matters much more than is suggested by the advocates of structuration theory (Pred 1986). For example, Hägerstrand (1984: 378) views 'the world as a fine-grained configuration of meeting places rather than as a system of regions or of aggregate categorical variables' while Pred (1984: 292) has taken this a stage further since he sees geography 'as a process whereby the reproduction of social and cultural forms, the formation of biographies, and the transformation of nature ceaselessly become one another, at the same time that time–space specific path–project intersections and power relations continuously become one another'. Emanating from these viewpoints has been a growing interest in reviving regional geography in a new, more structuralist form (Sayer 1989; Johnston *et al.* 1990).

In short, structuration theory has gone some way towards overcoming the view that Marxism is 'a model of mechanism in which the actors themselves have no say: they become puppets who dutifully act out the roles prepared for them by the theorists' (Ley 1980: 11). Although structuration suggests that there is more to the explanation of social relations than a few transcendental structures directing the course of history, many geographers would still argue that insignificant emphasis is given to the power of human consciousness and human action to redirect the course of events (see Harvey 1985b).

In summary, it is clear that the very same errors and shortcomings inherent in spatial interaction models are also evident in structural Marxism. As a result, neither a focus on statistically average behaviour nor an emphasis on Marxism provides a satisfactory basis for the study of human spatial behaviour. The shift towards the consideration of social structure and structuration theory, and in particular their fusion, does go some way to overcoming some of the deficiencies noted in earlier approaches based on empirical structuralism. However, many would still maintain that there is a continuing need for the incorporation and integration of a more cognitive and humanistic perspective into the analysis of people–environment interaction if a deeper understanding of the human condition is to be achieved. These issues are taken up in Chapters 4 and 5.

4 Information and Choice

A wide variety of what might be termed 'behavioural' approaches has come to characterize the study of people–environment interaction in the last three decades. Although varied, this work has in common a focus on:

1. How individuals come to know the environments with which they interact;
2. How they differentiate and choose between different locations as places to live, work, play, and shop;
3. How they develop simplified images of environments which are too big and too complex for them to be known in their entirety;
4. How knowledge, decision-making strategies, and overall images influence overt (or acted out) behaviour;
5. How certain parts of the environment come to take on meaning for the people who interact with it.

This chapter examines the first two of these issues. The remaining three are dealt with in Chapter 5.

The build-up of knowledge

It is a fundamental proposition of behavioural approaches in human geography that humans derive information from their environment and process that information in such a way as to provide a basis for overt behaviour. Environmental information is, in other words, fundamental to survival, to everyday living, and to the maintenance of a sense of well-being. In short, it is central to most basic human needs (Maslow 1954). The study of the acquisition of information, and of the build-up of knowledge generally, therefore underpins behavioural approaches in human geography.

Investigation of how humans use information about their environment has generated an enormous volume of literature (Moore 1979). Much of the early work spoke in terms of 'perception' and focused overwhelmingly on the notions of 'perceived environment' and 'mental map'. Indeed much of this early work was concerned with the 'perception' of natural hazards such as floods (Kates 1962) and drought (Saarinen 1966). The aim was to explore the extent to which humans are rational in their actions and to investigate how environmental perception influences the way in which people cope with risk and uncertainty (Jackson 1981). Thus attention came to focus on alternative ways of 'perceiving' hazard events: some people tried to impose order where none really existed (e.g. they fell back on folk sayings about climate to 'explain' droughts and floods), while others denied the possibility that hazards could be understood at all, preferring instead to regard them as unique, unpredictable 'acts of God'. This use of the term 'perception' created confusion in that it differed markedly from the usage that had evolved in experimental psychology where perception was usually taken to mean the impinging of environmental stimuli on the human sense organs.

Geographical use of the term 'perception' covered both this sort of immediate response to a stimulus and the more general awareness that is not contingent upon an immediate stimulus and to which the term 'cognition' is more appropriately applied. In this sense, then, geographers were adopting an 'ecological' approach to perception, similar to that advocated by the psychologist Gibson (1979). From this perspective, 'perception of the world is not direct ... what we perceive is a mental representation of the world, which is the product of cognitive operations on sensory input' (Heft 1988a: 327). As a result, it may in fact be best to think of both perception and cognition as contributing jointly to knowledge of the environment, the former providing *figurative knowledge* (images resulting from direct contact) and the latter *operative knowledge* (information that has been structured through a variety of mental operations) (Moore and Golledge 1976b). Certainly environmental information is gained through all the senses (Stea and Blaut 1973) and is stored in some way in the memory, as is demonstrated by the ability of humans to recall knowledge of environments that are not immediately present (Gould and White 1974).

The question of how people come to know the environment is a fundamental epistemological problem in so far as it raises queries about the source, nature, and form of knowledge. At issue are not only people's awareness and information but also the impressions, images, and beliefs that they have about environment (Moore and Golledge 1976a). In terms of environmental knowing, there are few objective facts. Rather information about the environment is sought and collected in a subjective and purposeful way that reflects the needs and values of the individual concerned. In other words, how an individual approaches the environment determines what he or she finds. Thus, environmental psychology has shifted the emphasis of much of behavioural science from the dictum that 'seeing is believing' to the point of view that 'believing is seeing' (Proshansky *et al.* 1970a: 101–2). As a result of that shift, the environment is not seen just as another variable to be thrown into a multiple regression model but rather as the *raison d'être* of a whole class of behaviour (Downs 1976: 74). In this connection Ittelson (1978) has suggested three foci for the study of the urban environment: the city as a source of information; the psychological processes involved in the use of this information; and the resulting varieties of urban experience. This threefold classification can be extended, through inclusion of the environment itself, into a fourfold listing of the interrelated elements that need to be considered in any examination of how humans use environmental information:

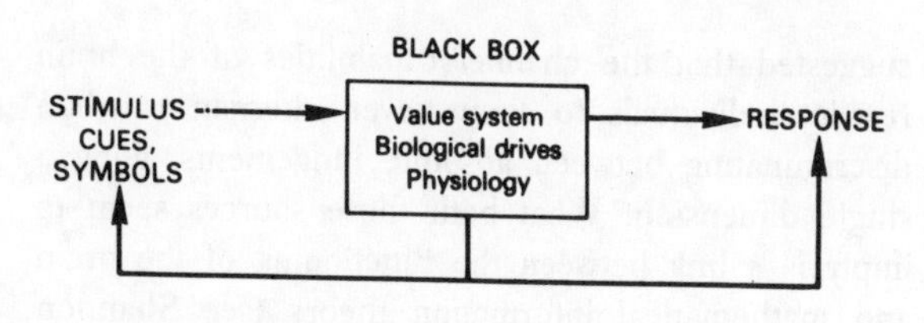

4.1 The mind as a 'black box'

1. The nature of the *individual*, and in particular the psychological processes that are used to cope with information.
2. The nature of the *environment* from which information is drawn.
3. The manner in which the individual seeks out *information* in the environment.
4. The way in which the environment provides an arena within which *overt behaviour* is acted out (Walmsley 1972a: 43–9).

Each of these is worth examining in detail.

Very little is known about how individuals cope with environmental information. In many studies the mind has been looked upon as a sort of black box: the environment has been manipulated, resultant behaviour observed, and imputations made as to how the mind works (Burnett 1976) (Fig. 4.1). More commonly in geographical studies it has been assumed that individuals are rational economic beings who seek to optimize their overt behaviour. Clearly such a proposition is unreal in that it ignores unpredictable change in the environment and makes wholly unacceptable claims about the ability of humans to handle information (Wolpert 1964). Unfortunately, the substitution of 'bounded rationality' for omniscient rationality does little to improve the situation in that it leaves unspecified the information-processing capacity of the individual (a point explored at greater length later in this chapter). In this connection Sagan (1977) has argued that the human brain is characterized by 10^{13} synapses capable of carrying 'yes/no' information and that the brain is quite able to process 100 such bits of information per second. Conversely, Miller (1956) has suggested that the channel capabilities of the brain restrict individuals to about seven alternatives when discriminating between absolute judgements along a single dimension. What both these sources seem to imply is a link between the functioning of the brain and mathematical information theory (see Shannon and Weaver 1949).

Although frequently cited, such theory has no relevance to the development of environmental knowledge because it is concerned with how much information a *message* has, not with how much information an *individual* has (Ackoff and Emery 1972: 145). In short, little is known about how environmental information is stored in the brain. Milgram (1973) has put forward the idea that size distortions in mental representations of the environment may in some way be proportional to the molecular storage units used for different realms of the environment, but this remains no more than a suggestion despite some supporting evidence from Sadalla and Staplin (1980b) to the effect that the length of routes with more remembered attributes (and hence more stored information) were consistently overestimated relative to simpler routes. Of course such neurophysiological explanations overlook the fact that how individuals handle environmental information is influenced to at least some extent by their needs and the degree of need arousal (Ittelson *et al.* 1974: 84), their 'cognitive set' (the way they think about the environment) (Leff and Gordon 1979), and by the extent to which they are 'space sitters' or 'space searchers' (Gould 1975).

The nature of the environment from which information is drawn has been studied in depth by Ittelson *et al.* (1974: 105–9) and their conclusions can be summarized briefly:

1. Environments have no fixed or given boundaries in space or time, the boundary at any one point in time being a product of the information generated through the interaction of the individual with her or his surroundings.
2. Environments provide information through all the senses (although sight is probably dominant, see Pocock 1981a).
3. Environments include peripheral as well as central information.
4. Environments provide much more information than can be handled adequately, largely because much information is either redundant, ambiguous, or contradictory.
5. Environments are defined by, and experienced through, action. That is to say, environmental cognition can never be completely passive because individuals are always part of the situation about which they are gathering information.
6. Environments have symbolic meanings. In particular certain environments are imbued with social meaning and behavioural expectations.
7. Environmental experience always takes on the systematic quality of a coherent and predictable whole.

Although these seven points have a definitive appearance, they should be regarded as no more than generalized assumptions.

No such generalizations have yet been made about the other two elements (information seeking and overt behaviour) that were suggested above as being central to the study of human use of environmental information, although Downs and Stea (1973a) have pointed out that there are three key questions: What do people need to know? What do they know? And how do they get their knowledge? Obviously people need to know something of absolute and relative location, something of distance and direction, and something of the attributes of the phenomena that exist at different locations. This knowledge tends to be incomplete, distorted, and schematized. Moreover it is based on both direct experience (learning by doing) and vicarious experience (learning through someone else's actions). This much is known. But according to Downs and Stea (1973b: 82) 'we have no idea how spatial information is processed (although we know something about the products of the transformation), how environments are learned (although we know something of the development of spatial cognition), and of inter- and intracultural differences in representation'. This is partly because the concept of information is 'one of those intuitively obvious, but extremely slippery, ideas which appear difficult to pin down and measure' (Gould and Lafond 1979: 386). It may well be that much everyday activity is carried out in a 'mindless' state and according to pre-existing routines of behaviour with little processing of new information (Langer and Piper 1987). Unfortunately we do not know. This is partly because human geographers have frequently conducted their research without any thought being given to the philosophy of mind on which it is based.

Philosophies of mind

Geographers have frequently made inferences about what happens in the mind but very rarely have they stopped to think what is meant by 'mind'. An exception is a review by Burnett (1976: 25–6) which outlined nine beliefs about the mind that are common in the literature of human geography:

1. Minds are valid objects for scientific enquiry.
2. Minds can be described in psychological rather than neurophysiological terms.
3. Stimuli exist outside the mind.
4. Minds observe, select, and structure information about the environment.
5. Minds evaluate environmental information.
6. Minds are the seat of sensations such as stress.
7. Spatial choices are made in the mind according to decision rules.
8. Spatial choice is the cause of overt behaviour.
9. The mental states of different people are comparable.

Burnett considered these beliefs in relation to three philosophies of mind: the *behaviourist* view that mental states can be defined only in terms of overt action, the *materialist* stance that mental states are purely neurophysiological events in the brain, and the *neodualist* position that a description of the mind must take account of both physical and peculiarly mental (and therefore *non*-physical) phenomena. Although some geographical work on cognitive representation has tended towards a materialist standpoint, the vast majority of researchers have emphasized imagery and therefore adopted a neodualist point of view.

How individuals come to know an environment is of course a special case of the more general epistemological problem of how a subject comes to know an object. In this sense, there are at least three ways of looking at the relationship between behaviour and environment. These have been described in detail by Moore and Golledge (1976b). The first position – *empiricism and environmental determinism* – rests on the assumption that knowledge is built up by encountering the world through the senses. From this point of view the world is seen as acting on people and the environment is viewed as *real* in that it has an objective existence independent of a subject. The second position – *rationalism and nativism* – rests on the contrary assumption that knowledge is innate and that it precedes experience. In other words, this viewpoint holds that the person acts on the world in that the form and meaning of the environment are given a priori. The environment is thought of as *ideal* in that it exists in the form of innate ideas. The third position – *interactionalism and constructivism* – flows from the work of Kant who distinguished between the *matter* of knowledge (which is acquired through experience) and the *form* or way in which knowledge is organized (which is given a priori). To Kant, there were only two ways of knowing: scientific method (with its emphasis on realism, materialism, and the denial of mind) and the exploration of the processes by which reality is formed in the mind (with its emphasis on idealism, phenomenology, and the primacy of mind). However, neo-Kantians accept that reality exists independently of mind but argue that it 'can only be grasped through the effort of particular minds' (Moore and Golledge 1976b: 14). As a result, from this point of view, it is only possible to study reality through studying people and, even then, the study is not of reality *per se* but of reality as a product of the act of knowing. Reality, in other words, is thought of as being *constructed*. Until relatively recently this constructivist position had little impact on geographical studies of environment and behaviour (Livingstone and Harrison 1981). In recent times, however, transactional constructivism has provided a philosophical basis for the study of environmental learning.

Environmental learning

The transactional constructivist point of view has come to the study of people–environment interaction through the work of environmental psychologists (see Altman and Rogoff 1987). It differs from a simple interactional perspective in that it refutes any existential and methodological distinction between people and environment and postulates instead that the two are linked in an indivisible relationship (Patricios 1978). No distinction is made between cognition and behaviour; instead the focus is on the person–environment whole (Aitken and Bjorklund 1988). The origins of transactional constructivism can be traced in *Gestalt* psychology and in the proposition that the environment is a complex and organized stimulus field that provides the context for human behaviour (Ittelson *et al.* 1974: 81–2). The thrust of the transactional constructivist position has been summarized by Moore and Golledge (1976b: 14) albeit with an element of jargon:

> experience and behaviour are assumed to be influenced by intraorganismic and extraorganismic factors operating in the context of ongoing transactions of the organism-in-environment. Transactions between the organism and the environment are viewed as mediated by knowledge or cognitive representations of the environment; but these representations are treated as constructed by an active organism through an interaction between inner organismic factors and external situational factors in the context of particular organism-in-environment transactions.

In essence, the transactional constructivist position assumes that humans are goal-directed animals whose actions take place in a social and historical context and are based on how past environmental experience is construed (Wapner *et al.* 1973). The idea that individuals 'construct' their own versions of reality is an idea with strong intuitive appeal to geographers, largely because of their interest in 'mental maps'. Nevertheless, it is probably transactional theory generally, rather than transactional constructivism *per se*, that has attracted interest.

An example of a transactional perspective in people–environment research is to be found in Oxley *et al.*'s (1986) description of 'Christmas Street'. In this study, a transactional approach was adopted as a way of analysing social networks among families on a city street at two times in the year. Underlying the work was the contention that the social networks and behaviours under investigation took shape through the inseparable interplay of people, psychological processes, environment, and time. Thus a holistic, temporally linked perspective was needed to highlight both stability and change, and the presence of both unique and general events. Such a perspective can, of course, be challenged on the grounds that the transactions between people and environment are difficult to measure (Pick 1976) and on the grounds that it is sometimes difficult to assess whether findings are valuable (in the sense that they are able to be generalized) or spurious (on account of their uniqueness) (Kaplan 1987).

Nevertheless, a transactional perspective does provide insight into why people interact with the environment in the way that they do. For example, Aitken (1991b) used a transactional framework as a basis for studying the way in which the film-maker Bill Forsyth uses 'image-events' to penetrate the extraordinary in Scottish everyday life. Indeed, this study sets out the fundamentals of transactional theory in some detail. It emphasizes, for instance, that individuals do not attain a stable adaptation to, and integration with, their environment. Rather the relationship is a *changing* one where change is triggered by *events* that create imbalance in pre-existing behaviours and understandings. In this sense, 'events are a nexus of behavioural, environmental and temporal features and, as such, it is important not to fragment a person-in-environment whole artificially by studying behaviours or environments separately' (Aitken 1991b: 107). Transactional approaches in geography have also been reviewed by Aitken and Bjorklund (1988). They argue that, in adopting transactional constructivism, human geographers have tended to place too much emphasis on constructivism and not enough on transactionalism. Instead of looking at the outcome of transactions in the form of constructed 'images', they believe that geographers should direct attention to understanding the nature of the transactions that take place. Obviously some sort of methodology is needed for such a study and Aitken and Bjorklund (1988) lay much store by the six principles of transactional methodology outlined by Altman and Rogoff (1987):

1. No context can be assumed to be generalizable and researchers should therefore remember that behaviour always reflects the specific context in which it occurs.
2. Events need to be interpreted from the perspective of

the participants.
3. Researchers should recognize that they are not independent of the situations they study and that, as a result, what they choose to focus on may reflect their value system.
4. Researchers should study change in people–environment interaction.
5. Consideration of the unique can sometimes provide as much insight into behaviour as can the scientific pursuit of generalities.
6. Eclecticism should characterize the choice of research techniques and research designs should be tailored to the questions that are being asked.

In short, a transactional perspective has a good deal to offer the student of people–environment interaction, not least in terms of methodological guidance. However, the impact of transactionalism is probably most in evidence in the way in which it provides insights into both the evolutionary adaptation of humans to the environment and the ways in which conceptions of the environment change with intellectual development.

Evolutionary adaptation to environment

Kaplan (1973, 1976) has argued that environmental information processing was a necessary skill for the survival of the species in that it enabled humans to adapt to uncertain and dangerous environments. In this sense survival depends on four types of knowledge and four associated responses: where one is (*perception and representation*); what is likely to happen (*prediction*); whether it will be good or bad (*evaluation*); and what to do about it (*action*) (Stea 1976: 108). The fact that humans were able to recognize, anticipate, generalize, and innovate ensured their survival. In particular, an ability to cope with *specific* information (such as that relating to features in a particular landscape) and to abstract from this in order to get *generic* information (which provides a way of understanding new environments based on experience of reasonably similar situations) gave humans an evolutionary advantage. In other words, generic information processing was the key to going from *perception* to *prediction*. Kaplan (1976) saw a neurophysiological basis to this process and suggested that the structures in the brain by which information was handled were very similar to the 'neural nets' hypothesized by Hebb (1949, 1966). Briefly, the idea is that there exists a set of elements in the brain that correspond to real world features and that these are linked in such a way that arousal of one element can lead to arousal of a collection of elements. The real world, in other words, becomes a representation in the brain.

Although undoubtedly based on people's transactions with their environment and on their construction of that environment in their minds, the neural net hypothesis gives Kaplan's work a materialist flavour that has perhaps inhibited further work in the field, particularly because it is a hypothesis that is very difficult either to prove or disprove. Little is known, for example, about where the concept of space is located in the brain, although the most likely place seems to be the parietal lobe of the right hemisphere (O'Keefe and Nadel 1978). Likewise little is known about whether the ability of males to perform better than females at spatial orientation tasks is linked in any way to brain structure (McGee 1979). Of course, such differences in performance mainly relate to the manipulation of geometrical shapes in laboratory exercises and differences in environmental knowledge and wayfinding in real world situations are very rare despite some evidence that males might use distance estimates and cardinal directions rather more than females in wayfinding experiments (Ward *et al.* 1986).

Although undoubtedly contentious in that they attempt a simple explanation of millennia of human history without a great deal of supporting evidence, Kaplan's views are complemented by Blaut's (1991: 55) argument that 'mapping behaviour is carried out by all ages and all cultures; it is therefore a natural ability, or habit, or faculty, "natural" in a sense very close to the way language acquisition is "natural"'. Blaut's view is that mapping is psychologically primitive, culturally universal, and 'natural' in that it forms the basis for a specific form of human activity, what he terms 'macroenvironmental' or 'place' behaviour (what might be more widely labelled 'spatial behaviour'). In other words, mapping developed in humans as a way of communicating information that could not be communicated, except with difficulty, in the spoken word. In fact Blaut (1991: 58) sees the origins of mapping in the 'pictorial representations of landscapes, incised in or painted on various kinds of surfaces'. He goes on to argue that, just as Chomsky (1965) and others have suggested the existence of a 'language acquisition device' (or 'LAD'), so there might exist in humans a 'mapping acquisition device' (or 'MAD'). Such a view is of course highly speculative and contentious but Blaut cites a lot of evidence, not least the protomapping skills widely found in young children (see also Matthews 1992). It is appropriate, then, to turn attention to research on the intellectual development of the young child.

Intellectual development

The argument that mapping is 'natural' and that environmental information processing plays an important role in evolutionary adaptation (that is, in *phylogenetic* development) has a parallel in the argument that similar processes are important in the development of the individual (that is, in *ontogenetic* development) (Stea 1976: 111). In both cases humans are seen as interacting with their environments, in the first place ensuring the survival of the species and in the second providing for intellectual growth. Intellectual development in this sense implies 'qualitative changes in the organization of behaviour' (Hart and Moore 1973: 250). Although there is a certain amount of literature on how environmental perception and cognition change with old age (see Gold and Goodey 1989; Rowles 1978), changes in perception and cognition have been studied most extensively in relation to child development where there is ample evidence that environmental awareness and understanding improve with age (Siegel *et al.* 1979). At issue are three fundamental problems: how environmental knowledge comes about; whether intelligence is the awakening of a priori structures or the gradual accumulation of bits of information; and the nature of the interaction between knowledge and behaviour. In addressing himself to these problems, Moore (1976) proposed six postulates to cover the development of environmental knowing.

Moore's first postulate states that environmental knowledge is *constructed* in that individuals invent structures in order to enable them to cope with reality. Thus, in the study of environmental knowledge it is impossible to separate the *objects known* from the *process of knowing*, in much the same way that it is impossible to separate a description of the world from the language used in that description (see Whorf 1956). The fact that individuals construct their own realities implies of course an element of subjectivism in the resultant knowledge. In practice, however, it is usually assumed that similar people have similar constructions. The second postulate is that environmental knowledge results from the interaction between factors internal to the individual (e.g. needs, personality) and the demands made of the individual by the situation (e.g. constraints on the appropriate form of behaviour in certain environments). This is closely related to the third postulate which states that individuals adapt to their environment in that they actively seek out and assimilate information. From this perspective the individual and the situation should be viewed as a whole: people imbue the environment with meaning and that meaning then influences their behaviour.

Moore's fourth postulate is that transactions with the environment are mediated by previously constructed conceptions. There is, in other words, an element of learning involved in the processing of environmental information to the extent that cognition takes stock of experience. The fifth postulate is that environmental knowledge has different structures at different stages of development. More specifically, the cognitive structure '*precedes*, *selects*, and *orders* the specific parts of environmental experience and behaviour that will be attended to and assimilated' (Moore 1976: 144) with the result that a change of structure will mean a change in the type of knowledge that the individual has. The final postulate argues that the nature of these structures can only be studied by taking account of their genesis and transformation. Such transformation can take several forms, two of the more important of which are *ontogenesis* (developmental changes associated with the life cycle) and *microgenesis* (short-term adaptation to environmental change which can take place over periods as short as a few days) (Pearce 1977).

The idea of developmental changes has been studied mainly in relation to young children. For example, Stea (1976) has suggested that children think of their environment in a *Gestalt* manner in that they do not memorize routes so much as *understand* the way in which paths are all related. He ties this in with the structure of the brain. The left-hand hemisphere controls language, writing, and other linear processes (*propositional functions*) while the right-hand hemisphere controls perception and spatial knowledge (*appositional functions*) and Stea suggests that, as the child develops, there is a shift in dominance from appositional to propositional functions (see Ornstein 1972). As a result Stea believes that an ability to cognize large-scale environments may be developed in children well before they reach school (see Blaut and Stea 1971). Unfortunately the evidence for this is somewhat equivocal (see Fishbein 1976). The idea of hemispheric differences in the brain remains, however, an important one that is potentially capable of accounting for what people find attractive in urban design (P.F. Smith 1976) and the adoption of either egocentric or domicentric systems of orientation (Sonnenfeld 1982).

From a developmental point of view, the study of brain structure has never been as popular as the work of Piaget (see Matthews 1992). In simple terms Piaget argued that there are four general levels of spatial knowing. First comes the *sensorimotor* level based on representation through action. At this stage, lasting for about the first two years of life, individuals have very limited mental representations of their environments and what they do know tends to be based directly on experience and on the manipulation of objects. Individuals have very little ability to conceptualize the environment. Young children, for instance, do not realize that objects that roll under chairs can be retrieved by walking around the chairs. Next comes the *pre-operational* level where knowledge tends to be based on simple mental constructs and intuitive, uncoordinated images of the world derived from the memory of previously perceived objects. At this stage (lasting from about 2 to between 5 and 8) individuals have little ability to differentiate between the particular and the general (Walmsley 1988a: 18). The *concrete operational* level follows. At this stage (which lasts until about the age of 11) simple logical thought permits mental representations of the environment that are symbolized and systematic. It is at this stage, for example, that individuals realize that the volume of a fluid remains the same when poured between differently shaped containers. Finally, individuals enter the *formal operational* stage during which they are capable of hypothetico-deductive reasoning and can handle abstract spatial notions that are not dependent upon real actions, objects, and spaces (Moore 1976: 148). At this stage, then, children are freed from relying on experience for knowledge (Walmsley 1988a: 18).

These levels are sequential in that a child passes through each of them as that child develops (Piaget and Inhelder 1956). Moreover, although nodes (e.g. the home) are critical to the environmental learning of all young children (not least because they facilitate orientation and route selection, Golledge *et al.* 1985), the Piagetian stages are each associated with an ability to handle different types of spatial relations; first come *topological* relations (e.g. proximity, separation), then *projective* relations (e.g. objects are thought of in relation to perspectives like straight lines), and finally *Euclidean* or *metric* relations (based on a system of axes and coordinates). Figure 4.2 shows that an ability to cope with these sets of spatial relations emerges during the pre-operational period. Figure 4.2 also illustrates Hart and Moore's (1973) hypothesis that children's understanding of the spatial layout of the environment falls into three stages: an *undifferentiated egocentric reference system* organized about the child's own position and actions; *partially co-ordinated, fixed reference systems* organized around specific places like a home or a town centre; and *operationally co-ordinated and hierarchically integrated reference systems* organized in terms of abstract geometric patterns (Moore 1976: 150). In simple terms, there is a sequence 'from action-in-space to perception-of-space to conceptions-about-space as a function of increasing differentiation, distancing, and reintegration between the organism and its environment' (Hart and Moore 1973: 255). Only at the third level can movement be described in relation to an abstract reference system. Below that level, orientation is either in terms of egocentric perspectives associated with an individual's own specific location, or in terms of fixed and familiar locations (e.g. home, school, playspace) (Fig. 4.3).

In short, Hart and Moore follow Piaget's argument that adaptation is the key to knowledge and that intelligence is the key to adaptation. Intelligence itself cannot be inherited but what is inherited is a variety of modes of intellectual functioning, in particular *assimilation* (the incorporation of new information into existing schemata or forms of organization in the brain) and *accommodation* (the readjustment of schemata,

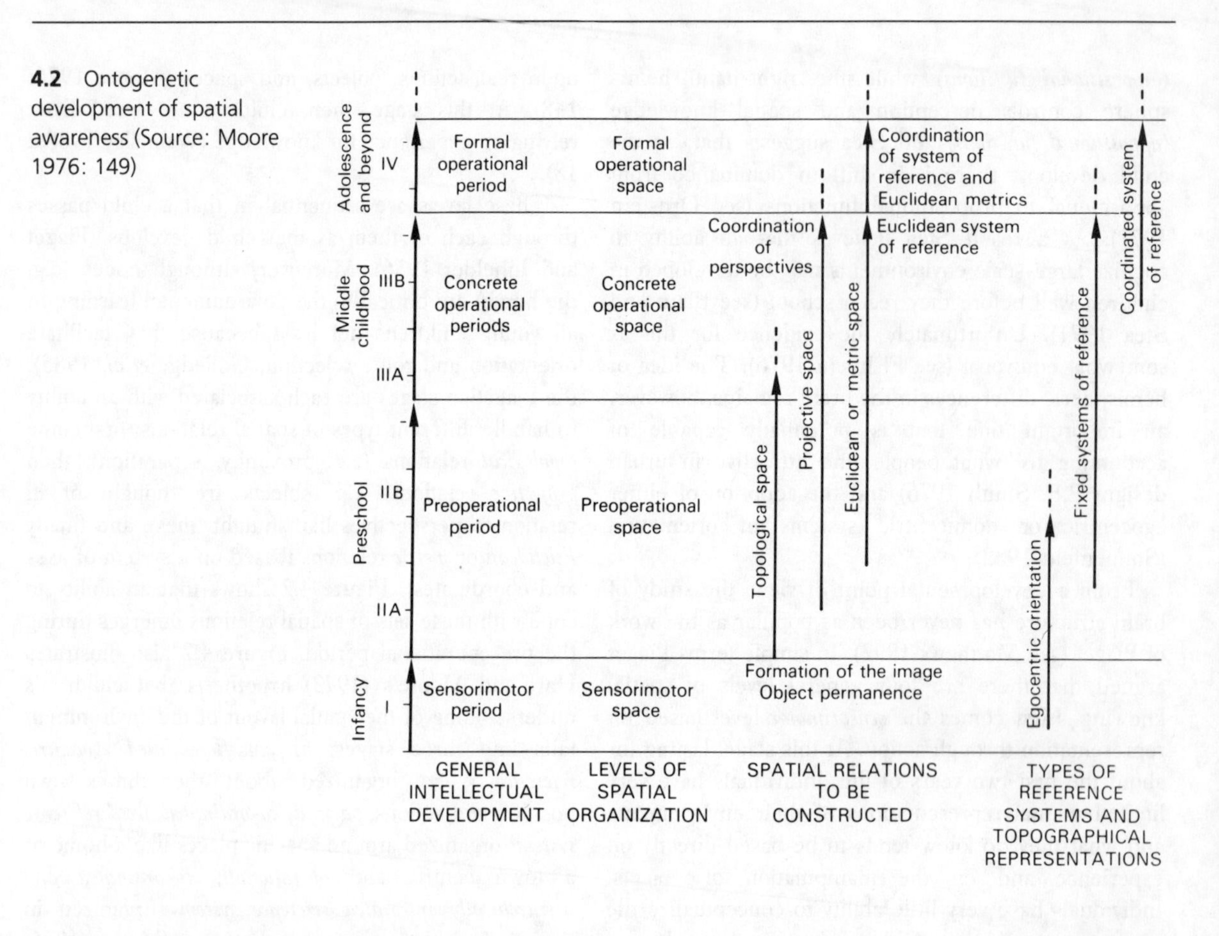

4.2 Ontogenetic development of spatial awareness (Source: Moore 1976: 149)

or means of organizing information in the brain, in the light of assimilation). Of course, not all authorities agree that Piaget's stages can be easily recognized in the field of environmental learning. Thus, although some researchers advocate using Piaget's work as a basis for geographical education (see Downs *et al.* 1988), others have pointed out that Piaget's various stages are only apparent in certain circumstances and, even then, might be rather more complex than Fig. 4.2 suggests (Spencer *et al.* 1989). Matthews (1984b, 1992), for instance, has pointed out that there are two different ways of looking at the development of the ability of young children to draw sketch maps of the areas with which they are familiar. The first is to discern qualitative changes in the way in which children externalize their mental representation of the environment and the second is to study the quantitative accretion of knowledge. The first approach rests on Piaget's structuralist proposition that, as a result of assimilation and accommodation, a child will pass through an evolving sequence of stages in its understanding of space. The second approach is much simpler and holds that the environmental information-processing potential of young children should not be underestimated because it may be that all individuals share an innate ability to comprehend spatial relationships (e.g. the relative position of various features in the landscape) and that this ability simply opens up with experience (Matthews 1984b: 89).

When he presented children in Coventry (England) with a blank piece of paper and asked them to draw their journey to school and their home area (explained to children as 'the area around your home'), Matthews was able to discern clear stages of development. Overall the range of the home area increased from a radius of 180 m at 6 to 870 m at 11 with three stages clearly evident (at ages 6–8, 9–10, and 11), corresponding approximately to the egocentric, fixed, and

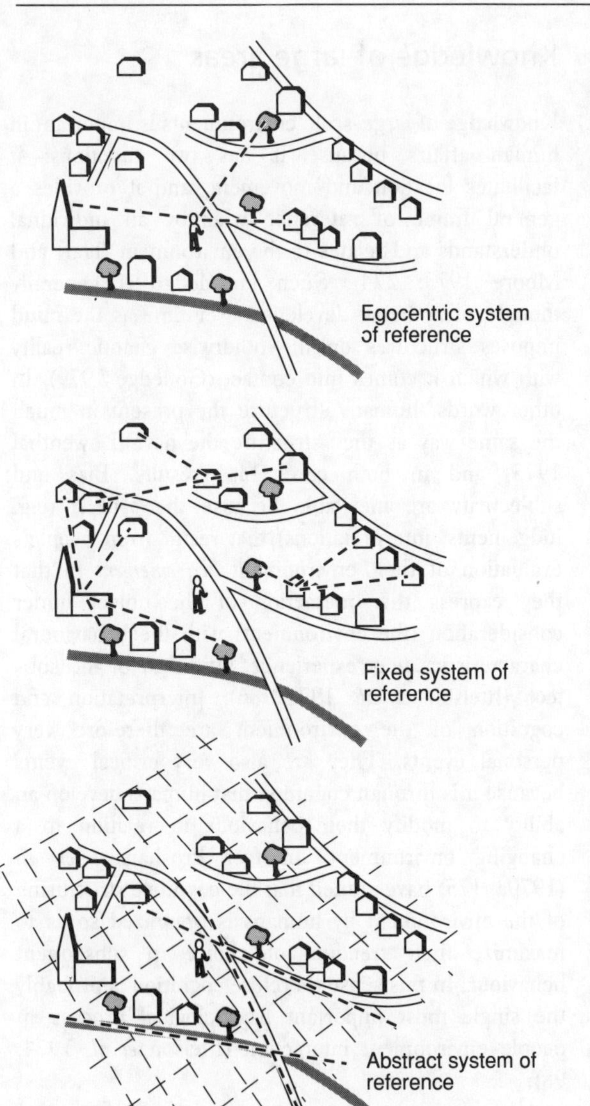

4.3 The orientation systems of young children

abstract systems of reference. However, the maps of the journey to school and home area were rather different in style. Maps of the journey to school emphasized paths whereas maps of the home area emphasized landmarks. This suggests that the type of knowledge acquired might be a function of the purpose for which that knowledge is needed. Furthermore, the timing of the stages was not always clear cut. Some individuals as young as 6 proved to have relatively objective and accurate impressions of very familiar environments. In other words, the relationship between age and cognition is very complex. More importantly, Matthews (1984a) found that the type of environmental knowledge displayed by children tended to be an artefact of the experimental method used. When children were given air photographs and maps of the area surrounding their homes and asked to indicate the extent of their home area, the task was very readily achieved with most children outlining areas much larger than was the case with their sketch maps. At the same time, the distinct stages disappeared. Instead there seemed to be an incremental increase in knowledge with age.

The implication to be drawn from Matthews's (1984a) work is that the existence of stages in the development of environmental cognition might be an artefact of the mode of presentation of material. Several authorities have argued along similar lines (see Piche 1981) and suggested that, in order to combat this possibility, observations of the environmental knowledge of young children need to focus on real life behaviour rather than laboratory experiments (Sonnenfeld 1985; Spencer and Darvizeh 1981). One such view has come from Blades (1989). He argued that most tests of children's environmental knowledge rely on *recall memory* (as when individuals make a verbal description, or a map, of an area). In contrast, real world behaviour may be more influenced by *recognition memory* (e.g. the ability to recognize scenes). It may be, for example, that recognizing a scene triggers appropriate navigational action. In this regard, it is interesting to note that there is evidence that children can remember a route after walking along it just once and that rudimentary environmental knowledge can be acquired as young as 3 years of age. In Blades's view, individuals come to terms with unfamiliar environments by using three sorts of information:

1. Extrapolations from experiences with similar environments;
2. Information within the environment (e.g. signposts);
3. Information external to the environment (e.g. maps, verbal instructions).

Environmental learning is, in other words, a complex process and it may be that, in terms of processes like scene recognition, young children can sometimes outperform adults (Doherty *et al.* 1989). It would be wrong therefore to assume that children simplistically follow Piaget's stages.

It would also be somewhat simplistic to assume that all children are alike in their environmental learning because important differences have been found between the sexes. Saegert and Hart (1978), for example, have pointed out that the home range of girls is usually more restricted than that of boys. A similar conclusion was reached by Matthews (1987) who also found discernible sex differences in both the quantatitive accretion of knowledge and in the way in which children externalize their mental imagery. The boys in his sample, for instance, drew maps that were more detailed and broader in conception. The cause of these differences is almost certainly to be found in the different socialization histories of boys and girls (see Moore 1986). In other words, the differences are gender-based rather than sex-based. Boys tend to be given more freedom and are encouraged to explore more than girls who are often rather protected. It is not surprising therefore that a study of the after-school playspace of children aged 7–12 in Armidale (Australia) showed that boys, on average, ranged up to a distance of 500 m from home whereas girls were limited to 200 m. Indeed only 30 per cent of girls played away from home in contrast to 55 per cent of boys (Cunningham and Jones 1991). Clearly, then, child-rearing practices seem to be important in influencing environmental learning. This may, however, only relate to Western society because a study of Kenyan children found no sex-based differences in spatial ability (Munroe *et al.* 1984).

In summary, the views of Piaget are both stimulating and somewhat contentious. And, of course, it needs to be remembered that studies of environmental learning in children throw little light on how individuals who are already at the formal operational level cope with new environments or with environmental change. There may be, as Moore (1976: 163) suggests, a 'conceptual parallel' between ontogenesis and microgenesis but as yet its nature has not been outlined.

Knowledge of large areas

Knowledge of large-scale environments is important in human affairs because it has two functions: it facilitates location and movement, and it provides a general frame of reference whereby an individual understands and relates to the environment (Hart and Moore 1973: 274). Such knowledge is generally thought of as being developed over time as the mind imposes structures on the otherwise chaotic reality with which it comes into contact (Golledge 1979). In other words, humans structure the present in much the same way as they structure the past (Lowenthal 1975) and in both cases bias results. Bias and subjectivity are inevitable because the *percepts* (e.g. judgements, interpretations) that result from people's evaluation of their environment are *emergent* in that they express the interaction of the object under consideration (the environment) with the behavioural characteristics (e.g. experience, attitudes) of the subject (Ittelson *et al.* 1974: 86). Interpretation and cognition of the environment are therefore very personal events. They are also very critical events because it is through cognition that humans develop an ability to modify their behaviour in relation to a changing environment. In fact Proshansky *et al.* (1970a: 175) have argued that the cognitive structuring of the environment by humans is organized so as to maximize their freedom of choice in subsequent behaviour. In this sense therefore cognition is probably the single most important psychological process in people–environment interaction (Ittelson *et al.* 1974: 98).

Cognition is based on the extraction of information from the environment. Different people will of course extract different information from the same environment. This does not mean, however, that it is impossible to generalize about environmental cognition. After all, the collection of information about large-scale environments is seriously constrained by distance with the result that people living close together tend to have similar information fields (Hanson 1976; Hudson 1975) and very similar motives for acquiring information (Gold 1980: 48). Thus an understanding of what information the mind processes and uses is central to any investigation of people–environment interaction (Golledge and Rushton 1976:

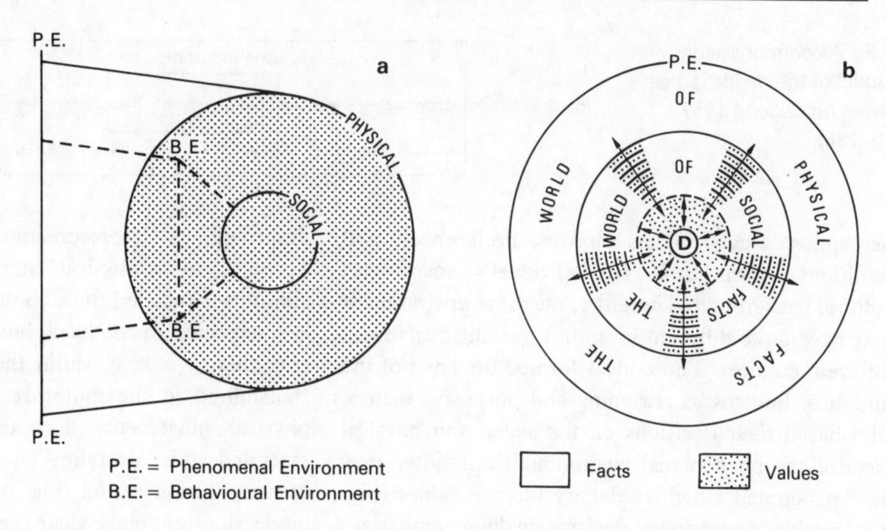

4.4 The filtering of environmental information: (a) the behavioural environment, (b) the behavioural environment of a decision-maker (Source: Kirk 1963: 363, 367)

viii). For the most part research in this direction has been theoretical and, despite attempts to look at information overload and urban complexity (Rapoport and Hawkes 1970), most authors have focused on the contribution of learning to cognition.

There are a great many references in the geographical literature that purport to show how psychological studies of learning can throw light on people–environment interaction. Prominent among these citations have been the stimulus–response (S–R) theories deriving in varying ways from the work of Hull (1952). For example, Stea (1976) has proposed a seven-stage learning model whereby (1) the environment is viewed as a complex of stimuli that (2) act on an organism resulting in (3) a cognitive process within the central nervous system such that (4) a response is learnt and (5) subsequently rewarded or reinforced subject to (6) the organism being disposed in such a way as to make the reward effective with a result that (7) learning can be said to have occurred.

Stea's argument clearly represents the *behaviourist* point of view. According to this, by constant repetition a suitable response to a stimulus is reinforced and learnt so that, whenever the particular stimulus is presented, the resultant behaviour is predictable (Wright *et al.* 1970). Although the early behaviourists (see Watson 1924) did most of their research with animals under laboratory conditions (see Canter and Kenny 1975; Stea 1973), there has been no lack of researchers ready, to apply S–R theory directly to human behaviour (see Skinner 1971; 1974). This experimental approach has, however, attracted stern criticism, particularly of the way in which behaviourism oversimplifies human behaviour. Koestler (1967) expressed this well when he pointed out that behaviourism was based on a 'ratomorphic' view of humans that was no better than the nineteenth-century anthropomorphic view of animals. In short, behaviourism overlooks the cultural and socio-economic context of learning (Lee 1976). It therefore ignores the transactional constructivist proposition that humans actively create their version of reality rather than simply learning an environment (Golledge 1979).

Partly as a response to these weaknesses of behaviourism, there emerged a *Gestalt* theory of learning which argued the need to incorporate the individual mind in any consideration of human decision-making. Basically, this theory postulates that perception forms a crucial intervening variable between stimulus and response, and that individuals organize the objects they observe in the environment into patterns, or *Gestalten*. Koffka's (1935) claim that behaviour was based on the environment as perceived rather than the environment as it is, was taken by Kirk (1952, 1963) and used as a means of developing a simple geographical decision-making model (see also Boal and Livingstone 1989). In this model Kirk argued that the decision-maker is embedded in a world of physical facts and a world of economic and social facts. Both worlds impinge on decision-makers through

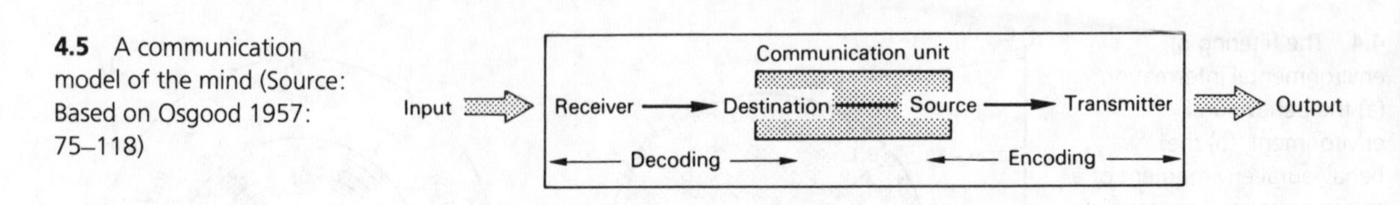

4.5 A communication model of the mind (Source: Based on Osgood 1957: 75–118)

perception which reflects motives, preferences, and traditions drawn from an individual's social and cultural background. Therefore, the same environment may have quite different meanings to individuals from different cultures. These ideas formed the basis of the notion of humans as reasoning and purposive beings, who based their decisions on the social and physical facts of the phenomenal environment only after they have penetrated a highly selective filter of values (Fig. 4.4). Such a conception of decision-making emphasizes the point that behaviour is guided, not by the external environment as such, but rather by a distorted psychological representation of it. Despite early suggestions that it was innate, this perceived environment has been shown to be a learned phenomenon (see Lewin 1936).

Both the S–R and *Gestalt* theories can be regarded as being too limited in their conception of decision-making since they fail to take into account what actually goes on in the mind when a decision is being made. This is an important issue but one that has traditionally been outside the realm of geography. The significance of the perceived environment is widely recognized but little is known about how information is actually processed in the mind. Instead, as has already been noted, the mind is viewed very much as a 'black box' which is the receiver of stimuli from the environment and an initiator of both innovative and habitual behaviour (see Fig. 4.1). The neurophysiological properties of the box are unknown and its supposed mode of operation is simply deduced by observing the way in which output varies as inputs are changed.

One refinement of this conception of decision-making is Osgood's (1957) adaptation of the basic S–R model, which he divided into two stages: an encoding and a decoding stage. Briefly, *decoding* is the term used to describe the association an individual has with, and her or his disposition to respond in a certain way to, a particular sign. The eventual response is known as *encoding*. Such a process involves what have been described as 'representational mediators'. During decoding some physical energy is expended when a stimulus is recoded through nervous impulses before it is interpreted in the black box. On the other hand, in encoding, a motive within the black box, after being transmitted to the motor responses, is recoded into physical movements that are the output. Osgood extended this decoding/encoding concept with a communication model (Fig. 4.5). Of course even this model does not make clear exactly what happens in the mind. Indeed the black box can be characterized as a 'hidden world'. This, of course, is a notion familiar to geographers who have long recognized that the *terrae incognitae* in the mind may influence behaviour to a much greater extent than actual environmental stimuli (Wright 1947). Essentially what is being argued in Osgood's variation of S–R theory is that it is not so much what we see, but how we feel about what we see, that is crucial to understanding behaviour (Lowenthal 1961). In other words, humans have 'the peculiar aptitude of being able to live by notions of reality which may be more real than reality itself' (Watson 1969: 14). In summary, what the black box concept emphasizes is that spatial behaviour involves not only stimuli and responses but also a complex process of relationship and feedback based on cognitive and physiological processes that are still only vaguely understood.

Anchor point theory

Dissatisfaction with S–R theories has encouraged geographers to look to the idea of *place learning* and the notion that individuals learn linkages between elements within the environment rather than simple associations between environmental stimuli and resultant behaviour (Tolman 1948). Although this approach is appealing in that it focuses on 'cognitive maps', doubts among psychologists about its logical status tend to have inhibited its adoption (Moore and Golledge 1976b: 9). Instead it is increasingly recog-

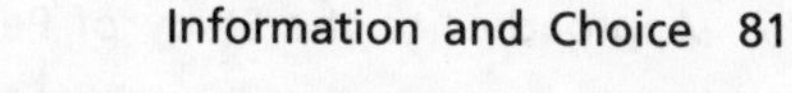

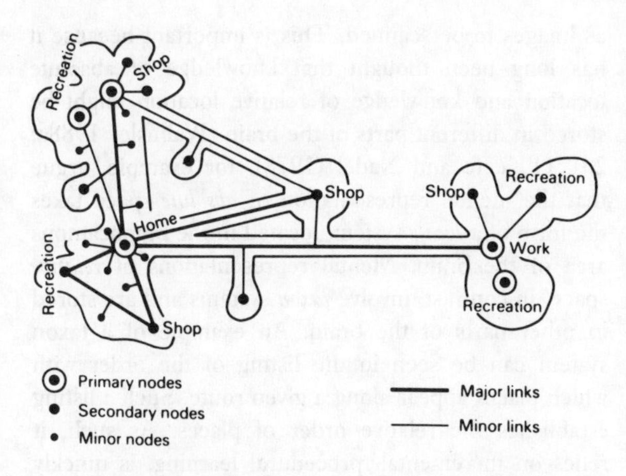

4.6 The anchor point theory of environment learning (Source: Golledge 1978: 80)

nized that the sort of problem that is involved in the learning of large-scale environments is one that has traditionally lain outside the scope of experimental psychology. As a result, researchers interested in large-scale people–environment interaction are likely to have to develop their own models of learning. In this context Appleyard (1973) has suggested that there are three types of knowledge that result from environmental learning:

1. *Operational knowledge* concerned with features that are critical to the functioning of the environment (e.g. location of shops, schools);
2. *Responsive knowledge* (how people react to features like landmarks);
3. *Inferential knowledge*, that is, an ability to generalize and to make inferences that extend beyond what is actually known.

Responsive knowledge is environment-dominant but operational and inferential knowledge are people-dominant and are the product of information derived from personal experience, from contacts, and from the mass media. Of these sources of information, direct experience has been studied most. For example, Golledge (1978) has proposed that learning about an environment takes place primarily as a result of interaction with that environment such that a cognitive representation of the environment is built up over time. First an individual comes to know locations, then links between locations, and then areas around groups of locations (Golledge and Spector 1978). The result is the pattern displayed in Fig. 4.6. This theory of learning is usually referred to as 'anchor point theory' because the nodes and landmarks to which an individual is linked in the urban environment (notably the home) serve as 'anchors' to which knowledge is attached (Couclelis *et al.* 1987). The theory therefore corresponds very closely with Siegel and White's (1975) view of how young children learn their way around the environment (namely, the idea that they learn landmarks, then paths, then areas).

Although there is a certain amount of evidence to support anchor point theory (Evans *et al.* 1984), the argument underlying it may be flawed. For one thing, although an individual's knowledge of an area does improve over time and although environmental learning can be very quick (Guy *et al.* 1990), it is possible that information is continually evaluated and some of what has been learned is discarded, at least in the sense that it does not show up in sketch maps but rather remains as somewhat vague 'background' information (Humphreys 1990). For example, Walmsley and Jenkins (1992a) found that the relative importance of landmarks, routes, and districts varies over time (Fig. 4.7). Increasingly, too, researchers are coming to recognize that there exist two different sorts of

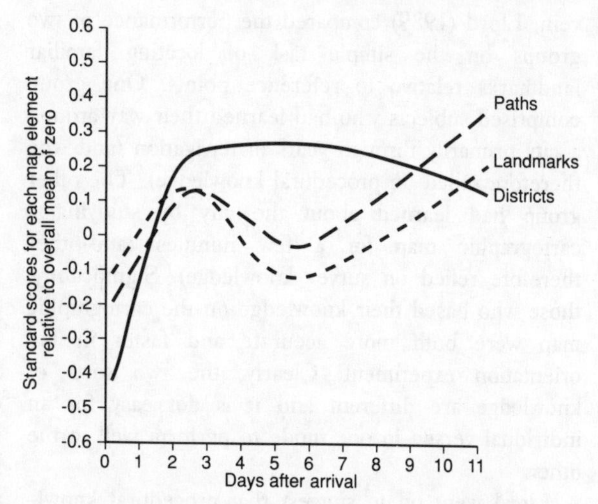

4.7 The relative importance of landmarks, paths, and districts over time (Source: Walmsley and Jenkins 1992a: 279)

environmental knowledge. Aitken and Prosser (1990), for example, accept that qualitative changes take place in environmental learning over time (an idea underlying anchor point theory) but contend that knowledge of a network of routes does not necessarily lead to knowledge of areas. Instead they argue that *procedural knowledge* (knowledge of routes and of how to get from one place to another) might be very different from *configurational knowledge* (the ability to traverse, without getting lost, complicated configurations of paths and nodes through the application of some overall frame of reference). In other words, there need be no necessary link between the two types of knowledge. Procedural knowledge *might* quickly lead to configurational knowledge in areas where there is a simple street plan but the link need not be there. As a result, people who gain their environmental knowledge in the form of procedural knowledge might experience genuine difficulty when called upon to draw a map of the area in which they live (Thorndyke and Hayes-Roth 1982).

Although the terminology varies from case to case (with configurational knowledge often being referred to as *survey knowledge*), several researchers agree with Aitken and Prosser (1990). Gale *et al.* (1990), for example, point out that route knowledge (i.e. procedural knowledge) can be quite parsimonious and that successful navigation along a route does not imply survey knowledge of the area in question. In a similar vein, Lloyd (1989) compared the performance of two groups on the simple task of locating familiar landmarks relative to reference points. One group comprised subjects who had learned their way around a city primarily through years of navigation (and who therefore relied on procedural knowledge). The other group had learned about the city by studying a cartographic map for a few minutes (and they therefore relied on survey knowledge). Significantly, those who based their knowledge on the cartographic map were both more accurate and faster in the orientation experiment. Clearly, the two sorts of knowledge are different and it is not easy for an individual versed in one mode to perform well in the other.

Lloyd went on to suggest that procedural knowledge is stored in the memory as conceptual propositions related to the process of going from one place to another, whereas survey knowledge is stored as images to be scanned. This is important because it has long been thought that knowledge of absolute location and knowledge of relative location might be stored in different parts of the brain (Walmsley 1988a: 24). O'Keefe and Nadel (1978), for example, argue that the mental representation of *absolute* space takes the form of a *locale* system, located in the hippocampus area of the brain. Mental representations of *relative* space, in contrast, involve *taxon* systems and are stored in other parts of the brain. An example of a taxon system can be seen in the listing of the order with which places appear along a given route. Such a listing establishes the relative order of places. As such, it relies on incremental, procedural learning, is quickly accomplished, but is relatively inflexible in that individuals frequently become lost once the known sequence is broken or missed. In contrast, the development of a locale system is motivated more by curiosity than by a need to remember routes. In this sense, a locale system leads to an overall mental representation of the city which can subsequently be used to aid understanding of how the urban environment is structured and functions (Russell and Ward 1982).

It may well be that much everyday life is dominated by procedural rather than survey knowledge. For example, at the micro-scale of an architecturally complex five-storey hospital, Moeser (1988) found that student nurses failed to develop survey knowledge even after traversing the building for 2 years. Similarly, at the scale of metropolitan living, Stern and Leiser (1988) have argued that environmental knowledge is procedural rather than spatial, on account of the limited activity spaces that most residents have.

The study of environmental learning is undoubtedly a fascinating area of research and one where a great many refinements in the understanding of people–environment interaction are being achieved. Recent research has taken many forms. Some, for example, is focusing on the simulation of environments whereby computer-operated cameras are used to provide highly realistic models of typical but fictitious urban environments (Evans *et al.* 1984). Other researchers are attempting to devise computer-based algorithms to simulate how learning occurs (Gopal and Smith 1990; Leiser and Zilbershatz 1989). Some researchers are even studying the visually impaired in order to see how such groups fare relative to the non-visually impaired. The results of this research are interesting in that they show that socio-cognitive competence can be acquired without vision. Indeed, congenitally blind subjects often outperform the adventitiously blind and the partially sighted (Passini and Proulx 1988; Passini *et al.* 1990). However, fascinating as all this work is, researchers interested in environmental learning should not lose sight of the fact that direct interaction with the environment does not account for all knowledge; social and historical factors are also important. Learning about the environment is, in essence, the process of compiling information which includes all the spatial relations among environmental elements as well as their socio-economic, cultural, and other meanings and significances (Golledge 1978: 81). It is important therefore to look at information flows and why they are important in people–environment interaction.

Information flows

The subject of information flows has been of interest to human geographers ever since Hägerstrand's (1952) seminal work on the propagation of innovation waves. For the most part this information-orientated research has focused on innovation and diffusion (see Brown 1982). It has usually been assumed, for example, that information about innovations is passed on only through interpersonal communication (the probability of which is influenced by the distance separating individuals), and that different individuals have a differing resistance to innovation. The result of this type of information flow is that diffusion describes an S-curve over time as first early adopters, then late adopters, and finally laggards succumb to the innovation (Fig. 4.8). In simple terms, diffusion can be likened to a wave that spreads outward from a point of innovation dissipating its energy as it goes (Morrill 1968). Even hierarchic diffusions, where innovations tend to follow the central place hierarchy, can be seen as a special case of the general diffusion process (Brown and Cox 1971; Morrill and Manninen 1975).

This classical view of diffusion has attracted a good deal of criticism in recent years (Brown 1981). Blaut (1977) has stressed that information is not something that lies around in discrete bits but rather something

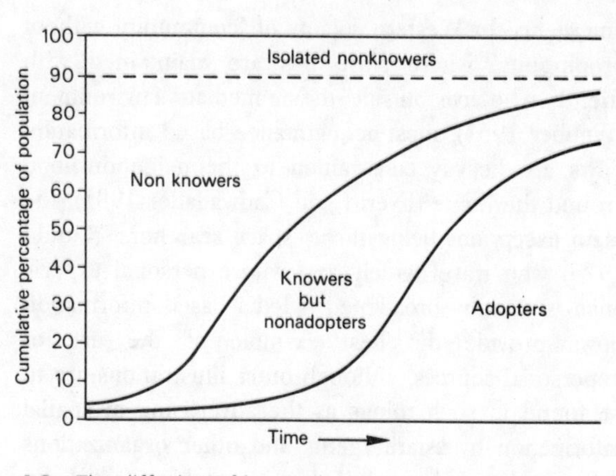

4.8 The diffusion of innovations

that is part and parcel of the structure of society with the result that diffusion should be seen as a force for cultural change. A similar theme was taken up by Blaikie (1978) in his argument that information transfer is class-specific and that a focus on regularities in the diffusion process results in a failure to ask vital questions about the ideological, political, and economic origins of innovations. On top of this, of course, many studies of diffusion fail to recognize that as time–space convergence continues and as the world becomes increasingly well connected in an informational sense, so the ability of information movement to shape patterns of innovation adoption weakens. In a world awash with information, it might be that place-to-place variations in the receptiveness of the population to innovation is the major determinant of diffusion (Ormrod 1990). In other words, studies of information flow can be criticized from an ideological viewpoint and because they ignore technological change. They can also be criticized in that most of them have concentrated on the macro-scale and only rarely has any attention been paid to the micro-scale and the way individual decision-making units identify and evaluate information sources (see Clark and Smith 1979).

Individuals acquire information from both personal and impersonal sources (Spector *et al.* 1976). Personal sources usually imply a point-to-point information flow and impersonal sources a point-to-area flow (Jakle *et al.* 1976: 121). Personal sources are best illustrated by communication between acquaintances. Despite the emergence in Western society of 'community without propinquity' where close ties are maintained with friends who live outside the immediate environment (Webber 1963), most acquaintance-based information flows are heavily constrained to the neighbourhood around the home (Everitt and Cadwallader 1981), the main exceptions being those 'space searchers' (Gould 1975) who travel widely and whose personal experience is thereby broadened. Media-based information flows provide the best example of the use of impersonal sources, although other illustrations are to be found in such things as the advertising of spatial information by estate agents and other organizations concerned with locational decision-making (see Palm 1976a). In all such cases the flow of information is controlled by *gatekeepers* (journalists, editors, agents) who decide what the public should see and hear. Predictably the use to which personal and impersonal information sources are put varies from individual to individual according to differences in their needs and outlook (Leff *et al.* 1974). Some generalizations can, however, be made.

Research into the effects of information on its users has traditionally focused on the question 'who says what, to whom, and how?' and this question can be used to differentiate private information flows from public information flows (Table 4.1). Private information is exchanged in personal contacts between neighbours on a direct basis usually at a small scale whereas public information involves communication

Table 4.1 Types of spatial information

	Public information	Private information
Who?	Skilled professionals (e.g. journalists)	Personal contacts (e.g. friends)
Says what?	Descriptions based on observations	Experiences based on recollections
To whom?	Mass society (e.g. TV viewers)	Primary contact groups (e.g. neighbours)
How?	Broadcasts and publications	Face-to-face meetings telephone contact

from skilled professionals to a mass society often at a reasonably large scale. This is not to say that public information flows in general, and the media in particular, encourage cultural homogeneity and uniform behaviour. In fact, the impact of public information flows on individuals is not at all clear and a variety of different positions have been adopted (Walmsley 1980). Behaviourists assume that exposure to things like the media influences both actions and attitudes. That is, the behaviourist position overlooks the fact that human behaviour is largely goal directed and assumes that information does something to the people it reaches in an almost deterministic fashion. For this reason, the behaviourist approach is sometimes known as the 'hypodermic model' of media influence; media exposure is thought to have a direct and observable impact on media users, much in the same way that drug use does. An opposite position is taken by functionalists who argue that public information fulfils a number of needs (e.g. orientation, reassurance, escape). In other words, public information sources are put to a number of uses from which gratification is derived. This approach is therefore known as 'uses and gratifications' research.

Yet a third perspective argues that public information flows have little *direct* impact on behaviour or attitudes but that they are important in that they tell information users what to think about. In this sense, public information sources set an *agenda* of issues for public consideration (Walmsley 1982b). This is an interesting proposition and it may well be that it has relevance to the distinction that is sometimes made between *specific* and *generic* information (Kaplan 1976). Private information flows are ideally suited to the provision of specific knowledge relating to what is where in a reasonably restricted small-scale environment. They are possibly less useful in providing generic knowledge and the ability to extrapolate from a known situation to an unknown situation because many private information flows contain little that is of use in the development of the schemata or mental structures that are necessary for the classification and organization of information. The agenda set by public information flows possibly fulfil this role of providing such schemata although the evidence for this is not yet definite (see Walmsley 1980, 1982b), partly because the nature of public information flows, like private information flows, is changing (Walmsley 1983, 1989).

Certainly the mass media are important in generating stereotypes. In this sense, then, the media are 'an integral part of a complex cultural process through which environmental meanings are produced and consumed' (Burgess 1990:139). Journalistic goals of objectivity, wide coverage, and 'the public interest' combine with geographical stereotypes and accessibility to create highly biased spatial patterns of news coverage (Brooker-Gross 1983), despite attempts by some media outlets to achieve balance in their coverage (Gans 1979). On top of this, of course, advances in communications technology are continually reducing the friction of distance by making information exchange quicker, cheaper, and more available (Abler and Falk 1981; Falk and Abler 1980). This is likely to affect both public and private information flows. For example, in the area of private information flows, more telephones and a better telephone system are likely to stimulate more social visiting and hence more travel and more private information exchange (Clark and Unwin 1981).

Choices between alternatives

Information flows, be they private or public, obviously play a key role in the build-up of environmental knowledge. They are important, too, in another critical component of people–environment interaction: the way in which individuals make choices between alternative locations, be they places to live, to shop, to work, or to play. Great numbers of people are continually faced with choices as to where to shop, and where to go during their leisure time. Others are more highly constrained; they may, for example, have little choice in where to live and their leisure time may be enforced rather than discretionary, possibly as a result of the high levels of unemployment that have come to characterize 'advanced' Western society. It is important, therefore, to understand how people make choices between competing alternative locations (what is sometimes called 'locational decision-making' or 'spatial decision-making'), and how constraints can curtail choice (see Clark and Flowerdew 1982). In this sense decision-making can be thought of as the translation of motives into overt action within the context of available information and subject to constraints. Without any motivation, an individual may be described as inactive and made up simply of a series of 'functional' characteristics (such as mental and physical abilities and value systems), a series of 'structural' characteristics (such as sex, age, and occupation), and a series of 'existence' variables (such as location). Decisions arise when an individual is motivated to act by changing needs or by changes in the environment (Huber 1980).

Decision-making can be classified into a number of types, ranging from highly deliberate problem-solving to habitual, subconscious decision behaviour (Moore and Thomas 1976). In general, most decisions tend towards the latter type, largely because of the inability of people to process large amounts of information and also their predisposition towards minimizing effort (Jarvis and Mann 1977). Simon (1952) has in fact argued that, in order to compensate for these traits, individuals are inclined to simplify the decision process by greatly reducing consideration of alternative courses of action. Certainly this type of decision-making process is commonly used in coping with the trials of everyday living. True problem-solving is, however, a very different type of decision making in that it involves confrontation with a problem which requires deliberate thought in a specified direction, whether it be searching or vicarious trial and error, and hence a choice from among a wide range of alternatives. Inevitably such a decision is a highly selective process and the resultant behaviour is often an abrupt change from what has happened before and involves a substantial alteration of life style. Many problem-solving decisions of this nature represent attempts by people to adapt to changing environmental conditions.

The core of most locational decisions is the notion of *choice*, which implies a search by individuals among alternatives within the environment (Golledge and Rushton 1976). The criteria for the ultimate choice are invariably relative, rather than absolute, and so the process of spatial choice involves, in principle, an individual in comparing each alternative with every other one in order to select the one which gives the greatest expected satisfaction. Such choice behaviour is, of course, part and parcel of an ongoing learning process that involves both correct and incorrect choices (Golledge 1969). For example, once overt action has taken place, individuals tend to restructure their decision processes in the light of the new and additional information so as to confirm and repeat a certain course of action, or lay the foundations for alternative actions (and this sort of feedback relationship between overt behaviour and decision-making was clearly shown in Fig. 1.3). Very often, too, choice behaviour has been conceptualized as involving a ranking procedure whereby all conceivable spatial alternatives are ordered on a scale of preferredness (see Rushton 1969b).

A major consequence of this focus on choice behaviour in human geography has been a plethora of studies of discretionary behaviour, such as shopping and recreational trips, rather than of those activities which are more space and time bound, such as journeys to work and to school (Pipkin 1981). This suggests that certain forms of behaviour involve much greater choices than others. Also, the range of choices for any one form of behaviour will of course vary for different individuals, with the wealthy having greater choices than the poor, and the young more mobility than the elderly. Inevitably realization of this fact forces attention upon the concept of choice within a wider environment and on the way in which social institutions may constrain individual choices by means of a whole series of 'entry rules' (Moore *et al.* 1976; Pred 1981). First, however, it is essential to appreciate the psychological basis for decision-making.

Maximizing and satisficing

Until the late 1960s, the overriding assumption in most studies of choice was that decision makers were *rational economic beings*. Briefly, this view assumes that humans have perfect knowledge of the environment, together with complete rationality, and are intent on undertaking the optimum economic activity. Applications of this idealized postulate frequently assumed an isotropic surface characterized by equal accessibility and peopled by identical human beings. Such an approach to decision-making in fact underpinned the scientific 'revolution' which swept through human geography from the early 1960s onwards. Despite its value in illustrating what should occur under certain conditions, this approach has many shortcomings (see Golledge and Timmermans 1990; Pred 1967). The criticisms have been many and varied (Barnes 1988). Some attack the postulate of rationality for being unrealistic. That is to say, it makes grossly unrealistic assumptions both about how much information decision-makers have and about their ability to handle that information. Others attack the rationality postulate because it is both reductionist (in that it seeks to reduce all choice to a matter of maximizing economic benefits) and determinist (in that it allows for only one logical outcome in any choice situation). Some attack the postulate because of what are perceived to be logical inconsistencies (Barnes 1988: 483).

The consequence of these criticisms has been a questioning of the value of the assumptions upon which the concept of rational economic behaviour is based (Simon 1959). For example, it has frequently been noted that, in reality, decision-makers have only limited information and are subject to factors that are simply not predictable. In addition the 'capacity' of the human mind for formulating and solving complex problems is very small compared with the size of the problems whose solution is required (Simon 1957: 198). As a result a certain amount of attention has shifted from the concept of rational economic behaviour to the principle of *bounded rationality*. According to this principle, individuals seek to be rational only after having greatly simplified the choice before them. This view is in line with the argument by psychologists that humans are 'cognitive misers'. Because it is not possible to process all available information, humans develop strategies (like 'bounded rationality') for dealing with complex situations (Taylor 1981). However, even this view may be unreal in that many individuals may not seek to optimize at all. Rather, given the enormous effort required for searching out and evaluating alternative courses of action, it may be that many decision-makers simply opt for a solution to decision problems that is *satisfactory* in relation to their aspiration levels.

March and Simon (1958: 256) have in fact claimed that most human decision-making, whether individual or organizational, is concerned with the discovery and selection of satisfactory alternatives; only in exceptional cases is it concerned with the discovery and selection of optimal alternatives. According to this view, optimization requires processes several orders of magnitude more complex than those in satisficing. This, then, is the postulate of human beings as *satisficers* (Simon 1952, 1957). The conditions for satisfaction are, however, not static, but are adjusted upwards or downwards on the basis of new experience and new aspiration levels. When the outcome of a decision falls short of the decision maker's level of aspiration, he or she either searches for new alternatives or adjusts the level of aspiration downwards, or both. The goal is to reach levels that are practically attainable. In other words, the satisficer may evaluate alternatives and assess likely outcomes on a scale that runs from optimistic to pessimistic projections (Table 4.2).

An early attempt to verify the satisficing model was Wolpert's (1964) study of actual and potential farm output. This showed that actual labour productivity was substantially less than its theoretical optimum. In fact less than half of the sample farms had labour productivity of more than 70 per cent of the optimal level and, in some areas, productivity fell to 40 per cent of optimum. In other words, the average farmer achieved less than two-thirds of the productivity that the farm's resources would have allowed. On this evidence Wolpert concluded that the concept of satisficing was much more appropriate than that of rational economic behaviour for explaining the observed behaviour of his sample. Farmers not only lacked knowledge but also, to some extent, were influenced by value systems that emphasized aversion to uncertainty. The question of how people cope with uncertainty has of course been studied most intensively in relation to natural hazards (see Walmsley and Lewis 1984: 109–17). Here the evidence seems to support the contention that behaviour is, at best, boundedly rational and, at worst, guided by mistaken notions of risk. Many individuals in hazard-prone situations, for example, display the sort of behaviour predicted by cognitive dissonance theory. According to this theory, individuals suppress information that is incongruent with their view of the world, no matter how inaccurate that view might be (Aitken 1991a: 186). Shippee *et al.* (1980) used the notion of cognitive dissonance to good effect to show how residents living in a hazard zone tend to deny the existence of the threat posed by the hazard whereas residents living near by, but outside the hazard zone, tend to have a heightened perception of the threat. In other words, how people cope with risk and uncertainty is an important topic in any consideration of decision-making.

Table 4.2 Types of decisions

Solution	Characteristics
Maximum	Optimum; maximum–minimum pay-off
Optimistic	Minimum regret at not optimizing
Satisfactory	Good enough within choice range
Pessimistic	Less than satisfactory
Worse off	Unsatisfactory

Probability, risk, and uncertainty

With the adoption of the concepts of satisficing and bounded rationality into their analysis of decision-making, human geographers began to develop an approach that conceived of spatial behaviour in probabilistic terms. An important contribution to this development was Pred's (1967) *behavioural matrix*. After a detailed critique of the concept of rational economic behaviour, Pred depicted decision-making outcomes as a function of the quantity and quality of information available in a given situation, and the individual decision-maker's ability to handle this information (Fig. 4.9). In turn each matrix is a function of the environmental system, since this governs both the availability of information and the perception of its usability. The amount of information

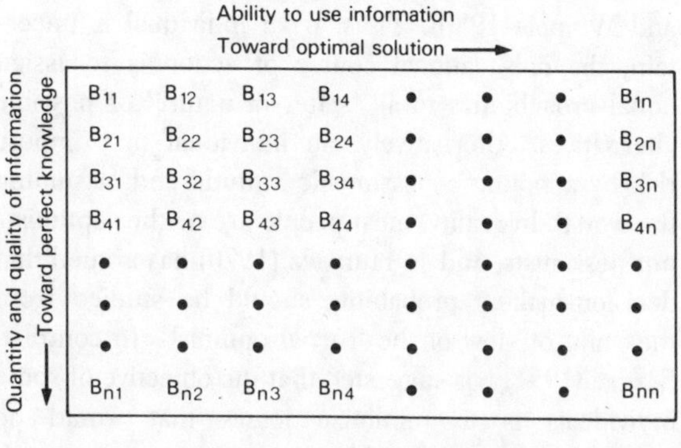

Ability to use information
Toward optimal solution ⟶

Quantity and quality of information
Toward perfect knowledge ↓

B_{11}	B_{12}	B_{13}	B_{14}	•	•	B_{1n}
B_{21}	B_{22}	B_{23}	B_{24}	•	•	B_{2n}
B_{31}	B_{32}	B_{33}	B_{34}	•	•	B_{3n}
B_{41}	B_{42}	B_{43}	B_{44}	•	•	B_{4n}
•	•	•	•	•	•	•
•	•	•	•	•	•	•
•	•	•	•	•	•	•
B_{n1}	B_{n2}	B_{n3}	B_{n4}	•	•	B_{nn}

4.9 The behavioural matrix (Source: Pred 1967: 25)

available can be related to an individual's exposure to the media, an individual's ability to pick out information, and the credibility attached to information sources. Four factors can be recognized as influencing an individual's ability to use information: experience of the user; intelligence; flexibility and adaptiveness; and desires, preferences, attitudes, and expectations. Generally speaking, placement at the bottom right corner of the matrix results in 'better' locational decisions than placement at the top left-hand corner. However, the matrix does not preclude an individual with little skill and information from making a better locational solution than one who possesses greater ability. Such an eventuality is of course relatively unlikely, hence the emphasis upon probability. In short, although it still incorporates the idea of rational economic behaviour, Pred's matrix affords a permissive framework which allows for the fact that many real world decisions fail to come anywhere near optimality.

Since the impetus given by Pred's model, human geographers have attempted to focus more directly on to the varying ability of people to predict future events. Inevitably, this involves the incorporation of the concept of *risk*, that is the notion that all estimates of likely future states are prone to error (Webber 1972). If the probable error is small, then the risk is slight. If the probable error is great, the situation may be described as being 'fairly risky'. Where an individual is unable to make any estimate of future events, that individual can be viewed as operating in conditions of *uncertainty*. Numerous researchers have suggested that, when decision-makers are faced with uncertainty, more often than not they estimate or impute risk in line with their own psychological characteristics (Starr and Whipple 1980). Thus, if an individual is uncertain, the only rational course of action is to assign equal probabilities to all 'states of nature' or possible alternatives. Alternatively, an individual may impute risk by adopting a pessimistic attitude and 'assuming the worst'. In reality, most people are neither optimists nor pessimists, and so Hurwicz (1970) has argued that decision-making probability should be studied from the point of view of the 'partial optimist'. In contrast, Savage (1951) has suggested that the objective of some individuals is to minimize losses that would be incurred if an unfavourable 'state of nature' occurred. In other words, such people are less concerned about the actual gains than with what they would stand to lose in cases of disaster.

In considering spatial decision-making under conditions of risk and uncertainty, several geographers have adopted formal game theory as a means of simulating the decision process (Alwan and Parisi 1974). Formal *game theory* should not be confused with informal gaming, which is a general technique of simulating real-life processes in the form of games, with individuals playing the role of real-life participants (Pred and Kibel 1970; Wolpert 1970; Cohen *et al.* 1973). Instead, formal gaming involves the simulation of the probable strategies of various opponents, each trying to obtain certain objectives without any knowledge about the actions of others (Berkman 1965; Gould 1963; Chapman 1974; Amedeo and Golledge 1975). In such games opponents formulate strategies which yield differing pay-offs, depending on the strategies of the others in the game. The strategies and resultant pay-offs are normally indicated in a pay-off matrix. At issue, in other words, are differing chances of success. A similar perspective has been adopted in the use of information theory as a conceptual framework within which to assess the significance of information upon which decisions are based (Marchand 1972; Louviere 1976; Walsh and Webber 1977; Webber 1978).

All of these attempts to incorporate uncertainty into decision-making are essentially probabilistic in their formulation. As an analytic device, probability does increase the predictive powers of the normative perspective. However, as a means of aiding the understanding of how decisions are made, it leaves much to be desired. For example, an emphasis on probability fails to unravel the dynamics of spatial decision-making. In particular, it tends to treat the environment as 'given' and tends to ignore the way in which people can modify their environment. Much of the probabilistic strategy relies in fact on what Lipsey (1975) has called the 'law of large numbers'. According to this view, the inconsistencies that are apparent in behaviour at the individual level cancel each other out when behaviour is viewed in the aggregate with a result that uniformity and generality can be observed. Although such a perspective may appear to be reasonable, it has the weakness that it implies that individual decision makers lack any common or unifying objective. It therefore runs counter to the assumption on which much behavioural research in human geography is founded, namely the notion that general principles guide the way in which individuals come to know and interpret their environments.

Utility and preferences

The introduction of the notion of boundedly rational satisficers necessitates the abandonment of the idea of perfect information. It also means a retreat from the position where decision makers are thought to be economic optimizers. The concept of satisficing remains vague, however, unless some attempt is made to define exactly what is being 'satisfied'. In this regard some guidance may be found in the concept of *utility* that was developed by economists as a common denominator for comparative value. In a series of pioneering papers, Wolpert (1964, 1965, 1966) adopted this concept within a geographical context. Specifically he developed the idea of *place utility* defined as the 'net composite of utilities which are derived from the individual's integration at some point in space' (Wolpert 1965: 60). Basically, then, place utility is a measure of the appeal of a place to an individual for whatever activity is under consideration. Place utility may be positive or negative, thereby expressing the individual's satisfaction or dissatisfaction with respect to a given place. The operation of place utility in an analysis of spatial behaviour may be illustrated with reference to migration. Where migration is intended and not forced, the argument goes, the migrant will tend to resettle at a destination which offers a relatively higher level of utility than both the place of origin and the alternative places to which the individual might migrate. This higher utility may be expressed in terms of actual characteristics of the place, or it may be the potential of the place as perceived by the migrant. Thus place utility enters the migration decision in two ways: first, in the potential migrant's decision to seek a new residential site arising from dissatisfaction with present utility; and second, as part of the decision as to actually where to search for a new residence. In this sense, then, application of the concept of place utility encompasses the 'subjective expected utility' which individuals ascribe to the different outcomes of decisions (see Graves 1966).

Despite its elegance, the utility concept is extremely difficult to apply. One popular method of attempting to measure utility has been to get individuals to rank order alternative choices in order to arrive at a transitive preference structure (Edwards and Tversky 1967; Lieber 1976). Unfortunately one of the problems with this technique is that it often uncovers intransitive preferences. Such intransitivity may be illustrated by reference to three hypothetical alternatives, A, B, and C: if A is preferred to B, and B to C, and C to A, intransitivity is said to exist. Such intransitivity is particularly common between multidimensional alternatives where individual alternatives are often assessed on different dimensions (May 1954). However, the extent of such intransitivity may sometimes be exaggerated because Walmsley (1977) has shown how a majority of a sample of Sydney women were quite capable of ordering competing shopping centres into a transitive preference ranking. Another problem in applying the concept of utility surrounds the question of whether an individual's preferences are stable through time (Rushton 1976). In particular, the concept of utility is difficult to apply when changes occur in the environment or when an individual's knowledge of alternatives changes markedly. In short, changing utility values may be attributable to any one of three factors (changes in the distribution of alternatives, changes in the action space of individuals, and changes in the individual's preference structure) and it is not always clear which combination of these should be the focus of attention.

Models of decision-making

There have been a number of attempts to formulate a general and comprehensive model of decision-making. A significant initial framework was Herbst's (1964) simple behaviour system, which postulated that every decision-making unit has a *need set*. In geographical terms, this means that individuals located at a given place have needs which are satisfied by that location (e.g. housing needs are satisfied by the dwelling in which the individual resides). If a location fails to fulfil all the needs required of it, stress is generated. Such stress may eventually become transformed into strain, which may be thought of as a force impelling a change in behaviour. According to Herbst, an individual confronted with intolerable levels of strain can adopt one of three courses of action: (1) a change in overt spatial behaviour; (2) an adjustment in the nature of the need set (e.g. reduced aspirations); or (3) an adjustment of the environment (e.g. building extensions to an inadequately small house). Unfortunately, Herbst's framework, although attractive through simplicity, does little to suggest how individuals actually learn about the environment in which they live (see Walmsley 1973).

A similar focus on aspiration levels is contained in Fig. 4.10, which illustrates decision-making in terms of single and multiple choices, within the context of aspiration levels and satisfactory solutions (Wilson and Alexis 1962). This formulation suggests that when an individual faces only a single choice, three phases are involved in reaching a decision: first, an aspiration level is determined; second, after some searching a limited number of outcomes or alternatives are defined; and third, the alternatives are assessed with a view to finding a solution consistent with the aspiration level. In multiple choice situations where behaviour is repeated at relatively short intervals (e.g. shopping, leisure time activities), the position becomes much more complex as the satisficer attempts to balance aspiration and action. In such cases each new decision is the product of earlier ones. For example, a positive discrepancy between behaviour and aspiration (the case where behaviour is actually better than what was aspired to) (Stage 1 in Fig. 4.10) leads to an upward adjustment of aspirations and a renewal of search produces Stage 2 (again with three phases). In contrast, a negative discrepancy (where behaviour falls short of aspirations) leads to a downward adjustment of aspirations and a commitment to search over a wider range of alternatives (Stage 3 of Fig. 4.10). Of course in neither case does the model say anything about the myriad of factors that influence aspirations.

A somewhat different framework for the study of locational decision-making, and one which gives some clues as to what influences aspirations, is Huff's (1960) model of consumer space preference. Although developed to analyse consumer movements, this model has much wider appeal because its holistic perspective highlights major components of the decision process.

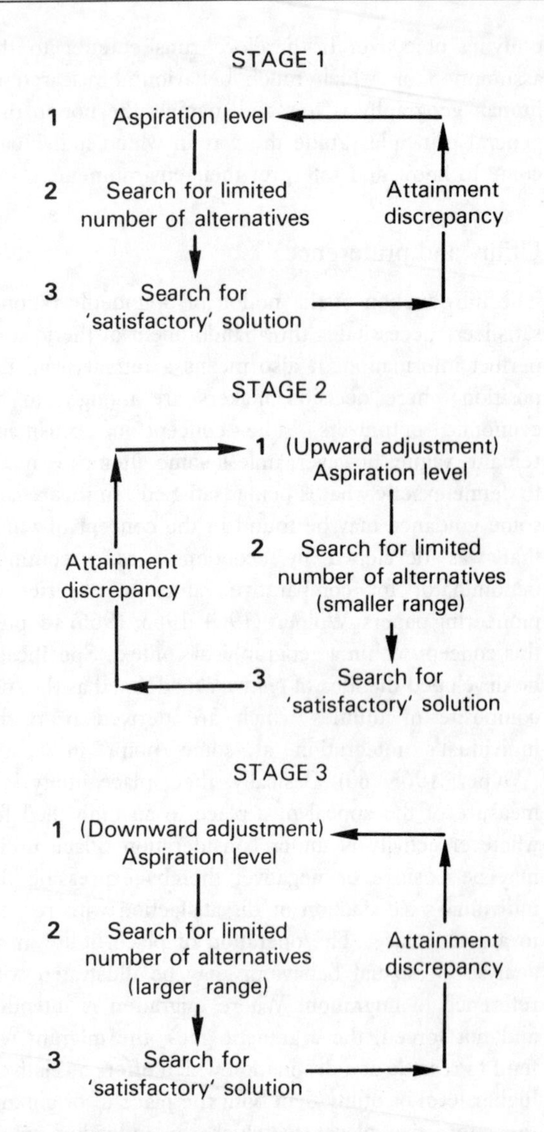

4.10 Single and multiple choice decision-making (Source: Based on Wilson and Alexis 1962: 24)

Briefly, five major components are identified (see also Fig. 9.2):

1. A *desideratum*, or a desire for a particular objective, which results from physiological needs and a stimulus drive;
2. A *value system* which determines how values operate as a filter between an individual's objectives and an eventual decision;

3. *Perception*, or the filtering process through which an individual selects and distorts information about the location and relative accessibility of different alternative choices;
4. *Movement imagery* which focuses on how individuals perceive their ability to overcome the frictional effect of distance;
5. The *overt behaviour* or movement which results from the evaluation of alternatives.

No matter what behaviour results, there is always a feedback of information which causes a restructuring of perception activities and movement imagery. Huff's model is, in other words, a very broad one. In fact it is probably better thought of as an organizing framework rather than a strict model in that its main function is to highlight the critical elements to be considered in a study of spatial decision-making. The same can be said of course of the work of Herbst (1964) and Wilson and Alexis (1962). Indeed it was dissatisfaction with these general frameworks and a desire for greater precision in the analysis of decision-making that encouraged many researchers to turn to mathematically based models of decision-making.

Most mathematical approaches to decision-making in human geography follow the work that has already been cited in acknowledging that decision-makers limit the number of environmental characteristics that are taken into consideration so as to avoid situations of information overload (see Louviere and Timmermans 1990b). These characteristics are then either implicitly or explicitly evaluated so as to yield subjective impressions of the overall attractiveness of the different choice alternatives in an individual's choice set (which is, of course, likely to be a subset of all available choice alternatives) (Timmermans and Golledge 1990: 312). In other words, individuals are assumed to make some subjective evaluation of how well each choice alternative scores on each characteristic under consideration, before combining these evaluations into some measure of overall utility (see Couclelis 1986). This means that in any choice situation there will emerge a preference structure or ranking of alternatives, with the most preferred option being the one selected (subject to whatever constraints might be operating). Given this overall framework, a great deal of effort has gone into modelling the way in which evaluations of the different dimensions of individual alternatives are combined into overall utility scores. Two approaches have dominated: a focus on *discrete choice models* and a focus on *decompositional multi-attribute preference models* (Timmermans and Golledge 1990).

Discrete choice models derive mainly from microeconomics (see Wrigley 1982) in that they assume that a decision-maker's preferences among a set of choice alternatives can be described with a utility function and that the decision-maker opts for the alternative with the highest utility (see Henscher and Johnson 1981). 'The utility function typically consists of two components: 1) a deterministic component which represents the systematic effects of the selected variables influencing choice behaviour, and 2) a random component which accounts for the effects of unobserved variables, measurement error and heterogeneity' (Golledge and Timmermans 1988: *xx*). Obviously, one of the critical considerations is how evaluations of the different dimensions of a choice alternative are combined into an overall measure of utility. Timmermans (1984b) has studied this problem and explored a wide range of compensatory combination rules (where a good score on one dimension can compensate for a poor score on another dimension) and non-compensatory combination rules. Notwithstanding this issue, depending on the type of probability distribution that is used, discrete choice models can take the form of either multinomial logit models or multinomial probit models (Timmermans 1984b). Discrete choice models are usually developed in situations where the dependent variable is observed in the form of real-world behaviour and where real-world evaluations of the various characteristics of the choice alternatives are also available. In other words, the models are usually calibrated in terms of observed choice probabilities and to this extent they can be considered to be models of 'revealed preference' (Timmermans 1984a).

Decompositional multi-attribute preference models are somewhat different in that they have the following structure (see Timmermans and Golledge 1990: 314). First, the characteristics thought to be influencing choice behaviour are listed. Second, these characteristics are combined to provide a set of hypothetical choice alternatives. Third, an individual is asked to express a preference for each of these alternatives. Fourth, these preferences are then 'decomposed', that is to say, techniques like regression analysis and analysis of variance are used to discover what combination rules can be used to blend evaluations of individual dimensions so that the overall preferences are reproduced. Finally, the preference structure is linked to overt behaviour by specifying some decision rule (e.g. an individual may be deemed to select the alternative with the highest utility, or with the highest utility subject to certain budgetary constraints) (see Timmermans 1984a). In other words, decompositional multi-attribute preference models are similar to discrete choice models to the extent that both assume that decision-makers arrive at an overall preference 'by cognitively integrating the part-worth utilities associated with the attributes of the choice alternatives according to some mathematical function' (Golledge and Timmermans 1988: *xx*). They differ, however, in so far as they rely on experimental data rather than real-world data. Strictly speaking, then, decompositional multi-attribute preference models relate to 'expressed choice' rather than 'revealed choice' (Timmermans 1984a). If they have an advantage over discrete choice models it is that they can model alternatives beyond the realm of the decision-maker's direct experience (i.e. they can handle combinations of characteristics that may not exist in the choice sets with which the decision-maker is usually confronted). As against this, it is unclear whether experimental estimates made by hypothetical decision-makers are at all congruent with real world behaviour (Timmermans 1984a: 215).

There are problems with both discrete choice models and decompositional multi-attribute preference models and some important criticisms have been made by Pirie (1988). Pirie believes that the pursuit of elegance in model building has been conducted to the point where the usefulness of the models might be diminished. For example, the goal of global transferability (the idea that the models should be applicable to all people and all places) means that they overlook the way in which specific local environments may be inextricably implicated in decisions. Likewise, in Pirie's view, the models overemphasize choice and therefore underplay the significance of constraints, a point also made strongly by Fischer and Nijkamp (1985). They also commonly fail to consider whether preferences are stable over time (Golledge and Timmermans 1988: xxii). A further telling criticism is that the models tend to focus on simple choices and single-purpose trips (Timmermans and Golledge 1990: 331). This is at odds with a real world in which multi-stop, multi-purpose behaviour is common. To overcome this weakness, Hanson (1980) has suggested that models should focus on the way in which trips are 'chained' (in the sense that different activities are linked together on a single journey). However, mathematical modelling of such 'chaining' has yet to impact significantly in human geography. Indeed, mathematical modelling generally, for all its elegance, still seems somewhat removed from real-world situations. Perhaps this is because real-world situations are often so incredibly complex that they defy description in simple models. Louviere and Timmermans (1990a), for example, point out that a situation with nine three-level attributes would not be unreal and yet, in terms of decompositional multi-attribute preference models, it presents almost 20 000 possible combinations. Nowhere is the inability of mathematical models to capture this sort of complexity more apparent than in organizational decision-making.

Organizational decision-making

Organizational decision-making warrants separate study from decision-making by individuals in that organizational decisions are characteristically of a different order of magnitude and a different degree of complexity from individual decisions (Castles *et al.* 1977; Bartlett 1982). As a result the behaviour of business enterprises, large corporations, and government has increasingly become the centre of explicit attention from human geographers (e.g. Mounfield *et al.* 1982; Hamilton 1974; Hamilton and Linge 1979; Flowerdew 1982) (see Chapter 7).

Within the business world there are at least two groups of important decision-makers: the owner-manager and the manager, both of whom tend to have different motivations and outlook (Harrison 1975). According to Florence (1964), such motivations may be economic, psychological, sociological, or 'alogical' and in each case the entrepreneur differs from the salaried manager (Table 4.3). A similar distinction has

Table 4.3 Business decisons and motivations

Motives	Entrepreneur	Salaried manager
Economic A	Maximum profits	Choice of greater salary
Economic B	Satisfaction – balance of profit and other utilities	Satisfaction – balance of pleasing the shareholder and other utilities
Psychological hobby, boss, and free man	Power seeking, empire building, love of work	Power seeking, fear of others' power
Sociological	Prestige, status seeking in industry	Identification with firm, prestige from successful firm
Alogical	'His own money to do as he wishes', tradition	No initiative or ideas from outside, tradition

Source: Adapted from Florence (1964)

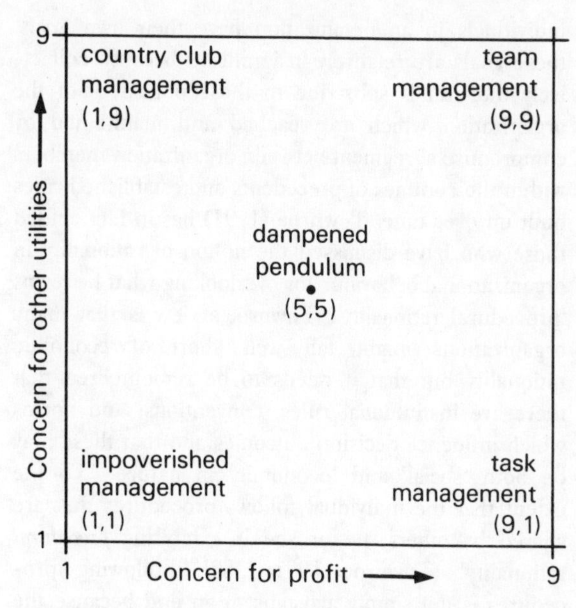

4.11 The managerial matrix (Source: Based on Blake *et al.* 1962: 13)

been drawn by Blake *et al.* (1962) in terms of a schematic 'managerial matrix' (Fig. 4.11). The axes of this matrix are represented by a concern for profit and a concern for other utilities, and different types of managerial approach can be identified within the matrix. Thus the 'task manager' seeks to maximize profit while the so-called 'country club' management places greater reliance on satisfaction of non-profit-orientated concerns. Likewise, 'impoverished management' de-emphasizes profit and other utilities while 'team management' attempts to maximize both its own satisfaction and the monetary success of the business. In the middle of the matrix is to be found what is aptly labelled the 'dampened pendulum'; in this area are the managers whose decisions are made partly to achieve personal satisfaction and partly to ensure the business's survival. Of course it would be wrong to place too much emphasis on this type of matrix because, in many respects, the role of the manager is declining in advanced economies as the scale of organizational activity increases and diversifies (Brown *et al.* 1974). It is important therefore to look not just at managers but also at the processes through which decisions are reached.

Some clue as to the nature of organizational decision processes can be gained from organization theory (see Chapter 7). In general terms this theory argues that all organizations possess similar features, namely, a structure of status and responsibility, a network of communication and delegation, and a set of rules and procedures aimed at achieving a group goal (Castles *et al.* 1977). Such goals may vary from one organization to another. In all cases, however, they form the basis of organizational structure and the guiding principle of organizational policy. Essentially, then, an organization is a grouping of individuals, each with different tasks and authority: in the case of the corporate organizations these individuals include managers, stockholders, suppliers, and employees while in a government organization the key individuals are politicians, administrators, and planners (Simon 1976). Cyert and March (1963) in fact described an organization as being a coalition of mutually dependent, but often conflicting participants. Although individuals in an organization have their own goals, these goals are relatively insignificant because collectively individuals subscribe to those objectives of the organization which are reached and maintained by compromise agreement between organization members within the confines of precedents and established rules built up over time. Townroe (1991) has in fact chided those who have dismissed the notion of rationality in organizational behaviour for overlooking what he terms 'procedural rationality'. Townroe's view is that many organizations might fall well short of economic rationality but that it needs to be remembered that there are institutional rules, conventions, and norms which influence decision outcomes and that these may be both 'social' and 'economic' in nature. 'To the extent that the individual follows procedures that are shared by others, he or she is exhibiting *procedural* rationality' (Townroe 1991: 388). Following procedures is not simply a means to an end because the procedures themselves may contribute to the definition of ends (e.g. consultation might help the minimization of staff turnover and the enhancement of a corporate image).

In general, then, corporations produce collective decisions rather than individual ones. Some of these collective decisions are derived democratically in that they are based on a consensus; others emerge from the workings of a small clique. The development of cliques owes much to the fact that the personal goals of the manager, such as personal prestige or advancement, often diverge from the general, collective, company goal of profit making (McNee 1960). As a result there often emerges what might be called a 'managerial sub-culture' which acts as the fulcrum of corporate decision-making. Thus the perceptions and feelings of managerial groups within corporations are frequently accompanied by a degree of corporate autonomy, a particular tradition of planning, and a particular relationship with other corporations and trade unions, as well as with government. The workings of such managerial groups are extremely complex and individualistic and it is not surprising therefore that there are as yet few general models of organizational decision-making. However, the work to date suggests that a detailed understanding of organizational behaviour necessitates an appreciation of how the organization in question construes the world in which it is operating.

5 Image, Behaviour, and Meaning

It is widely acknowledged in the study of people–environment interaction that the acquisition of environmental information, and the use of that information in some form of decision-making process, serves as a prelude to overt or 'acted out' behaviour. In many cases, however, the processing and evaluation of environmental information do not influence overt behaviour and human activities directly. Rather these processes operate to change how the mind *construes* the environment, very much in the way proposed in the transactional constructivist approach to environmental awareness. Thus it is the changed mental construction of the environment that most immediately influences overt behaviour. This point is shown clearly in Pocock's (1973) model of how a perceiver interacts with an environment to create a mental image of that environment: the information status of the environment influences what the individual comes to know about the real world while the individual's psychological, physiological, and cultural make-up determines how that knowledge contributes to the development of an *environmental image* (Fig. 5.1). Although individual formulations vary from author to author, this model is characteristic of a large volume of research that can be summarized by Gould's (1973a: 183) contention that 'the manner in which men [*sic*] view their spatial matrix impinges upon and affects their judgements to some degree'. This emphasis on people's 'view' of the environment and the resultant mental image encouraged human geographers, particularly in the early 1970s, to think in terms of 'environmental perception' and 'mental maps' (see Wood 1970). For example, Downs (1970a) drew attention to 'geographic space perception' as part of the 'behavioural revolution' in the discipline and suggested that research should focus on how people store spatial information and on the way in which they develop a preference for different activities and locations within the environment.

The idea that an individual's behaviour is influenced by that person's construing the environment as some form of perceived mental map is not an entirely new one because in some ways it has antecedents in Lewin's (1951) proposition that behaviour is a function of a person's 'life-space' (see Chapter 1). Nor is the notion of a mental map an idea that has escaped criticism: Graham (1976) has argued that mental maps have made little contribution to the development of theory in geography because of the incoherence of the central concept, and Lowenthal (1972) has suggested that work on environmental perception generally is weakened through a lack of commonly accepted definitions, objectives, and mechanisms for applying research results in the realm of environmental design. In part, these problems arise because studies of perception have been conducted at a variety of scales that range from 'room geography', through architectural space, neighbourhoods, and cities, to large conceptual spaces such as nation states (Saarinen 1976). Moreover, space can be thought of in a number of ways depending on whether action, orientation, or imagery is the central issue (Wheatley 1976).

The term 'perception' is used loosely by geographers. In fact much geographical research is concerned with spaces that are so large that they cannot be perceived in the strict sense of stimuli directly impinging on the senses but rather must be conceived

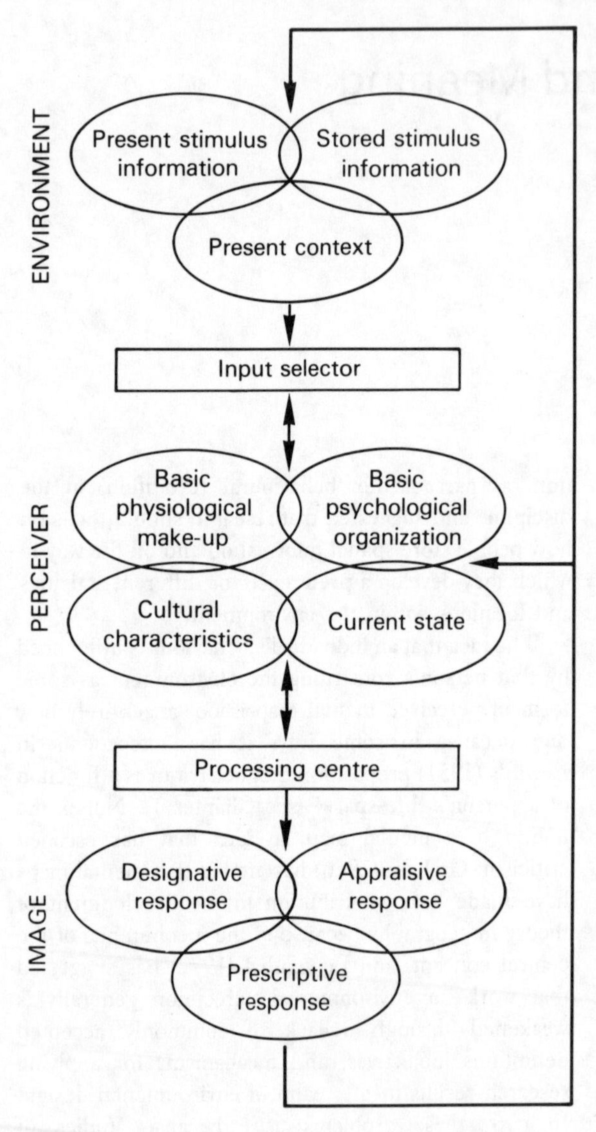

5.1 A conceptual model of people–environment interaction (Source: Pocock 1973: Environmental perception: process and product, *Tijdschrift voor Economische en Sociale Geographie* **69**, 252)

or cognized in the absence of direct stimuli (Stea 1969). The distinction between perception and cognition is, however, a heuristic device rather than a fundamental dichotomy because, in many senses, the latter subsumes the former and both are mediated by experience, beliefs, values, attitudes, and personality such that, in interacting with their environment, humans see only what they want to see. Perception and cognition are therefore purposive acts and the mind is very far from being an empty container ready for facts. Rather knowledge of the environment is something that is created in the mind of an individual as that individual interacts with the environment. The end product of the act of perception and cognition has been given a variety of labels: mental map; image; cognitive representation; schemata. The idea of 'mental maps' is, as will be shown later, really nothing more than a metaphor, while the notion of an 'image' conjures up mental pictures, and the term 'cognitive representation' is rather vague and all embracing. As a result the idea of *schemata* as frameworks for coding and structuring environmental information is probably the most useful concept in helping to understand the way in which information processing and spatial decision-making influence overt behaviour. In order to appreciate why this is so it is necessary to look at the origins and evolution of studies of image and environment.

The concept of image

Environmental images can be thought of as learned and stable mental conceptions that summarize an individual's environmental knowledge, evaluations, and preferences (Pocock and Hudson 1978: 3). In this sense an image is a partial, simplified, idiosyncratic, and distorted representation that is not necessarily isomorphic to the real-world environment (Pocock and Hudson 1978: 33).

Academic interest in the concept of image first reached prominence with the work of Boulding (1956) who proposed that all behaviour is dependent upon an image built up of information derived from the social and physical milieu (Powell 1978: 17). In this sense an image can be thought of as part of the culture in which it develops. That is to say, an image is both an individual phenomenon and a cultural phenomenon to the extent that similar individuals in similar milieux are likely to have similar images in their minds and hence they are likely to exhibit similar forms of behaviour. Boulding's idea of an image was in fact very much in accordance with the spirit of the times. For example, in many ways it complemented Simon's (1957) idea of boundedly rational human beings arriving at decisions on the basis of simplified models of the real world. As a result, Boulding's work has had a great impact on subsequent research, at least to the extent that his book is cited repeatedly. Powell (1978: 17–18) is, however, of the view that, despite the citations, Boulding's ideas have not been thoroughly understood. He has therefore drawn attention to the fact that an image is not the simple concept that it is sometimes thought to be but rather is comprised of at least ten main features:

1. A spatial component accounting for an individual's location in the world;
2. A personal component relating the individual to other people and organizations;
3. A temporal component concerned with the flow of time;
4. A relational component concerned with the individual's picture of the universe as a system of regularities;
5. Conscious, subconscious, and unconscious elements;
6. A blend of certainty and uncertainty;
7. A mixture of reality and unreality;
8. A public and private component expressing the degree to which an image is shared;
9. A value component that orders parts of the image according to whether they are good or bad;
10. An affectional component whereby the image is imbued with feeling.

Of these features it is the spatial component which has generated most interest from geographers, not only because it is perhaps the most important part of the total structure but also because it appears to be accessible in a way that other parts are not (Boulding 1973: viii).

Geographical studies of spatial and environmental imagery have drawn considerable inspiration from the pioneering work of Lynch (see Appleyard 1978) and in particular from his book on the image of the city (Lynch 1960). In this sense, then, the geographer's interest in environmental imagery parallels that of psychologists and designers (see Krupat 1985). In his book Lynch asked respondents in Boston, Jersey City, and Los Angeles to draw a sketch map of their respective cities. The resultant drawings *simplified* the spatial structure of the environment by omitting numerous minor details and straightening out complex geometrical shapes into more manageable patterns of straight lines and right angles (Day 1976). This simplification and rotation of images is a topic that has subsequently been studied in some depth. Tversky and Kahneman (1974), for example, have shown that a great many human judgements derive from inferences based on what are perceived to be particularly salient pieces of information. In terms of urban living, individuals often build up a simplified image of reality by treating particularly vivid components of the landscape as characteristic of entire areas. Thus a neighbourhood image might be based on one or two buildings that are notable for their architecture, disrepair, or occupants. Similarly, the straightening of angles in the mental rearrangement of street patterns might be simply one application of a 'rotational heuristic' used by people called upon to remember complex configurations (Tversky 1981). There is certainly ample evidence that humans are far better at remembering angles in pathways if those angles approximate 90° or 180°, and evidence that people estimate all turns in routeways as more like 90° than they actually are (Byrne 1979; Sadalla and Montello 1989). There is also evidence of city dwellers 'rotating' directions in orientation experiments such that perceived cardinal points are made to fit the alignment of major transportation routes (e.g. road systems might be interpreted as running N–S when in reality they run NW–SE) (Lloyd and Heivly 1987).

With Lynch's subjects, the simplication process was carried on to such an extent that he was able to argue that the image of the city was organized in terms of five elements:

1. *Paths* which are the channels along which individuals move;
2. *Edges* which are barriers (e.g. rivers) or lines separating one region from another;
3. *Districts* which are medium-to-large sections of the city with an identifiable character;
4. *Nodes* which are the strategic points in a city which the individual can enter and which serve as foci for travel;
5. *Landmarks* which are points of reference used in navigation and wayfinding, into which an individual cannot enter (Lynch 1960).

According to Lynch, an image of a city based on these elements serves as a basis for interpreting information, as a guide to action, as a frame of reference for organizing activity, as a basis for individual growth, and as a provider of a sense of emotional security (Downs and Stea 1973b: 80). More particularly, Lynch argued that an individual's knowledge of the city is a function of that city's *imageability*, that is the extent to which it makes a strong impression on the individual concerned. Imageability is, in turn, closely related to *legibility* by which is meant the extent to which parts of the city can be recognized and interpreted by an individual as belonging to a coherent pattern. Thus a legible city would be one where the paths, edges, districts, nodes, and landmarks are both clearly identifiable and clearly positioned relative to each other. In such a case, the overall structure of a city would be able to be 'read' by a city dweller in much the same way as a text can be read. Of course neither prominence nor architectural detail guarantee imageability because places and spaces only become significant when they are given meaning through a combination of usage, emotional attachment, and symbolism (Pocock and Hudson 1978: 31–2). In other words, an image acts back on the people in whose mind it forms by influencing what they see.

The five elements of the Lynchean landscape are by no means as clear-cut as is sometimes implied. For instance, a church may be a node to one person (because it is visited regularly) but a landmark to another person (because it is not visited at all but used for directional wayfinding). Despite this, Lynch's approach has been used with a reasonable degree of success in a wide variety of towns. Figure 5.2, for example, interprets the structure of an Australian country town in terms of Lynch's image elements. A wide range of similar studies have been undertaken and only edges have proved to be an elusive characteristic (see Pocock and Hudson 1978: 51). There are none the less a number of important criticisms to be made. To begin with, the free-hand mapping technique may not be a good method of eliciting the image that people hold of an environment because it demands a certain level of education, training, and aptitude. Moreover the entire thrust of Lynch's work concentrates overwhelmingly on the visual element of the image at the expense of considerations like sound and smell. This means that not enough attention is given to the functional and symbolic meaning of urban space (Steinitz 1968), largely because Lynch's approach focuses on the passive encountering of the environment rather than on interaction.

A more serious criticism of geographical work on imagery generally is that it has been characterized by 'borrowed methodology, a pot-pourri of concepts, and liberal doses of borrowed theory' (Stea and Downs 1970: 3). As a result, in the eyes of some commentators, there has been little advance in knowledge. In fact Stea and Downs (1970: 6) have highlighted five key issues on which research needs to focus:

1. *Elements of an image*. What are the main elements of an image? How do these interrelate? Is there a common set or do they vary with the particular environment?
2. *Relations between elements*. What are the distance metrics relating elements in cognitive representations? What are the orientation frameworks used to relate elements?
3. *Surfaces*. Do the interrelationships between the elements lead to a surface? Is the surface a continuous or a discontinuous phenomenon?
4. *Temporal nature*. How stable is the image over time? How far backwards or forwards in time do images extend, i.e. what is the temporal rate of information decay? How do images respond to environmental changes? Are there lags in response time?
5. *Covariations with other features*. What is the spatial extent of the image, i.e. what is the spatial rate of information decay? Is there a hierarchy of images according to the particular desired behaviour? How do all these factors vary with cultural, socio-economic, and personality correlates?

Additionally, Canter (1977: 16) has argued that images represent only one type of cognitive experience and that cognition more frequently takes the form of *schemata* that exert an important influence (of which individuals might not be fully aware) on interaction with the environment. The idea of schemata stems initially from Bartlett's (1932) work on memory and has been used by Lee (1968) in his study of how people define and know urban neighbourhoods. Basically, a schema is a framework on which people can 'hang' information. In other words, an individual

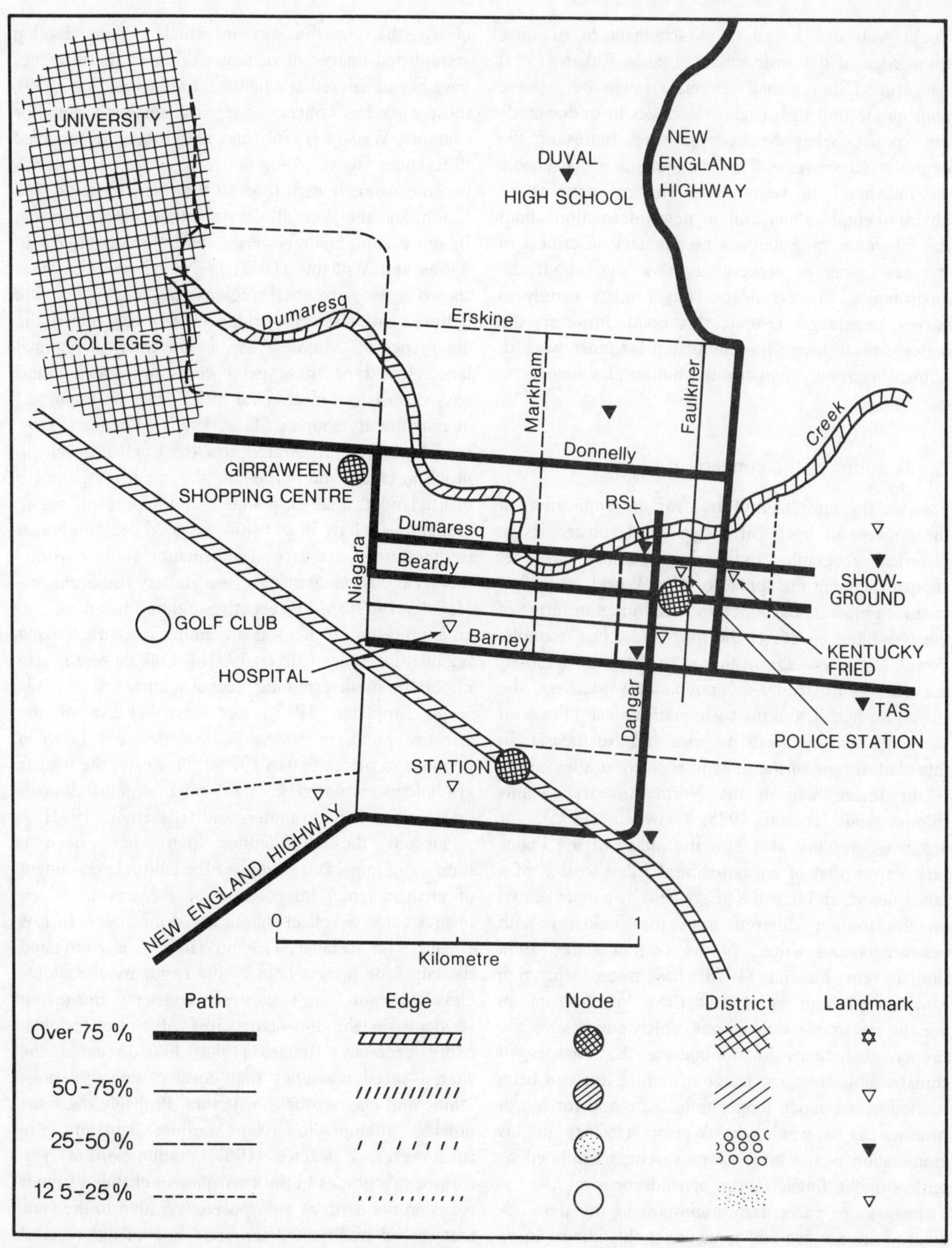

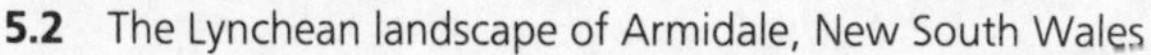
5.2 The Lynchean landscape of Armidale, New South Wales

might well use a variety of schemata to organize knowledge of the environment. Indeed Aitken (1987) has argued that mental schemata might be a more appropriate tool than cognitive images in understanding spatial behaviour because they represent the cognitive structures and coding systems which enable an individual to respond to shifting patterns of environmental stimuli and to new information about the environment. Schemata may in fact be critical in the processing of procedural knowledge about the environment whereas images might relate mainly to survey knowledge. Despite this point, however, the concept of an image has stimulated far more work in human geography than has the notion of schemata.

Applications of the concept of image

Possibly the most fruitful area for the application of the concept of image in geographical research is in historical geography and various authors have set about explaining the spread of New World settlement in the eighteenth and nineteenth centuries in terms of the prevailing images of the new lands. For example, Powell (1978) has shown that at one stage or another, the quest for Arcady, the notion of wilderness, the search for health, and the co-operative ideal influenced the settlement of North America and Australasia. In this context, one of the most intensively studied topics is the image held of the North American Plains (Blouet and Lawson 1975). Bowden (1975), for instance, has illustrated how the image of the Plains varied from that of an agricultural Eden to that of a sandy desert, and how the image held by people settled on the frontier differed from that held by both newcomers and writers in the eastern states. In a similar vein Jackson (1978) has traced changing Mormon perceptions of Salt Lake Valley from an initially favourable view to one which emphasizes the agency of humans in overcoming the vagaries of climate. This changing image of nature has also been studied over a much longer time span, and for North America as a whole, by Watson (1976a) in his examination of the increasing awareness exhibited by settlers of the fragile nature of wilderness.

Images are particularly important in the development of stereotypes and nowhere is this shown more clearly than in the way in which people develop stereotyped images of regions. Although these images have been analysed at a national level (Boulding 1959), most work has concentrated on a smaller scale. For example, Watson (1976b) has noted that the sectional differences that arose in early colonial America became stronger with time so that New England, the South, and the West all developed a distinctive image. In more contemporary terms, both Jordan (1978) in Texas and Williams (1977) in New Brunswick have shown how perceptual regions are a part of folk culture. Likewise, Shortridge (1984) has traced the emergence of 'Middle West' as an Americal regional label. Just how these perceptions are created and passed on is not clear. Oral history may be important as may literary sources (Tuan 1976a). Pocock (1979), for instance, demonstrated how the English novel did much to create and perpetuate the image of the north of England as a harsh, polluted, and depressed region. Geographers have in fact shown a good deal of interest in literature as a source of information about environmental and regional images (see Mallory and Simpson-Housley 1989). More recently, attention has turned to an appreciation of films as a medium in characterizing regional life styles (Aitken 1991b). Folk culture is also important in determining people's attitudes to landscape (Appleton 1975), not least because of the affective bond, or *topophilia*, that develops between people and places (Tuan 1974a). However, the insular symbology employed in many environmental images makes them hard to understand (Blakemore 1981).

Despite these difficulties, there have been a number of important studies of the cultural component of environmental images and of the way in which images come to reflect shared social meanings. In Los Angeles, for instance, Orleans (1973) has contrasted the city-wide images held by the residents of middle-class Westwood with the very restricted images of residents in the low-status area of Boyle Heights. More generally, Reitzes (1983) has discussed the shared social meanings that develop among various ethnic and class groups in Atlanta. Probably the most notable attempt to relate culture and imagery is, however, Glacken's (1967) examination of how European attitudes to the environment changed from a focus on the earth as a purposeful creation to the view that the environment determined human nature, and subsequently to the idea that humans altered the face of the earth.

At a smaller scale, environmental images of cities have attracted a good deal of attention, reflecting a close relationship between the study of images and that of urban design (Goodey and Gold 1987). Although in some senses this work began with Firey's (1945) study of sentiment and symbolism in locational decision-making and planning in Boston, for the most part it has been inspired by Lynch (1960). Buildings, paths, and areas have been the most commonly identified elements of the urban image (Pocock 1975). The topic of paths has attracted particular attention and studies have been made of the way in which the image associated with a path varies with the mode of transport (Pocock 1975). For example, Cullen (1961) and Appleyard *et al.* (1964) have looked at the sequential and almost linear image of cities held by car travellers. Underlying much of this work is the idea that the image of a city is very like a text that can be read by city dwellers. This has led geographers to explore the field of *semiotics* (see Walmsley 1988a: 83–6). In essence, semiotics (sometimes also called *semiology*) is the study of signs and of the way in which elements within a landscape can act as signs and symbols for the society occupying that landscape. It will therefore be discussed later when attention is directed to the way in which meaning is ascribed to the environment. For the moment, enthusiasm for the application of the concept of 'image' needs to be tempered by an appreciation of some of the methodological difficulties encountered in trying to measure imagery.

Methodological difficulties in the study of images

There are three critical methodological problems in the study of environmental imagery: the question of how to *extract* meaningful information from individuals about large scale environments; the question of how this information can be *manifest* so as to make obvious the extent of an individual's cognitions; and the question of how to *analyse* this material (Golledge 1976: 300–1). The range of available techniques varies from drawing exercises (e.g. sketch maps) to interviews (which provide verbal descriptions), and from natural environments to controlled situations (e.g. manipulation of models). However, by far and away the most common approach has been the sketch map of a real-world environment. After all this technique is a revealing one because it provides information on the sequence in which elements of the environment were drawn, the connectivities and gaps between these elements, and the style, detail, and scale distortions of the image, as well as evidence of emotional attachment or indifference to different parts of the environment (Lynch 1976: vi). There are nevertheless a number of difficulties with the approach. To begin with, it is very time consuming. Moreover it remains unknown to what extent the technique relies on artistic prowess; it may well be that more complete images are drawn by better artists rather than by individuals who are more aware of their environment. This raises the issue of *validity*, that is whether sketch maps actually provide a true measure of environmental images. In this context the actual instructions given to map drawers may be important in that Lynch (1960) noted a significant difference between sketch maps and maps built up from verbal descriptions. *Reliability*, or the extent to which the results of a sketch-mapping exercise are replicable, is another problem. Sketch mapping is a reactive technique in the sense that the drawing of one map of a particular environment will influence the drawing of subsequent maps of the same environment. As a result it is very difficult to replicate a study or, for that matter, to examine the change in an image over time. The best that can be attempted is an inference based on cross-sectional studies (Norberg-Schultz 1971).

In view of these difficulties, the attitude of geographers to 'the world in the head' has changed somewhat. Many view a mental map not as 'an explanatory construct' but as 'a potentially dangerous nostrum' (Downs 1981). As a result, the argument goes, the process of 'mapping' is probably a more important focus of attention than the finished product or 'map'. Given this, it is somewhat surprising that human geographers have not paid more attention to *projective techniques* whereby the mapping process can be uncovered. According to Saarinen (1973), there are five such techniques:

1. Association techniques (like word association tests

where a response is given to a stimulus);
2. Construction techniques (like the thematic apperception test where a subject goes beyond a stimulus such as a photograph and constructs a story about what is happening in the photograph);
3. Cloze procedures (where individuals fill in the missing parts of a stimulus, the term coming from the *Gestalt* psychology notion of 'clozure' which is itself a term used to describe the tendency of humans to perceive objects as wholes (or *Gestalten*) and to thereby fill in any gaps that are present in a stimulus);
4. Choice techniques (such as drawing lines on an existing map to show the perceived extent of a neighbourhood or other area with which an individual identifies);
5. Free expression techniques (exemplified by sketch maps).

Of these only the free expression techniques have been widely used, possibly because they are well suited to the Lynch-type studies with which many human geographers have been preoccupied. Sketch mapping, in other words, has perhaps been given too much emphasis in geographical methodology because it may well be that it is only appropriate to a small portion of the total number of approaches that geographers have developed in the study of how people come to know their environment. The full range of these geographical approaches has been classified by Briggs (1973) into four categories: studies of cognitive distance; examination of designative images of what is known to be where in the environment; analysis of appraisive images that express preference for different parts of the environment; and investigation of activity patterns. Each of these approaches has attracted a good deal of attention.

Cognitive distance

Briggs (1973) has suggested that an image of a city is built up on the basis of public and private information, from four types of knowledge: knowledge of nodes; knowledge about the closeness of nodes; knowledge of the relative location of nodes; and knowledge of sets of nodes and their interlinking paths. Central to each of these is the concept of *distance*. One of the first researchers to realize that the notion of distance that people hold in their mind is different from objective distance was Thompson (1963) who discovered that mile and time estimates to shopping centres in San Francisco were overestimated relative to real-world distances (see Chapter 9). This finding has since been corroborated in a number of studies, thereby leading to the suggestion that cognitive distance is overestimated relative to objective distance irrespective of the mode of transport, city size, or the distance metric used (time, road, or straight line distance) (Pocock and Hudson 1978: 53). The significance of this lies in the fact that cognitive distance is important in influencing three types of trip-making decisions: whether to stay or go; where to go; and what route to take (Cadwallader 1976).

Unfortunately the results of experiments on cognitive distance, although generally pointing to overestimation, are less than clear-cut in detail. For example, Pocock and Hudson (1978: 53) have noted that cognitive distance is *underestimated* relative to objective distance beyond a distance of about 11 k in London and 5.5 k in Dundee. Similarly, Day (1976) has found slight underestimation beyond distances of 1100 m in central Sydney, a fact which he attributes to such distances being beyond walking range. Lowrey (1973) has interpreted this sort of marked overestimation of very short distances in terms of the extra effort required to get started on a journey. He has also noted the very considerable range that exists in the regression equations between cognitive and objective distance. These can vary greatly from person to person. A high degree of variance in distance estimates is in fact characteristic of all experiments on cognitive distance. According to Briggs (1976), this idiosyncratic response comes about because cognitive distance estimates are likely to be influenced by factors relating to the subject, to the stimulus, and to the interaction between the subject and the stimulus.

Among subject-centred factors, it has sometimes been suggested that familiarity with the study area influences the accuracy of distance estimation (Lee 1970), as does age in the case of schoolchildren (Matthews 1981). For example, in a comparison of tourists and permanent residents, Walmsley and Jenkins (1992b) found that permanent residents were more accurate than tourists in estimating inter-city distances up to a distance of about 160 k. Interestingly, this corresponds approximately with the upper limit of the range of day trips from their study area. However, despite such evidence that length of exposure to an area improves the accuracy of distance estimations (Golledge *et al.* 1969), the relationship is unclear (Cadwallader 1976). Likewise the relationship between distance cognition and gender is unclear: some work suggests that females perform less well than males (Brown and Broadway 1981) whereas other research has shown that females are better able to reappraise distance following the opening of a new road system (Antes *et al.* 1988).

In terms of stimulus-centred factors, the most common finding has been that, in intra-urban settings, downtown distances are exaggerated relative to out-of-town distances, a fact which may be associated with the denser packing of land uses – and hence the greater number of stimuli – around the central city (Golledge *et al.* 1969). Once more, however, the evidence is equivocal because, although the exaggeration has been corroborated in other areas of dense land use (e.g. Byrne 1979; Ericksen 1975), for time estimates (Burnett 1978a), and in rural areas (Walmsley 1978), Lee (1970) has produced a contrary finding to the effect that out-of-town distances are relatively more exaggerated in Dundee. This may conceivably be because out-of-town journeys are considered by some to be boring and therefore appear longer. In other words, the differences in the results might stem from differences in the road networks used. Similarly, an urban area might be represented in the mind as a 'network map' (derived from procedural knowledge) and, depending on the regularity of the street pattern, distances might be accurately cognized or greatly distorted (Byrne 1979).

In terms of factors relating to the interaction of the subject with the stimulus, two considerations stand out: the attractiveness or salience of the end point of the distance to be estimated; and the nature of the connecting path. Thompson's (1963) pioneering work on cognitive distance established that a positive evaluation of a shop tended to shorten the subjective distance between a consumer and that shop. Similar observations of the way in which the distance to desirable facilities is underestimated relative to the distance to undesirable facilities have been made by Lowrey (1973) and Eyles (1968). Likewise C.D. Smith (1984) found that the accuracy of distance estimates increases with the 'pleasingness' of a place. In fact such is the pervasiveness of this type of finding that Ekman and Bratfisch (1965) postulated an 'inverse square root law' to relate emotional involvement (*EI*) with a place to estimates of the subjective distance (*SD*) to that place:

$$EI = \frac{b}{\sqrt{SD}}$$

where the value of *b* is determined by the arbitrary units of measurement. Unfortunately this 'law', despite claims to universality, is simply an artefact of the range of cities considered in the experiments on which it was based, and different ranges of stimuli give very different equations (Walmsley 1974a). As a result the nature of the connecting path may be a more fruitful topic of research than the attractiveness of the end point. The nature of urban paths has in fact attracted a lot of attention despite Tuan's (1975a: 207) observation that some people can travel a route without remembering certain parts of it. It seems that the more turns a path has, the longer it is estimated to be (Sadalla and Magel 1980), and the greater the number of intersections, the longer it is thought to be (Sadalla and Staplin 1980a), although the generality of this finding has been challenged by Antes *et al.* (1988). Sadalla's findings can be interpreted in terms of a greater volume of information requiring a greater storage area in the brain (Sadalla and Staplin 1980b). However, such a view does not explain why cognitive distances are non-commutative (in the sense of being different lengths when viewed in different directions) (Pocock 1978). Nor does it explain why distances across the River Thames are more overestimated than distances along the river (Canter 1977: 92), or why cities with a central river tend to have less exaggerated cognitive distances than cities without such central features (Canter and Tagg 1975). Perhaps the problem is that Sadalla tended to focus on micro-scale, contrived, model environments rather than real-world situations and extrapolations between the two are very difficult (Heft 1988b).

In short, there is no unanimity in the results of experiments on cognitive distance. One possible explanation is that some studies have used straight line distances, others have used road or time distances.

Likewise, distance estimates have been elicited in a variety of ways: simple and direct questions; scales graduated in miles; ratio estimates; and inferences from sketch maps. Day (1976) is of the opinion that the nature of the technique makes no significant difference to the results, but this finding has been challenged by both Phipps (1979) and Cadwallader (1979). The latter has stressed that the distance estimates provided by a subject are not independent of the way in which these estimates are sought. Moreover, he has suggested that there should exist no simple relationship between cognitive and objective distance because people do not possess internalized spatial representations of the world that are based on Euclidean geometry. This is in agreement with Robinson's (1982: 284) view that humans think of distance in absolute, ratio, and relative terms: 'most of us may rely largely upon relative judgements, resort to ratio judgements only when finer discrimination becomes a matter of significance, and elect to employ absolute judgement only rarely and, even then, with hesitation, uncertainty and qualification'.

Designative images

Distance is but one element of an environmental image. Individuals also know something of the direction and relative positioning of places, as was shown by Lynch (1960). This type of knowledge of what is where in the environment has been described by Pocock and Hudson (1978) as a *designative image*. There is of course no one type of designative image. Ladd (1970), for example, has identified four sorts of maps among the responses produced by adolescents when asked to draw their neighbourhood: pictorial maps representing buildings as part of a street scene; schematic maps showing lines and areas in a rather unconnected way; images that resemble a map in that they can be used for orientation; and true maps with identifiable landmarks. Appleyard (1969, 1970) has likewise produced a classification of sketch map images. Basically two sorts were identified: *sequential* (focusing on linkages between places) and *spatial* (concentrating on landmarks and areas rather than paths). Each of these two categories was divided into four types on the basis of increasing complexity (Fig. 5.3). According to Appleyard (1973) these types of cognitive representations are built up on the basis of interaction with the environment, imagery, and symbolization of the environment through language labels, much in the manner suggested by Bruner *et al.* (1966). The classification has been validated in both Western (Pocock 1976b) and non-Western cities (Wong 1979) and in both cases sequential maps dominate with low social status being reflected in simple maps. Appleyard (1970) has also suggested that over time there occurs a shift from sequential to spatial maps and an associated shift from topology to Euclidean geometry. This point has been corroborated by Moore (1979) but challenged by Humphreys (1990), by Lee and Schmidt (1988), and by Spencer and Weetman (1981) who made two contrary findings: first, neighbourhoods tend to be represented in spatial form irrespective of when the maps are drawn; and second, the choice between spatial and sequential maps at the city scale is dictated by respondent preference for mapping style and not by the timing of the map-drawing exercise.

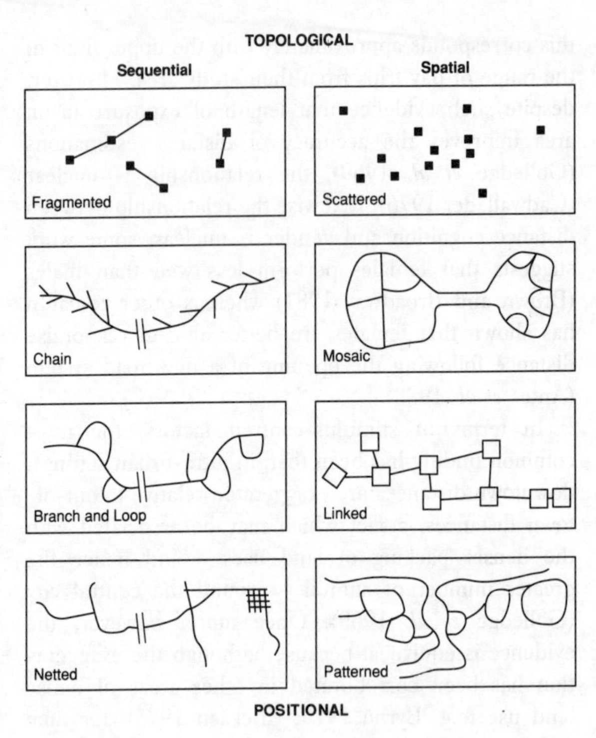

5.3 Styles of cognitive mapping (Source: Appleyard 1969: 437)

The sort of exercise in which Lynch and Appleyard were engaged is frequently referred to as *cognitive mapping*. The idea of maps in the mind has a long history (see Trowbridge 1913; Tolman 1948). It is, however, a term with 'tremendous surplus meaning' (Stea 1976: 107). Basically what are referred to as cognitive maps are convenient sets of shorthand symbols that can be used to describe the environment (Downs and Stea 1973a: 9). In this sense, cognitive mapping is a way of simplifying environments that are too big and complex to know entirely through the process of acquiring, coding, storing, recalling, and manipulating spatial information (Downs and Stea 1977: 6). The term 'map' is of course a *metaphor* rather than an analogy because the form in which information is stored in the brain is not known (Downs and Meyer 1978). Thus maps can be elicited by sketches but this does not necessarily imply that information is stored in terms of spatial coordinates. The metaphor is nevertheless a useful one, despite adverse criticism (Kuipers 1982), because it focuses attention on three critical issues in the relationship that people have with their environment: what people need to know (location, distance, direction, attributes); what people actually know (incomplete, distorted, and schematized information); and how people get their information (public and private sources) (Downs and Stea 1973a). The implicit assumption behind work on cognitive maps is that the nature of the map influences real-world behaviour: in the words of Downs and Stea (1977: 36), 'the most powerful, flexible, and reliable method of spatial problem solving is cognitive mapping'. However, not all writers have been so enthusiastic. Boyle and Robinson (1979: 64), for example, argue that cognitive maps play only a minor and intermittent role in influencing behaviour and that, as a result, it would be wrong to attach any great significance to cognitive maps in the co-ordination of spatial activities. To Boyle and Robinson, maps are best thought of as part of a general, learned cognitive structure or schema that enables the completion of commonplace tasks without resort to careful deliberation (see also Tuan 1975a).

Similar people, operating in the same environment, tend to have similar maps, partly because they have similar physiological information-processing capacities and partly because their interaction with the environment is constrained to similar origins and destinations (Downs and Stea 1973a). There have nevertheless been many attempts to treat cognitive maps as a dependent variable and to identify significant independent variables that account for differences in maps. One of the most studied has been age, possibly because Gould (1973b) has demonstrated that children of different ages have different amounts of spatial information. Curiously, however, neither age nor IQ has proved to be a really important variable in the analysis of children's maps (Maurer and Baxter 1972; Moore 1975). Instead the familiarity which individuals have with their environment seems to be the critical factor. This suggests that individuals imbue an environment with meaning and that this meaning influences the nature of the environmental image (Harrison and Howard 1972). Social status may well influence this process of attributing meaning to the environment (Orleans 1973), and familiarity generally seems to be determined by mobility and travel behaviour (Murray and Spencer 1979; Beck and Wood 1976).

The morphology, or physical structure, of cognitive maps is another topic that has attracted a lot of attention. Much of this work has assessed the relative merits of sketch maps and multidimensional scaling as ways of uncovering structure. The results are, however, less than clear-cut: although sketch maps tend to offer better representations of how subjects image reality (Mackay 1976), it is as yet unclear whether such representations are better interpreted through Euclidean or non-Euclidean geometry (Golledge and Hubert 1982). Despite this methodological dilemma certain generalizations can be made about the structure of sketch maps. Pacione (1978) has found, for example, that maps of Great Britain tend to be more uniform and less haphazard than reality and tend to assume the shape of a triangle. More specifically, districts, nodes, and edges accounted for 95 per cent of the information in a set of maps of Britain and the coastline appeared to act as a cue in the structuring of the maps (Pacione 1976). A similar tendency for maps to use the coastline rather than cardinal points as a frame of reference was apparent in maps of Los Angeles (Cadwallader 1977). In this case the influence of the coast as a frame of reference was weakest for inland locations and, moreover, appeared to decline with increasing length of residence. The shape and salience of landforms also seem to be important in maps at a continental scale (Saarinen 1988; Walmsley *et al.* 1990) and at an urban scale where images are most easily formed when there is a regular street plan, characteristic nodes, and unique landmarks (de Jonge 1962). In other words, there is ample evidence that individuals orientate themselves to conspicuous urban features rather than to cardinal directions (Pocock and Hudson 1978: 62). Furthermore, this applies not only to physical land use but also to the social structure of cities because there is evidence that people build up expectations about residential segregation and neighbourhood characteristics (Cox *et al.* 1979; Lloyd and Hooper 1991).

Cognitive maps are of course influenced by the technique on which they depend, whether it be drawing, describing, or modelling (Sherman *et al.* 1979; Pocock 1976a) and this presents problems in analysis. One way around these problems is to shift attention from getting subjects to *construct* a map to getting them to *complete* a map that has certain parts missing. Such completion tests are easy to analyse, comparable from person to person, and relatively unambiguous. The *cloze* procedure is one prominent technique. It derives from the *Gestalt* notion of 'clozure', that is the tendency for a subject to go beyond a specific stimulus and to fill in missing information. It has been used in cognitive mapping by Robinson and Dicken (1979) in a study that imposed a grid over a map of Britain and then asked students to provide the name of a settlement in each of 65 grid squares for which information had been omitted at the rate of 1 in 5. Three samples of students achieved success in between 15 and 30 per cent of cases. Obviously the grid size and omission frequency are critical in this sort of experiment. Moreover, the naming of a place does not necessarily imply knowledge of other characteristics of the place. The technique nevertheless offers an alternative to the more widely used constructive approaches, especially where the background information gives cues as to the positioning of localities (see Tversky 1981).

There is no doubt that, despite the lack of much really significant progress in recent years, cognitive maps will continue to be studied, perhaps because of their 'lucky-dip' element and the genuine intellectual challenge of bringing order to the apparent chaos which they present (Boyle and Robinson 1979: 73). Some authorities will continue to claim that cognitive maps are reflected in real-world activities (e.g. Holahan and Dobrowolny 1978), whereas others will stress that images and maps do not play any essential role in spatial behaviour and will argue that attention needs to turn from maps to the schemata wherein public and private, specific and general information is organized (Tuan 1975a). In this sense, cognitive maps can be thought of as 'plans' (Canter 1977: 51), much in the same way that Miller *et al.* (1960) used the psychological concept of a plan as a construct that helped structure behaviour. From this viewpoint, a cognitive map needs to be concerned not only with what is where in the environment (the designative element) but also with what emotions individuals feel towards the attributes of different locations (the appraisive element).

Appraisive images

The designative aspects of the image of an environment may be less important in contributing to the understanding of behaviour than are the *appraisive* aspects of the image, that is the meanings evoked by the physical form (Pocock and Hudson 1978). This appraisive response can itself be differentiated into two components: the *evaluative* (concerned with the expression of an opinion) and the *affective* (concerned with the specification of a preference). Although in many ways it underlies the work of Ley (1983) in differentiating parts of the city in terms of scales such as stress–security, status–stigma, and stimulation–ennui, the evaluative component has been most widely studied in relation to verbal descriptions of environmental quality where an individual is required to state an opinion of a particular place by, for example, completing a semantic differential questionnaire. The difficulty with this approach is that the opinions offered tend to vary from case to case and it is therefore difficult to arrive at a common set of evaluative dimensions (see Kasmar 1970). In contrast, the affective component of appraisive images is more easily compared from case to case because it is

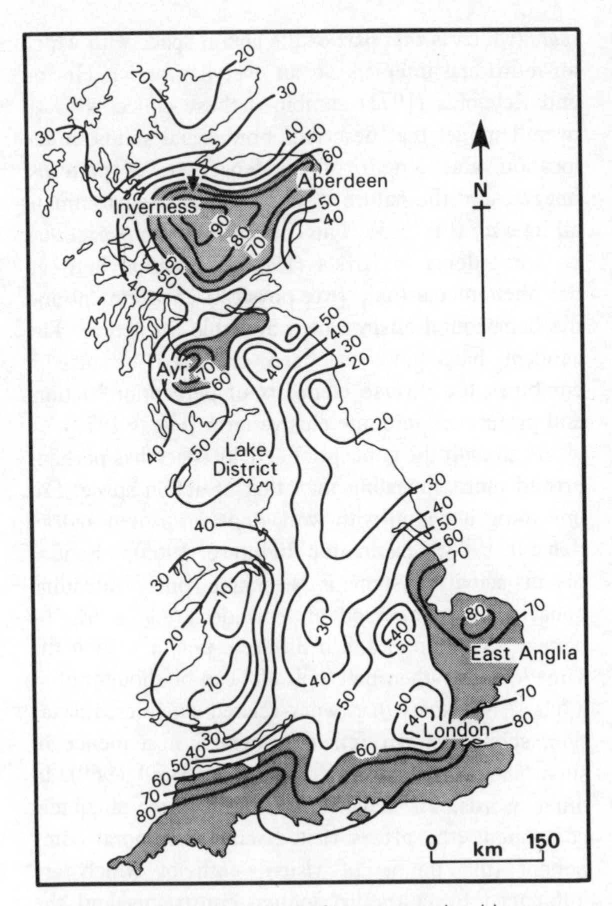

5.4 The preference surface of Inverness school-leavers (Source: Gould and White 1968: 170)

possible to measure the response to stimuli on relatively straightforward ordinal or interval scales. For instance, people can be asked how much they would like to live in various regions within a country and their answers can be calibrated on a scale that ranges from a strong desire to live in a certain place through to absolute opposition to living in that place. Given the predominance of such studies of affective imagery, it is not surprising that the terms 'appraisive' and 'affective' have come to be used synonymously.

One of the first people to explore the affective component of appraisive images was Gould in a paper (1973a) concerned with residential desirability in the United States, Europe, and West Africa. He concluded that space preferences seem to be related to the information that people have about places which in turn seems to be related to the population size of the places in question. The methodology, whereby individuals score regions in terms of residential desirability to provide a data matrix which can be reduced by factor analysis to a component representing preference, has been applied widely (Gould and White 1974). The sort of picture of preference that emerges, with component scores standardized to 100-point scale, is shown in Fig. 5.4.

The main problem with this approach is that it elicits preferences in a relatively unconstrained context. In other words, subjects state how much they want to live in an area irrespective of the real likelihood of living there. It is not surprising therefore that preferences vary widely in apparently homogeneous subject groups. Jackson and Johnston (1972) have illustrated this variability and have attributed it to the fact that individuals have different environmental images and different criteria for desirability. This suggests that personality variables may be important, thereby corroborating Proshansky's (1978) view that 'place identity' is a significant but much ignored component of overall self-identity. The evidence for the impact of personality in people–environment interaction is as yet limited although Walmsley (1982a) has shown how introverts and extroverts differ in residential preferences, with the former preferring quiet country towns to both the cities and the outback of Australia. Similarly, Aitken (1990) has used the 'locus of control' personality scale to analyse evaluations of neighbourhood change. This measure differentiates people in terms of whether they feel in control of their fate (an 'internal' locus of control) or powerless to change the course of events (an 'external' locus of control). It may be, in other words, that individuals develop distinctive styles of dealing with the environment in the same way that they develop characteristic styles of social interaction. Aitken (1991a: 182–4) has explored such 'environmental disposition theory' and speculated on which personality traits might influence environmental attitudes and behaviour. This work therefore builds on the earlier work of McKechnie (1977) who developed the multi-scale 'environmental response inventory' as a tool for assessing environmental dispositions.

Of course it may be that appraisive images will become more important as 'footloose' employment develops in the tertiary and quaternary sectors of the economy (Pacione 1982a), with the result that public authorities will have to place increasing emphasis on the manipulation, through advertising and promotion, of stereotyped images (Burgess 1982). To some extent this is already happening: Burgess and Wood (1988), for instance, have examined the 'advertising' used to promote the redevelopment of the London 'Docklands' and shown that the campaign was much more effective in reassuring those who had already relocated to the area than it was in attracting new firms. For the moment, however, it is probably true to say that unconstrained choice has little bearing on real-world behaviour (Massam and Bouchard 1976). It is important therefore to look specifically at overt behaviour rather than at the images that might predispose individuals to certain forms of activity.

Activity and constraints

There are two major methodological choices to be faced in any study of overt behaviour: whether to do experimental research with controlled variables or whether to conduct holistic research that looks at the real-world environment and behaviour as a whole; and whether to rely on recall and self-report techniques or whether to adopt the logistically difficult approach of looking at behaviour as it happens (Ittelson *et al.* 1974: 208). In many ways these choices reflect what Craik (1968) believes are the key issues in the study of everyday environments: how to identify and present the pertinent characteristics of the environment; what behavioural reactions to record; and what groups to study.

Early research into overt behaviour by human geographers was strongly inductive in approach in that it sought to generalize about observed empirical generalities. Thus Adams (1969) studied movement patterns within a city and concluded that individuals have wedge-shaped mental maps that take in both the central city and the individual's residence. From such empirical observations arose the concepts of *action space* (which refers to that part of the environment which has a place utility to the individual and with which the individual is therefore familiar) and *activity space* (which is that part of the action space with which an individual interacts on an everyday basis). Horton and Reynolds (1971) combined these concepts in an overall model that describes how social status, home location and length of residence, travel patterns, images, and the nature of the real-world environment all interact (Fig. 5.5). The concept of action space has its antecedents in Kirk's (1963) distinction between the phenomenal (objective, physical) environment and the behavioural environment available for action. The concept has, however, been criticized because it combines the diverse elements of perception, action, and preference into one elusive term (Higgs 1975).

As a result the concept of activity space has perhaps proved more appealing than that of action space. For one thing it ties in with work done on *activity systems* (Chapin 1968; Chapin and Hightower 1965). The idea of an activity system is a simple one. Individual behaviour is comprised of 'episodes' (e.g. a trip by person *a* to shop *b*) and if different people exhibit the same 'episode' then that 'episode' can be thought of as a 'class' of activity (e.g. work-related activities, relaxation, shopping). An activity system is a sequence of such 'classes' of activity (Chapin and Brail 1969). In other words, an activity system has both a spatial component (the places visited) and a temporal component (the timing of visits), both of which are influenced by culturally defined constraints and the nature of the geographical environment (e.g. working hours and land-use zoning determine the nature and timing of the journey to work). An activity system is, in this sense, a manifestation of a space–time budget (Anderson 1971). Obviously, the activity system of one person is likely to be very different from that of another person because of the complex nature of travel trips (see Daniels and Warnes 1980). Similarly, there can be considerable variability over time in the activity system of a single person (Hanson and Huff 1988).

There are a number of problems with activity systems research, notably how to get detailed records of behaviour, how to cope with multi-purpose activities, how to study socially unacceptable behaviour, and how to overcome the faulty recall of respondents. For the most part these problems are overcome by using either diaries or large samples at one point in time and these approaches have both yielded a great deal of descriptive data. For example, women have

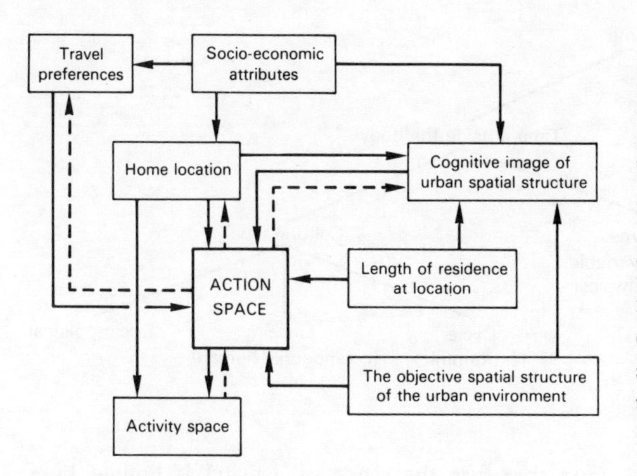

5.5 Influences on action space (Source: Horton and Reynolds 1971: 41)

been shown to have different activity patterns from men (Hanson and Hanson 1980) and the elderly to have different activity patterns from those of working age (see Rowles 1978; Smith 1991). Such studies remain, however, essentially descriptive and activity systems research has made little progress towards the explanation and prediction of travel patterns.

Time and behavioural constraints

One reason why studies of activity systems have not achieved as much as some researchers had hoped lies in the fact that they have sometimes tended to ignore time as a dimension of human behaviour. This is in fact a failing that is common to much social science: all too often time has been approached from the point of view of comparative statics, that is to say behaviour has been compared at various points in time without due emphasis being given to consideration of what time means to the actors involved and to the way in which this meaning influences behaviour. Yet, in a fundamental sense, space, time, and place are irresolvably linked in experience (Tuan 1977). According to Lynch, what people think of time is crucial for their well-being and a desirable environmental image is 'one that celebrates and enlarges the present while making connections with past and future' (Lynch 1972: 1).

Recognition of the importance of time has led to pleas for a *chronogeographical* perspective in the study of human behaviour (Parkes and Thrift 1980). It has also generated a vast literature (see Carlstein *et al.* 1978a, b, c). Time is of course a multi-faceted concept and it is easy to identify many different meanings. For example, in addition to the personal images of time hinted at by Lynch, it is possible to differentiate 'time inside the body' (biological rhythms) from 'time outside the body' (Walmsley 1979) (Fig. 5.6). The two are often in conflict as when shift workers are required to change established patterns of eating and sleeping. 'Time outside the body' usually manifests itself in cultural time because human beings everywhere tend to make very similar temporal judgements using natural cycles and culturally determined social events (e.g. public holidays) (Doob 1978). This cultural time may be cyclical (e.g. lunar cycles), linear (birth to death), or imposed by the socio-political system (e.g. daylight saving).

Just as it is possible to identify different types of time, so it is also possible to identify different approaches to the study of the importance of time. Three approaches are particularly relevant to human geography: the Lund School of time geography; time budget analyses; and work on the sequencing of activities (Parkes and Wallis 1978).

The Lund School of *time geography* began with the work of Hägerstrand in the mid–1960s but did not become widely known in the English-speaking world until much later (see Pred 1973b). Hägerstrand (1978: 123) has stated his value position very eloquently: 'we need a geography today which helps us to see ourselves, our fellow-passengers and our total environment in a more coherent way than we are presently capable of doing . . . the answer seems to lie in the study of the interwoven distribution of states and events in coherent blocks of space–time'. The development of time geography is predicated on eight basic assumptions (Hägerstrand 1975) which state that:

1. The human being is indivisible;
2. The length of each human life is limited;
3. The ability of a human to undertake several tasks at once is limited;
4. Every task has a duration;
5. Movement between points in space consumes time;

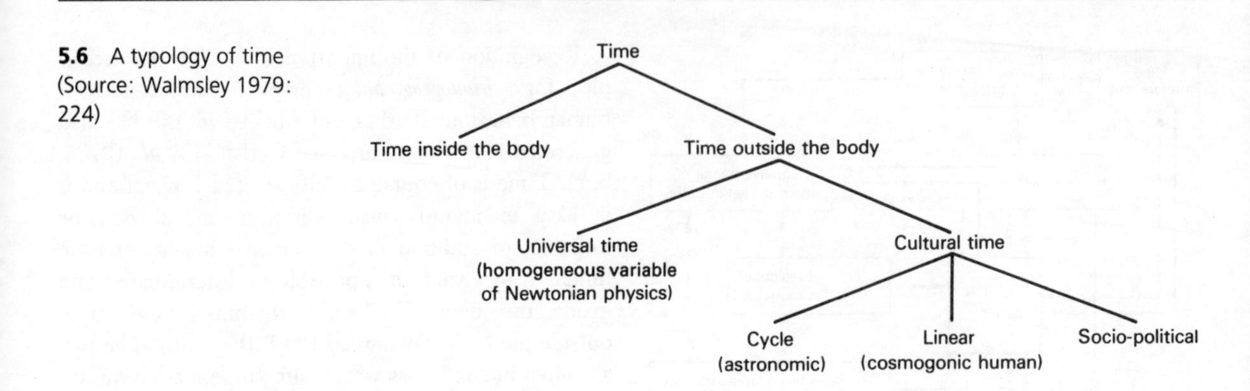

5.6 A typology of time (Source: Walmsley 1979: 224)

6. There is a limit to the number of individuals that can be packed into one space;
7. There is a limited space available for one activity at any one time;
8. Every situation is inevitably rooted in past situations.

The force of these assumptions is most clearly demonstrated when an individual is thought of as describing a *life-path* through *time–space*. This conceptualization can operate at a variety of scales (day-path, year-path, etc.). For example, Fig. 5.7(a) shows the day-path of the author of this chapter on the day that the chapter was completed. Places (work, home) that are essentially points (in that movement within them is inconsequential) are usually termed 'stations' and are displayed as vertical lines (Lenntorp 1978). Travel is displayed as a diagonal line because it consumes both time and space.

In addition to describing what actually happens, time geography can be used to investigate *possible* behaviours. It does this by introducing the idea of a *time–space prism* to delineate the scope of activity open to an individual. What this means can be illustrated by returning to Fig. 5.7(b). If it is assumed that the author cannot leave home before 0800 hours and must return by 1815 hours and that work commitments demand his attention between 0815 and 1800 hours (except for a one-hour lunch break), then some indication of his scope for other activities can be gained by drawing prisms to indicate how far and how long he can go between prearranged commitments. Clearly, the range of behaviour is more restricted at lunch-time when the author must travel on foot than in the evening when he may travel by car. In other words, the size of the prism (and therefore the scope of activity) is limited by a variety of *constraints* which reflect such things as social considerations, environmental opportunities, and the attitudes of the individual concerned (Parkes and Thrift 1978). According to Hägerstrand (1970), these constraints can be differentiated into three types:

1. *Capability constraints* limit the activities of individuals because of their biological make-up (and hence the need for sleep and food) and because of restrictions on the facilities at their disposal (such as limited transport networks and limited access to cars);
2. *Coupling constraints*, which are often determined by a clock, refer to the need for individuals to be at the same place at the same time as other individuals (as, for instance, when schoolteachers and pupils need to coincide);
3. *Authority constraints* relate to the tendency for some activities to be available only at the times when the person in charge of that activity deems appropriate (for example, child care facilities are open only during those hours that the authorities specify).

These and other ideas inherent in time geography are, at one and the same time, disarmingly simple and inspiringly powerful (Parkes and Thrift 1980: xiv) and the approach has generated a great deal of research that ranges over questions of how the activity organization of a community influences child development (Mårtensson 1977), where to locate single-parent housing so as to provide access to services (Robertson 1984), and how neighbourhoods differ in terms of what they offer within given time–space prisms (Forer and Kivell 1981).

Time budget studies and research on the sequen-

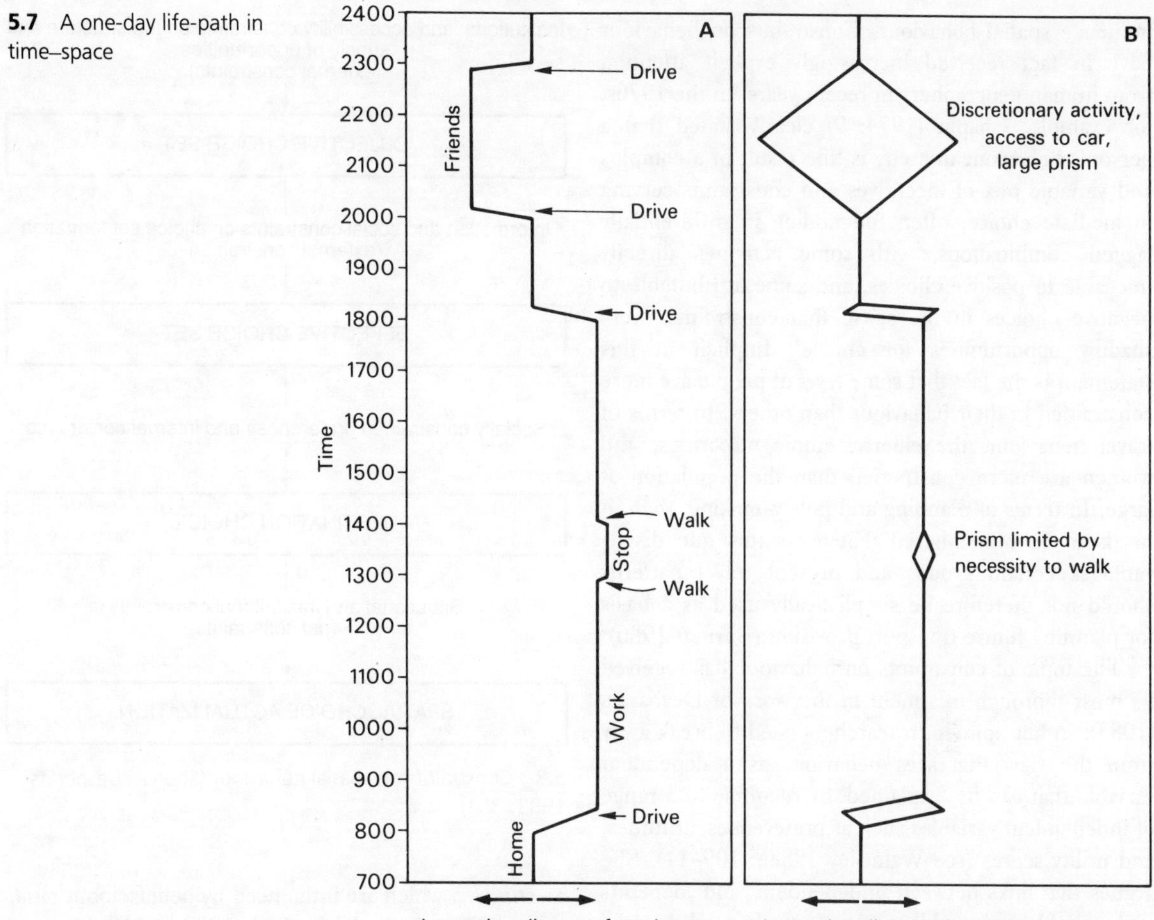

5.7 A one-day life-path in time–space

cing of activities have attracted less attention from human geographers than the work of the Lund School, possibly because they lack any major theoretical component and at times seem overly concerned with the measurement of actual rather than possible behaviour patterns. The first time budget studies were conducted in the 1930s but the approach experienced something of a revival in the 1970s, largely as a result of the work of Chapin (1974, 1978) and a major international study of the use of time involving 30 000 people in 12 countries (Szalai 1972). Generally a distinction is drawn between *obligatory* and *discretionary* activities and the amount of time devoted to each is ascertained by means of either diaries or questionnaires. This results in a methodological approach that is fraught with problems: the sampling designs adopted are often of debatable value; a focus on 'average' days raises questions of validity; and the general *ex post* emphasis means that alternative behaviours (and therefore constraints) are seldom considered (Carlstein and Thrift 1978). Above all, the approach pays little attention to the sequence in which activities are undertaken. This has been rectified to a certain extent by Cullen and Godson (1975) who isolated structural elements in activity profiles by calculating transition probabilities for individuals moving from one activity to another. However, even this work does not really do justice to the temporal constraints that influence human behaviour.

Time is, of course, not the only constraint to

influence spatial behaviour. Constraints on behaviour have in fact received increasingly explicit attention from human geographers in recent years. In the 1970s, for example, Chapin (1974: 9) clearly stated that a person's behaviour in a city is 'the result of a complex and variable mix of incentives and constraints serving to mediate choice, often functioning in differentially lagged combinations, with some activities directly traceable to positive choices, and some attributable to negative choices in the sense that constraints overshadow opportunities for choice'. Implicit in this statement is the fact that some *types* of people are more constrained in their behaviour than others. In terms of travel behaviour, the elderly, ethnic minorities, and women are more constrained than the population at large. In terms of planning and policy-making, then, it needs to be remembered that the status quo disadvantages certain groups and present travel patterns should not therefore be simplistically used as a basis for planning future transport provision (Burnett 1980).

The topic of constraints on behaviour has received its most thorough treatment in the work of Desbarats (1983). In her opinion, researchers need to break away from the view that sees behaviour as a dependent variable that can be 'explained' by recourse to a range of independent variables such as preferences, attitudes, and utility scores (see Walmsley 1988a: 109–11). She argues that links between 'independent' and 'dependent' variables are seldom clear-cut. Instead, people arrive at decisions as to what to do by considering the feasibility of alternative courses of action, as assessed in terms of what society expects of them (i.e. what is considered important) and in terms of what the environment actually permits. In other words, 'spatial choice may bear little relation to movement behaviour and . . . movement behaviour, in turn, may say more about social structure than about individual preference and choice' (Desbarats 1983: 347). That is to say, a great many individuals find themselves hampered by socially generated constraints on what they can do. The mode of operation of such constraints is set out in Fig. 5.8. The content of this diagram goes a long way towards explaining why there is often a discrepancy between thought and action. Desbarats's inspiration came from the work of Ajzen and Fishbein (1980) on the *theory of reasoned action*. According to this view, *behaviour* is influenced by *intentions*, which are influenced by *attitudes*, which are influenced by beliefs about *social norms* (Fig. 5.9). Walmsley (1988a: 110–11) has explained the operation of this theory in the following manner:

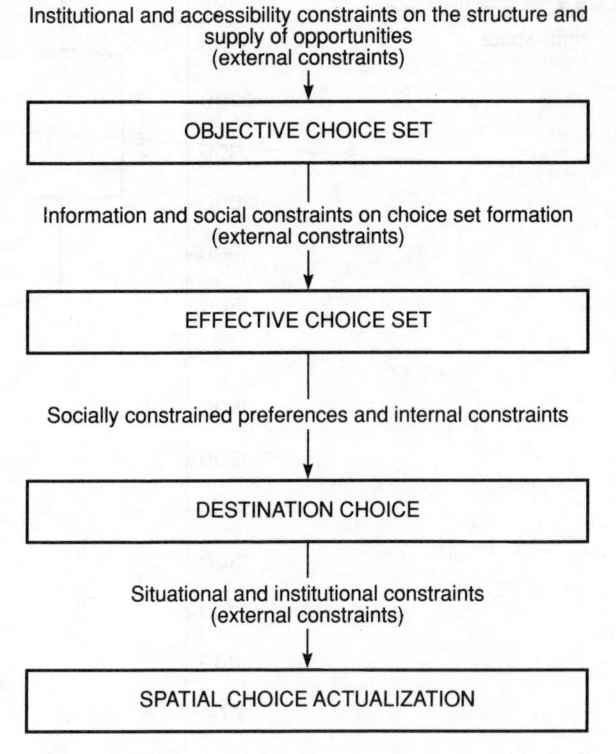

5.8 Constraints on spatial behaviour (Source: Desbarats 1983: 351)

> the theory proposes that individuals have beliefs about the environment (e.g. a belief that suburb X is pleasant). Evaluation of such beliefs leads to the formation of attitudes (e.g. favourably disposed to living in X). Such attitudes form the basis for intentions (e.g. intention to buy a house in X), but only after stock has been taken of both society's notions of what is proper behaviour (e.g. would the prospective migrant be acceptable in X?) and the extent of the individual's desire to comply with society's norms.

In short, constraints influence behaviour in four ways: (1) they restrict the 'opportunity set'; (2) they mould attitudes and preferences; (3) they bring

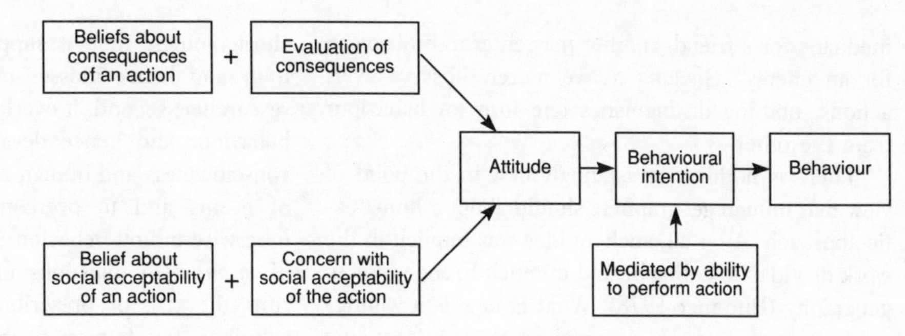

5.9 The theory of reasoned action (Source: After Ajzen and Fishbein 1980)

about choices that do not conform to preferences; and (4) they prevent choices from being realized (Desbarats 1983).

Desbarats's work provides a stimulating framework for considering how socially generated constraints impact on all sorts of groups. Nowhere is the importance of socially generated constraints more clearly shown than in relation to *women*. For example, there might be very real differences in the way in which men and women experience city living because of the way in which gender-based roles lie at the heart of the organization of society (see Holcomb 1986; Tivers 1985). Cities, for example, are very much man-made structures in the sense that they are created in accordance with a set of ideas about how society works, about who does what, and about who goes where (Matrix 1984). These ideas centre around the male breadwinner and the male head of household. They incorporate sexist ideas about the role of women, perhaps displayed most clearly in the implicit (and sometimes explicit) notion that suburbia is a woman's place. Although there are undoubtedly pleasant features to suburban living (space, quiet), negative features (loneliness, immobility) tend to have been overlooked. Sadly, then, there has until recently been no widespread recognition of the way in which social organization can 'trap' some women in the home, notably those who have to care for young children (Popenoe 1981). The impact of such constraints can be striking, especially when compounded by location in public housing estates, the existence of single-parent households, and poor public transport. An Australian survey of a peripheral housing estate in Sydney found, for instance, that 47 per cent of women felt lonely and that 30 per cent had no friends within walking distance (Sarkissian and Doherty 1984). In short, suburban living can constrain women in ways that it does not constrain men (Wekerle 1980). Society is, of course, changing and the stereotypical suburban family is now very much a minority as women participate much more in the paid workforce and as the birth rate drops. The fact remains, however, that socially derived constraints still impinge significantly on women. It is important therefore to explore how the environment is experienced and how meaning is ascribed to it in such a way that meaning influences subsequent behaviour.

Ascribing meaning to the environment

An appreciation of the importance of constraints and of the need to understand what the environment 'means' to those who interact with it has led to the adoption of a *hermeneutic* perspective which argues that people–environment interaction is best understood by focusing on the way in which humans experience and interpret their surroundings. Because this approach rests on a non-positivist philosophical framework that encourages a holistic, experiential view of people–environment interaction rather than analysis by way of mainstream scientific method and 'independent' and 'dependent' variables, it has frequently been labelled 'humanistic' (see Buttimer and Seamon 1980; S.J. Smith 1984). The central argument is that in order to understand behaviour we need to get inside the mind of the person undertaking the behaviour. Tuan (1989: 77) makes the point vividly by quoting Barker (1968: 29): 'One person may enter a drugstore to buy medicine for a friend, another may enter to buy poison for an enemy.' So long as we merely observe overt actions, nothing distinguishes one form of behaviour from the other.

There is nothing devastatingly new in the point of view that human geographers should adopt a humanistic approach. After all, such an idea was implicit in the work of Vidal de la Blache and in much French human geography (Buttimer 1978). What is new is a willingness on the part of geographers to confront the philosophical basis of humanistic thinking directly rather than merely by implication (see Harrison and Livingstone 1982). Thus in the 1960s human geographers, like other social scientists, began to become aware of the strengths and weaknesses of humanistic principles over and against the prevailing preoccupation with techniques of scientific rationality (Ley and Samuels 1978a: 1). In simple terms, humanism in its broadest sense offers a critique of scientific method in that it stresses that humans must be viewed as complete beings with due emphasis given to their creativity, individuality, and to the subjective way in which they interpret the world around them. It is therefore very critical of quantitative social science wherein:

> some shun consideration of a person's experience of situations as both private and qualitative, and therefore falling outside the scope of 'objective science'. Others, even more dogmatic, reject experience as essentially 'epiphenomenal', having no causal role in behaviour or action, and hence no explanatory value in accounting for how individuals deal with their surroundings (Wapner *et al.* 1980: 224).

Such a view is unacceptable to humanists because they see people as 'determined not by stimuli as such, but by definitions of situations partly enjoined upon them through their participation in a particular society and cultural milieu' (Wapner *et al.* 1980: 225). To persist with a fragmented view of human beings that highlights particular reactions to particular environmental stimuli (e.g. distance cognition experiments) is, it is argued, dehumanizing. In the view of Ley (1980), a geography that fails to give adequate attention to people as complete human beings has four significant shortcomings: first, it suppresses the issue of *subjectivity* and hence glosses over the subjectivity of the researcher; second, it overlooks *intentionality* in human behaviour and hence devalues the power of human consciousness and human action to redirect the course of events and to overcome constraints that would otherwise inhibit behaviour; third, it reduces questions of *meaning* to questions of technique and therefore runs the risk of prescribing inappropriate technical solutions for human problems; and finally, it lays altogether too much emphasis on human *passivity*.

A geography that adopts a humanistic perspective and does justice to a person's distinctly human characteristics would emphasize that action is a response to the meanings that individuals attribute to the physical and social environment (Cullen 1976). Such a geography would therefore begin with the premise that the study of human spatial behaviour is different in kind from the sort of study conducted in the natural sciences in so far as humans exhibit characteristics (creativity, values, individuality) that are not generally evident in the sort of phenomena studied in natural science. Couclelis (1983: 32), for example, has argued that 'much social science to this day has been too busy mimicking the surface features of last century's experimental physics' to appreciate that it is quite wrong to assume that researchers are detached observers objectively recording an external world. It follows from this that the techniques of natural science are inappropriate for the study of human behaviour and need to be replaced by an approach that gives much more attention to *values* (Buttimer 1974; Tuan 1974c) and to people's day-to-day *experience* of the environment (Tuan 1976b). Of course it is one thing to advocate a new value-oriented, interpretive approach but quite another to spell out what is involved because the adoption of a humanistic perspective poses problems concerned with philosophy, methodology, and the degree to which different people share the same experience and attribute the same meaning to their surroundings (Ley and Samuels 1978a: 9).

To distinguish between philosophy (issues of belief) and methodology (a concern for logical procedures) in relation to humanism is, however, to extend a separation made in 'science' to a field where it does not really apply. That is to say, advocates of a humanistic perspective contend that philosophy and methodology cannot be differentiated because the *means of analysis* are intimately related to the *meaning of analysis* (Ley and Samuels 1978c). To be specific, the philosophical stance associated with humanism stresses *anthropocentrism*, *holism*, and *intersubjectivity* and none of these can be investigated other than from the point of view of human experience. In short, there is no way that human experience can be studied other than through human experience (Ley and Samuels 1978c: 50–1).

Anthropocentric views of the environment have been recognized by geographers for a long time. For example, Lowenthal (1961) has stressed that everyone interacting with an environment builds up a geographical epistemology which is founded on personal geographies composed of direct experience, memory, and fantasy. This world view is influenced by both unique variables such as personality and social variables like culture. Moreover, although shared, this world view is transient: 'it is neither the world our parents knew nor the one our children will know' (Lowenthal 1961: 245). In a very fundamental sense these personal geographies are 'mirrors' for human beings in that they reflect and reveal human nature by emphasizing the way that people seek order and meaning in their experiences of the world (Tuan 1971). It would be simplistic, however, to assume that it is an easy matter to 'uncover' these personal geographies. This is because people's attitudes towards the environment are sometimes ambivalent and ambiguous (and therefore difficult to interpret) (Tuan 1973) and because a researcher can only understand a subject's behaviour to the extent that they share common cognitive categories and conceptualizations of the environment (Ley and Samuels 1978b). Interpretative knowledge is, after all, 'mediated through the categories of the interpreter's pre-understanding' (Buttimer 1979: 22).

A further difficulty in interpreting the world as viewed by an individual stems from the fact that an individual's view is *holistic* and needs to be comprehended in its entirety. Furthermore, humans continually interact with their environment and are therefore continually confronted by new experiences that may necessitate changes in their world view. 'There is no room in a humanistic perspective for a passive concept of man [*sic*] dutifully acquiescing to an overbearing environment. But neither is man fully free, for he inherits given structural conditions and, indeed, may be unaware of the full extent of his bondage' (Ley and Samuels 1978a: 12). The need for humanistic geographers to appreciate the context of behaviour becomes particularly acute when it is realized that humans operate to a very significant degree in a *symbolic* environment (see Smith 1974; Tuan 1978b). Symbolism in this sense exists in the mind. It is an element of consciousness. To focus on symbolism is therefore to concentrate on consciousness and the world of ideas. Of course, no matter how intellectually fascinating such research might be, it does run the risk of diverting the humanist's interest from *intersubjectivity* in the world of experience and material conditions (Ley 1978: 45).

Although humanism puts great emphasis on the existence of intersubjectively shared experiences and meanings, it studiously rejects all types of aggregate, statistical measurement. Instead a more particularistic and qualitative methodology is advocated whereby the meanings that individuals or groups attribute to an environment are uncovered through such techniques as participant observation (see Jackson 1983), the use of unobtrusive measures, and various forms of encounter and in-depth interviewing (Burgess *et al.* 1988a, b; Ley and Samuels 1978c: 121). It is not surprising therefore that there has been a surge of interest in qualitative methodology in human geography (see Eyles and Smith 1988). By adopting qualitative techniques, the humanistic perspective avoids the idealized and contrived laboratory conditions that are common in much of social science. After all, participant observation allows the researcher to become an involved insider rather than a detached outsider and thereby to gain an insight into the everyday world of her or his subjects, as was well illustrated in Symanski's (1979) study of tramps and freight trains. However, qualitative research 'involves more than negotiating an entry into communal life and participating in the drama. ... Experiential research differs from the process of living as a non-researcher in that it requires not just participation, observation and description, but also abstraction, contemplation and selective communication to academic (and other) peers' (Smith 1988: 33). There are of course problems with this sort of interpretive approach. To begin with, it runs the risk of becoming preoccupied with the unique and the esoteric to the extent that description becomes an end in itself. Research, in such cases, becomes non-incremental (Ley and Samuels 1978c). Conversely, there is always the danger that the researcher will see order where none exists and will assume a concordance between mind and behaviour that is perhaps not justified (Tuan 1976b).

Humanism itself is an umbrella term that covers a wide variety of approaches. These approaches are perhaps best classified according to their underlying philosophy. It is therefore important to look at the main philosophies. Each of these tends to be characterized by an arcane vocabulary (Billinge 1983) and what follows therefore is but an introduction to each issue.

Phenomenology

Phenomenology is a philosophy developed largely in continental Europe and, as a result, it made very little impact until relatively recently on the world of English-speaking geographers (see Pickles 1985). The philosophy is usually said to have begun with Husserl (1931) and his attempts to break away from an obsession with facts in the study of human behaviour. In its simplest form, it affords a radical method of enquiry that proceeds from pure consciousness without presupposing an existent world (Walmsley 1974b). Its aim, in other words, is to shed inquiry of any preconceptions. Pickles (1988: 237) in fact sets out four principles which are often ignored in social science but which guide phenomenology:

1. The objects of inquiry cannot be specified a priori;
2. All knowing is intentional;
3. The task of inquiry is the precise and accurate description and account of the phenomena we encounter in the world;
4. The task of phenomenology is to permit description without the distorting influence of a priori and unclarified assumptions.

The focus of attention is on the link between experience and meaning in an individual's interaction with the environment, and its underlying value position holds that social scientists change an object of study while they study it and because they study it, with the result that they need to know as much about the eye that sees as about the object that is seen (Strasser 1963). Phenomenologists argue that people cannot be independent of the world, that people come to know the world through their own consciousness, and that social scientists must therefore study how people experience the world (see Seamon 1987). Thus 'if scientific method is a way of thinking that realises itself as a way of doing, phenomenology is a way of thinking that reveals itself in a *way of being*' (Relph 1981a: 101). In order to concentrate on 'being', phenomenologists focus on the lived world or *Lebenswelt*. Their goal is to identify the *essences* (necessary and invariant features of objects) that provide the structure of being that underlies all relationships in this *Lebenswelt*. In simple terms this goal is achieved by adopting the method of *verstehen*, that is to say, the researchers attempt to understand the behaviour of the person under study by empathising with that person and imagining themselves in that person's position in order to comprehend the intentions that motivated the person to act in a particular way.

There is no one brand of phenomenology. Rather the philosophy has changed and developed over time (Walmsley 1974b). Probably, it is the work of Schutz (1967) that has had the greatest impact on social science, largely because he developed phenomenology as an academic method rather than as pure philosophy. Schutz argued that 'phenomenology is empirical because it is based on observation, it is systematic because it is concerned with the organization of the phenomena of experience, and it is rigorous because it is reflexive, that is, it subjects its own procedures to critical appraisal' (Relph 1981a: 103). In this sense phenomenology redefines the subject matter of social inquiry and presents new issues for investigation. It is not, however, simply a new perspective on old problems. The new focus is on lived experience. In the words of Relph (1981a: 109):

> We live in a world of buildings, streets, sunshine and rainfall and other people with all their sufferings and joys, and we know intersubjectively the meanings of these things and events. This pre-intellectual world, or life-world, we experience not as a set of objects

> somehow apart from us and fixed in time and space, but as a set of meaningful and dynamic relations.

This pre-intellectual life-world manifests itself in a sense of belonging to a place and in the existence of a *taken-for-granted world.* Phenomenologists are therefore interested in how a sense of place develops and in how the character of a place can transcend the behaviour of the particular people living there at a particular time (Seamon 1984). Sadly, despite this being an important topic and phenomenology being an important philosophy, phenomenological approaches in human geography have made relatively little progress in recent years, possibly because the true nature of the philosophy is as yet poorly understood (Pickles 1988).

Existentialism

Existentialism and phenomenology are similar to the extent that both seek 'to define the relationship between *being* (existence, reality, and material condition) and *consciousness* (mind, idea, image)' (Samuels 1978: 23). Both philosophies are therefore anthropocentric and both offer a chance to break away from quantitative geography's emphasis on space and to look instead at the concept of *place.* In simple terms, as it has been interpreted by geographers, existentialism is a philosophy that is concerned with an individual's attachment to, and alienation from, place. Existentialism thus attempts to restore the concrete, immediate experience of existence to a position of prominence in the study of people–environment interaction. The philosophy begins with the subjective in the sense that the human subject is seen as firmly grounded in irreducibly concrete historical and geographical facts of existence. From this point the philosophy argues that people set themselves apart from the space around them in order to enter into relations with the objects and phenomena in that environment (Samuels 1981: 117–19). In other words, people define the environment as opposite to and separate from themselves. This means that an individual sets the world at a distance thereby *alienating* herself or himself from it. This alienation, or setting apart of people and environment, is overcome by individuals developing *relationships* with objects within the environment. 'What distance necessitates (detachment), relation fulfils (belonging)' (Samuels 1981: 119). In short, relations with the environment give meaning to people's existence. People's distancing of themselves from their environment and their relations with that environment are therefore dialectical concomitants of one another.

Despite an attempt to look at such things as the 'uprootedness' of alcoholics from an existential viewpoint (Godkin 1980), the philosophy has had only a limited impact on human geography. Samuels (1981: 129), the staunchest advocate of an existential viewpoint in geography, sees existential geography as 'a type of historical geography that endeavours to reconstruct a landscape in the eyes of its occupants, users, explorers and students in the light of historical situations that condition, modify, or change relationships'. Whether this is an attainable goal is open to debate. Many authorities still see existentialism as anti-intellectual philosophy on the grounds that one of its central tenets holds that reality and existence can only be experienced through living and cannot therefore be made the object of thought. Perhaps ultimately then the significance of existentialism to geographers is simply that it offers a perspective that stresses the quality and meaning of human life in the concrete everyday world (Buttimer 1979).

Idealism

The philosophical position of idealism holds that the activity of the mind is the foundation of human existence and knowledge. The world, in other words, can only be known indirectly through ideas (Guelke 1981: 133). There is no 'real' world that can be known independently of mind because each person acts on the world, and the 'world' is formed and structured through consciousness (Moore and Golledge 1976b: 13). In other words, 'the real world, or what we think of as the real world, is a mental construct' (Guelke 1989: 295). The corollary to this is that the social sciences are necessarily different from the natural sciences: the former look at the world in the head and the latter at the world outside the head. The idealist thesis is that knowledge 'gives birth to a world which for us is the only world' (Moore and Golledge 1976b:

13). To strict idealists, nothing exists apart from what is in the mind and the contents of the mind therefore comprise reality (Guelke 1974). 'The interpretation of the phenomena we experience and perceive depends entirely on the ideas and beliefs of the perceiver' (Guelke 1989: 295). Order, in other words, is present in all rational action. The ideas inherent in an action comprise the order. From this point of view, a human action or its product is understood – that is to say its rationality is discovered – by a scholar rethinking or reconstructing the thought contained in the action. The researcher has therefore no need of theories which impose order because order is already inherent in action. As a result, the key concern of the idealist geographer is not to provide a causal explanation of an event (in the sense of listing all the necessary and sufficient conditions for its occurrence) but rather to elucidate its meaning and significance (Guelke 1981: 134–6).

This process of 'rethinking' is virtually identical with the *verstehen* of the phenomenologists. Idealism differs from phenomenology, however, in that it goes beyond an interest in the subjectively experienced life-world and seeks out verifiable knowledge (Guelke 1981). Idealists, for example, pay a great deal of attention to the concept of culture in so far as this comprises an idea in the mind of people which influences their behaviour and which is capable of verification by means of shared meanings which are manifest in societal norms, institutions, and customs. The task of the human geographer is seen then to involve the relation of human behaviour to the total cultural context in which it is embedded (Guelke 1989: 296). However, as with so many interpretative approaches, idealism suffers from inherent weaknesses. There is, for instance, no way of knowing whether the 'rethinking' of two or more idealists would produce the same results. Nor is there any way that the process of 'rethinking' can take complete account of emotions and feelings without researchers imputing their own values to the subject under study (Curry 1982). Above all, idealism as adopted by geographers tends to suffer from a rather cavalier use of language (Harrison and Livingstone 1979).

A sense of place, territory, and personal space

Profound psychological and emotional links develop between people and the place they live in and experience (Tuan 1977). Place, for example, can serve as a context for action, as a source of identity, and as a focus of environmental meaning (Entrikin 1990). For the most part, however, people's attachment to place has not attracted the attention it merits from human geographers who have been very largely preoccupied with the abstract, geometrical, and objective concept of space (Tuan 1974b). Places are none the less important phenomena in the lived world of everyday experience. Norberg-Schultz (1980) in fact talks of *genius loci*, meaning the unmeasurable, unobjectifiable character of a place and its importance to humans. In short, experience and consciousness of the world centre on places. Thus, although contemporary society is highly mobile, most people retain a strong sense of attachment and belonging to key places (Lenz-Romeiss 1973). Despite this, little is known about how places acquire meaning. In the words of Relph (1976: 6), 'we live, act and orient ourselves in a world that is richly and profoundly differentiated into places, yet at the same time we seem to have a meagre understanding of the constitution of places and the ways in which we experience them'.

Most of a person's experience of the everyday world is unselfconscious and not clearly structured (Eyles 1989). The experiential concept of place is particularly difficult to grasp, largely because of its complexity (Relph 1976). The term 'place' implies a location and an integration of nature and culture. As a result places are unique and yet they are connected together by the circulation of people within the environment. In this sense places are localized in that they are part of larger areas. Moreover, the nature of places changes over time as the nature of culture and society changes. Above all, places have meanings. A place is in fact a centre of meaning and it ranges in size from a rocking chair, through an urban neighbourhood, to a nation state. Similarly, experience of places can occur in different modes, from relatively passive ones like smell (Porteous 1985) to active ones like seeing and thinking. Small places may be known directly and intimately through the senses whereas large places may be known as a result of indirect experiences gained through concepts and symbols (Tuan 1975b). And the net result of this process of getting to know a place is often a strong sense of attachment which, in terms of its contribution to a person's well-being, may be every bit as important as close relationships with other people. Places might in fact be integral to a person's psychological development. 'Place-identity' may foster well-being. For example, the definition of 'self' might reflect relations with places just as much as it reflects relations with people (Proshansky *et al.* 1983). For those who are highly mobile and therefore unable to bond with particular places, it might be that certain types of places foster 'settlement-identity' (Feldman 1990).

By conceptualizing place as a multi-faceted phenomenon of experience and by examining properties of place such as location, landscape, and personal involvement, some assessment can be made of the extent to which these properties are essential to people's experience and sense of place (Relph 1976: 29). The task is not, however, an easy one. Although places usually have a fixed location and possess features which persist in an identifiable form, the meaning of a place comes neither from location, nor from the trivial function that places serve, nor from the community that occupies it, nor from superficial and mundane experiences (Relph 1976: 43). The meaning of places is far deeper and can only really be uncovered by studying the intentionality that underpins human behaviour. Interpreting such a sense of place implies *verstehen*. It therefore presents problems of validity and verification. It should not be imagined, however, that these problems are any greater than those surrounding positivistic scientific method because quantitative, survey-based attempts to delimit a sense of place in north-east England have also met with methodological detractors (see Townsend and Taylor 1975; Taylor and Townsend 1976; Cornish *et al.* 1977). Perhaps the way to go therefore in identifying 'a sense of place' is to use both positivistic and humanistic approaches, as in Eyles's (1985) study of Towcester.

At the centre of methodological debates about the measurement of a sense of place lies the question of the extent to which the meaning that attaches to a place is a shared meaning. This was a problem addressed by Tuan (1974a) when he used the term *topophilia* to refer to the affective bond that develops between people and place or setting. (The opposite of topophilia is *topocide*, the destruction of the bond between people and place, and Porteous (1988) has written passionately about the demise of the village of Howendyke in the north of England.) Topophilic sentiments range in variety and intensity and cover aesthetic and tactile responses as well as a sense of belonging. The development of these sentiments has, in Tuan's view, a lot to do with the role of groups in the transmission of culture. For example, Tuan argues that, although humans are biological organisms with a number of senses, it is culture which tends to determine which sense is favoured (as in the case of vision being dominant in most Western societies, Pocock 1981a). Similarly culture serves to encourage the development of symbols. It does this partly through the medium of literature which provides a means whereby shared meanings can be passed on from individual to individual (Mallory and Simpson-Housley 1989; Pocock 1988). For example, the rise of the English regional novel in the nineteenth century (Pocock 1981b) did much to perpetuate the sense of place that attaches to areas such as Hardy's Wessex (Birch 1981) and geographers generally can learn a good deal about people and place through close scrutiny of how literary sources treat the environment (Pocock 1981c).

Much behavioural research into people–environment interaction avoids issues of symbolism and topophilia and instead simplifies the world into easily represented structures or models that ignore much of the subtlety and significance of everyday experience (Relph 1976). This is an unfortunate bias, not only because it means that a sense of place tends to be overlooked in academic research, but also because the simplified models often serve as a basis for the design of new environmental settings and the manipulation of people and places. Uniform planning and environmental design is in fact doing away with localism and creating homogeneous landscapes (clearly shown by the tendency for international hotels to look alike, irrespective of their location). Planners, in other words, are creating a placeless geography and fostering a sense of *placelessness*. Increasingly, people have no sense of awareness of the deep and symbolic significance of places and no appreciation of the role of places in their own identity (see Relph 1976, 1981b).

Of course not all researchers would agree that the bonding of people to place is best studied through humanistic approaches and an emphasis on experience. An alternative view is to be found in the argument that an individual's relations with the environment are in some way innate. Several popular books have explored the theme of humans as territorial animals (e.g. Ardrey 1967; Morris 1968) and have pointed towards an important issue, namely the extent to which ethological studies of the biological bases for behaviour in the natural environment provide clues for the understanding of human behaviour. Other studies have shown that humans might have innate and little understood skills such as the ability to sense direction (Walmsley and Epps 1988a, b). To ethologists, the concept of territory is linked to the instinct of aggression. A territory is, in other words, a place to be defended on account of its role in providing food and a place for mateship. As such it serves four functions:

1. It ensures propagation of the species by controlling population density;
2. It keeps animals within communication distance so that food and danger can be signalled;
3. It facilitates group activity and therefore group bonding;
4. It provides individuals with a known terrain within which to learn and play (Eibl-Eibesfeldt 1970).

It is difficult to see simple parallels with this in human behaviour (Malmberg 1980) and the idea has even attracted opposition in biology (see Montague 1968). However, if a territory is thought of as the space around an individual or group which that individual or group thinks of as in some way its own, and which therefore distinguishes it from other individuals or groups, then the concept has at least intuitive appeal in the study of human affairs (Gold 1982). Territory in this sense is characterized by a feeling of possessiveness and by attempts to control the appearance and use of space (Brower 1980). This is a point made forcefully by Sack (1986) who sees territoriality as a culturally derived means of either enhancing or impeding social interaction. Its existence can be interpreted as an attempt 'to affect, influence, or control actions, interactions, or access by asserting and attempting to enforce control over a specific geographic area' (Sack 1983: 55). Sack, in other words, is not interested in whether territoriality is innate so much as in the way in which it contributes to human social organization. Indeed he has attempted to build a theory of the way in which territoriality reifies power and both communicates and enforces social control. A somewhat similar view, but with less emphasis on social control, has come from Taylor (1988) who coined the term 'territorial functioning' to refer to a complex system of sentiments, cognitions, and behaviours that are very much place specific and socially and culturally determined. Place, to Taylor, is central to the existence of small groups and 'territorial functioning' serves to minimize social conflict by clarifying spatial relationships, strengthening social bonds, and enhancing control.

Similarly, Edney (1976a) has stressed that human territoriality bears little relationship to aggression but that it is important as a way of organizing behaviour. In his view territoriality is characterized as the continuous association of a person or persons with a specific place and it operates at three levels: the community; the small group; and the individual. At each level the main benefit from territoriality is reduced randomness, added order, and hence predictability in the environment. In this way territory is something that is very much taken for granted with a result that formidable problems are encountered when attempts are made to measure territories (see Walmsley 1976). In fact 'the functions and benefits of a territory would be drastically reduced if it could not be taken for granted' (Edney 1976a: 43). Thus territoriality operates at the community level to encourage group identity and bonding, exemplified in the feelings of loyalty that urban dwellers display towards their suburb or town (Hall 1982; Norcliffe 1974) and in the way in which teenage gangs associate with particular 'turfs' (Porteous 1973). At the level of the small group territoriality operates to encourage congruency between behaviour and setting. Certain behaviours are deemed appropriate to certain places and territory holders are deemed to have certain rights. At an individual level territory provides security, stimulation, and identity. It can also be important in distinguishing between social classes and in providing cues that lead to an individual's behaviour being integrated or 'chained' into sequences that in turn encourage routine and order (Edney 1976b).

An alternative to this community–group–individual interpretation of territory is Altman's (1975) differentiation of primary, secondary, and public territories. In primary territories, occupation is permanent and invasion is resented. In secondary territories, individuals have some control over their surroundings but others have access (typified perhaps by seating patterns in a common room). Public territories are those occupied for only a short period (e.g. park benches). No matter what classification is used, however, human territoriality serves the purpose of regulating social interaction by establishing rules of behaviour and protocol (Brower 1980).

The significance of territory to the individual is shown most clearly in relation to the home (Porteous 1976, 1977). The home provides for physical security, psychic security, and a sense of well-being; it provides a threat-free environment in which the territory holder can control and manipulate sensory stimulation; and it provides both an identity and a way of communicating that identity to the outside world. Cooper (1974) has even suggested that the home is important to people in that it symbolized 'self' in a very deep psychological way. Space at a variety of scales, in other words, is assigned meaning and is thereby used as a communication system. Hall (1959, 1966) referred to this non-verbal communication, which differs from culture to culture and of which people are rarely aware, as 'the silent language'. He introduced the term *proxemics* to refer to the human use of space as an elaboration of culture and suggested that research should focus on three things: fixed feature space (the realm of architectural design); semi-fixed features (such as furniture); and informal space (the distances maintained by an individual in encounters with others). Informal space can be resolved into *intimate space* (up to 0.6 m); *personal space* (0.6–1.5 m), *social–consultative space* (1.5–3.6 m), and *public space* (beyond 3.6 m). Of course these spaces are not spherical about the individual and the distances quoted apply specifically to studies of white, American, middle-class males. There is none the less a good deal of support for the existence of these spaces (Sommer 1969). In particular personal space seems to be related to considerations of privacy and freedom (see Westin 1967) with a result that invasion of personal space, either deliberately (e.g. interrogation techniques) or accidentally (e.g. overcrowding), can have discomforting effects.

The taken-for-granted world

Attempts to identify a sense of place have met with criticism. All too often research has been cast in obscurantist terms, that either parrot the views of artists or rely on anecdotes which convey mood but little message (Ley 1978: 45). Too much attention has been devoted to cognitive processes and too little to concrete situations. Moreover, too much research has focused on trivial subject matter (Ley 1981). Ley (1977: 498–501) claims that these criticisms arise because human geography has failed to draw upon an appropriate philosophical underpinning to engage the distinctive epistemological issues of subjectivity and has relied instead on the general positivist viewpoint that the subjective is metaphysical and therefore unknowable, irrational, private, and beyond the range of theory. In the view of Ley, the solution to this problem is for human geography to break away from the type of *psychologism* that, in imitating the natural sciences, destroys the situational aspects which are integral to the meaning of experience. Instead researchers should recognize that people–environment interaction is characterized by relations which are fuzzy and ambiguous (Rapoport 1977). In particular, Ley advocates a focus on the world of experience, and especially the *taken-for-granted world*, as a means of developing a holistic view of the dialectical relations between people and their surroundings. This is an important point because, as has been shown, people's sense of place and territorial bonding is often taken for granted and therefore difficult to study by means of quantitative, positivistic social science. In short, people are rarely aware of how they experience the environment (see Kaplan and Kaplan 1989). To Ley, Schutz's (1960) phenomenological theory of social action provides an appropriate philosophical underpinning for a focus on this taken-for-granted world in that it entails the researcher in asking the question 'what does this social world mean for the observed actor within this world and what did he [*sic*] mean by his acting within it?' (Ley 1977: 505). Obviously, such a question can be applied to individuals, to groups, and to organizations.

Ley's argument about the taken-for-granted world is closely related to Seamon's (1979) humanistic study of three key elements in people–environment interaction: *movement* (which varies in scale but is often habitual), *rest* (which involves belonging to somewhere where energies can be renewed), and *encounters* (situations of heightened consciousness when individuals are particularly aware of environments). It is also related to Buttimer's (1976) earlier plea for geographers to explore the concept of life-world. Buttimer argued that geographical studies of overt behaviour and its cognitive foundations often fail to speak in categories appropriate for the elucidation of lived experience to the extent that they make artificial distinctions between subjects and objects, people and environments. Buttimer (1976: 278) implies that phenomenology may provide a starting point for exploration of the life-world but cautions that the philosophy provides neither ready-made solutions to the epistemological problems facing social science today nor clear operational procedures to guide the empirical investigator. However, in Buttimer's view, phenomenology and positivism are not opposing views, and she appeals for a dialogue between the two approaches. Although there is only a little evidence of such 'dialogue' taking place, there does seem to be an increased willingness on the part of many researchers to be eclectic in the choice of research strategies and to choose techniques that seem to offer insight into the way in which people ascribe meaning to the environment. One such technique is personal construct theory.

Personal construct theory

The key problems in coming to terms with 'meaning' involve operational definition of the term, its measurement, and consideration of representivity (Harrison and Sarre 1971). One approach which has appealed to many people as a potentially fruitful way of examining the *meaning* that the environment has for people is Kelly's (1955) *personal construct theory*. This approach, derived from clinical psychology, assumes that humans are proto-scientists who try to understand the workings of the world around them (Downs 1976). The theory in fact derives from Brunswik's (1956) suggestion that individuals sample their environment perceptually and then test the accuracy of their perceptions by trying out the environment through their actions (Ittelson *et al.* 1974: 110). Because sampling is never perfect, environmental knowledge is never perfect. The fundamental postulate of personal construct theory is that a person's processes are psychologically channelled by the ways in which he or she anticipates events. From this, eleven corollaries are deduced to show how constructs help in understanding and ascribing meaning to the world (Table 5.1). In this sense, then, personal construct theory is very different from traditional learning theories. Indeed, it fits very comfortably with the notion of *schemata*, or frameworks by which knowledge is organized (Aitken 1991a: 185). Constructs can in fact be thought of as the evaluative dimensions by which the environment is interpreted. The difference between personal construct theory and traditional learning theory is perhaps best seen by reflecting on Kelly's distinction between *accumulative fragmentalism* and *constructive alternativism*: under the former perspective, knowledge is obtained by piecing together bits of information whereas under the latter perspective, exemplified by personal construct theory, the world is not seen as fact-filled but rather as a place where individuals construct and test alternative assumptions as to how things work (Downs 1976: 79–80).

The constructs that individuals have about the environment can be uncovered in a variety of ways. Perhaps the best known of these is the *minimum context form* whereby items are presented in triads to respondents who are asked to state what differentiates two members of the triad from the third (Fransella and Bannister 1977: 14–19). The resultant constructs are usually expressed as bipolar scales. These constructs can then be listed on one axis of a matrix, the other axis being a listing of the environmental features under study. For example, a group of shopping centres can be evaluated in terms of a number of constructs. Values in the matrix are scores for each element (shopping centre) on each bipolar scale (construct). In other words, the matrix is a grid of the repertoire

Table 5.1 The formal content of personal contact theory

Fundamental postule A person's processes are psychologically channelled by the ways in which he anticipates events

Construction corollary A person anticipates events by construing their replications

Individuality corollary Persons differ from each other in their constructions of events

Organization corollary Each person characteristically evolves for his convenience in anticipating events, a construction system embracing ordinal relationships between constructs

Dichotomy corollary A person's construction system is composed of a finite number of dichotomous constructs

Choice corollary A person chooses for himself that alternative in a dichotomized construct through which he anticipates the greatest possibility for the elaboration of his system

Range corollary A construct is convenient for the anticipation of a finite range of events only

Experience corollary A person's construction system varies as he successively construes the replications of events

Modulation corollary The variation in a person's construction system is limited by the permeability of the constructs within whose range of convenience the variants lie

Fragmentation corollary A person may successively employ a variety of construction systems which are inferentially incompatible with each other

Commonality corollary To the extent that one person employs a construction of experience which is similar to that employed by another, his processes are psychologically similar to those of the other person

Sociality corollary To the extent that one person construes the construction process of another he may play a role in a social process involving the other person

Source: Adapted from Bannister and Fransella (1971: 202–3)

(rarely more than ten to fifteen) of constructs necessary to describe a given type of environmental feature, hence the term *repertory grid technique*. The technique is not a test because the form and content vary from case to case (Bannister and Fransella 1971). It is simply a method that puts the least possible number of constraints on individuals as they try to communicate their understanding of the environment and what it means to them. In fact the flexibility and open-endedness of the technique have led to criticism. Downs (1976) has pointed out, for instance, that personal construct theory has often been adopted in geography because it affords a ready package rather than because it is deemed theoretically attractive. Moreover, the repertory grid technique only provides for interpersonal comparisons when a common set of constructs and elements is used, and yet such a move runs counter to the idiosyncratic focus for which personal construct theory was originally developed. Additionally the theory says little about feedback and its influence on behaviour. Such criticism may, however, be harsh because the theory has the potential to be developed in the light of application and it has been used with success in the study of housing (Aitken 1987), environmental imagery (Harrison and Sarre 1975), shopping (Hudson 1974), tourism (Pearce 1982b), and holiday-making (Riley and Palmer 1976).

Shared environmental meanings

Personal construct theory can be very effective in describing how an individual interprets the environment. It is less effective in indicating the extent to which that interpretation is shared. Nor does it give any clues as to how *shared meanings* come into being, whether they are contested, and, if so, whose views dominate. The focus of personal construct theory, in other words, is on the individual. However, as has been pointed out in relation to territoriality and a sense of place, group and community views are also important. It is not surprising, then, that there is a rapidly growing body of research in human geography concerned with shared meanings and their role in people–environment interaction. In large measure, these shared meanings comprise the taken-for-granted world at any point in time. Anderson (1988), for example, has examined the notions of 'race' in the structuring of society and space in Vancouver in the last century, paying special attention to the social construction of the racial category 'Chinese' in white

European culture. The definition of what comprises 'Chineseness' reflects the power of some groups over others, notably the European community over the Chinese in Vancouver. 'The race idea has been a unifying concept in the evolution of white European global hegemony. With more or less force in different colonial settings, racial ideology was adopted by white communities whose members (from all classes) indulged it for the definition and privilege it afforded them as insiders' (Anderson 1988: 131). In this process, the state has been instrumental in creating and perpetuating racial classifications. In contemporary society, for example, where multiculturalism is the dominant ideology, the commercialization of 'Chinatown' has been fostered by the state, not least as a tourist attraction.

In addition to its intrinsic value, Anderson's study is also important in that it is indicative of a growing interest in *locality studies*. Such studies focus on the way in which 'larger processes in part construct, and are themselves constructed through, the locality' (Anderson 1988: 146). In this way, what is happening in society at large can be *interrelated* with what is happening in specific places; to Anderson, what happens at the local level (e.g. the definition of the geographical extent of Chinatown) can both grow out of, and feed back into, a politically divisive system of racial discourse that justifies domination over people of Chinese origin. Locality studies have been conducted in a number of areas, often as a way of emphasizing geographical variations in the impact of global economic restructuring (see Cooke 1989a). Their growth points to a resurgence of interest in social science in *case studies* as opposed to huge surveys (Mitchell 1983). In some respects, a growing interest in locality was foreshadowed by Pred (1984: 279) in his 'conceptualization of place as a constantly becoming human product as well as a set of features visible upon the landscape'. To Pred, the development of 'place' involves the reproduction of social and cultural forms, the formation of biographies, and the transformation of nature, in the manner shown in Fig. 5.10. Pred's work therefore integrates parts of humanistic geography, work on time geography, and a materialist interest in economic production in an overview of how locations develop their distinctiveness.

One of the things perhaps not given sufficient emphasis in Pred's work is the role of the mass media in developing images of localities among those who do not have first-hand knowledge of the places in question. The mass media are, for example, notorious for trading in *stereotypes*. Often these have little foundation in fact, as demonstrated by Burgess's (1985) study of the way in which the media 'described' a largely hypothetical 'inner city' in their coverage of the 1981 rioting in British cities. The media, in other words, are an integral part of a very complex cultural process by which environmental meanings are produced and consumed, and attention therefore needs to be paid to the way in which meaning is encoded in different forms of media text and decoded by different groups among media users (Burgess 1990).

Integral to locality studies is an appreciation of 'cultural' matters (Cosgrove 1989a). *Culture* has in fact become so important in geographical inquiry that some authorities speak of a 'new agenda' for cultural geography; instead of emphasizing the evolution of the cultural landscape (as did traditional cultural geography), new cultural geography is concerned with the way people make sense of the world (Jackson 1989). This new agenda obviously encompasses popular culture and hence the mass media. It also poses the question of what comprises 'culture'. As Thrift (1989) has pointed out, culture is a difficult concept to pin down because it is so all-embracing: it includes material artefacts and socially constructed meanings, as well as beliefs and attitudes and a wide variety of symbols. One way of investigating the symbolism inherent in culture is through landscape (Cosgrove 1989b; Cosgrove and Daniels 1988). Others have chosen to explore symbolism and culture through *semiotics*, placing the emphasis on how humanly designed landscape features can be imbued with meaning (Duncan 1987). Semiotics in this sense can be divided into three types: the *pragmatic* deals with the origins, uses, and effects of signs within the landscape; the *semantic* deals with the way in which signs actually carry meaning; and the *syntactic* deals with the overall development of 'sign-systems', as in the organization of an urban fabric that can be read as a 'text' (Duncan and Duncan 1988; Broadbent 1976). At the heart of semiotics is the proposition that 'if architecture is a language then the city must be the text' (Krampen 1981: 32). Unfortunately, the reading

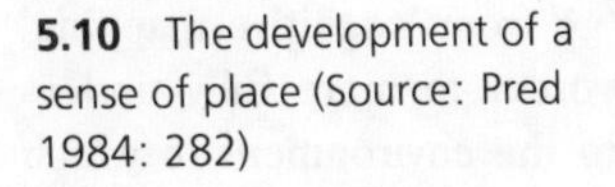

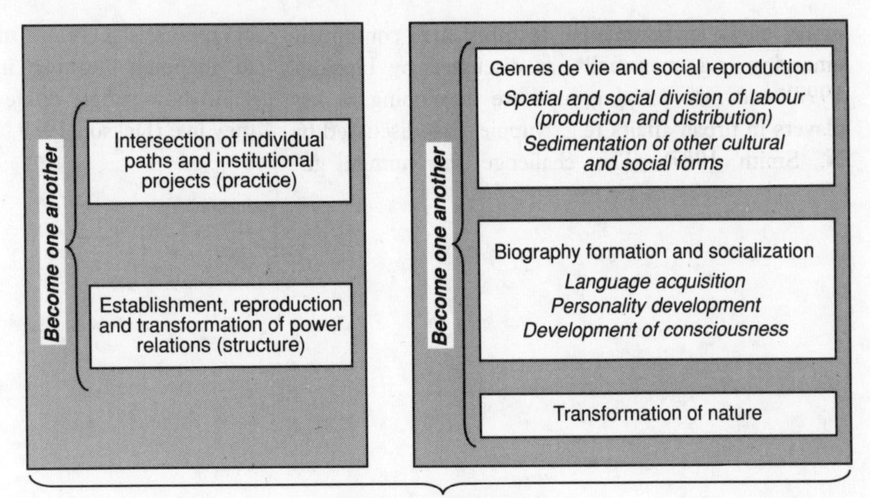

5.10 The development of a sense of place (Source: Pred 1984: 282)

of such 'texts' has been fraught with problems despite the best efforts of Rapoport (1982) in decoding vernacular architecture. In short, the metaphor of landscape as text might not be a useful one (Rowntree 1986). As a result, Eyles and Peace (1990) have looked instead at both *iconology* (the interpretation of symbols) and *iconography* (the description of symbols) in their study of Hamilton (Canada).

Any interest in culture and in the intersubjectivity of meaning on the part of human geographers necessarily involves a focus on *interpretation*. Eyles (1988), for instance, has discussed 'interpretative geography' and Ley (1988) has called for 'interpretive social research' in the inner city. Both these authors see human geography as being concerned to explain the nature of social reality. In advocating interpretation, Eyles (1988: 1) argues that 'order cannot simply be imposed on the world through the use of scientific methods and constructs because an "order" already exists'. Human geography is therefore in the business of revealing the taken-for-granted assumptions that people have about their world. The goal then is 'to make sense of the actions and intentions of people as knowledgeable agents . . . to make sense of their making sense of the events and opportunities confronting them in everyday life' (Ley 1988: 121). This is no easy task and Eyles (1988) has argued that it is best approached by using multiple investigators, multiple theories, and multiple data sets. Above all, it means striving for what Geertz (1980) has called 'thick description' (cf. 'thin description' which merely repeats the unexamined accounts of informants). It is not surprising then that human geography seems to be showing renewed interest in *ethnography*, particularly the detailed description of city life (Jackson 1985). Not since the ethnographies of the Chicago School of Human Ecology (see Hannerz 1980) have geographers shown so much interest in the subject. Eyles and Donovan (1986), for example, have used an ethnographic approach very effectively to examine individual perceptions of health and health care in the English Midlands.

In summary, a good deal of work is seeking to understand how meaning is ascribed to the environment by examining the mode of operation of social networks (Huckfeldt 1983) and culture generally (Jackson 1991). Social and contextual effects on behaviour have of course been of interest to geographers for a long time. For example, much of electoral geography has studied such effects (Walmsley and Lewis 1984: 145–51). What is new in contemporary work is the attempt to place the research in an overall context of social and cultural theory. The challenge is great. Even the language used to recount urban experiences and interpretations might vary between social groups (Goodchild 1974). Furthermore, new environmental features are continually emerging (e.g. megamalls, as discussed by Hopkins 1990) and new social groups are developing as key players in urban affairs (e.g. 'yuppies', as discussed by N. Smith 1987). The challenge for human geographers is therefore to go from 'cognitive mapping' to 'mapping meaning' in order to more fully understand how people relate to the environment in which they live (Jackson 1989, 1991).

6 Critique

The structural approaches to people–environment interaction described in Chapters 2 and 3 and the behavioural approaches outlined in Chapters 4 and 5 have a number of shortcomings. For example, spatial interaction models such as the gravity model and similar formulations may provide a reasonable description of the empirical regularity that is to be seen in the behaviour of aggregates of individuals but they offer little by way of explanation. They therefore give little insight into why individuals come to act in a given way. Aggregate statistics, in other words, frequently mask the variability in behaviour that exists between different individuals and between different groups and they thereby cloud the causes of such behaviour. Likewise, structural Marxist approaches adopt a holistic rather than an individualistic stance. They place great faith in transcendental structures as the motivating force in the functioning of society and they subjugate individuals to a passive role in a system dominated by the relationship between the mode of production and the associated superstructure. Spatial structural approaches, again reviewed in Chapter 2, fare little better. An emphasis on social space as a macro-scale phenomenon, and on the identification of social areas, runs the risk of directing inquiry at an artificial target in that many of the areas about which information is gathered (notably census districts of one sort or another) are merely administrative units designated for the convenience of census organizers. Many of them have little significance in terms of either behaviour or social belonging. Worse still, to focus on units such as census districts can encourage geographers to commit the *ecological fallacy*. This involves two things:

1. Attributing the characteristics of an area to all the individuals living within that area even where there is ample evidence to show that such 'diagnostic' characteristics are carried by only a minority of residents (Hamnett 1979);
2. Proceeding with a style of analysis whereby areal characteristics are correlated with each other, often creating fallacious impressions of causal relationships (Openshaw 1984).

For instance, a correlation of summary rates for unemployment and for crime, across a number of suburbs, might produce a statistically significant coefficient. This, in turn, might lead to the interpretation that unemployment 'causes' crime. However, based on suburb-level data alone, such a conclusion could be erroneous. Crime might be caused by any number of factors and might merely manifest itself in a geographical pattern that is coincidentally similar to the pattern of unemployment.

Given these problems with structural approaches, it is not surprising that many researchers have advocated the behavioural approaches described in Chapters 4 and 5. However, much of the behavioural emphasis in human geography over the years has failed to look at individual actors and has concentrated instead on combining individuals into groups and then studying the average behaviour of those groups. As a result, much of geography has tended to look upon a population as made up of 'dividuals' rather than 'individuals', meaning that there has often been a tendency to divide a population up into rather arbitrary groups rather than into actual decision-making units (Hägerstrand 1970). In other words, there has been a tendency to compartmentalize people into aggregate categories and to lose track of their existence as individual human beings. Very often these aggregate categories have been artificial in the sense that they are imposed by the researcher rather than reflecting the purposiveness of the actors concerned (Golledge and Couclelis 1984). For example, it might be convenient to divide up a population according to occupational status but the exercise can be somewhat irrelevant unless some sort of occupational identification is significant in the lives of the people themselves.

Obviously a balance has to be struck between focusing on potentially misleading aggregate statistics on the one hand and being weighed down by a mass of individual statistics on the other. Likewise, it has to be realized that individual actors are not sovereign decision-makers but rather are often highly constrained in terms of what they can and cannot do. Again, a balance has to be struck between choice and constraint. Moreover, it has to be realized that many of the constraints impinging on human behaviour are themselves socially created. It is not surprising, therefore, that much of the debate about the status and form of contemporary human geography centres on a few key issues: how to relate individual human behaviour to the structures within which it takes place; how to demonstrate that the structure and spatial organization of society are both a creation of individual human beings and a constraint on their activity; and how to describe aggregate patterns of human activity in such a way as to be useful to planners and policy-makers without losing sight of the richness and variety of individual experience.

Hägerstrand (1970) has suggested that the middle ground between biography and aggregate statistics can be found by studying the individual's 'path' through time and space (see Chapter 5). In this way, the needs and wants inherent in individual life-paths can be related to geographical constraints in the provision and accessibility of opportunities for need fulfilment. Moreover, similarities between life-paths can be identified and links thereby provided between what individuals do and what entire populations do. In this sense, a focus on what might be thought of as the *micro-scale* (i.e. the activities of individual human beings) complements the study of what might be thought of as the *macro-scale* (i.e. the overall geographical context within which behaviour is conducted).

Hägerstrand is not of course the only person to have addressed this problem. In fact a great many geographers have wrestled with the problem of relating the macro-scale and the micro-scale. Olsson (1969), for instance, coined the expression 'geographical inference problem' to refer to the difficulties involved in trying to relate macro-scale structures (e.g. the built environment of cities) to micro-scale processes (e.g. individual decisions on trip-making) at a satisfactory level of aggregation (in other words, in such a way that one can be seen as arising from the other). Predictably, several methodologies have been put forward for tackling this problem. Gould (1963) has advocated a game theoretic approach for investigating how individuals develop strategies for combating environmental uncertainty and how such strategies manifest themselves in macro-scale features such as land use patterns. By way of contrast, Burnett (1978b) has investigated Markov models of how individuals adapt to environmental change in an attempt to describe aggregate travel patterns within urban spatial structures. In other words, both Gould and Burnett sought an answer to the geographical inference problem in logico-mathematical methodologies. Neither of them was particularly successful. Nor were others working in a similar vein.

Micro- and macro-scale inquiry

Watson (1978) has attributed this failure of micro- and macro-scale studies to complement each other, and therefore the failure of researchers to solve the geographical inference problem, to the fact micro- and macro-scale approaches are based on different methodological and philosophical positions. Researchers interested in the macro-scale of spatial structures see individual spatial behaviour as so complex as to be indeterminate and therefore impossible to study. In the view of these researchers, it is more efficient to study the generalized patterns that emerge in aggregate behaviour than it is to look at the individual processes that underlie people–environment interaction. Studies of individual behaviour, in other words, are deemed to be impossible, unnecessary, misleading, and ultimately unsuccessful since what is important to society can be seen in aggregate patterns (Watson 1978). The opposite view is of course that macro-scale studies are unsuccessful. The fact that such studies ignore individual behaviour means that they cannot hope to establish causal relationships between the environment and human actions. They are therefore of limited explanatory and predictive power. Moreover, they tend to ignore phenomena such as cognition which are obviously important but which cannot be measured at the macro-scale.

The choice which researchers make between micro- and macro-scale inquiry is characteristically very much a private act of faith. 'Aggregate analyses are economical when human behaviour is indeterminate, or highly repetitious, or highly constrained by space, time and the environment' (Watson 1978: 44). Moreover, they form a necessary first step in the study of any form of people–environment interaction. However, they are insufficient in themselves because aggregate patterns are only truly meaningful if their interpretation is soundly based on knowledge of individual behaviour. 'Perhaps it is time to consider a distinction between macrogeography and microgeography (along the lines of macro- and microeconomics). . . . Perhaps this would open the door to the interdisciplinary co-operation (with a micro-scale oriented environmental psychology) that always seems to beckon but never arrives' (Walmsley 1988b: 311). Despite the attractiveness of such a proposition, it raises a problem: most studies of individual behaviour have borrowed the research methodology of macro-scale spatial analysis and have therefore placed undue reliance on the approach to problem formulation and hypothesis testing that is handed down indirectly from the natural sciences. In other words, many geographers have failed to realize that the central assumption of natural science – that study can proceed in such a way as to be context-independent – is untenable in the social sciences (Eyles 1989: 104). They have naïvely assumed that laws can be found in people–environment interaction in the same way that they can be found in physics (see Guelke 1977). Couclelis (1983: 32) expressed this eloquently when she observed that 'much of social science to this day has been too busy mimicking the surface features of last century's experimental physics to grasp the deeper structure of innovative, creative, contemporary scientific thought'.

An emphasis on the methodology of the natural sciences in much of human geography has helped preclude consideration of a number of alternative approaches to people–environment interaction and has led to a situation where what is sometimes recognized as 'behavioural geography', instead of being an integral and fundamental part of the study of fields like urban planning, is something of an intellectual backwater (Cullen 1976). Far too much attention has been paid to generalization through aggregation and to what might be termed 'analytical behavioural geography' (see Golledge and Stimson 1987). This is a rather narrow field of inquiry characterized by an emphasis on logico-mathematical thinking and language, a quest for generalization, and the desire for public verifiability of results through statistical testing (Golledge and Couclelis 1984). For example, it has been standard practice to take an element in the environment (e.g. a shopping centre) and to try to measure people's image of that element in order to arrive at some summary description of that image. This presupposes that images can be committed to paper in experimental situations and that individual images can be aggregated to a group view. In short, the generality of this approach is one of its strengths. It is also its main failing: 'it can be applied, as indeed it has been applied, to a whole variety of different situations because it is almost wholly lacking in content' (Cullen 1976: 399). It may be no coincidence, then, that the impact on the discipline of geography of behavioural approaches to the study of people–environment interaction has been somewhat less than its early advocates predicted (Gold and Goodey 1984). Inadequate theoretical conceptualization of people–environment relations and inappropriate methodologies have done much to characterize 'behavioural geography' as an eclectic and incoherent sub-field of the discipline as a whole (Aitken and Bjorklund 1988).

One corollary to the widespread adoption of the methodology of natural science has been a deterministic flavour to some of the research that has been conducted. For example, individuals are commonly viewed as organisms whose behaviour can be fully explained by reference to the combined effects of a set of environmental stimuli to which the individual passively and predictably responds in a manner similar to the way in which a Newtonian mass responds to the gravitational forces exerted upon it (Cullen 1976: 401). As a result, much of the behavioural emphasis in human geography has tended to shy away from focusing on the *meaning* that an environment has for an individual. Likewise, much research has avoided studying the way in which meaning can influence the use to which an environment is put. Subjective meanings simply do not lend themselves to 'scientific' methodology. As a result, despite pleas such as Hägerstrand's (1970), human geography has been very much preoccupied with macro-scale phenomena such as cities, shopping centres, and land use generally.

Of course this changed to a certain extent with the growth of interest in humanistic approaches (see Chapter 5). Buttimer (1990) has observed that there are four distinct strands to Western humanism:

1. A focus on the nature of 'humanness' that encompasses such diverse themes as individuality and sociality, freedom and responsibility, and rationality and hedonism;
2. An inquiry into humanistic modes of knowing that opposes the way in which natural science attempts to reduce the real world to simple relationships, and seeks instead to elucidate rather than explain, thereby highlighting the subjectivity of consciousness;
3. The study of the humanities as a means of nurturing the cultivated mind;
4. The advocacy of humanitarianism as an approach to life that encourages social responsibility and liberal politics.

All of these strands have had some impact on geography. However, the nature of that influence is sometimes difficult to discern. Indeed, in the view of Buttimer, the impact of humanism is not to be seen in particular techniques but rather in an awareness of the following: geosophy (the way in which individuals construe nature, resources, space, and time into their own personal geographies); temporality (recognition of the inextricable bonds between time and space); relativity (in particular, an appreciation of how the 'inevitable ripple' between an observer and the observed has heightened awareness of how research instruments and techniques can selectively focus inquiry); and hermeneutics (for example, recognition of cultural bias in the design and interpretation of research has forced geographers into consideration of issues such as ideology and power). It has also focused attention on the dialectical relationship between structure and agency. Sadly, however, the language in which humanistic ideas are expressed has sometimes been their undoing. This is because, in the opinion of Billinge (1983: 400), human geography has witnessed the emergence of a particular style of writing which, although claiming to capture the richness and subtlety of human consciousness, has 'in fact served different and more covert purposes: the perversion of meaning, the disguise of mediocrity of sentiment, the inflation of the authors' self-regard and the representation as profound of ideas which are in reality clichéd or banal'. Although this might be a harsh judgement, there are several authorities of the opinion that humanistic approaches 'constitute a confused melange of subjectivist geographies' which together comprise 'little more than introspective esotericism, lacking a cohesive framework for substantive research' (Harrison and Livingstone 1982: 2).

Overarching models of human behaviour

A reluctance to give up the methodology of natural science and a penchant for the arcane vocabulary of humanism are both factors to have militated against the development of behavioural approaches in human geography. Underlying both these factors, and inhibiting consideration by human geographers of the individual *qua* individual, has been a reluctance to think about the question of what overall model of human behaviour should be incorporated into geographical research. In many ways, this situation was rectified by Agnew and Duncan (1981). They pointed out that all social scientists, whether they engage in the explanation of human behaviour *per se* or the results of human activity (e.g. the form and structure of cities), necessarily adopt – albeit sometimes only implicitly – one of several alternative models of human behaviour.

These models can be differentiated from each other by reference to three dimensions: *holism* versus *individualism*; *structure* versus *action*; and *determinism* versus *free will*. Although these dimensions are really continua, the difference between them is perhaps best explored by looking at the polar ends of the scales in question.

The individualistic viewpoint holds that large-scale social phenomena are simply the sum of the behaviours of the many individuals that participate in these phenomena. In contrast, holists argue that such phenomena are to be explained in terms of their own autonomous, macroscopic level of analysis. Put simply, the distinction is one between looking at individuals and looking at overarching concepts such as 'culture' and 'mode of production' (Harrison and Livingstone 1982). Likewise the distinction between a structure-oriented approach and an action-oriented approach amounts to a choice between explanation based on autonomous and reified concepts and explanation based on the actions of individual people. In support of their view, structuralists point to the regularities which often occur in social group behaviour and which can be observed even when group membership changes. The persistence of such regularities precludes, it is argued, explanation of behaviour in terms of the attitudes and intentions of particular individuals. What matters, and what is therefore seen as the cause of behaviour, is the overall structure of the group and the way in which this impinges on the individual. In this context, of course, structure means something more than just 'organization' in the sense of composition or constitution. It also means more than spatial structure in the sense of the physical environment surrounding an individual. In a sense it incorporates all these things. What it refers to therefore is the overarching context of behaviour, encompassing such unobservable features as ideology. At the centre of attention, in other words, are *transcendental structures*, so called because they exist prior to, and as a necessary condition for, experience.

Structuralism in this sense occurs in a wide variety of forms. Transcendental structures, in particular, are exemplified in social science in at least three ways (Agnew and Duncan 1981). First, they are to be seen in the structural–functional view that there are self-regulating social structures (such as organizations) that seek an equilibrium that is entirely separate from the wishes of the people comprising the structure. To study structure in this sense is analogous to studying the physiology of an organism (Harrison and Livingstone 1982: 4). Second, transcendental structures are to be seen in the anthropological notion that cultures continually evolve towards a state of greater complexity. Finally, transcendental structures are very evident in the Marxist idea that the social structure of advanced capitalism is inexorably driven along a preordained path. In the words of Harrison and Livingstone (1982: 7):

> With the central conceptualizations in the language of structural Marxism revolving around such key terms as 'the logic of capitalism', 'social formations', and 'the capitalist mode of production' (all of which refer to structures explicitly defined as holistic, autonomous, and self-determining entities displaying regularities which cannot be accounted for in terms of individual and collective action), the resulting model of man [*sic*] subordinates human creativity, individuality, and agency to the operation of transcendental social structures as the only truly active forces.

Each of these three examples of transcendent structures in social science involves an element of teleological argument to the extent that they all focus on the ends as a means of explaining the cause. In each case human action is subordinate to transcendent social structures. In short, the action of individuals is seen as the *product* of transcendent structures, perhaps most clearly seen in the idea that the 'logic of capitalism' supposedly influences individual behaviour. Opponents of this view argue that structures should not be reified and treated as objects with causal properties; rather opponents of structuralism argue that regularities in social life are ultimately caused by individual and collective action. People are seen as the active forces behind structures which, therefore, have no power other than that created and endowed by people (Agnew and Duncan 1981: 48). Taken to its extreme, of course, such an argument would lead to the view that inquiry should focus exclusively on the attitudes, ideas, and behaviour of individuals and therefore overlook the social origins of intentions and desires. Fortunately such an extreme point of view is rare except perhaps in those branches of economic geography that have fêted the concept of rational economic behaviour and held to the view that, where the real world does not fit the economic model, the real world is wrong! Similarly, the extreme positions of free will and determinism are now very rare. This is not surprising. Free will is a patently implausible proposition. And Franck (1984) has effectively demolished determinism by pointing out that:

1. It exaggerates the influence of the physical environment by ignoring or underestimating the importance of other factors;
2. It unrealistically assumes that people are passive in their relations with the environment and it therefore ignores the importance of human choice and goals;
3. Because it assumes that the environment is a given and immutable entity, it neglects processes of environmental change and modification.

In general, then, social science can be characterized by less extreme positions that either see humans as active creators or sustainers of the social order, or as people imprisoned by norms, values, and roles determined by the social and cultural system in which they exist (Agnew and Duncan 1981: 48).

Human geography has, at various times, taken different positions along the three dimensions described by Agnew and Duncan (1981). For instance, much of North American cultural geography can be thought of as holistic, structural, and deterministic to the extent that, in the eyes of many researchers, it was a culture, viewed as being something greater than the sum of its component parts, that was the focus of attention and not the actions of individual persons carrying that culture. Indeed, for many, it was the impact of culture on a macro-scale phenomenon – the landscape – that was at the heart of their research. Similarly, structural Marxism, as was shown in Chapter 3, can be characterized as holistic, structural, and deterministic: the structure determines the places and functions occupied and adopted by the agents of production who themselves are never anything more than the occupants of those places and functions. Of course, not everyone would agree with this characterization of Marxism. Chouinard and Fincher (1983), for example, have claimed that much criticism of Marxism is unjust in that it fails to recognize that much of the contemporary writing on Marxism plays down the reification of structures, is far from 'economistic', and pays a good deal of attention to the consciousness of the actors involved. Despite this, many human geographers would accept that both cultural geography and Marxist geography reify structures and therefore put the human world outside the power of human agency (Agnew and Duncan 1981: 51).

In contrast, early behavioural initiatives in human geography, involving the postulate of rational economic behaviour (Pred 1967: 5–10), can be thought of as individualistic, action-oriented, and deterministic. The focus of attention was on the overt behaviour of individual actors. Unfortunately, however, the omniscience implied in economic rationality meant that individuals could act in only one way (the most rational way) in any given situation. Deep down, then, the approach was deterministic. A similar deterministic flavour marred the stimulus–response (S–R) theories in which geographers were interested for a time (Golledge 1969): human cognition was regarded very much as a 'black box' and humans were viewed as animals unable to transcend their reflexes and therefore at the mercy of environmental stimuli (Agnew and Duncan 1981: 51). Downs (1976: 72), for instance, quotes personal correspondence with Golledge (one of the pioneers of geographical explorations of learning theory) to the effect that learning theory is '*used* but may not be very *useful* in understanding things in any but a strict normative sense'.

What seems to be needed is an approach that is individualistic, action-oriented, and which allows individuals a modicum of free will and a degree of latitude in interpreting, and ascribing meaning to, the environment. At the same time, the model should take stock of the fact that much human behaviour is constrained. It should therefore incorporate the study of constraints, particularly the way in which constraints are socially generated (in the sense of arising from the mode of operation of particular societies in particular places at particular times). Unfortunately the development of such a comprehensive model of human behaviour has not been an easy task. In many instances, geographers have shown a preference for empirical description over theorizing. Where theorizing and model building have been attempted, a number of problems have characterized the research that has been undertaken: for example, many model builders have been unclear as to which aspects of behaviour should be measured, many have failed to consider a full range of explanatory variables, many have used vague operational definitions or surrogate variables for rather vague concepts, and many have overlooked the extent to which behaviour is constrained by the environment in which it occurs (Burnett 1981).

Moreover, many geographers – as is amply demonstrated by discussion in this chapter so far – have been overly concerned with identifying dualisms or binary oppositions. Sayer (1991) has explored this tendency in detail and shown how the simple labelling of polar opposites serves to mask the complexity of the issues involved and the existence of continua rather than contrasts. Particularly revealing is his discussion of the contrast between *contextualizing* and *nomological* approaches. Sayer points out that it is not uncommon nowadays to find people arguing that the social sciences have a 'contextualizing' character rather than a law-seeking, law-invoking, or nomological one. The argument is that social science explains things 'by reference to their context rather than nomologically by reference to universal laws or principles' (Sayer 1991: 287). This is not a new dichotomy for geographers interested in behaviour because a distinction was drawn some time ago between the study of 'behaviour in space' (where behaviour is described and understood in relation to the context in which it occurs) and the study of 'spatial behaviour' (where the focus is on finding the general in the particular so as to arrive at rules, principles, and laws that describe behaviour independently of the context in which it occurs). 'Nomological approaches are most successful in relation to objects which are highly context-*in*dependent, whose nature and behaviour is the same whatever the context' (Sayer 1991: 287). They are therefore ideally suited to the natural sciences. The law of gravity, for instance, shows what will happen to apples falling from trees, irrespective of the type of apple and the location and ownership of the tree. They may be less suited to the social sciences where many of the phenomena under study are highly context-dependent. Sayer gives the example of political ideology, a phenomenon that varies markedly from place to place despite overall descriptions like 'democracy' and 'socialism'. Of course, 'law making' and 'contextualizing' might not be opposites. It may be that there is a middle ground: the 'laws' of social science might have a rather greater degree of specificity than the laws of natural science. Geographers therefore need to note that when they move from the study of *non-sensate behaviour* (e.g. the type of macro-scale population flow studied in gravity models) to the study of *sensate* behaviour (which implies goal-directed activity based on processes like cognition, perception, memory, recall, thinking, problem-solving, and decision-making, Golledge 1985: 113), they are moving into a realm where the context of behaviour becomes very important.

In the eyes of many researchers, the foundations for individualistic, action-oriented models that allow people to come to terms with their environment in a non-deterministic way, and in a way that takes full cognizance of the impact of constraints on behaviour, are to be seen in behavioural studies based on transactional–constructivist philosophy. The main features of this approach have been outlined by Aitken and Bjorklund (1988) who drew attention to the six principles of transactional methodology, as identified by Altman and Rogoff (1987) and outlined in Chapter 4. The force of these principles is to focus attention on the fact that behaviour (1) always reflects a specific context, (2) is best understood from the perspective of the actor, and (3) is subject to change. It follows from this that the study of people–environment interaction needs to recognize the uniqueness of phenomena and the necessity for eclecticism in the choice of methodology. The major tenets of transactional–constructivist philosophy are therefore very congruent with the agenda of new cultural geography (see Jackson 1989) to the extent that both are concerned with the way environmental meaning originates, is contested, and is negotiated (Chapter 5). However, enthusiasm for these approaches must be tempered by an appreciation of the fact that behavioural approaches generally have attracted a good deal of criticism. Sometimes this has come from within the ranks of practitioners, as when Aitken and Bjorklund observed that geographers have concentrated too much on constructivism (e.g. images constructed in the form of mental maps) and not enough on transactionalism (the processes whereby constructions of the world are brought into being). More commonly, behavioural approaches have been criticized by researchers who have chosen to parody such approaches as esoteric and ephemeral.

Golledge (1981) has dealt with the criticisms and misconceptions of behavioural approaches at some length. He observed that behavioural studies in geography began in the 1960s when researchers realized that, in order to exist in and comprehend an environment, people learn to select and organize critical subsets of information from the mass of experiences open to them. It is this information processing that is very much of interest in behavioural research. The goal of behavioural approaches to the study of people–environment interaction is not therefore to define a new sub-discipline within geography but rather to provide a new perspective on much of the geographical domain. Implicitly, this reorientation calls for links with other disciplines studying the relationship of people to their environment. The focus of attention is very much on sensate as opposed to non-sensate behaviour. Therefore to confuse behavioural approaches with macro-scale model building (e.g. gravity models), as some have done, is to grossly misrepresent the approach. Likewise, to argue that behavioural studies in geography are 'behaviouristic' (in the Watsonian sense of wanting to reduce everything to stimuli and responses) is again to grossly misrepresent reality because the fact is that the rationale for behavioural approaches seeks to avoid deterministic interpretations of people–environment interaction. Similarly, it is wrong to suggest that geographers inclined towards behavioural approaches are only really concerned with overt behaviour because a great deal of effort has gone into studying the decision processes from which overt behaviour results.

Of course, this is not to say that behavioural approaches in geography are free from weaknesses. This is far from the case: there is still too facile an equation drawn between 'behavioural geography' and 'environmental perception'; there is still insufficient attention paid to epistemology; there are still problems with the interdisciplinary transfer of concepts and methods; there is still too little attention given to investigating the degree of congruence between image and behaviour; and there is still an element of conflict between practitioners, notably between the positivists and the humanists, and between those who focus on individual decision-makers and those who prefer to look at a social and cultural agenda.

Criticisms of behavioural approaches in human geography should not be exaggerated, nor should the weaknesses of the approach. The reasons why behavioural approaches developed in the first place are still as valid as ever: a desire to break away from simple yet deterministic propositions about human behaviour and to recognize the importance of human consciousness in influencing the way in which individuals respond to the environment; a desire to foster multidisciplinary links with other researchers interested in people–environment interaction; a desire to contribute to issues of social and moral concern, and to policy-making; and a wish to enhance the geographer's role as an environmental educator by showing a heightened appreciation of the critical nature of many environments to the people who live within them (Gold and Goodey 1984). Moreover, there are many signs that behavioural approaches are becoming widely accepted as an integral part of the discipline of geography, and that the differences between the various behavioural approaches are lessening. In part this situation is coming about because of a recognition of the rapidity of environmental change in contemporary society and of the need for behaviour to change as a consequence.

It is obviously imperative that human geographers study such behavioural adaptation (see Aitken and Bjorklund 1988). In part, too, increased interest in behavioural approaches stems from the fact that the differences between the various behavioural approaches might not be as great as some researchers have suggested in the past. In terms of micro- and macro-scale inquiries, for example, Cadwallader (1989b) has attempted to study migration from both perspectives and to show how they complement each other. Similarly, Aitken (1991a) has observed that the vehemence in the exchange between advocates of the humanist and positivist world views in geography has ebbed to the point where there are some signs of convergence. In his view, there is a new interest in theory development in relation to people–environment interaction, reflecting a desire to 'reflexively account for both spatial behaviour (the rules that govern decision-making) and behaviour in space (the contexts of, and constraints on, behaviour)' (Aitken 1991a: 181). In other words, those geographers interested in behavioural approaches are breaking away from the interest in comparative statics which has hampered the discipline in the past. They are no longer prepared to look at behaviour at different points in time and to make inferences as to why those behaviours might be different. Instead they are prepared to confront the dynamic nature of environmental relations much more directly and to examine what sort of transactions people make with the environment and how those transactions influence both the development of mental imagery and the conduct of overt behaviour.

Similarly, geographers interested in behavioural approaches are moving away from tautological arguments that see preference structures as a determinant of behaviour and behaviour as a reflection of preference structures (Golledge and Rushton 1984). Instead they are asking how preference structures come about, how they are socially influenced, the extent to which they are context-specific, and the extent to which preferences are repressed in the face of constraints. To some extent a lead in this direction can be taken from French geographers interested in behaviour, for they have never been as obsessed with models of individual behaviour as have their anglophone counterparts; rather, francophone geographers tend always to have set events within a framework of social values (Bailly and Greer-Wootten 1983). In short, those geographers interested in behavioural approaches are showing a willingness to break out of the straightjacket of positivism and to countenance an exchange of ideas with those of humanist or Marxist persuasions. To some authorities, it is becoming clear that behavioural approaches in geography 'must entail a conception of "cognition" broad enough to encompass most of what we understand as human consciousness' because 'human beings are somehow the primary, most important reality and deserve to be treated as such' (Couclelis and Golledge 1983: 333, 337).

An appreciation of the complexity of causation in human affairs and of the need for a multi-faceted approach to the study of people–environment interaction has caused many geographers to look to that branch of philosophy known as *realism*, and in particular to the work of Bhaskar (1975). The argument here is that there are influences on individual behaviour which are not able to be observed but which are nevertheless very real. This is of course a point of view that is very similar to the Marxist argument that 'the logic of capital' cannot be observed directly but yet it underpins the behaviour of economic entities in the capitalist world. Bhaskar's views have been summarized very clearly by Johnston (1989). According to Johnston, Bhaskar argues that there exists a three-tiered stratification of 'reality':

1. *Mechanisms* which are the 'causal powers' underlying specific structures;
2. *Events* which are the realizations of the mechanisms;
3. *Experiences* which represent the ways in which individuals appreciate the outcome of events.

Basic mechanisms cannot be observed because they exist in an underlying structure that can be thought of as the *real* domain. What can be observed is the manifestation of those mechanisms as events in what can be thought of as the *actual* domain. The domain in which people actually experience events is known as the *empirical* domain. Johnston (1989: 57) gives a vivid example of how the three tiers are linked: an unemployed miner in Durham *experiences* an *event* (the closure of a pit) which itself can be interpreted as a manifestation of an underlying *mechanism* (the operation of the 'laws' of capital). Clearly, followers of Bhaskar are likely to argue that social scientists need to study all three levels of 'reality'. In this sense, there might be a *rapprochement* between such diverse fields as Marxist approaches and humanistic approaches. The former might be primarily interested in fundamental mechanisms while the latter might be mainly interested in the world of experience. If this is so, adopting both approaches might be sensible in that they could be conceived of as complementary. Of course, some people do not accept this proposition. Fincher (1983), for example, has claimed that the epistemological and political characteristics of Marxism distinguish it from other forms of inquiry and preclude its combination, on a piecemeal basis, with other theoretical perspectives. Nevertheless, there is a genuine interest on the part of many human geographers in philosophical debates with the rest of social science and a willingness at least to consider a wide variety of views. Golledge (1983: 58) summed up the point well when he pointed out that 'perhaps the fundamental mistake of many who search for theory in social science is to assume that all aspects of social science are subject to formalization as theory'. Theory building in social science might be very different from theory building in the natural sciences.

The interrelatedness of behaviour and structure

One of the theoretical approaches in social science that has been of considerable interest to geographers is Giddens's (1979, 1981, 1984, 1987) idea of *structuration* (see Chapter 3). Gregory (1978) did much to turn geography's attention in this direction when he argued that spatial structures (the way we organize the physical environment) and social structures (the way society is organized) are inextricably linked. To Gregory, the spatial structure in an area is not merely the container within which social forces (e.g. class conflict) are played out. It is also a means by which those social forces are actually constituted. Class conflict, for example, often finds expression through spatial structures (e.g. the segregation of land uses within the physical environment). Therefore, 'spatial structures cannot be *theorized* without social structures, *and vice versa*, and . . . social structures cannot be *practised* without spatial structures, *and vice versa*' (Gregory 1978: 121, quoted in Soja 1989). In short, it is impossible to differentiate social relations from spatial tructures: the one moulds the other (Johnston 1991).

Since Gregory's book, the relevance of structuration to geography has been explored by Dear and Moos (1986), Moos and Dear (1986), and Thrift (1983, 1985). The basic features of structuration have been set out by Dyck (1990). Basically structuration theory is concerned with the problem of how to theorize about human agency in the analysis of economic and social change. In a fundamental sense, then, structuration theory is concerned with the interplay of structure and agency. 'Structuration theory suggests an inherent spatiality to social life and permits context . . . to be placed at the centre of analysis' (Dyck 1990: 461). To advocates of structuration theory, society exists neither independently of human activity nor as a product of human activity. Rather the two are linked in a recursive manner such that a change in one triggers a change in the other, followed by a subsequent change in the former, and so on.

The *context* of behaviour, both in a temporal and a geographical sense, is therefore very important to any explanation of that behaviour. Moreover, structuration theory holds that the context that matters is very much a social construction. To understand that construction, researchers therefore need to adopt a hermeneutic approach so as to see the world through the eyes of those whose behaviour is under study. In the words of Dyck (1990: 461):

> Structure, rather than a model constructed by observers, is understood as consisting of 'rules and resources', which only exist temporally when 'presenced' by actors; that is, when drawn upon as stocks of knowledge in day-to-day activity. Thus structure only exists through the concrete practices of human agents, who reproduce social life through their routinized day-to-day encounters.

In other words, humanly constructed structures constrain behaviour which itself subsequently alters the form of those structures and their interpretation. In this context, human agency is not simply the behaviour that people undertake but their very ability to do things. It is therefore a broad concept. It covers potential as well as activity. It is human agency that underpins much of the taken-for-granted world; what is taken for granted is very much influenced by what people are capable of doing, irrespective of whether they actually do it. Many constraints (e.g. social mores) do not have to be spelled out to be effective. Their impact is simply accepted. In other words, it is human agency that influences how the meaning of the environment and of social structures generally is built up, how it is challenged, and how it is negotiated. Although individual actors may be able to make a difference in the world, the existence of social structures and the meaning given to those structures, constrains the actors, not just as individuals but as totalities (Dyck 1990: 462).

All human activity occurs in a context and any understanding of that behaviour therefore demands an appreciation of the context. 'The contextualization of human action in time and space is central to the linking of structure and agency, for it is context which delimits and shapes the parameters of social action' (Dyck 1990: 462). Deep down, the notion of context has a certain resonance with the ideas of time geography, not least because both sets of writings stress the interrelated nature of time and space. However, what sets structuration apart is the notion of human agency and the recursive way in which agency and structure modify each other.

One of the things that makes Giddens's writings appealing to geographers is the fact that they place 'space' at the centre of social theory (see Chapter 3). No longer are spatial structures merely a by-product of essentially social, economic, and political forces. Above all, structuration theory encourages the study of *locales*. Locale in this sense does not simply mean 'place'; rather it refers to the common awareness that actors have of the meaning of specific social and spatial structures.

Structuration theory is a very ambitious theory. It is a very broad theory (see Cohen 1989; Held and Thompson 1989). It is also relatively new. Its appeal to geographers is self-evident. Dyck (1990), for instance, has used it to advantage to study the development of the notion of motherhood in a Canadian suburb and to show how the practices of motherhood were built up, interpreted, and negotiated in response to economic and social change. Others have been more critical. Gregson (1987), for example, has argued that it is very difficult to translate structuration theory into propositions that can be analysed empirically. Similarly, Johnston (1991: 238) has observed that structuration theory is more concerned with the nature of social science than with providing methods by which facts can be obtained to elucidate the nature of social life. Certainly, structuration theory does seem to stimulate debate that is one step removed from empirical investigation.

There is of course nothing wrong with reflecting on the general nature of social science inquiry, especially if such reflection reveals how human geography can be brought into a more central position in the social sciences. Moreover, such speculation is not a new activity. For example, for several years geographers have shown an interest in critical theory as developed by Habermas (1972). According to this, the study of social change should proceed by reconstructing the conditions that made the activity in question possible. This, in turn, necessitates consideration of the way in which structures, including spatial structures within the physical environment, constrain human activity. More recently, Agnew and Duncan (1989) have looked at the centrality of human geography within the social sciences. In particular, they examined the possibility and importance of bringing together what can be called the geographical and sociological 'imaginations'. 'The geographical imagination is a concrete and descriptive one, concerned with determining the nature of and classifying places and the links between them. The sociological imagination aspires to the explanation of human behaviour and activities in terms of social-process abstractly and, often, nationally construed' (Agnew and Duncan 1989: 1).

There is no reason why these two 'imaginations' should be kept separate. Indeed, their separation is merely an artefact of the institutionalization of knowledge (the existence of different geography and sociology departments within universities). Agnew and Duncan argue that *place* can provide a focal point for the coming together of these two 'imaginations'. The importance of looking at how the meaning of place develops is of course very high on the agenda of the new cultural geography (Jackson 1989; see Chapter 5). According to this school of thought, there is a pressing need to look at how environmental meanings are contested and negotiated. This is obviously an integral step in the examination of how structures are reproduced through everyday action, often through things as basic as the use of language (Pile 1990). However, the need to understand socially constructed meanings extends beyond place. Harvey (1990), for example, has discussed how general notions of space and time are themselves socially constructed. In Harvey's (1990: 418) view, conceptions of space and time are inevitably 'contested as part and parcel of social change, no matter whether that change is superimposed from without (as in imperialist domination) or generated from within (as in the conflict between environmentalist and economic standards of decision making)'.

Postmodernism

Predictably, given his leaning towards Marxian analysis, Harvey (1990) sees the roots of conceptions of space and time as lying in the mode of production and the concomitant social relations. Others have suggested that we may need a completely new approach to thinking that takes account of the fact that social change might be leading advanced Western society into a new epoch. Of particular interest here is the idea of *postmodernism*. Western society has undergone major transformations in the past. Tuan (1982), for instance, has drawn attention to the transformation that occurred about the time of the Renaissance when society changed from a community orientation (where group behaviour was rewarded) to an individual orientation (involving a substitution of competition between individuals for the community allegiance that had existed hitherto). The change was a profound one. Many authors have suggested that a similarly profound change is currently taking place (Featherstone 1988; Lash 1990; Turner 1990).

The term used by many of these writers is *postmodernism*. Unfortunately the term is a very vague one that is open to a variety of interpretations. Jackson (1989: 175) cites Featherstone's (1988: 15) quotation of a dictionary definition of postmodernism: 'The word has no meaning. Use it as often as possible.' The term 'postmodernism' was probably first used widely in architecture and design (see Jencks 1985). In this sense, 'Post-modern design is anti-elitist, aiming to appeal to a range of tastes including the vernacular. It employs a complex language, with heavy metaphoric content, drawing on previous styles and subverting architectural conventions by self-conscious use of allusions, visual parody, and wit' (Jackson 1989: 176). However interesting such design fashions might be, it was the attempt to link these developments to wider changes within society that really drew the attention of social scientists to postmodernism. If there was a new fundamentally different style emerging in architecture, then maybe that style reflected, in some way, the nature of late capitalism (Jackson 1989: 176).

In many ways, postmodernism began as a critique of the modernist movement in architecture and planning. In the view of Beauregard (1989: 381), modernism was characterized by a desire to '(a) bring reason and democracy to bear on capitalist urbanization, (b) guide state decision-making with technical rather than political rationality, (c) produce a coordinated and functional urban form organized around collective social goals, and (d) use economic growth to create a middle-class society'. As capitalism ran into problems in the 1980s, and as the need for economic restructuring became very urgent throughout the advanced Western world, so the nature of modernism came in for criticism (see Gregory 1989a, b). Quite simply, modernism seemed ill equipped to handle the crises of capitalism that were becoming apparent. The origins of postmodernism were therefore radical in the sense of looking at the roots of social problems and trying to discern what approach would increase understanding of those problems (and thereby their solution).

The features that many social scientists deemed were characteristic of advanced Western society in the 1980s (hypermobile capital, concentrations of advanced services, juxtaposition of wealth and poverty, decentralized cities, and customized production complexes, see Beauregard 1989) tended to be bracketed together as the hallmarks of postmodernism. The search was therefore on for a theory of postmodernism that would account for all these phenomena. Given this ambitious and very general goal, it is not surprising that postmodernism has, at times, appeared to be all things to all commentators. Dear (1986), for example, pointed out that postmodernism could be viewed as style, as method, or as epoch: as style it refers to what came after modern architecture (designs including paradox and irony and the juxtaposition of contrasts); as method it involves *deconstruction* (a method of analysis of literary texts that questions whether there are deep structures in the mind that determine the limits of intelligibility); and as epoch it signifies a radical break with the world as it has developed in the last couple of decades.

Fundamentally, postmodernism is against the generalizing approaches of modernism and against the search for universal truths (see Dear 1988; Johnston 1991). Instead, postmodernism celebrates variety and variability. Supporters of the idea of postmodernism argue that there can be no overarching theory of how society works because how society works will vary from place to place. An interest in postmodernism therefore puts the local dimension back on the geographical agenda (Cooke 1990). In other words, both social structures and human behaviour are context dependent. Postmodernist thinking in geography has probably been taken furthest by Soja (1989). He argued that geographical patterns should not be seen merely as a reflection of social processes. Geography itself matters. Place matters. Soja therefore focused very much on the *spatiality* of social life. In Soja's (1989: 79) words, 'It is necessary to begin by making as clear as possible the distinction between place *per se*, space as a contextual given, and socially-based spatiality, the created space of social organization and production.'

Spatiality is not just context; it is a social construction that covers the way in which meaning is invested in particular geographical locations. Soja (1989: 80) drew an important corollary from this: 'Once it becomes accepted that the organization of space is a social product – that it arises from purposeful social practice – then there is no longer a question of its being a separate structure with rules of construction and transformation that are independent from the wider social framework.' He then went on to formulate a very interesting argument:

1. Spatiality is a . . . recognizable social product . . . which incorporates as it socializes. . . .
2. As a social product, spatiality is simultaneously the medium and outcome . . . of social action and relationship.
3. The spatio-temporal structuring of social life defines how social action . . . [is] . . . made concrete.
4. The . . . concretization process is problematic, filled with contradiction and struggle. . . .
5. Contradictions arise primarily from the duality of produced space as both outcome/embodiment/ product and medium/presupposition/producer of social activity.
6. Concrete spatiality – actual human geography – is thus a competitive arena for struggles . . . (Soja 1989: 129–30).

Soja's argument amounts to an attack on *historicism* (the idea that prediction is achieved by looking at the history of events). Instead, Soja wants social scientists to look at spatiality. In his view, we need to recognize that social processes are differently constituted in different places (see Johnston 1991: 249–51). Predictably, there have been criticisms of this approach. For example, Massey (1991) has criticized Soja for overlooking women, ethnic minorities, and non-Western societies, and Curry (1991) has argued that there is much in postmodernism that is not really new. Others have taken a more equivocal position. Harvey (1989), for instance, has shown an interest in postmodernity but does not go all the way with arguments like Soja's, preferring instead to see postmodernism as linked with the flexible accumulation stage of the capitalist mode of production. In his view, the changes pointed to by proponents of postmodernism, 'when set against the basic rules of capitalistic accumulation, appear more as shifts in surface appearance rather than as signs of the emergence of some entirely new postcapitalist or even postindustrial society' (Harvey 1989: vii).

Despite all the theorizing that has gone on in social science generally, and human geography in particular, inquiry is still troubled by certain enduring problems. Many of these relate to the question of how to study people–environment interaction in such a way as to integrate (1) choice and constraint, (2) process and pattern, (3) the micro-scale and the macro-scale, and (4) behaviour and structure. Moreover these need to be integrated in such a way that the dialectical links become apparent. In the past, friction has existed between the proponents of different viewpoints: the positivists against the humanists, and the behaviouralists against the structuralists. Perhaps the way ahead is not to be disparaging and not to compartmentalize. Rather, more may be gained from attempting to see a plurality of approaches as a basis for complementary perspectives on people–environment interaction. In Part Three of this book we therefore focus on a few issues and explore the interplay of choice and constraint. Although claims for the existence of 'behavioural geography' as a distinct sub-discipline may not have much credence, behavioural approaches generally have wide acceptance in the study of people–environment interaction (Golledge 1985). Such approaches highlight constraints just as much as activities. To give but one very simple example, the built environment is a social construction that constrains the behaviour of the people who live within it to a significant degree (Knox 1987). In order to explore the interplay between choice and constraint, social science needs theories that explain 'paths taken' and 'paths foreclosed' and theories that show how small events are implicated in the creation of large social processes which, in turn, are 'the stuff out of which social structures are ultimately made or broken' (Storper 1988: 165). While not explicitly trying to develop such a theory, the chapters in Part Three seek to advance understanding of selected issues to the point where theory might be better articulated and better informed.

Part 3 Fields of Study

The general principles and character of people–environment interaction have been outlined in Part Two. Part Three looks at how what is known about people–environment interaction can be applied to gain insight into specific aspects of contemporary life in advanced Western society. Part Three is not, however, a 'cookbook' on how to apply any one type of behavioural approach. In line with the overall thrust of this book, Part Three takes a broad view of 'behavioural approaches' and uses the term to denote the entire variety of ways in which humans come to know and interpret the environments in which they operate. Indeed, it is worth re-emphasizing the sentiment that underlay Part Two: a plurality of approaches to the study of people–environment interaction in human geography is not a weakness but a strength of the discipline because such a plurality provides for multiple insights into why the world is as it is.

Just as no claim can be made for the primacy of behavioural approaches over other approaches in human geography, so no claim can be made for the primacy of any one behavioural approach over other behavioural approaches. In what follows, therefore, there is no one theme in the sense of a technique or philosophical orientation that is advocated above all others. In fact, the behavioural approaches to be considered range over both positivistic research and humanistic attempts to understand and interpret the experiential environments within which people find themselves living.

In order to appreciate the significance of behavioural approaches in understanding people–environment interaction, several fields of study have been selected for detailed examination. Inevitably, this selection process was subjective. However, we have chosen to focus on what we consider to be five fundamentally important components of human existence. These aspects of life are particularly worthy of attention at a time of rapid social, economic, and technological change. The five topics are jobs and work (Chapter 7), housing and migration (Chapter 8), shops and shopping (Chapter 9), leisure and recreation (Chapter 10), and belonging and well-being (Chapter 11). In other words, Part Three can be thought of as relevance-led rather than literature-driven. We have chosen topics because of their intrinsic importance, not because of the amount that has been written about them.

Much of the attention is the ensuing five chapters is on evaluating the relative autonomy of people in their everyday lives and on teasing out the nature of the balance between choice and constraints in everyday activity. Obviously, the format of each chapter varies a little, reflecting the varied nature of the research that has been conducted to date in each field of study. Nevertheless, there is common ground among the five chapters in that each looks at four things: (1) the salience in everyday life of the particular issues under consideration; (2) the key actors and decision-makers involved and the role of choice in their behaviour; (3) the constraints that limit behavioural freedom; and (4) the gaps that exist in knowledge and the research questions that still need to be answered.

In addition to this common focus for each chapter, there are certain engaging themes that recur continually throughout Part Three. First, all the chapters stress the very rapid rate at which change is occurring. Examples are to be seen in the speed of onset of economic restructuring (and the associated rise in unemployment), housing crises and the emerging problem of declining house affordability, the relatively recent emergence of new retail facilities (e.g. retail parks, hypermarkets), growth in the amount of leisure time ('enforced' leisure in the case of the unemployed), and the rapidly changing social, economic, and political character of cities.

A second theme running throughout Part Three concerns the need for multiple perspectives on why things are as they are. The traditional approaches adopted by human geographers in the study of people–environment interaction (e.g. the gravity model and formulations based on the notion of economically rational behaviour) are shown to be inadequate. Likewise the study of spatial structures (e.g. factorial ecologies of cities) are not enough of themselves to explain why certain geographical patterns have emerged. Structuralism, too, does not take cognizance of the rich variety of human behaviour. To understand fully why things are as they are, researchers need to take stock of all these approaches but also need to pay attention to people's actual (cf. idealized or assumed) behaviour.

This raises the third general theme. In looking at why people interact with the environment in the way that they do, it is necessary to ask where people get information about the environment, how they evaluate that information, how they build up environmental images, what motivates them, how they learn from their own experiences and those of others, and how they actually go about making decisions as to where to live, work, or play. It is also necessary to appreciate that individuals are not sovereign decision-makers and that the choices that they make are very often highly constrained choices. Moreover, the relationship between people and their environment is a reciprocal one: human beings make the environment what it is and then the nature of that environment acts back on them to influence their subsequent behaviour. Many constraints are of course socially constructed. As a result, an implicit theme in much of Part Three is that any concern for people–environment interaction must pay due attention to social groups and to the way in which intersubjectively shared meanings are developed. For instance, jobs can influence social status, housing may be important in self-identity, retail developments sometimes take on the role of 'spectacle', leisure is often a social construct whose meaning is contested and negotiated, and 'place' can serve as a retreat and haven in a turbulent world, or as an enclosure and control mechanism in human affairs. The way in which such meanings develop is influenced significantly by the social groups and sub-cultures to which people belong.

The final theme that runs through Part Three relates to the fact that the study of people–environment interaction is not merely an academic exercise but rather is an activity with very real value in the policy arena. Regional development policy, housing programmes, retail provision, recreation and tourism planning, and neighbourhood design, urban renewal, and community development are all areas of activity where success rests upon a sound appreciation of the complex nature of human behaviour.

Ultimately, the challenge that confronts human geographers is to understand why patterns of societal development take the form that they do at particular points in time and space. In order to achieve this understanding, human geographers need to know a lot more than they presently do about the detail of people–environment interaction.

7 Jobs and Work

One of the most important aspects of everyday life is that related to paid work. Such work contributes much more than money to a person's existence. For example, the type of job held by an individual has often been taken in social science as an indicator of overall well-being. In other words, the holding of a certain job or position can contribute significantly to the definition of 'self' and to the way in which the community at large treats an individual. For this reason, and because of basic economic factors, jobs and work are topics well worthy of study by human geographers. This is particularly the case at a time of rapid economic change when alterations in both the nature and availability of work are having profound effects on the lives of many people throughout the world. In order to appreciate the importance of jobs and work, it is essential to appreciate the nature of contemporary economic change because the economic behaviour of individual actors is ultimately very dependent on the context in which they find themselves operating.

The literature on what is widely referred to as 'industrial restructuring' is enormous. The nature of the changes involved has been clearly outlined by Van der Knaap and Linge (1989). The period from the 1944 Bretton Woods agreement on exchange rates to about 1970 is often known as 'the Long Boom' because it was characterized by considerable overall economic growth in advanced Western economies. This economic growth reached its peak in the 1960s, a time characterized by the rise of multi-plant, multi-product firms. This was a time therefore when the organizational structure of firms became an important influence on where various activities of the firm were located (Van der Knaap and Linge 1989). In the 1970s things began to change. Mergers were common as firms sought either to diversify or increase market share. In the late 1980s and 1990s this process ended. The trend nowadays is for firms to concentrate on so-called core activities in which they have long-standing experience. Moreover, the onset of a global economic recession in the mid–1970s encouraged firms to adopt risk-avoidance and risk-sharing behaviour. This fostered extensive subcontracting arrangements whereby firms shed what were deemed to be non-core activities.

One of the most obvious manifestations of industrial restructuring in countries like the UK, the USA, and Australia, has been job loss and unemployment. Economic recessions, like the one that began in the mid-1970s, bring downturns in demand. Firms cope with this through a variety of short-term strategies (e.g. stockpiling, laying off labour) and long-term strategies (e.g. reorganizing production), the end result of which is job loss (Fig. 7.1). This has led to very high levels of officially recorded unemployment. Official unemployment rates are of course notoriously unreliable: discouraged job seekers may not register as unemployed with the result that official statistics underestimate the 'true' level of joblessness; conversely, many workers may have paid employment in the so-called 'black economy' that evades taxation, government legislation, and official notice. The Spanish Ministry of Economy, for instance, has estimated that 30 per cent of the workforce has jobs in Spain's 'black economy' (Van der Knaap and Linge 1989). Informal work is therefore a topic well worthy of investigation.

Sharpe (1988) has suggested a typology of work

7.1 Strategies of multi-plant firms faced with falling demand and/or non-profitability (Source: Thrift 1979: 147)

Reduce labour force, non-replacement of labour, call in sub-contract work, etc.

Stockpile

Fall in demand and/or non-profitability

JOB LOSS

Short term

Long term

Numerical control, automation, standardization, etc.

Yes

Closure

No

Product concentration, relocation, concentration of production

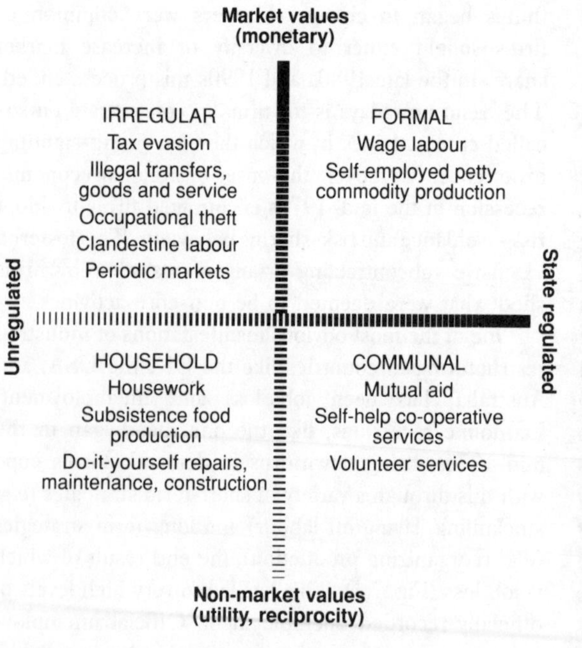

7.2 Categories of work (Source: Sharpe 1988: 317)

that stresses unpaid activities as well as the wage economy (Fig. 7.2). The key dimensions that Sharpe adopts for the typology are (1) whether the work is state regulated or unregulated, and (2) whether the focus is on market (i.e. monetary) values or non-market (e.g. reciprocal help) values. On this basis, a distinction can be drawn between formal work, irregular work (which may encompass up to 35 per cent of the labour force and account for up to 40 per cent of GDP), household work (including 'housework' which is still undertaken largely by women, and do-it-yourself activities, see Pahl 1984), and communal work (especially common in rural and ethnic communities). Sharpe's view is that informal work may be no bad thing if it fosters self-reliance, personal autonomy, self-identity, and the development of personal skills. The fact remains, however, that many people seek formal work and are 'trapped' in unemployment as industry either closes or moves elsewhere (Thrift 1979). Although this entrapment is very evident in inner-city areas, there is a wider dimension that is clearly obvious. Green (1986), for example, has shown that the chance of joining the ranks of the long-term unemployed varies between the different regions of Britain. In particular, a relatively active labour market in the South in the mid-1980s contrasted with a sluggish labour market in the North.

De-industrialization lies behind the high levels of unemployment. The term 'de-industrialization' is, however, used in a variety of ways (see Massey 1988). In its simplest form, it is best defined as a relative or absolute decline in manufacturing employment. In this sense, de-industrialization may well be associated with 'jobless growth', the situation where production is increased without an increase in employment. The precise nature of the job loss resulting from de-industrialization has been explored by Massey and Meegan (1982). They found that there were distinct variations in job loss between the regions of Britain and that four factors influenced the regional shifts in employment: labour costs; access to local markets;

corporate form and size of enterprise; and the age of the capital equipment. Firms caught in a cost–price squeeze and recession can respond in one of three ways: intensification (increase labour productivity without new technology), investment in technological change, and rationalization (reduction in capacity). The significance of this lies in the fact that the type of strategy adopted influences the resultant geographical impacts. Intensification leads to unemployment but no plant closure. Rationalization involves both. Investment, too, often has distinct regional patterns. For example, investment in the 1970s in mass production techniques in the iron-castings industry in Wales saw productivity per worker increase by 50 per cent. This contrasts with the North where there were many complete plant closures on account of the preponderance of small jobbing foundries that provided little scope for increased productivity through mass production (Massey and Meegan 1982). Other industries to have undergone rationalization are cotton weaving, machine tools, and electrical machinery making. Intensification is characteristic of the footwear industry and textile finishing.

It would be wrong to paint too bleak a picture of the industrial performance of advanced economies in the 1980s. Certainly some growth has been evident (see Harris 1988). New small firms have continued to be created, admittedly with high failure rates. In particular there have emerged new forms of small business activity (franchising, co-operatives, networks, management buy-outs), often involving entrepreneurs from groups not traditionally heavily involved in business (e.g. ethnic minorities and women) (Mason and Harrison 1985). Keeble (1990) has identified five reasons for the development of new small firms:

1. Recession-push mechanisms whereby entrepreneurship is provoked by unemployment;
2. The fragmentation of large firms and the tendency for firms to subcontract business services that were hitherto provided internally;
3. The customization and specialization of high-income consumer demand and the emergence of 'niche' markets;
4. The emergence of innovation opportunities resulting from technical advance;
5. The promotion of an 'enterprise culture' on the part of government.

This growth has, of course, been concentrated on certain sectors of the economy. In the USA, for example, 80 per cent of new jobs created in the 1980s were in retailing, health, and business services. Moreover, more than 70 per cent of the US workforce are employed in 'services' generally (Christopherson 1989).

Reasons for economic change

Advanced economies have obviously changed considerably in the 1970s and 1980s. It is not surprising therefore that a number of attempts have been made to explain these changes (see Harris 1988; M.J. Taylor 1984). Indeed, according to Van der Knaap and Linge (1989: 18), 'It is not only the case of change which is of interest but also the way people react to, cope with and manage its impacts. Little is known about whether people have explicit strategies and how these develop and what their spatial implications are.'

Some researchers are proponents of *long-wave theories*, most notably the 50-year cycles of expansion and contraction associated with Kondratiev. Hall and Markusen (1985), for example, interpret the develop of hi-tech industry in areas like Silicon Valley as a manifestation of the 'fifth Kondratiev' (see also Hall *et al.* 1987). In this sense, then, the growth of industry centred around information processing will mark the fifth upswing of the international business cycle (the original one being the Industrial Revolution in Britain in the 1780s). This approach has, of course, critics. Some argue that its treatment of time is simplistic (i.e. it reduces everything to neat cycles) with the result that the behaviour of individual firms and industrialists is overlooked. Instead the people actually involved in economic enterprises are 'captured by time' (Barnes and Curry 1988).

Other writers have promulgated *evolutionary models of the nature of economy and society*. The best known here is Bell's (1973) argument that advanced Western society is developing into a 'post-industrial society' as service employment comes to dominate. To be specific, Bell argued that as incomes increase, so the demand for services will increase, leading to this sector being the dominant form of employment. The argument has been extended by B. Jones (1982) to include the notion of a 'post-service society' beyond 'post-industrial society' but challenged by Gershuny (1978) who argued that the growth of service employment is not a sign of an emergent post-industrial society but an indication that, as the technical and organizational structure of production becomes more complex, an increasing number of service workers are required to maintain goods production. It must also be remembered that Bell wrote during the era of post-war economic prosperity and did not foresee the collapse of work in the Western world in the 1970s and beyond (Allen 1988).

A third group of researchers has focused on *world system theories*, notably the idea that there has emerged a *new international division of labour* (Fröbel *et al.* 1980). The argument here is that the economic order that prevailed throughout the post-1945 boom has disintegrated (Thrift 1986). As a result, multinational corporations have tended to concentrate their top management (those engaged in goal determination and corporate planning) in core metropolitan areas, often known as 'world cities'. In contrast, production plants tend to be spread globally and focused on locations where access to raw materials, cheap labour, or government incentives guarantees low production costs. This trend is very evident in industries such as footwear, apparel, and electrical goods (see Graham *et al.* 1988). Such an international division of labour has been facilitated by the emergence of multi-product, multinational corporations and by technological change (especially improved communications) which has permitted the fragmentation and separation of different stages of production in different places (Pinch 1989). The argument is, then, that an international division of labour develops on account of the ability of multinational corporations to command capital and technology and to rationalize their use on a global scale (Morris 1988b: 337). Reality, is course, more complex than simple models. Teulings (1984), for instance, has shown that, in the case of the electronics giant Philips, the flow of capital to the Third World is partly compensated for by a flow of capital *to* countries in the core of the world economy, largely to ensure access to big markets and as a strategic response to the increasing dependency of multinational corporations on the 'backing' of government. Despite this, the notion of a new international division of labour remains a powerful one.

A fourth group of researchers adopts a Marxist perspective on the causes of economic change and attempts to translate ideas on capital into the specifics of everyday reality (M.J. Taylor 1984). This work has concentrated very much on labour markets (see Braverman 1974) and on what are known as *regulationist theories*. Briefly, these focus on the way in which the labour market is regulated. Usually capitalist economies are interpreted as being marked by a succession of distinct historical stages which differ from each other in terms of the organization of production (Harris 1988). The clearest division is usually that between mass production, assembly line techniques (commonly termed 'Fordist' production after Henry Ford's pioneering initiatives in the motor vehicle industry) and contemporary moves to flexible production and customized (i.e. short-run) products (commonly termed 'post-Fordism'). These terms (Fordism and post-Fordism) have in fact gained wide acceptance and are used nowadays by many researchers who do not subscribe to the regulationist school of thought.

Finally, there is a group of researchers who argue that the precise nature of industrial change can only be understood by looking at *the behaviour of individual firms*. This may necessitate the descriptive monitoring of trends in industry and conjecture about what underlies change, or it may involve consideration of how various enterprises and organizations come to terms with the overall political, social, and economic environments in which they operate. These researchers aim to show that 'individual decision-making units with some degree of discrimination still lie behind the overarching laws of capital accumulation, notwithstanding the imperatives of profit making and the prevailing economic climate' (M.J. Taylor 1984: 264). Clearly, it is impossible to explain behaviour without consideration of the environment within which behaviour is conducted. However, it is the contention of this last school of thought that behaviour is highly variable and cannot be reduced to simple assumptions.

No one mode of explanation has 'a monopoly of insight' into the nature of economic change (Chapman and Walker 1987: 28). Indeed, some of the most engaging and intellectually stimulating debates in the whole of human geography have occurred in what is known as 'industrial geography' in recent years. Among other things, these debates have focused on the relative autonomy of actors and on the influence of environment and economic structure on behaviour. The debates also relate to the linkage between industrial change and regional economic growth and to the vexed question of what can be done to arrest economic decline and to ameliorate its detrimental consequences. In short, geographical studies of jobs and work are of theoretical and applied value. To some extent, of course, this has always been the case because it has been in relation to the location of industry that geographical thinking has often reached its most sophisticated level.

Industrial location theory

Location theory has traditionally been the cornerstone of geography's respectability as a social science in so far as its nomothetic emphasis stood out for many years in stark contrast to the rest of an essentially descriptive discipline. Predictably, models of location have taken many forms that vary from the least transport cost solution advocated by Weber (1929) and subsequently elaborated upon by Hoover (1937), to the market area models of Lösch (1954), and the locational interdependence emphasis of Hotelling (1929). In essence all these models are normative and draw on the neo-classical micro-economic theory of the firm (Pocock and Hudson 1978: 11). Indeed, Moses (1958) and Smith (1971) did much to integrate location theory with marginalist economics, despite the fact that, unlike transport costs (which are regularly if not linearly related to distance), the costs of many factors of production (e.g. interest rates on borrowings) do not vary between different locations to any significant degree (Lever 1985).

Central to all these economic models has been the notion of profit maximization: to Weber maximum profit could be equated with minimum cost while to Hotelling and Lösch maximum profit equalled maximum sales. However, a major problem in applying the models occurs whenever an attempt is made to transpose the theoretical ideal of profit maximization to some real-world profit function. This is because some firms do not appear to attempt to maximize (see Wolpert 1964) while others consider non-economic factors such as technological supremacy (see Thomas 1980) and, in any case, make mistakes (Webber 1969). In other words, there is a great gulf between the theoretical ideal of location models and the variety of real-world conditions (Lever 1985). Location theory models, for instance, tend to overlook a number of major developments that characterized the world economy in recent years: the adoption of mass production and the consequent expansion in the scale of plant; the increase in the capitalization of industry; the great growth in the size of firms; the evolution of multi-plant, multifunctional and often multinational corporations; the agglomeration of industry in major city systems; inflation; technological advance; and international trading agreements (Clarke 1985; Hamilton 1974, 1978). Additionally, traditional models make the unreal assumption that spatial behaviour can be differentiated from general economic behaviour.

Coupled with this disquiet about the lack of reality of traditional location theory has been a concern with whether or not the theory is internally consistent and with the fundamental epistemological problem of whether human behaviour can be reduced to the form of an abstract model (Massey 1979). It is not surprising therefore that calls have frequently been made for a form of location theory that emphasizes real behaviour, decision-making, and the organization of business enterprises (Hamilton 1974). Lever (1985) has in fact spelt out five components to the rationale for such a behavioural approach:

1. Dissatisfaction with the assumptions of rationality on which traditional theory is based;
2. Recognition that firms are often big enough to control the price of some inputs (e.g. labour), thereby putting a new slant on the minimization of costs;
3. Awareness that prices and costs vary so slightly from place to place as to allow scope for consideration of non-pecuniary factors in location decisions;
4. Acceptance that government policy has a critical influence on industrial location but is at the same time a variable outside the scope of classical theory;
5. Familiarity with a growing body of management studies that suggest that firms pursue goals other than profit maximization.

Behavioural approaches to the study of industrial location

One of the earliest and most cogent criticisms of normative location theory came from Pred (1967) who disputed the logical consistency of the assumptions involved in traditional formulations, questioned the motives of optimizing behaviour, and rejected the knowledge levels and mental acumen required in economic behaviour. In particular, Pred argued that the concept of rational economic beings ignores satisficing behaviour and non-economic considerations. He also demonstrated that the requirement that firms outguess competitors but not be outguessed is unworkable under conditions of imperfect competition because, under such conditions, what is optimal for one firm depends on the actions of other firms. Likewise, other writers have suggested that, instead of maximizing profits, firms may tend to minimize uncertainty by repeating previous behaviour, by locating in areas with which they are familiar, and by imitating the behaviour of competitors (Lloyd and Dicken 1972: 157). All this implies that firms should be viewed as organizations existing in, and adapting to, economic and socio-political environments from which they derive information (Gold 1980: 219–20). Thus, in Pred's view, every decision-making unit concerned with economic activity can be placed conceptually within a behavioural matrix, the axes of which measure the quality and quantity of information available to the decision-maker and that decision-maker's ability to handle information. Within such a matrix, traditional economic rationality represents but one out of a wide range of positions (see Fig. 4.9).

Of course not all authorities subscribe to the view that a new behavioural location theory is needed. Some support modifications of the classical approach. For example, Webber (1971, 1972) has argued that much criticism of location theory is harsh because traditional formulations like central place theory have never been tested empirically in that no attempt has ever been made to control the environmental assumptions of the theory. Perhaps the theory is untestable by empirical means because its assumptions scarcely even approximate thc real world. Similarly, Dicken (1977) has pointed out that least-cost location is still important for large multi-plant business enterprises, albeit on a much enlarged geographical and organizational scale. In a somewhat different vein, Smith (1979a: 54) has suggested that a proper perspective on industrial location requires 'a broad view of the production process that stresses the satisfaction of human needs in a broad societal context and recognises the distribution of those things on which human well-being or ill-being depends'. A similar sentiment has been expressed by Linge (1988) who argued that it is increasingly anachronistic for researchers to focus solely on manufacturing industry because a coherent view of economic change also requires consideration of governmental, social, cultural, individual, and household impacts and responses.

Despite such initiatives it remains true that traditional location theory-based approaches ignore the way firms interact with their uncertain environment, overlook considerations of information availability, and neglect the fact that organizational goals are many and varied (Keeble 1976: 3). Indeed, in the view of Townroe (1975: 33), the neo-classical value tradition in economics, on which much location theory is based, has afforded researchers 'with opportunities for both increasing abstraction and mathematical elegance as well as for increasing irrelevance and a widening gulf between theoretical reasoning and empirical progress'. To Townroe, the alternatives to traditional theory appear to lie in the use of operations research techniques that rely on optimizing procedures like linear programming, a managerial approach that emphasizes organizational structure and good management practice, and a behavioural approach that highlights the theory of the firm, information processing, decision-making, and perception. The last two of these alternatives have attracted the greatest attention, no doubt reflecting the fact that the majority of industrial location decisions are both planned and non-optimal (Stafford 1972: 205). However, for the most part empirical investigations of real-world decisions have taken place after the event and have been structured around standard economic variables. Only occasionally has attention turned to the simulation of location decisions (see Hart 1980) and to in-depth interviews designed to uncover decision processes (Stafford 1972). The behavioural approach has nevertheless considerable potential, particularly if location decision are studied holistically as part of

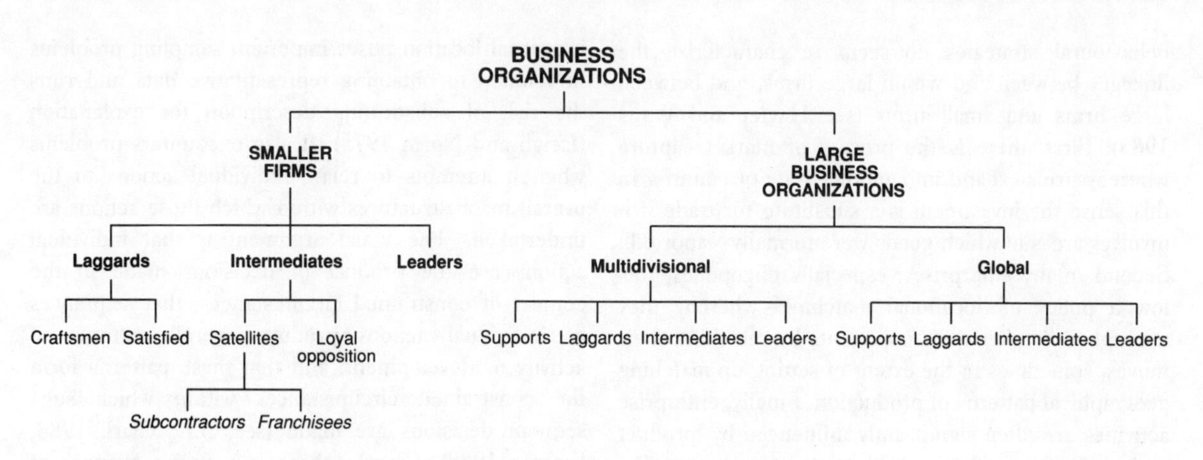

7.3 The segmentation of business organizations (Source: Taylor and Thrift 1983b: 452)

overall investment strategies and if they are treated in such a way as to recognize the importance of considerations such as liquidity, solvency, profitability, and the availability of capital (Carr 1983).

A geography of enterprise

A plea to understand how firms come to terms with the environment in which they operate has led to the development of what is known as a 'geography of enterprise'. This field of study focuses on the power relations between large and small firms (Chapman and Walker 1987; Watts 1987). This is an important issue because industrial geography 'has never dealt adequately with the competitive and controlling relationships of business organizations' environments and the asymmetry of both inter- and intra-organization relationships has never been recognized, let alone specified' (M.J. Taylor 1984: 269). Fundamentally, an enterprise approach to industrial geography seeks to re-establish business organizations 'as *active* agencies of change and not simply as *reactive* cogs in the machinery of macroeconomy' (Taylor and Thrift 1983a: 1287). In order to gain insight into the power relations between enterprises, Taylor and Thrift (1983b) have devised a typology of business organizations so as to point out that different sorts of organizations behave in different ways (Fig. 7.3). Their argument is that, since the mid-nineteenth century, the main changes to have occurred in capitalist economies have arisen as a result of a process of internal differentiation or 'fission' as firms responded to new competitive pressures (Taylor and Thrift 1983b: 450). Thus, particularly since the 1960s, there have developed multi-product, multinational corporations.

Exactly how these divide up their activities and where they locate their various undertakings has a major influence on the geography of economic development. Because such activities are not evenly spread, multinational corporations can contribute significantly to uneven development (Hymer 1976). However, care must be taken not to generalize too widely. The differences between multinationals are significant. Dicken (1988), for example, has looked at Japanese direct foreign investment and shown that the organizational structure of such investment tends to vary considerably between Asia and North America; in Asia complex intrafirm divisions of labour have developed whereas in North America, Japanese plants tend to be market-oriented and established primarily to overcome trading problems.

The typology developed by Taylor and Thrift is a relatively simple one. And of course a simple typology does not capture the dynamic nature of inter-firm linkages. To some extent this is inevitable. After all, large organizations comprise large numbers of people in different positions with different responsibilities, abilities, and powers, all contributing to the amalgam of corporate activity. Despite this variability, certain behavioural strategies do seem to characterize the linkages between and within large firms, and between large firms and small firms (see Hayter and Watts 1983). First, there is the process of market capture whereby firms expand into another state or country. In this sense the investment is a substitute for trade if it involves areas to which goods were formally 'exported'. Second, many enterprises, especially oligopolies, follow a policy of 'locational matching' whereby they attempt to counter their opponents by matching their moves, sometimes to the extent of setting up matching geographical patterns of production. Finally, enterprise activities are often significantly influenced by 'product cycles'. For example, as products 'mature', there is a tendency for their production to become more capital intensive. This can heighten corporate rivalry and lead firms to attempt to reduce their costs below the level of their competitors by developing locations where labour costs are low and labour productivity high (Hayter and Watts 1983: 165).

In pursuit of these overall strategies, big firms are powerful but not always unified. Trade unions may oppose some moves. So, too, might shareholders, especially institutional investors whose holdings might be sufficiently large to exert a powerful influence. There is no denying, however, that 'the growth of multiregional and multinational firms over the last few decades has been an important and possibly dominating force moulding the nature of interregional and international interdependencies' with the result that intra-corporation flows of technology, investment, services, information, and goods comprise the ties that bind places together (Hayter and Watts 1983: 170).

Clearly, the behaviour of enterprises demands attention. To geographers a behavioural interpretation of enterprises offers a number of attractions: it is more flexible than the traditional approach and it is process-oriented (Leigh and North 1975); it considers both explicit spatial behaviour (e.g. opening a new plant) and implicit spatial behaviour (e.g. altering the scale of existing plant) (Hamilton 1978); it is dynamic and holistic and focuses on organizational adaptation to differing environments (Wood 1975); and it highlights the importance of risk and uncertainty, internally generated constraints on growth and performance, information availability, and images and perception. As against this a behavioural approach to the study of industrial location poses important sampling problems in relation to obtaining representative data and runs the risk of substituting description for explanation (Leigh and North 1975). It also encounters problems when it attempts to relate individual actions to the overall macrostructures within which those actions are undertaken. The usual argument is that individual actions are the product of decisions made in the context of constrained circumstances, that sequences of individual actions produce overall patterns of activity or development, and that these patterns form the constrained circumstances within which subsequent decisions are made (see G.L. Clark 1988; Storper 1988). Clearly, there is a strong element of circularity in such an argument. More fundamentally, the behavioural approach still holds to the view that highly complex human behaviour can be represented by an abstract model at the level of the individual decision-maker. This is a proposition that some would question (Massey 1975a). Despite these problems, the behavioural approach, particularly as it focuses on organizations, remains very strong in industrial geography (Martin 1981).

Organization theory

Within the behavioural approach, organization theory appears to offer a means of understanding non-normative decision-making and learning by firms (Britton 1974). This is because organization theory focuses attention on the very issues that classical location theory has failed to accommodate: the persistent growth of large corporations comprising a multiplicity of branch plants (Watts 1981); imperfect competition, monopoly, and oligopoly; the fact that corporate strategy reacts to things other than the external environment; and the fact that much economic behaviour is non-spatial (Hamilton 1978). In other words, in adopting organization theory geographers recognize that large enterprises in different product markets often have more in common with each other than they have with smaller enterprises in the same market because their sheer size offers them similar amounts of power to influence their external environments (Lever 1985: 24). In this sense, then, a focus on organizations draws attention to the power and dependence relations within as well as between firms (Fredriksson and Lindmark 1979) and therefore goes some way to answering the pleas of those writers who have argued that a full understanding of industrial location can only be achieved if attention is paid to the structure of the political economy within which firms operate (Massey 1975a, 1979). What is required, then, from an organizational perspective, is 'analysis of the evolution, structure, adjustment and goals of organizations engaged in industrial activities to establish how these elements affect perception of the spatial variable in location decisions of any kind and how that spatial dimension has a feedback effect upon organizational structures and their functioning' (Hamilton 1978: 12). In the eyes of many, this end is best achieved through the adoption of a systems theory framework (Hamilton and Linge 1979) because such a framework enables a number of important questions to be asked about organizational behaviour: How and why did the system emerge? Why did it grow more in some directions than in others? What holds it together? And what tension and conflict are there within the system (McNee 1974: 49).

The fruitfulness of this systems framework has been demonstrated both in relation to hypothetical organizations like 'International Gismo, Inc.' (McNee 1974) and in relation to the harshly real world of the British plastics industry (North 1974). The latter study showed how locational decisions range across *in situ* expansion, complete transfer, the establishment of branches, acquisition or take-over, and closure, thereby corroborating a behavioural pattern that has been found elsewhere (Fig. 7.4) (see Rees 1974). The study also showed how the type of decision made tends to reflect the stresses impinging on the organization: for example, unplanned growth tends to lead to relocation, planned growth to the opening of branches, diversification to acquisition and take-over, and downturn to closure.

A major influence on what type of locational decision a firm takes in response to stress is the type of linkage that the firm has with other firms within its environment. Basically such linkages reflect economic considerations such as the type of product, the size of the production centre, the size of the market, the distance between producers and consumers, variations in pricing strategies, ownership ties, and the availability of local external economies (Taylor 1973). However, often overriding these monetary considerations are the behavioural benefits to be gained from the agglomeration of production units, notably the reduction in risk and uncertainty (Taylor 1978a). Likewise, the growth of the firm, the share of the market, the diversification of interests, entrepreneurial satisfaction, and perhaps even simple survival, are behavioural considerations that may weigh just as heavily in decision-makers' minds as more conventional economic considerations.

Indeed, locational decisions may be just as much a response to short-term stress as a part of a long-term economic plan (Hamilton 1974). Growth is, after all, both a way of achieving a further utilization of resources and a way of reducing uncertainty and mitigating the managerial, financial, and locational restraints that influence a firm's behaviour (Hakanson 1979). As a result, an organization theory perspective on industrial location needs to consider the organizational forms which shape the perception, images, and actions of decision-makers, and the interpretation of organizational goals and functions by key individuals within the firm's management structure (Hamilton 1978: 13). In this regard it is important to recognize that the organizational structure of a firm is dependent upon the firm's own technological core on the one hand and the constraints and contingencies of the environment (including technological opportunities and threats) on the other (McDermott and Taylor 1982). Sadly, most industrial geographers have not appreciated this and have adopted concepts from the organizational sciences without sufficient consideration of the specific context in which they were developed. For example, much of the borrowing by human geographers from organization theory has displayed a management bias (Marshall 1982). One such aspect is the idea of the 'flexible firm' (Atkinson 1985).

The flexible firm

The notion of 'flexibility' is a very broad one that covers everything from computer-aided design (CAD) to flexible manufacturing systems and changes in work practices. Moreover, although the pressure for change

7.4 A model of industrial location decision-making (Source: Rees 1974: 191)

Decision-making environment
Cognitive processes
Learning behaviour
External demand potential in sales regions
Internal expansion difficulties
(stress threshold)
Problem recognition
growth, demand
Short-term response
in situ expansion
Long-term response
Relocation
Acquisition
New plant
External search for location alternatives:
1. **Regional** delimitation of macro-scale areas due to **demand** needs.
2. **Community** evaluation by **comparative cost** approach.
3. Selection of few **location sites** on **judgemental** grounds.
Final evaluation for decision
Decision ratification
Post facto rationale
Outcome: resource allocation
Feedback

often has a common origin in market segmentation, job loss, and technological innovation, the way in which different firms adapt to such change, and the degree to which they become 'flexible', tend to reflect pragmatism and opportunism rather than a conscious strategy (Atkinson 1985: 14). Nevertheless, Atkinson has suggested what the structure of the archetypal flexible firm would be like (Fig. 7.5). Briefly, the model proposes a horizontal segmentation of a core workforce (undertaking key firm-specific activities) and a cluster of peripheral groups (which (1) provide extra labour as and when needed for the core (and thereby prevent the need for major numerical fluctuations in the size of the core workforce) and (2) undertake the kind of non-specific and subsidiary activities that all firms generate (e.g. typing, cleaning)). Clearly, the quality of working life may differ significantly between the core and the peripheral workforce to the extent that the former is likely to have more secure employment and better promotion prospects. Although many people would question the generality of Atkinson's model, there is certainly evidence of flexibility in

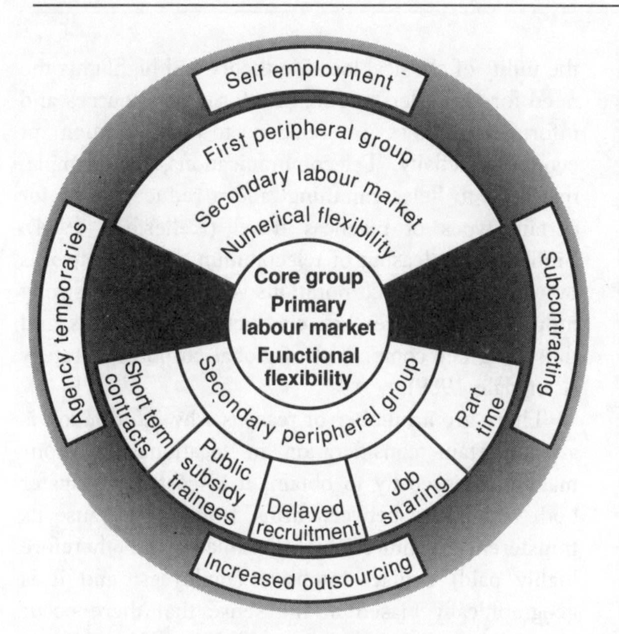

7.5 The flexible firm (Source: Atkinson 1985: 19)

the workplace. In the US economy, for instance, there has been a marked increase in part-time and temporary employment and in subcontracting: 25 per cent of the US workforce is now engaged in such activities (Christopherson 1989). The corollary to this notion of flexible structures is that the decision-making processes of firms must be complex, taking into consideration not only their own 'core' requirements but also their linkages with other firms on which they depend for the undertaking of 'peripheral' functions.

Decision processes within the firm

Rees (1974) has suggested that the process by which firms arrive at locational decisions can be reduced to a number of generalizations. For example, it seems that a firm's first action in accommodating stress at a particular location is to consider *in situ* adjustment. This has been corroborated by Bade's (1983) study of manufacturing in Germany. This found that the profitability of a plant is influenced significantly by factors which hardly vary from place to place. Furthermore, a large number of location-specific considerations are hard to assess. Thus decision-makers tend to approach their task in a defensive and risk-aversive manner which often leads them to see relocation as a last resort. If a move is inevitable, alternatives are narrowed down by a cost analysis of specific sites, although the final decision usually combines both economic and intangible considerations. In this context a decision to set up a new plant is invariably a long-term investment solution to corporate growth and planning problems. Predictably, such decisions are more common in large organizations than in small ones. Of course, in addition to moving, firms can reduce the uncertainty and stress of their current situation by repeating past behaviour or by imitating competitors.

Not all researchers agree, however, that behaviour can be simplified to such rational terms. Stafford (1972: 210–11), for example, has argued that decision-makers within business enterprises very rapidly and very drastically transform the infinite complexity of the optimal location problem to simple, intellectually manageable proportions by not indulging in difficult modes of analysis and, whenever possible, by avoiding arduous negotiation with groups such as unions and governmental regulatory agencies. On occasions information may be withheld from unions as a 'strategic secret' so as to subvert what might otherwise be the legitimate expectations of employees (Clark 1990). Stafford has in fact suggested that the techniques used and the location principles applied in locational decision-making vary with the stage of locational choice, from a simple emphasis on economies of scale through to the maximization of 'psychic' income (Table 7.1). A similar observation that decision-makers are aware of neither the full extent of the relevant information field nor the applicability of criteria for choice has been made by Townroe (1974) who concluded that, although large investments tend towards economic rationality, most firms choose a feasible rather than the optimal alternative out of the choices with which they are confronted. Knowledge and perception of the environment do not seem to be influenced, however, by the size of a firm (Barr and Fairbairn 1978), possibly because small firms can get to know their limited environments very well while big firms can often internalize much of their environment (e.g. having their own research and transport facilities) and thereby reduce uncertainty (Hamilton 1974).

Table 7.1 Location principles

Stage	Appropriate techniques (*most commonly used)	Location principle
I *In situ* expansion	—	Economies of scale
II Delimitation of region for location of new production facility	Regional demand projections*; regional and inter-regional input–output tables; linear programming	Maximum demand
III Selection of finite number of feasible sites	Comparative costs*; linear programming; extrapolation from past experience*	Least cost
IV Final site selection	Comparative costs; judgemental integration of data and attitudes*	Maximize 'psychic' income; minimize 'difficulties' (e.g. unions, governmental regulatory agencies)

Source: Based on Stafford (1972: 212)

Information flows

A dynamic perspective is needed in economic geography because environments change (Gertler 1988a). The goals of firms change. As a result knowledge of the environment can never be complete and comprehensive. New information continually becomes available. Information flows, and communication generally, are in fact central to modern economies (McDaniel 1975). Törnqvist (1977), for example, has suggested that the present-day economy comprises three communications planes: first, the transport of goods; second, face-to-face communications; and third, telecommunications. The current prominence of the second and third communications planes diminishes the utility of classical location theory and highlights the need for an understanding of information sources and information flows in relation to the location of economic activity. Telecommunications, for example, may lead to 'telecommuting' and a reduced need for certain types of business travel (Kellerman 1984). Similarly, the leasing of telecommunications networks by multinational corporations can exert a strong influence on where they conduct their business and thus be a key component of global corporate strategy (Langdale 1989).

There are a number of reasons why information is an important consideration in locational decision-making: it is costly to obtain; it is costly to transfer both within and between firms, primarily because its transference requires highly skilled (and therefore highly paid) contact intensive employees; and it is geographically biased in the sense that there occur locational variations in its availability. In addition, information circulation patterns and organizational locational patterns interact with each other, and feed back upon one another, to influence the nature of urban and economic growth through the creation of non-local multipliers in economic activity and the determination of diffusion paths for growth-inducing innovations (Pred 1973a). As a result, information is important not only in relation to explicit locational decisions regarding the setting up of new plant but also in relation to the sort of implicit locational decisions that are exemplified by such routine tasks as reordering materials when stocks are low. However, it is in relation to explicit decisions that the importance of information is most evident because place-to-place variations in the availability and accessibility of information are increasingly considered as a critical factor of location for administrative headquarters, banks, financial institutions, consultants, and data-processing centres (Pred 1977a: 24).

Törnqvist (1970) was one of the first geographers to stress the importance of information when he drew attention to the significance of contacts between expert personnel in general problem-solving, planning, and reconnaissance. To Törnqvist, many business organizations can be differentiated into operating units (e.g. manufacturing plant) and administrative units (e.g. offices) with the latter being responsible for both routine and non-routine decisions. A somewhat more

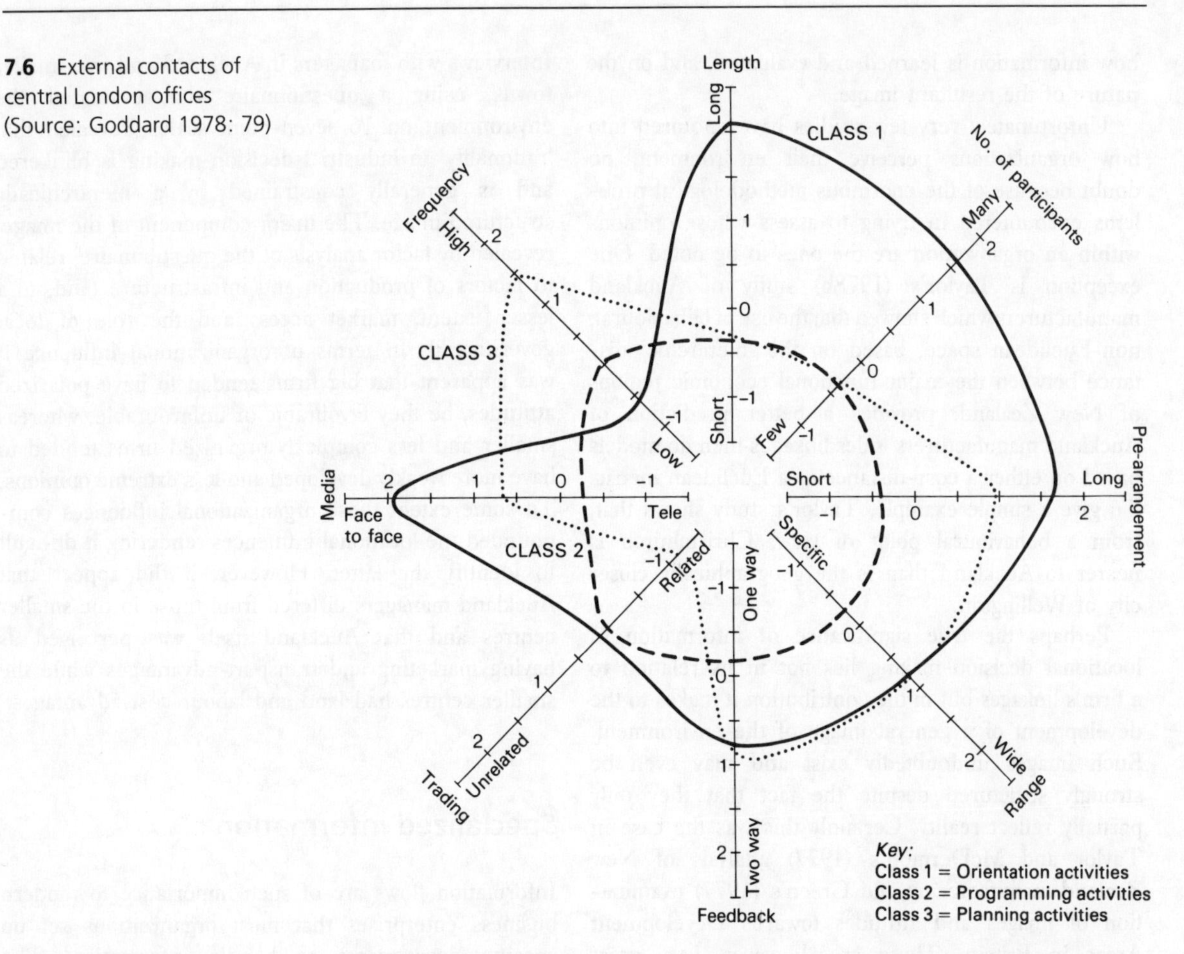

7.6 External contacts of central London offices (Source: Goddard 1978: 79)

detailed classification of the work of such administrative units was provided by Thorngren (1970, 1973) who argued that all organizations operate in a knowledge environment and a values environment and that they come to terms with each of these environments by considering three time horizons:

1. *Orientation activities* deal with the furthest time horizon, focus on things like changing ideologies and possible scientific advances, and usually involve face-to-face, prearranged meetings between key personnel (Class 1 in Fig. 7.6);
2. *Planning activities* are concerned with a medium term time span, and involve middle management in relatively frequent contact (often by telecommunication) in regard to such matters as potential technological innovations, administration, production, and sales (Class 3);
3. *Programming activities* centre around the most immediate time span and encompass largely routine tasks concerned with the day-to-day running of business organizations (see Frey 1981) (Class 2).

Empirical support for this threefold classification has come from Goddard's (1978) study of the external contacts of employees in a sample of central London offices (Fig. 7.6). Clearly, then, information processing is fundamental to the functioning and survival of business organizations. Moreover, the conceptualization of information, employed by Thorngren, Törnqvist, and others, adopts an open systems perspective (whereby there is a continuous input of new information) rather than the closed system perspective of classical location theory. It is important therefore that research focus on how information is handled and specifically on how the environment is perceived, on how information is learned and evaluated, and on the nature of the resultant image.

Unfortunately very few studies have enquired into how organizations perceive their environment, no doubt because of the enormous methodological problems encountered in trying to assess whose opinions within an organization are the ones to be noted. One exception is Taylor's (1978b) study of Auckland manufacturers which showed that the use of behavioural, non-Euclidean space, based on the 'hierarchic' distance between the major functional economic regions of New Zealand, provides a better prediction of Auckland manufacturers' sales linkages than do models based on either a cost–distance or a Euclidean metric. To give a simple example, Taylor's study shows that, from a behavioural point of view, Christchurch is nearer to Auckland than is the geographically closer city of Wellington.

Perhaps the true significance of information in locational decision-making lies not in its relation to a firm's linkages but in the contribution it makes to the development of a general image of the environment. Such images undoubtedly exist and may even be strongly structured despite the fact that they only partially reflect reality. Certainly this was the case in Taylor and McDermott's (1977) analysis of New Zealand manufacturing and Green's (1977) examination of images and attitudes towards Development Areas in Britain. There are, however, very great problems in trying to elicit such images from firms. The principal components methodology that works so well in the description of individual preference surfaces (see Gould and White 1974) is not truly applicable to the delimitation of organizational images because it does not really allow for the weighting of responses that is necessary in order to reflect how much each contributes to any overall view. Instead a simple psychological scaling technique may be appropriate, providing that the list of alternatives between which respondents are to state a preference is kept to relatively limited proportions (Taylor 1977).

Despite these difficulties, McDermott and Taylor (1976) and Taylor (1977) have come up with some interesting conclusions about the general image of the economic environment held by New Zealand industrialists and about the way in which locational and organizational attributes influence the image. Interviews with managers in Auckland and five smaller towns, using a questionnaire that evaluated the environment on 16 seven-point scales, revealed that 'rationality' in industrial decision-making is blinkered and is generally constrained by a hierarchically structured image. The major component of the image, revealed by factor analysis of the questionnaire, related to factors of production and infrastructure (and, to a lesser extent, market access and the role of local government). In terms of organizational influence it was apparent that big firms tended to have polarized attitudes, be they favourable or unfavourable, whereas smaller and less complexly organized firms tended to have more weakly developed and less extreme opinions. To some extent these organizational influences compounded the locational influences rendering it difficult to identify the latter. However, it did appear that Auckland managers differed from those in the smaller centres and that Auckland itself was perceived as having marketing and transport advantages while the smaller centres had land and labour cost advantages.

Specialized information

Information flows are of such importance to modern business enterprises that most organizations set up specific departments to handle information. The departments concerned are invariably offices of one sort or another and, as such, they have come under increasing geographical attention (see Goddard 1973; Alexander 1979). The economic activities undertaken in offices comprise part of what is generally referred to as 'services' and the 'geography of services' (transport and communications, retail and wholesale distribution, insurance, banking, finance, business services, professional services) is becoming an increasingly important topic of study (Daniels 1985). To date, however, research has tended to look at headquarters offices or at individual firms in preference to looking at offices within the overall spatial structure of corporate organizations (Goddard 1975a: 11). This bias is important because there may well be a significant difference in the way in which offices make locational decisions depending on their size. Edwards (1983), for example, has argued that as one proceeds down the office hierarchy, so incremental decision making becomes increasingly important. Large corporations may strive for optimality in office location patterns but small firms tend to opt for incremental changes that minimize disruptions to day-to-day operations. Despite this, there is no denying that offices and services generally are vitally important in contemporary economies. After all they contribute significantly to employment growth and they can have a major influence on a nation's balance of payments (through the 'invisibles' component of trade) (Daniels 1987). They also represent the base for those contact-intensive employees engaged in information exchange. The location of offices is therefore highly significant, particularly because it is not unusual for 70–80 per cent of office contacts to be within 30 minutes travel time (Goddard 1975a: 16). Contact, in short, is a prime consideration in office location.

It is unrealistic, of course, to talk about 'offices' in general, or about 'services' as if they comprise a homogeneous sector of an economy. Daniels (1985) has made an important distinction between 'consumer services' and 'producer services'. The former cater directly for final demand or end-consumers. Examples are to be seen in retailing, public transport, and sport. The latter comprise those activities whose output serves as a specialist input for other industries. Examples are to be seen in business services, research and development, advertising, market research, and banking. These are critically important in the modern economy because they provide 'a lifeline to other economic activities wishing to adapt to new opportunities for improved productivity or diversified product development made possible by the technological revolution' (Daniels 1985: 115). In short, they are a key element in the 'flexibility' that characterizes contemporary economies (Wood 1991). Surprisingly, given this importance, producer services have been the subject of relatively little attention. Their importance and the scale of some of their activities cannot, however, be overstated. The 'Big Eight' accountancy firms employed more than 300 000 people world-wide in 1986 and offered their services in at least 30–40 countries (Daniels 1987). Nor can the role of producer services in creating an international urban system focused on a small number of global cities be overlooked. For example, one-third of all the world's international loans go through London (Daniels 1986).

The reasons why office-based service employment concentrates on certain locations are complex. The choice of location by service establishments is rarely based simply on the availability of suitable premises. Rather it is also influenced by the residential distribution of key personnel, local initiatives to attract employment, and possibilities for improved access to existing and potential clients (Daniels 1985: 121). Generalizations about office location are therefore fraught with danger. In particular, different considerations are likely to dominate for consumer and producer services. Some would argue of course that the distinction between consumer services and producer services is itself a little simplistic. Britton (1990), for instance, has argued that service employment should be seen as complementary to, rather than competitive with, manufacturing employment. There are clearly interdependencies between the two sectors and, on this basis, Britton has suggested a fivefold division of labour within production (Table 7.2). In many ways, this ties in with Atkinson's (1985) writings on the flexible firm (Fig. 7.5). Obviously a firm's 'production' could encompass both consumer and producer services undertaken as either core or peripheral activities. In all cases, however, information flows are critical.

Contact surfaces

What contact-intensive office employees handle is not information in general so much as *specialized information*. The essence of specialized information is that it is private information (see Walmsley 1982b). It is therefore not uniformly available but rather is geographically biased in that it reflects the action space and contact patterns of the firm concerned. Predictably, the greatest number of office-orientated contacts occurs in metropolitan areas. Such areas also have advantages in accessibility to non-local specialized information because of their dominance of transport and communications networks (Pred 1973a: 43).

Contact-intensive employees are an important part of the workforce. Not only are they high-income earners, and therefore presumably high spenders, but they also tend to attract other personnel and business

Table 7.2 An extended division of labour

1. Primary (direct) labour	(a) Immediate workplace – processing, transfer, assembly (b) Supplementary – transport, repairs, etc.
2. Secondary (indirect) labour	(a) Auxiliary labour to immediate primary labour – inventory and clerical staff, quality controllers, supervisors, engineers, etc. (b) Pre-production or preparatory labour – product development (design, testing, research), cleaners, clerical (c) Post-production labour – packaging, wholesaling, transport, installation, maintenance, repair
3. Complementary labour	Construction labour – buildings, land improvements, infrastructure (overlaps 2 and 4)
4. Tertiary labour	Circulation and management – trade, management, advertising, banking, leasing, renting, some transport, communications, insurance, etc.
5. Quaternary labour	Knowledge production (pure research), labour reproduction scientists, educational and health workers, etc.

Source: Britton (1990: 533)

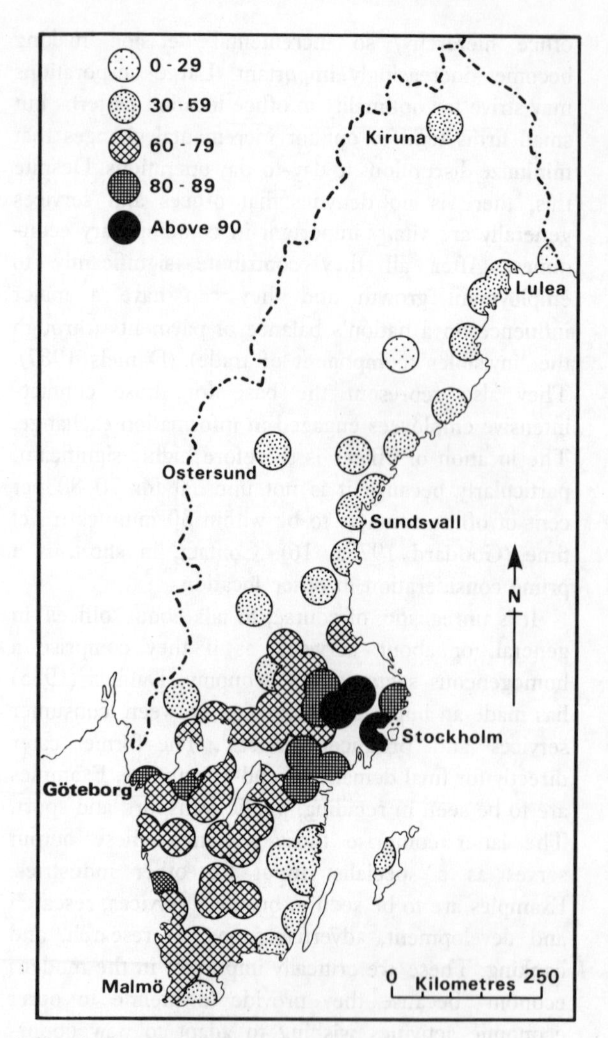

7.7 The Swedish contact surface in 1970 (Source: Törnqvist 1973: 97)

services (Törnqvist 1978). As a result there have been a number of attempts to measure the information flows, contact patterns, and awareness levels of these employees. Törnqvist (1970), for instance, has stressed the importance of contact-intensive face-to-face information exchange in post-industrial economies and has noted how the need for such exchanges encourages urbanization and metropolitan growth. In an attempt to illustrate the nature of the interdependence between information flows and city growth, attention was focused on the Swedish contact landscape (Törnqvist 1973). Given that there are groups within many business enterprises that are engaged in extensive contact activity, Törnqvist argued that it is possible to depict a contact landscape that assesses the contact potential of a given set of points. Values in the contact surface reflect the balance between the contact possibilities available at a point and the time and cost constraints involved in visiting that point. A sample contact landscape for Sweden in 1970, standardized to a 0–100 scale where the highest scores indicate the most contact, is given in Fig. 7.7. Clearly the Stockholm region dominates the contact surface.

The point of reference for contact surfaces is of

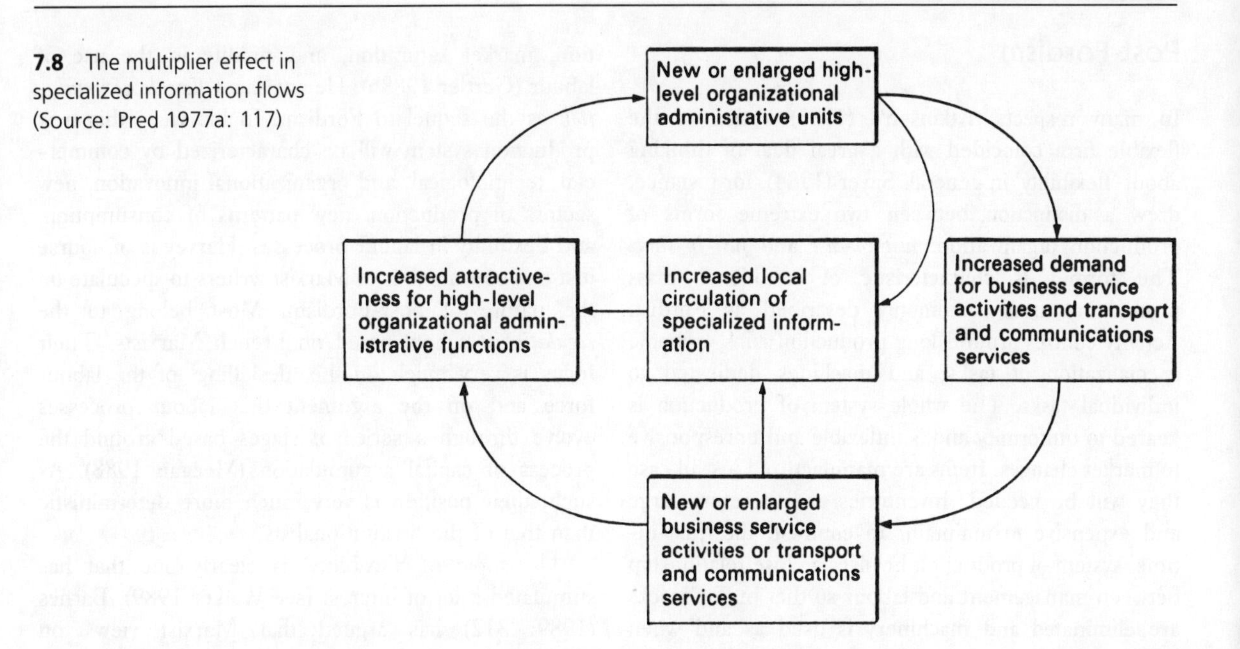

7.8 The multiplier effect in specialized information flows (Source: Pred 1977a: 117)

course the region rather than the firm, and their significance lies in their ability to model different transport and mobility constraints at a time when the circulation and availability of specialized information are mounting on account of ongoing structural shifts in the occupational composition of the workforce (Pred 1977a: 25). Of particular importance in this context is the increasing prominence of multilocational organizations, often resulting from mergers, in post-industrial societies (Pred 1977a: 99). This importance derives from the fact that the flow of information between the headquarters of such organizations and their subordinate units can potentially have a major impact on the pattern of interdependence that develops between cities. For example, because headquarters are invariably in metropolitan areas, and because these headquarters dominate specialized information flows, then it may follow that, regardless of where it occurs within a system of cities, investment by a firm may be likely to cause a multiplier effect at the firm's metropolitan headquarters in so far as new information processing requirements are almost certain to result from new investment (Pred 1977a: 116). This may further enhance the dominance of metropolitan areas within the contact and information landscape.

Metropolitan growth, therefore, may take on the appearance of a circular and cumulative feedback process as new high-level employment (Fig. 7.8). In short, 'interurban information circulation and organizational location patterns feedback upon one another to influence the process of city-system development' (Pred 1973a: 58). This means that information flows within the spatial structure of corporate organizations can underlie and steer such urban and regional development processes as the spatial diffusion of innovation and the development of regional external economies (Goddard 1975b: 88). This has obvious implications for regional policy in general, and for the elimination of interregional inequalities in employment in particular. As yet, however, there has been little research into the perceptions and images that industrialists have of this growth transmission process. It may well be that managerial attitudes are both malleable (Taylor and Neville 1980) and based on very limited information (Green 1977). It may also be that such images are grossly distorted and derive from stereotype, prejudice, and myth. It may even be that reality is much more complex than Fig. 7.8 suggests. For example, it may be that the nature of advanced economies is changing to the extent that new geographical patterns of economic activity are emerging, based around what is widely referred to as 'post-Fordism'.

Post-Fordism

In many respects, Atkinson's (1985) notion of the flexible firm coincided with a great deal of thinking about 'flexibility' in general. Sayer (1984), for instance, drew a distinction between two extreme forms of production organization: *just-in-case* and *just-in-time*. The former is characteristic of traditional mass production firms, commonly described as Fordist. Activity centres around long production runs, extreme specialization of tasks, and machines dedicated to individual tasks. The whole system of production is geared to uniformity and is inflexible and unresponsive to market changes. Items are manufactured just in case they will be needed. Inventories are therefore large and expensive to maintain. In contrast, the 'just-in-time' system of production involves a close relationship between management and labour so that buffer stocks are eliminated and machinery is used as and when necessary to produce required stock. The key feature of this system is obviously flexibility, based on close monitoring of market demand and an ability to redeploy resources within a firm at short notice.

Other writers have taken a more fundamental approach and sought to identify the basic parameters of the production system that many see as taking over from Fordism. Lash and Urry (1987), for instance, have suggested that distinct regional economies organized around a handful of industries in major urban centres (characteristic of what they term 'organized capitalism') are giving way to a more dispersed geographical pattern of employment (in an era they term 'disorganized capitalism') (see Allen and Massey 1988). Piore and Sabel (1984) have gone even further. They argue that, in volatile markets such as existed in the 1980s, firms have to experiment with flexible forms of organization. Mass markets have become saturated and consumers are now demanding specialized and differentiated goods which the mass production system is ill-equipped to provide. A crisis thus exists for mass production, the consequence of which will be a process of *flexible specialization* (see Amin and Robins 1990). This is commonly referred to as the *institutionalist* view of post-Fordism (Meegan 1988). Harvey (1987a) takes a similar view in his analysis of capital's response to a crisis of flagging productivity gains, increasingly international competition, market saturation, and rigidity in the use of labour (Gertler 1988b). He envisages *flexible accumulation* as the sequel to Fordism. In his view, this new production system will be characterized by commercial, technological, and organizational innovation, new sectors of production, new patterns of consumption, and flexibility in labour processes. Harvey is of course just one of a number of Marxist writers to speculate on the nature of post-Fordism. Most belong to the *regulationist* school based on French Marxists. Their focus is very much on the 'deskilling' of the labour force and on the argument that labour processes evolve through a series of stages based around the process of capital accumulation (Meegan 1988). As such, their position is very much more deterministic than that of the 'institutionalists'.

The topic of 'flexibility' is clearly one that has stimulated a lot of interest (see Walker 1989). Barnes (1989: 312) has argued that Marxist views on accumulation, like neo-classical economic views on location theory, are seductive in that 'clear lines of determination provide clear answers to difficult questions'. They are, however, equally unsatisfactory in that the real world is messier than the theory suggests. Barnes's view is that this 'messiness' should be celebrated, not packaged into grand theories and absolute conclusions. In short, it is essential to eschew grand theory and to look at why behaviour takes the form that it does in particular places. Such a goal means rethinking how human geographers view time and space. As G.L. Clark (1988) points out, theorists in geography have traditionally held time and space in abeyance so as to simplify analysis and concentrate on the economic processes at work. This approach is no longer tenable. In Clark's view, we need to move beyond empirical description of how geographical variations in technological innovation, local employment characteristics, and patterns of industrial specialization influence growth and decline. We need to theorize explicitly about time and space. To do this, we may need to forget about equilibrium models of reality and focus instead on why disequilibrium characterizes most real-world situations. It is not enough to argue that time and space are being annihilated as the global economy becomes internationalized because such an argument does not square with empirical evidence that locality is becoming much more important in influencing the nature of the industrial landscape (see Swyngedouw 1989).

Clark's views are reinforced by Soja (1987) who makes the point that 'Sweeping changes in the technology of production and communication; unexpected shifts in long-standing patterns of cultural expression, economic development, and political power; equally unanticipated revisions in the immediate social relations of class, gender, kinship, and allegiance have had a shattering effect on the experience and comprehension of modernity' (Soja 1987: 290). What is happening in relation to industrial production is, in other words, perhaps only part of a major shift in the nature of modern life. Soja advocates 'deconstruction' as one approach to theorizing about this changing world, that is a reinterpretation of events in terms of broader movements and structures. We need to see 'how the spatial fabric and texture of human societies derives from the organizational structures and social practices of the societies themselves' and how 'the unevenly developed geographical landscape . . . is not only a social product but also rebounds back to shape social relations, social practices, and social life itself' (Soja 1987: 291). At the heart of such pleas for a new bout of theorizing is recognition of the fact that the landscape of contemporary capitalism is littered with burgeoning industrial districts comprised mainly of small to medium-sized firms. At the same time, large corporations are being raided, dismantled, and stripped down (Walker 1989: 45). There is clearly a need to understand this reorganization of production systems and the way in which industries develop in and through the places in which they grow up rather than descending from the heights of corporate boardrooms (what Walker 1989 refers to as 'geographical industrialization' as opposed to 'industrial location'). In short, there is a need to understand how the nature of industrial production might be changing as a result of changes in behaviour and how such changing production, in turn, influences behaviour.

New geographical patterns of production?

New geographical patterns of industrial organization have been explored in depth by Scott. The focus of much of Scott's attention has been on locational trends in industrial activity in metropolitan areas. He argued that there are two components to such trends: a tendency for labour-intensive firms to cluster together at the centre, and a tendency for capital intensive firms to establish themselves on cheap land at the periphery (Scott 1982). In other words, where production processes are labour intensive and vertically disintegrated (i.e. production is fragmented, the labour process specialized, plants small, output highly variable, and inter-plant linkages labyrinthine), locational convergence or clustering should occur. Conversely, where production is capital intensive and vertically integrated (i.e. plants are large, output standardized, and inter-plant linkages substantial in magnitude), dispersal should occur (Scott 1983). Given an additional and general tendency for capital to replace labour, this has meant a suburbanization of industrial production.

Scott (1986) took this argument further in a theoretical sense in showing how processes of vertical and horizontal disintegration lead to increasing external economies of scale and how these then translate into a basic urban dynamic of centralization. The same processes are of course susceptible to reversal and this, coupled with increases in the scale of plants and with deskilling of the labour force, can lead to dispersal. Scott (1986: 29) sees disintegration and clustering as typical of industries such as clothing, furniture, foundry work, business services, and research and development, not least because such industries often depend on local pools of skilled labour. 'As local labour markets come into operation, innumerable additional external economies are created and these also help to hold the entire complex together as a geographical entity' (Scott 1986: 32).

If labour becomes deskilled, then it may pay firms to seek out cheap labour locations (e.g. the Third World). Conversely, technological change can stimulate vertical reintegration (Scott 1986). Technological innovation is of course a hallmark of the 1980s and 1990s and hi-tech industries are growing worldwide. One of the clearest examples of such hi-tech industry is the semiconductor industry. It is interesting to note therefore that the US semiconductor industry is characterized by a twofold pattern of agglomeration and dispersal (Scott and Angel 1987). The 'centre' in this case is not a downtown area but rather Silicon Valley, a transaction-intensive environment where firms concentrate on state-of-the-art, low-volume, customized and semi-customized production using a many tiered and segmented local labour market. In contrast, routinized forms of semiconductor manufacturing tend to be undertaken away from Silicon Valley (i.e. at dispersed locations, often low wage countries in the Third World). In other words, the advent of hi-tech, coupled with trends to either concentration or dispersal, can lead to new industrial landscapes (Scott 1988). Perhaps an appropriate term for these new concentrations is 'flexible production complexes' (Storper and Scott 1989). In short, production flexibility can lead to the clustering of firms into localized geographical areas. Flexibility can therefore ultimately lead to the emergence of industrial districts. On this basis, the late-Fordist regime of capital accumulation (characterized by branch plants and the decentralization of production to peripheral regions) can be differentiated from its successor regime (with its strong agglomeration tendencies in flexible production sectors) (Amin and Robins 1990: 13).

What Scott is arguing is that Fordist production systems (characterized by mass production, assembly lines, labour specialization, and routinized labour relations) are being replaced by flexible forms of production (craft-artisanal and design-intensive activities in the hi-tech and business service sectors) (see Lovering 1990). 'The flexible firm, in contrast to its Fordist predecessor, survives through perpetual adaptability rather than perpetual costs reduction' (Lovering 1990: 161). Such firms rely heavily on links with other firms and, as the costs of transport, communication, information exchange, and search and scanning increase, so too do the forces leading to agglomeration. Such agglomeration focuses not on the city centre but on the places where the industry in question first developed (e.g. Silicon Valley).

Several researchers have claimed to find empirical evidence to support Scott's general argument, notably in relation to the semiconductor industry (Angel 1990) and the motion picture industry (Storper and Christopherson 1987). Others have been very critical. Sayer (1990), for example, has pointed out that reality is more complex than much post-Fordist writing suggests. This is corroborated by Morris (1988a) who questioned, in a study of multinational branch plants in South Wales, assertions about the deskilling of the labour force and about the absence of research and development in branch plants. Pinch *et al.* (1989), after a survey of manufacturing performance around Southampton (UK), are also sceptical about the idea that advanced economies are entering a new era of flexible specialization.

The sternest criticism has come, however, from Amin and Robins (1990). In their opinion, the views of people like Scott are simplistic and contentious because they overlook contradictory components of the restructuring process. They see Scott's ideas as 'a kind of anti-Fordist utopia characterized by flexibility, diversity, and, in spatial terms, localism' (Amin and Robins 1990: 10). In their view, Silicon Valley might be an exception rather than an example. To read too much into the Californian experience is to create a new orthodoxy where virtually any example of localized economic life that is new and thriving can be interpreted as an 'industrial district'. Amin and Robins (1990: 15) see a danger in Scott's writings whereby 'certain tendencies in the present period are absolutized, and then projected forward as the paradigm for a future regime of accumulation, for a new historical phase of capitalism; and then it is in the light of this reified, ideal-type model of post-Fordism that the events of the present period are assessed and explained'. Amin and Robins' view is that rather than there existing some fundamental shift from centralization/concentration/integration to decentralization/dissemination/disintegration, what is really happening is the evolution of organizational forms that are an extension of Fordist structures.

Lovering (1990) has also criticized Scott on many grounds, most notably for ignoring the role of nation states and of political economy generally. Lovering points out that the so-called evidence for Scott's thesis (Silicon Valley, Route 128, the 'Third Italy') is not unequivocal; for instance, Route 128 in Boston owes much of its development to the US military rather than 'flexibility'. What seems to be happening in these regions looks, to Lovering, remarkably like the Fordism that is supposed to be disappearing. In other words, flexible specialization, if it exists, may be but one of several coexisting forms in which capital accumulation can take place (Lovering 1990: 169). Many of the key characteristics of post-Fordist flexibility may in fact be illusory: new machinery never seems to be as good and flexible as claimed and there are often significant labour-based barriers to its use; 'just-in-time' programming is not easy to introduce because it might necessitate the unacceptable rupturing of long-standing linkages with suppliers; and flexible production systems are sometimes introduced to control labour rather than to put a firm in a better competitive position (e.g. many jobs may be deliberately reclassified as 'technicians' and therefore outside organized labour unions) (Gertler 1988b). The response to these criticisms has sometimes been spirited (see Schoenberger 1989) but, ultimately, the debate shows one thing: we simply do not know enough about the behaviour of individual firms (or individual unions) to be able to theorize safely about what is happening. The scope for behaviourally oriented research is endless. Not only would this help in the interpretation of contemporary trends, it would also provide a sounder foundation whenever attempts are made to harness whatever is known about 'industrial geography' in the pursuit of regional development policy.

Regional development policy

The restructuring of advanced economies since 1970 obviously has had a great impact on patterns of regional development (Amin and Goddard 1986), often through plant closures (Watts and Stafford 1986). More to the point, the manifestations of restructuring have resulted in winners and losers, particularly in a geographical sense. A spatial division of labour has resulted from economic and technological changes over the last 20 years such that control functions (e.g. research and development, policy formulation) tend to be located in company headquarters in metropolitan areas and mass production at decentralized, sometimes low-wage locations (Massey and Meegan 1979). The net result is that the most critical components of corporations tend to be located in metropolitan areas, much in the manner suggested by Fig. 7.8. There is, in other words, a pattern of 'uneven development' to be observed in most advanced economies with 'core' areas being favoured at the expense of the 'periphery' (see Walker 1989). Massey (1984) conceptualizes this as the 'layering' of different 'spatial structures of production' (variants of Fordism and post-Fordism), each complexly superimposed and combined with the ones that went before so as to reflect the unique qualities of the places in question and the organizational structures of large corporations (especially the linkages between corporate headquarters and branch plants). This raises interesting questions: What do branch plants contribute to the local economies of the (often 'peripheral') regions in which they are located? To what extent are there 'local labour markets' and what is the significance of these?

It is a widely held view that branch plants do little for local economies (see Hagey and Malecki 1986). Walker (1989) questions this by citing the importance of IBM at Montpellier (France), semiconductor firms in South Wales, and Japanese car factories in the American Midwest. The question therefore arises as to why some branch plants are significant and others not. Again behavioural studies are needed to provide answers. In relation to local labour markets, Peck (1989) has shown that such markets are generally conceptualized as comprising two components: a primary sector of better quality jobs with high wages, good working conditions and sound prospects, and a secondary sector characterized by low wages, poor working conditions and a lack of job security. It is in this secondary sector that many disadvantaged workers become trapped, despite the best efforts of governments through local employment initiatives and training packages. The net result is therefore the high unemployment levels noted at the start of this chapter.

What can be done to integrate branch plants into local economies and to free up rigidities in local labour markets depends on a range of factors as diverse as the availability of venture capital (Thompson 1989), the presence of premises for rent (given that the majority of manufacturers rent rather than build) (Bull 1985), and the type of industry (with material goods production tending to have more local linkages than service goods production) (Marshall 1979). These are therefore the issues that regional policy needs to address. An even more fundamental question is whether regional policy should encourage branch plants at all. Some would argue that such plants are counter-productive to regional development because they bring primarily unskilled jobs, limit local autonomy over investment, and increase 'import' dependency by relying on corporate rather than local linkages (Hayter and Watts 1983: 171). Given this conflict of opinion, there is an urgent need to discover what happens in particular circumstances and to appreciate why the behaviour of the actors involved takes the form that it does. In short, behavioural approaches to the study of jobs and work have a big contribution to make in human geography, not least by filling in gaps in existing knowledge as to why the world is as it is.

8 Housing and Migration

Migration, or the movement of people from one geographical location to another, has been taking place since the origin of the human species. Over time, it has not only increased in volume and diversity but has also involved steadily lengthening distances (Potts 1990). With the emergence of a world economy and the globalization of communications, migration, in turn, has 'exploded' at all geographical scales and become of major concern, thus justifying Goldstein's (1976: 424) observation that 'whereas the study of fertility dominated demographic research in the past several decades, migration may well become the most important branch of demography in the last quarter of the century'. Therefore, the necessity to identify and explain such population movements is vital, both in practical and academic terms, in many parts of the world (Clarke 1989).

Despite the burgeoning of migration research, as yet no coherent theory of human migration has been validated (Roseman 1971; Zelinsky 1983; Zolberg 1989). However, as long ago as the 1880s, Ravenstein (1885; 1889), using birthplace statistics contained in the 1881 census of England and Wales, suggested that migration could be generalized into seven 'laws':

1. The majority of migrants move only a short distance, and consequently there is a general displacement of persons producing currents of migration in the direction of the great centres of commerce and industry.
2. The process of absorption is created by movement from immediately around a great city, creating gaps which are filled from more remote areas. This also means that few migrants will be found in cities from areas progressively further away.
3. Dispersion has similar features and is the inverse of absorption.
4. Each main current of migration produces a compensating counter-current.
5. Long-distance migrants usually go to large cities.
6. Urban dwellers are less migratory than rural dwellers.
7. Females are more migratory than males.

Although this highly generalized framework refers very much to late Victorian Britain, it does emphasize that in any analysis of migration, it is necessary to consider at least three interrelated elements: the place of migration; its causes; and its selective nature. Since the publication of Ravenstein's 'laws' no one researcher has attempted to produce such a comprehensive framework, although individual parts of the migration process have been analysed in a more sophisticated manner (Lee 1960; Lewis 1982; Clark 1986a).

One of the few attempts in recent times to provide a theoretical framework for migration, albeit far less comprehensive than Ravenstein's, has been Zelinsky's hypothesis of a mobility transition (Fig. 8.1): 'there are definite patterned regularities in the growth of personal mobility through space-time during recorded history, and these regularities comprise an essential component of the modernization process' (Zelinsky 1971: 221–2). Thus, different types of migration are characteristic of different phases (I to IV) of modernization. This hypothesis has been taken a stage further by Rowland (1979) and Wardwell (1977), in their interpretation of the continuing high levels of

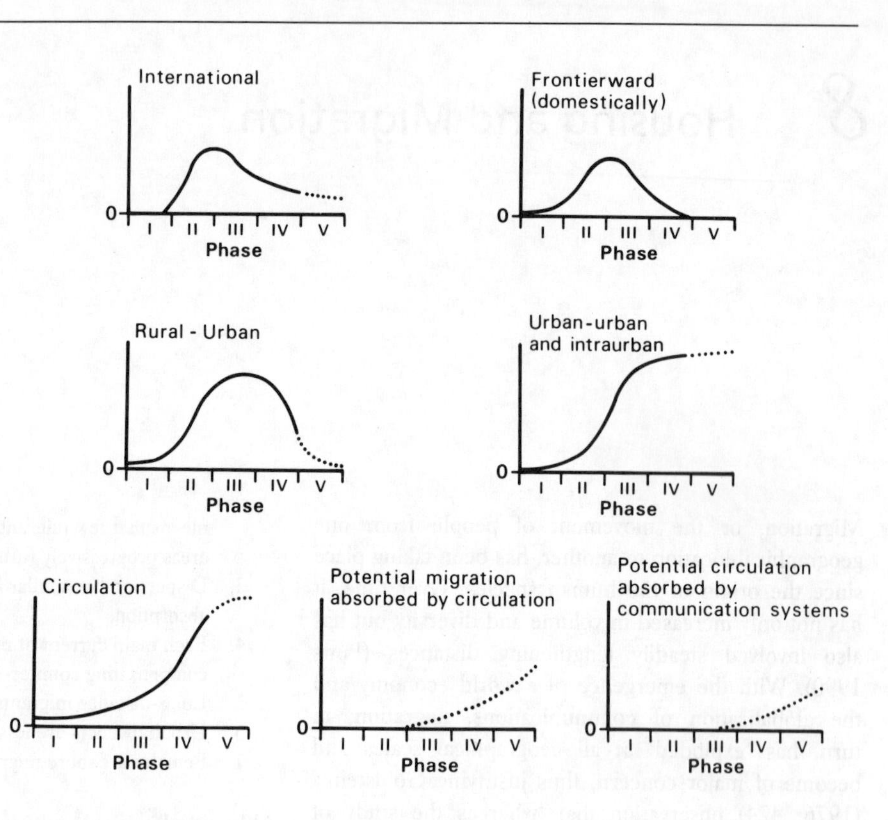

8.1 Comparative time profiles of spatial mobility (Source: Zelinsky 1971: from The hypothesis of the mobility transaction, *Geographical Review* **61**, 219–49 by permission of the American Geographical Society)

residential turnover and the emergence of a rural turnaround in Australia and the United States respectively, when they claimed that migration maintains a settlement system rather than modifying it. From this perspective, migration is therefore basically an equilibrium-seeking process.

> Just as the vital transition describes the process of fertility and mortality as tending from a semi-equilibrium characterized by high fertility and mortality, to a new semi-equilibrium with low fertility and mortality, so too a migration equilibrium may be approached towards the end of the mobility transition as all parts of the settlement system achieve a high level of modernization (Rowland 1979: 11).

Understandably, since the introduction of such 'laws' and 'hypotheses', considerable discussion has taken place as to the relationship between mobility and modernization and the underlying causes of migration; yet many have criticized the various interpretations for their Western overtones and value judgements (Brown and Lawson 1985; Zolberg 1989). The search is therefore still on for some coherent overall view of migration.

According to White (1980a), there have been two main approaches to the study of migration. The first is the 'macro' or structure-oriented approach. This initially involved the description and explanation of migratory flows in terms of distance decay, as expressed in the gravity model and its later derivative, the intervening opportunity model (see Chapter 2). However, even with the inclusion of additional socio-economic factors (such as income and unemployment), usually within a regression equation, as well as attempts to measure the interaction between the characteristics of places and their associated migration patterns, it was found that gravity model-type formulations achieved a fair degree of prediction only at the aggregate level (Willis 1974; Shaw 1975). In other words, such approaches provide little insight into the behaviour of individual migrants except in so far as such behaviour can be deterministically predicted by reference to variables such as distance and town size.

A similar deterministic flavour has influenced those researchers who have argued that migration can only be explained in terms of responses to the needs to capitalist production. Again, this is a structure-oriented approach. Access to employment and housing is interpreted as reflecting the power of the dominant capitalist class. Migration is therefore seen as a simple response to the structural constraints under which individuals operate (Castells 1978). From this perspective, changes in the system of constraints can only be achieved as a result of collective action rather than individual initiative. For example, in their analyses of the regional migration history of Wales and Scotland, Rees and Rees (1981) and Jones (1986) respectively emphasized the role of migration in the furtherance of systemic relations between regions which have been structured out of the process of capital accumulation. The focus of this type of political economy perspective has consequently been upon macro-economics and political structure rather than on individual responses.

Inevitably, the aggregate focus of both the gravity model and the political economy perspective, coupled with their failure to explain fully the diversity of migration, continues to be a source of much debate. For example, as long ago as 1970 Hägerstrand (1970: 8) made the telling point that 'nothing truly general can be said about aggregate regularities until it has been made clear how far they remain invariant with structural differences at the micro-level'. More recently, Thorns (1985) argued that structure-oriented approaches even fail to take into account the extent to which arenas are created within which individuals have opportunities to make (admittedly constrained) decisions as to their residential location.

In response to these criticisms, the last two decades have witnessed the emergence of a second, the 'micro' or behavioural, approach to the study of migration (Lewis 1982; Clark 1982a). Such an approach argues that people at different places and at different positions in the social structure have different degrees of knowledge about, different perceptions of, and are able to benefit to different degrees from, opportunities at places other than those in which they currently reside. For example, White (1980b) has shown that outmigration from Topeka, Kansas, to other cities was more closely related to awareness and preference indices than to variables such as size, distance, and

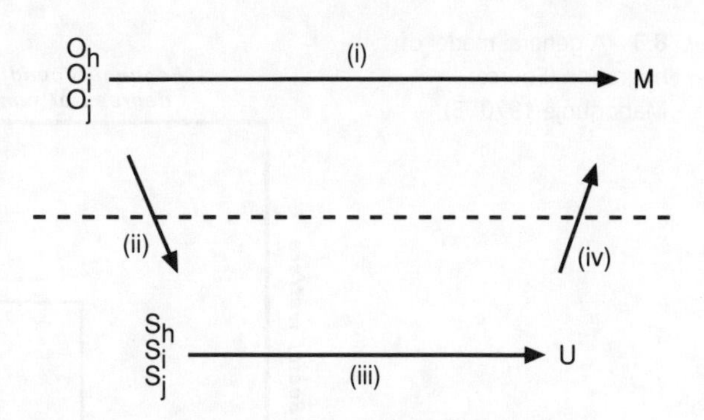

8.2 A conceptual framework for analysing migration behaviour (Source: Cadwallader 1989a: 496)

economic and demographic characteristics. Essentially, according to the behavioural perspective, it is these perceived differences in opportunities that are important rather than any response to some 'objective' factor such as wage rate fluctuations. To some, of course, such micro-scale approaches are not necessarily at odds with macro-scale approaches to the study of migration. Cadwallader (1989a), for instance, has suggested that the two approaches should be thought of as complementary rather than in opposition to each other. He argued, as illustrated in Fig. 8.2, that objective variables (O_{hij}) are transformed through the cognition of potential migrants into (S_{hij}) their subjective counterparts, which combine to form an overall measure of the attractiveness of an alternative destination (u) which, subject to constraints, may be translated into overt behaviour (m).

Despite this enthusiasm, by the early 1980s, there was criticism of the behavioural perspective just as earlier there had been criticism of the macro-scale approaches. Firstly, there was growing evidence to suggest that, in many instances, there was a discrepancy between preferences and residential mobility. Hwang and Albrecht (1987), for example, found among a sample of Texas homebuyers that this mismatch increased with the greater exclusivity of a neighbourhood. Secondly, there was also considerable disquiet over the emphasis that the behavioural approach put on choice and decision-making (the 'demand side' of migration) at the expense of constraints and opportunities within the labour and housing markets (the 'supply side') (Short 1978; Bassett and Short 1980). In other words, the view

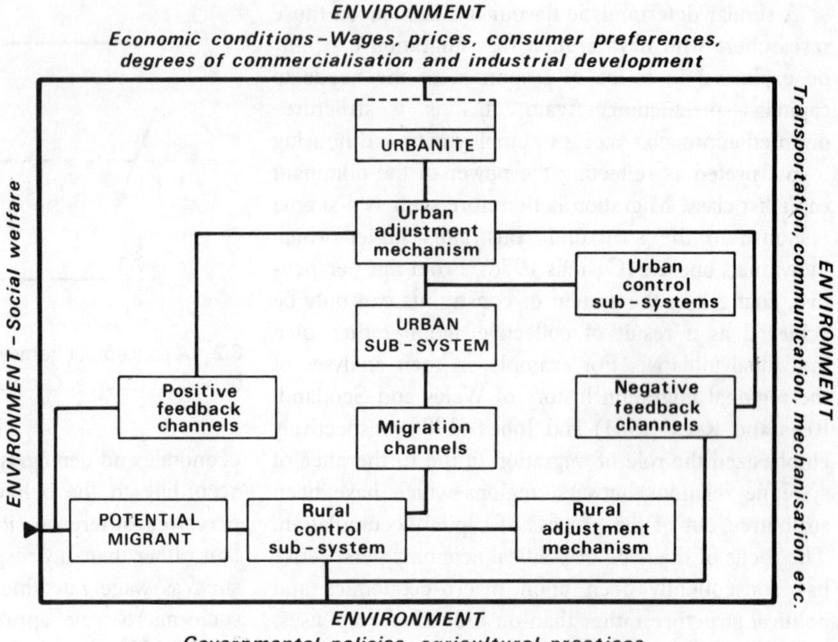

8.3 A general model of migration (Source: Mabogunje 1970: 3)

emerged that migration can only be understood within the context of the broader socio-economic system, though Maher (1984) has observed that the differing emphasis in the behavioural matrix between choice and constraint is only a matter of scale. That is to say, most studies by assuming away the societal and institutional context, have been able to focus on the decision framework and those independent variables influencing migration patterns. At a higher level of abstraction the change is, therefore, from variables with independent status to those which can be viewed as being dependent on a further set of factors.

So far, however, little research has actually integrated the study of mobility and the study of differential access to housing and employment, though a number have advocated such an approach. For example, Mabogunje (1970), in a study of rural–urban movements in West Africa, conceptualized migration as a circular, interdependent and self-modifying system in which the effects of changes in one part of the system have a ripple effect throughout the whole system (Fig. 8.3). Thus, the potential migrant operates within a changing environment with choices and constraints at both origin and destination. More specifically, Murie (1986) has suggested that the integration of mobility and opportunities can best be achieved by the adoption of a 'choice within constraints' framework which involves:

1. Household preferences (based on values, income, and occupation) which affect the interpretation of opportunities and their changing character;
2. Search and information gathering restrictions, which influence a household's perception and awareness;
3. Access to housing, which affects the eventual outcome since both the public and the private sectors have 'rules' of access;
4. Limited availability of the type of dwelling required.

In other words, what is being emphasized in this formulation is that all households make decisions about migration within a set of individual and societal constraints.

The decision to move

Over 30 years ago, Rossi (1955) suggested that the decision to change residence involved three, often interrelated, stages:

1. The decision to seek a new location;
2. The search for a new location;
3. The selection of a new location.

Several attempts have been made to model this process (Brown and Moore 1970; Brummell 1979; De Jong and Fawcett 1979; Sell 1983). The majority of these models were conceptualized in terms of a *stress–satisfaction formulation* derived from Wolpert's (1965, 1966) concept of *place utility*. Briefly, this viewpoint argues that the basis of any decision to migrate is the household's belief that the level of satisfaction obtainable elsewhere is greater than the present level of satisfaction. The difference between the two levels can, therefore, be regarded as a measure of *stress*. The major emphases within these models are on the nature of motivation, the preferences underlying a decision to look for a new dwelling, and on examination of the search and evaluation process.

Of the various attempts to model the migration process within a decision-making context, the most widely accepted has been that introduced by Brown and Moore (1970) (Fig. 8.4) (see also the critique of the model provided by Walmsley 1973). This model highlights changes in the needs and expectations of the household as well as changes in the characteristics of both the dwelling and the environment. Only when the place utility derived from the present dwelling and location is reduced below a certain threshold level, either by internal or external sources of stress (or both), will the household decide to look for an alternative. However, even this decision does not automatically entail migration because stress can sometimes be overcome by the household either reducing its aspirations or improving its existing circumstances (e.g. modifying the existing dwelling by building an extension). The decision as to where to move to, when it comes, involves searching information sources about likely vacancies and matching these with the household's aspirations. If an appreciably higher place utility is provided by a new location, then there is a decision to migrate. If not, a decision to remain in the present location, or continue the search, is made. The significance of the Brown and Moore model lies in its emphasis on the need to consider housing and environmental stress within the context of the needs, expectations, and aspirations of householders. For example, based on income, family status, religion, and ethnicity, the model suggests that each household has a frame of reference with respect to urban living in general and its housing situation in particular and that this frame of reference influences subsequent behaviour.

A similar viewpoint was evident when De Jong and Fawcett (1979) incorporated the concept of *value expectancy* with the migration decision model. They argued that a household's intention to act in a certain way depends on both the *expectancy* that the act will be followed by a given consequence (or goal) and on the *value* of that consequence (or goal) to the individual. In Fig. 8.5 it is therefore hypothesized that migration is the result 'of (1) the strength of the value expectancy derived intentions to move, (2) the direct influences of background individual and aggregate factors, and (3) the potential modifying effects of often unanticipated constraints and facilitators which may intervene between intentions and actual behaviour' (De Jong and Fawcett 1979: 59). In other words, the values and goals of migration, and the expectancy of attaining those values, interlink to form migration behavioural intentions which can lead, under some circumstances, to either a decision to migrate or to make *in situ* adjustment rather than to move.

In view of the fact that not all decisions are free, and the fact that a wide variety of constraints upon movement operate within all societies, many researchers have suggested that both the models of Brown and Moore and De Jong and Fawcett are rather unrealistic (Popp 1976; Thorns 1980). Essentially these critics argue that the relocation decision process may not come to fruition because of the existence of a set of behavioural constraints. To appreciate the nature of these constraints, it is necessary to understand the reasons behind migration.

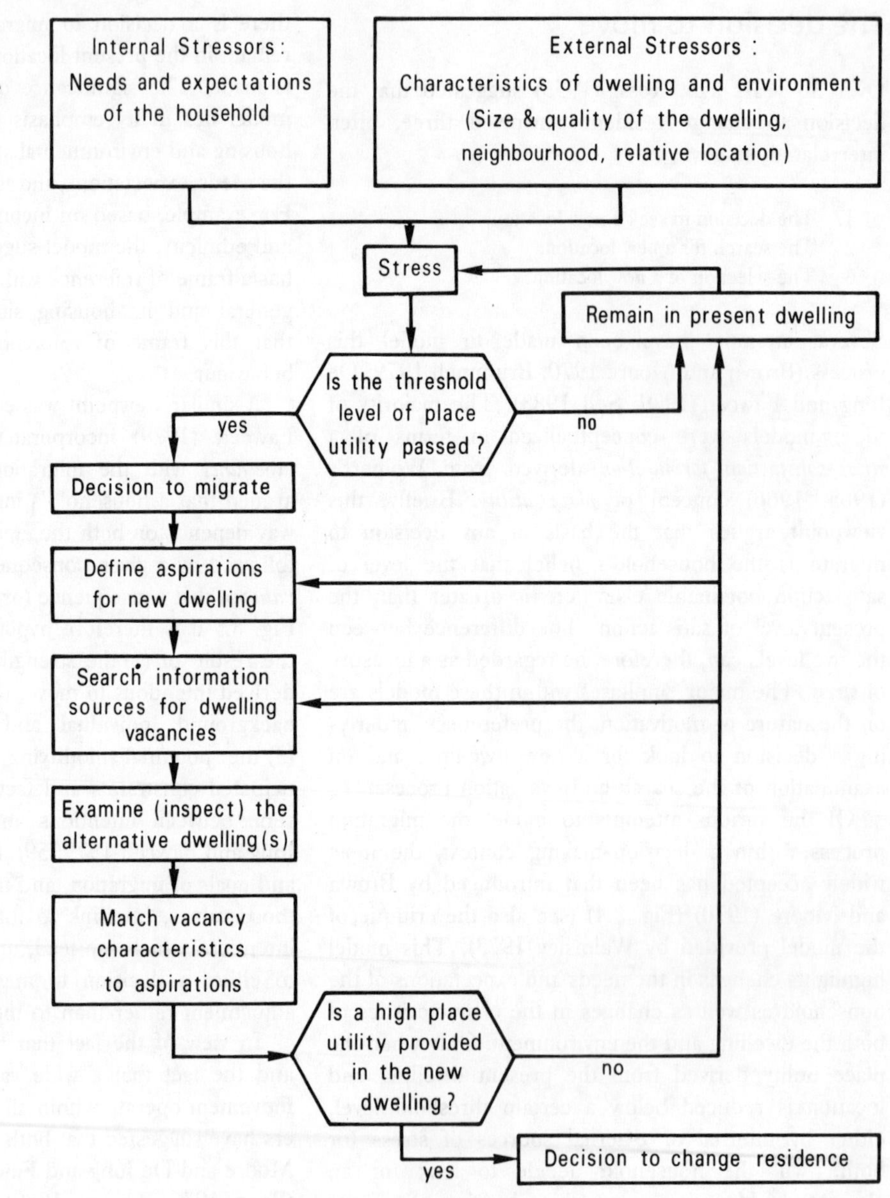

8.4 A model of relocation decision-making (Source: Based on Brown and Moore 1970: 1–13)

Reasons for migration

In any consideration of migration, it is necessary to distinguish between voluntary and involuntary moves (Cebula 1980). Although the latter are more often than not associated with wars, famines, and persecutions, there is growing evidence to suggest that, even within a city, something like 15 per cent of all moves are of a 'forced' nature, mostly associated with property eviction and demolition, retirement, ill health, and divorce (Kendig 1984). In terms of *voluntary* migrations, the literature suggests that long-distance moves are generally motivated by economic factors such as unemployment, income, and career advancement,

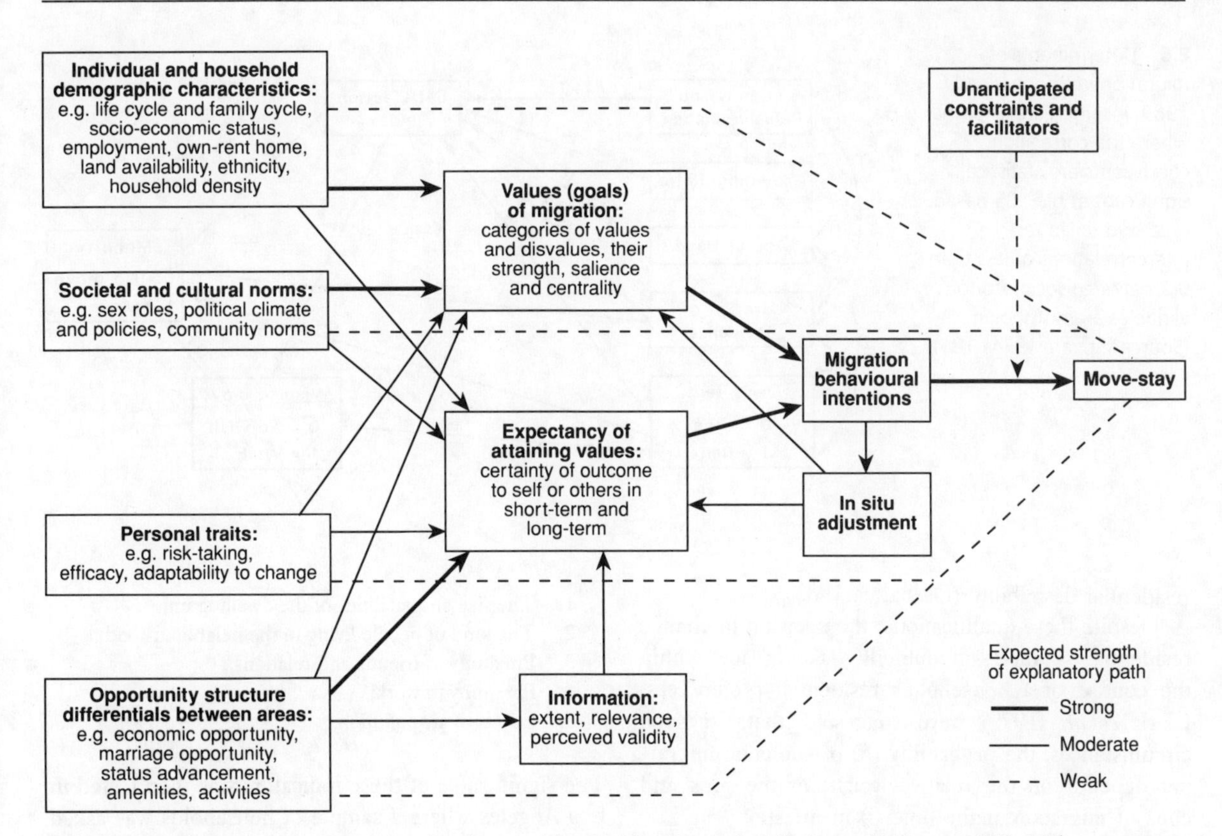

8.5 The importance of expectancy in migration decision-making (Source: Golledge and Stimson 1987: 261)

while short-distance moves are motivated by more social forces such as neighbourhood change, investment, marriage, and dwelling space requirements (De Jong and Fawcett 1979; Lewis 1982).

In any attempt to determine why a household changes its residential location, it is necessary to first elucidate the means by which environmental opportunities are evaluated. For example, Menchik (1972) employed regression analysis to look at the first- and second-order preferences of households by using sets of variables relating to natural and built environments, accessibility, housing characteristics, and familiarity with the residential environment. Similarly, from a questionnaire survey of Hamilton, Ontario, Preston (1987) suggested that residential areas are evaluated in terms of land use, lot size, social character and housing quality. Somewhat differently, Johnston (1973a) and Cadwallader (1979) argued that households rate neighbourhoods cognitively on three evaluative dimensions: the impersonal environment (mainly physical attributes), the interpersonal environment (mainly social attributes), and locational attributes. All these studies suggest that residential aspirations are complex phenomena that, to a certain extent, are expressed through market choice even though the salience of evaluative dimensions might vary across different types of neighbourhood (Preston 1987). Indeed, the identification by researchers of a relatively limited set of evaluative dimensions might suggest that neighbourhood perception is a fairly simple process which can be readily reduced to a simple overall utility value (Lieber 1979; Deurloo *et al.* 1988). This is not, however, the case because residential preferences seem to be strongly influenced by factors such as socio-economic status and stage in the life/family cycle. It also appears that households are constrained by economic considerations, plus individual taste and preference, when forming their aspirations and, moreover, in assessing

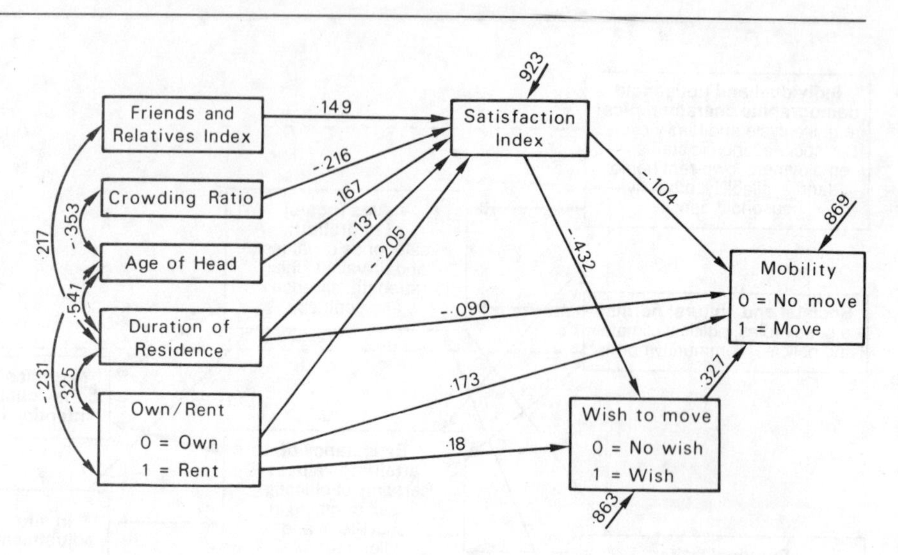

8.6 Determinants of migration in Rhode Island, 1969. Included are all paths where the correlation coefficients are statistically significant at $p > .05$ based on a two-tailed test. Intercorrelations of less than 0.2 between socio-economic variables are not shown (Source: Speare 1974: 181)

residential desirability (Desbarats 1983).

Despite these qualifications, the decision to change residential location undoubtedly takes place within the context of a household's residential preferences. Clark *et al.* (1979) have suggested that, in such circumstances, the probability (*P*) of someone migrating depends on the relative weight of the pros and cons of migration at the time (*t*) in question:

$$P(t) = c \text{ Stress } (t) - \text{Resistance } (t)$$

Therefore, the estimation by an individual at time *t* of his or her socio-spatial circumstances depends on weighing advantages and disadvantages and judging results within a set of tolerance limits:

$$P(t) = cA(t) - D(t)$$

where *A* stands for advantages, *D* for disadvantages, and *c* is a correction factor which is specific to each person's particular point in time and space.

Despite the simplicity of this stress–satisfaction formulation, it is difficult to apply. In fact it has been interpreted in a variety of ways including as a form of economic disequilibrium model, the disequilibrium in question being generated by the changing space needs of the household as well as by varying psychological phenomena (Salling and Harvey 1981). Nevertheless, Clark and Cadwallader (1973) have attempted to disentangle the processes by which stressors affect migration decisions. At the outset, five indicators of stress were chosen for analysis:

1. The size and facilities of the dwelling unit;
2. The kind of people living in the neighbourhood;
3. Proximity of friends and relations;
4. Proximity to work;
5. Levels of air pollution.

The significance of these indicators was then tested in Los Angeles where a sample of households was asked to rate, on the basis of the five indicators, how easy or difficult it would be to find a more desirable location elsewhere. They were also asked to evaluate their levels of satisfaction with their present residence and neighbourhood. A highly significant correlation of 0.384 was achieved between a desire to move and locational stress. Most stress was experienced by those households who thought they could find a better residence elsewhere and by those who were least satisfied with their present location. Apart from the pollution stressors, each of the other four indicators was important in creating household stress, with the size and facilities of the dwelling being the most important and proximity to work the least significant. Similarly, Speare (1974) developed an index of residential satisfaction based on housing and neighbourhood characteristics, and found, among a sample of households in Rhode Island, that the index provided a meaningful measure of whether the households wished to migrate and also whether they moved during the following year. It was also found that individual attributes acted indirectly through the residential

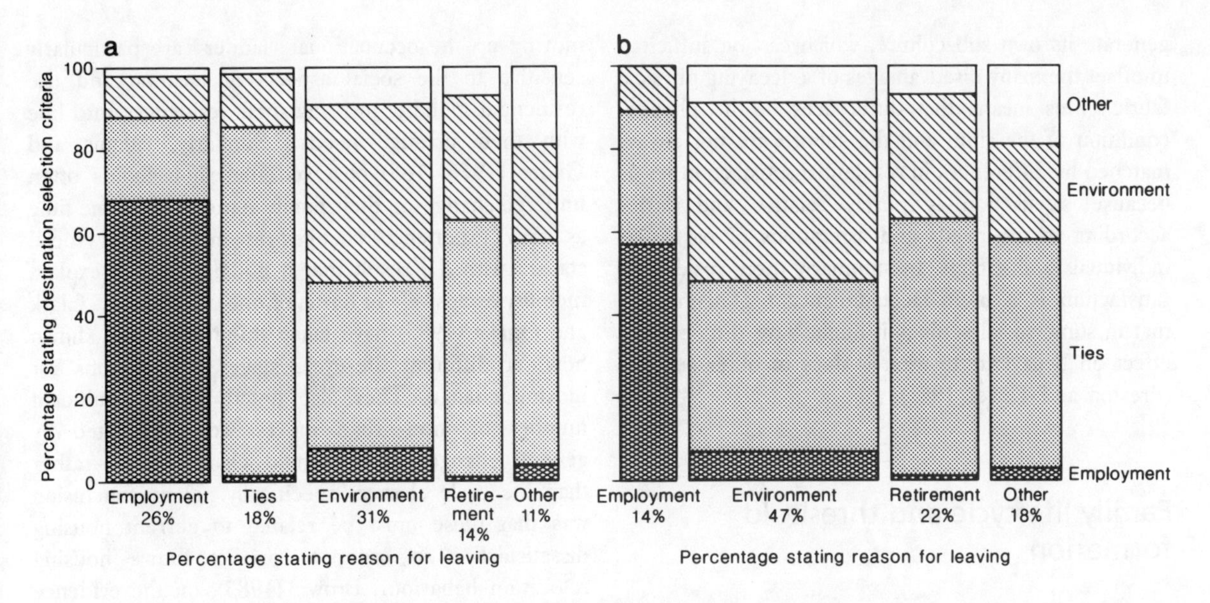

8.7 Relationship between destination selection criteria and reason for leaving origin: (a) all households; (b) households in which decisions were not simultaneous (Source: Williams and McMillen 1980: 199, 201)

satisfaction index to affect mobility potential (Fig. 8.6). Clearly, both of these studies identified housing as the critical variable in determining household stress and, therefore, they corroborate findings of studies more directly concerned with the causes of residential mobility. A contrary view has, however, come from Boehm and Ihlandfelt (1986) who suggested that, among inner-city residents, neighbourhood variables were the most significant predictors of mobility, irrespective of tenure. The implication of this study is that improvement policies can succeed in stabilizing inner city neighbourhoods.

Little research has been conducted with the specific goal of understanding the destination selection process as distinct from the decision to leave a place of origin. Williams and McMillen's (1983) study of migration decision-making among households relocating to non-metropolitan USA did, however, attempt to distinguish between those households that simultaneously chose a destination, given a certain reason for leaving (simultaneous decisions), from those that employed different criteria for each stage in the migration decision process (non-simultaneous decisions) (Fig. 8.7). Similarly, Morrill *et al.* (1986) found that non-metropolitan migrants were negative about city life and positive about the social and environmental characteristics of their new communities. Also these authors noted that the decision to move did not precede the choice of destination; more often a visit to the destination precipitated an evaluation of the origin and, thereby, the decision to move.

Clearly, these findings suggest that there is more to residential satisfaction than simply housing and family requirements. Individuals identify with their residential location and endow it with social meaning. For example, Feldman (1990) applied the concept of *self-identity* to explain cognitive and behavioural dispositions towards city and suburban settlements and suggested that this identity influenced future residential intentions. This confirms the view that both the suburb and the slum can be a source of satisfaction. It has long been considered in Western society that suburban living symbolizes all the values of 'good' living (Michelson 1977). In contrast, it might be thought that inner-city neighbourhoods with their decaying fabric, overcrowding, and poor living conditions, would be a source of considerable dissatisfaction. Surprisingly, this is not always the case. For example, in his classic study of the Adam district in Chicago, Suttles (1972) showed that the slum can generate its own sub-culture, which can be sufficient to offset the many disadvantages of a decaying physical fabric. This means that differences in the physical condition of the slum and the suburb are not always matched by differences in residential satisfaction levels because satisfaction is purely relative and varies according to the needs and aspirations of particular individuals and groups. In summary, then, residential satisfaction is a multi-faceted concept. This means that in some social settings it might have the opposite effect on behaviour to what it does in other settings (Preston and Taylor 1981).

Family life cycle and threshold formation

A major factor in determining residential mobility is the changing threshold of residential dissatisfaction that accompanies changes in the family life cycle. From the formation to the dissolution of the cycle, critical events can be identified which increase or decrease the propensity to migrate (e.g. marriage, birth of children, last child leaving home, and retirement) (Butler *et al.* 1969). The first and most significant interpretation of migration by means of the life-cycle approach was Rossi's (1955) classic study of residential mobility in Philadelphia. This concluded that the major function of mobility was as a process whereby families adjust their housing needs to match the shifts in family composition that accompany life-cycle changes. Despite several later researchers confirming this (Morgan 1976), some writers have revealed that life cycle changes are never perfectly correlated with mobility levels, with the result that a question mark must be placed over the universality of life cycle as a differentiator in migrant potential. In other words, although 'changes in household or environment may be a necessary condition for mobility, they are not however a sufficient condition' (Coupe and Morgan 1981: 213).

Clearly, the evidence provided by these studies points to the necessity of considering the selectivity of the life-cycle dimension within the broader context of social mobility and career patterns. Even Rossi (1955: 179) recognized this when he argued that 'families moving up the occupational "ladder" are particularly sensitive to the social aspects of location and use residential mobility to bring their residences into line with their prestige needs'. Similarly, Adams and Gilder (1976: 165) observed that 'households often undergo changes in their family status at the same time as they experience changes in income and social status', with the result that it is dangerous to explain mobility exclusively in terms of one or the other. Clark and Onaka (1983) have taken this further and shown how, at different life-cycle stages, the reasons for moving change. They also noted that a significant number of intra-urban moves were generated by general changes in household characteristics rather than life cycle changes specifically. Their conclusion was that these must be related to current housing dissatisfaction in order to predict future housing relocation behaviour. Brown (1983), on the evidence of a fourteen element typology of movers, suggested that job-, house-, and family-related causes of migration interrelate differently at different stages in the life cycle. For example, first-time buyers stress the need for access to such things as work, social facilities, and transport but, as incremental steps are taken to upgrade the quality of the house and location, there is a greater emphasis on requirements related to the immediate environment (e.g. access to the countryside), and increased spatial differentiation in the home (e.g. a desire for a study). As further incremental steps are taken, qualitative aspects such as 'atmosphere' and 'character' become increasingly important. Downgrading buyers, in contrast, stress the importance of the immediate environs of the house, and a dominant feature of their choice criteria is, as might be expected, price.

The growing significance of *organizations* in determining migratory flows has also led to a questioning of the relevance of the life cycle model. For instance, McKay and Whitelaw (1977) revealed that, in Australia, those seeking career advancement and employed in the health care, education, and local government bureaucracies, were generally restricted to *intra*-state moves, whereas those involved in the private sector (especially manufacturing and financial institutions) generally moved between metropolitan centres (and thus *inter*-state). Essentially, the latter are *spiralists*. An example is to be seen in 'a manager in a large corporation who moves around the country either at the corporation's bidding or while changing corporations . . . [and who] . . . is prepared to go anywhere to further his [*sic*] career' (Mann 1973: 11). Such individuals contrast markedly with *locals* who achieve career progression while staying within one particular locality.

Several other researchers have argued that this typology is too simplistic in that the pattern of household formation is becoming increasingly diverse (involving, as it does, the growth of single-parent and one-adult households). As a result of this diversity, any generalization about the life cycle and its influence on migration is becoming difficult (Stapleton 1980). It has been estimated, for example, that about 150 000 new households are formed annually in the Netherlands compared with about 75 000 per year vacating a property because of death or a move to special housing (e.g. aged accommodation) (Linde *et al.* 1986). Several studies assume that young people entering the housing market for the first time move into a dwelling in the less attractive parts of a city but there is growing evidence that this is not the case; rather they are involved in high levels of turnover even among the better housing (Goldschreider and Da Vanzo 1985; Linde *et al.* 1986). Similarly, the growing trend of retirement migration is diverse in nature in that it involves not only moves to coastal areas but also moves that reduce geographical separation from children as well as shifts into special retirement homes (Warnes 1986; Meyer and Speare 1985; Stapleton 1984). Of course, those households within the public rental market find it more difficult to adjust their housing needs with changes in their family status than do the rest of the population, given the limited availability of housing stock from which to choose (Ford and Smith 1990).

The limitations of looking within the life cycle of households for explanations of mobility has led some workers to argue that residential relocation should be seen not just as a reaction to family stress but also as a form of rationally motivated behaviour. In other words, the housing goals of individuals and families have become very important as a focus of study, as have the strategies they adopt to fulfil their goals. A more explicit concern with the long-term strategies of individuals has therefore led to the study of 'housing careers' in the sense of consciously planned housing–mobility strategies which allow the desired housing goals of the individual or household to be reached over time. From this perspective, households are seen to move in order to compensate for some perceived deficit or inadequacy in their current residential situation. However, once in their new location, the comparison of their housing with that of their peers begins again and a renewed desire to move is generated. The desire for mobility, therefore, can be a constant one until all perceived deficiencies in the current residential situation have been met. Interesting as such ideas are, it must be noted of course that the concept of a housing career is only realistic for those in a secure financial situation with good access to housing finance and income (Kendig 1984). It is perhaps more important therefore to examine how people in general search out information on possible places to move.

The search process

The housing market is a good illustration of a decision-making situation where there are many different channels of information, and where a good deal of effort is put into the acquisition of information, much of it spatial in nature. Thus the residential choice process can be conceived of as operating within the context of the housing market, migratory moves being the result of the interaction between the migrant and the market. This interaction involves *search*. The concept of search itself encompasses several interrelated characteristics: it is a goal-directed activity; it involves a complex process of information gathering; it takes place in a context of uncertainty; a point is reached where search ends and a choice is made; and it takes place within a set of constraints. The process of search may therefore be characterized by its duration, the number and type of information sources used, the number of neighbourhoods searched, the number of houses examined, and the radius of the area searched.

Several models have been developed to analyse the process of search (Clark 1982b; Clark and Smith 1982; Meyer 1980). Probably the most useful is the one suggested by T.R. Smith *et al.* (1979). This model

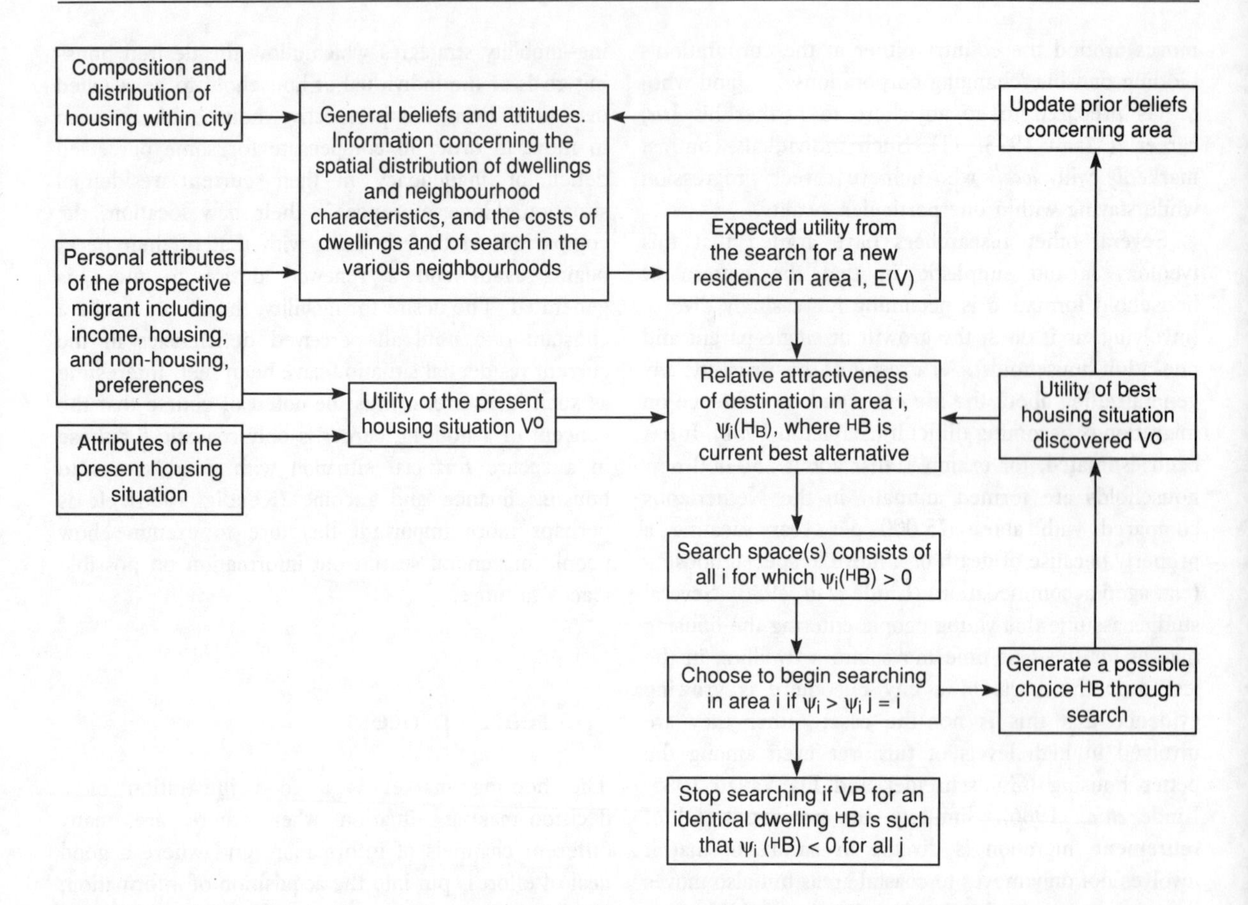

8.8 Schematic model of decision-making and searching (Source: T.R. Smith *et al.* 1979: 8)

proposes that the decision-maker attempts to reduce uncertainty by collecting information about the mean utility in each of several areas (in a way which can be compared to the financial problem of selecting a portfolio of investments). This process takes stock of:

1. The individual's perception of the decision-making environment;
2. The temporal-spatial distribution of vacancies by type and price;
3. The type of information sources available and used;
4. The perceived cost of using these information sources on the efficiency and sequential structure of the search process (Fig. 8.8).

Not surprisingly, several attempts have been made to refine this model and to evaluate its ideal characteristics against the real world (Clark and Smith 1985; Phipps and Meyer 1985; Phipps 1988; Jayet 1990).

In migration, the comparative evaluation of the present location and potential future locations is based on the knowledge a migrant has concerning each alternative. The procedure by which individuals gather such information is guided by the extent and content of their information fields, or the set of places about which they have knowledge. Such a field can be divided into two: an activity space and an indirect contact space. An *activity space* is made up of all those locations with which an individual has regular, almost day-to-day contact. It comprises, therefore, a fairly well-known, albeit geographically restricted, area. In contrast, *indirect contact space* lies beyond the area of the individual's day-to-day contacts, and familiarity with it partly depends upon information from the mass media and other public information sources. For both types of information field, there tends to be a distance decay in the accuracy and amount of information an individual possesses. Within cities, this bias is sectorally oriented, reflecting the regular movements of households from their home to the city centre. According to Adams (1969), this is why the majority of households move within sectors of the city.

The search for another house or a new job depends not only on the household's knowledge of alternative opportunities or locations but also on the type of available information (T.R. Smith *et al.* 1979). Generally, studies have shown that information sources used by potential migrants include newspapers, walking or driving around, friends and relatives, and estate agents. However, there is a high degree of variability in the usage rate (percentage of searches using each source) and the location rate (percentage of searches locating an acceptable alternative through a particular source). In a study of residential mobility in Baltimore, Rossi (1955) found that the most successfully used source was essentially informal. Similar conclusions were reached by Herbert (1973) in Swansea. For example, in the case of a high-cost neighbourhood, 39 per cent of respondents discovered the required vacancy 'by looking around' while a further 25 per cent found what they wanted through family and friends; only 11 per cent found vacancies through newspapers and only 17 per cent through estate agents. The picture in a low cost neighbourhood was even more striking in that the respective figures were 21 and 58 per cent as against 13 and 8 per cent. The importance which attaches to 'casual' information channels suggests it is more than likely that households will first search areas with which they are already familiar and, if this is so, then the action space of the households will play a not inconsiderable part in their selection of the area to which they eventually move. This was confirmed by Palm (1976b) in studies of estate agents as information mediators in San Francisco and Minneapolis; it was also concluded that searchers who depend heavily on estate agents make use of a highly structured information source which has the potential for creating localized imbalances between the supply and demand for housing (Palm 1985). Less is known about the sequence in which different information sources are adopted when a multiplicity of sources is used, and about the interrelationships between different information channels (Clark and Smith 1979), although Talarchek (1982) has suggested that the temporal sequence involves newspapers, agents, friends and relatives, and visiting, in that order.

Information about migration opportunities outside the 'local' environment tends to be provided from sources other than direct personal experience. One of the most significant of these channels is that formed by friends and relatives, who not only provide detailed information about their own location but also can assist the assimilation of the newcomer into a strange community. The effect of such information feedback is to create a distinctive migration stream between two places. For example, Hillery and Brown (1965: 47) have shown that southern Appalachia is not a region in the sense that its parts belong to the same migration system but rather it is a collection of 'backyards' which are connected to non-Appalachian areas, often distant cities, through migration. More recently, Roseman and Oldakowski (1984), in a study of long-distance migrants, and Williams and McMillen (1983), in an analysis of migration to non-metropolitan USA, confirmed the significance of such 'ties' in determining the migrant's choice of a locality. McHugh (1984) even went so far as to claim that, of a sample of metropolitan migrants, 95 per cent had at least one tie with the place selected and 72 per cent had two or more ties.

Information flows are not of course static. Rather they change in space and time. For example, individual learning occurs as activity spaces and indirect contact spaces develop and as new information sources come into prominence. This *learning*, in turn, can be reinforced by searching the environment, particularly the sort of purposeful searching that occurs in the choice of a new residential location. Unfortunately, the nature of environmental learning and searching is poorly understood. Of late, however, the changing information pattern available to individuals as a result of purposeful search has become a focus of investigation, particularly in relation to residential mobility in the city (Clark 1982b; Preston 1987). Generally, the evidence suggests that the search for a new residence in the city will be concentrated near to the home, or in other accessible areas, simply because an individual's activity and awareness spaces are so highly localized. Barrett (1973) and Golledge and Stimson (1987) in Toronto and Adelaide respectively found that the behaviour of the majority of their samples involved the purchase of a house after a short search which covered a few houses in a small area. Clearly, then, house searchers often consider only a small segment of the vacancy list. This fits in with the finding of Brown and Holmes (1971) that the initial stage of search tends to be *space covering* and generally involves a wide area, with the home area being the focal point. At the second stage, which is *space organizing*, the search focuses on a small area where a satisfactory vacancy is likely to be found. If a suitable vacancy is not found then a third stage is initiated, in which the intensity of the search is increased and, possibly, widened. Finally, as time runs out and still no suitable vacancy has been found, households redefine their strategy, either by changing house and neighbourhood requirements or even reversing the decision to move.

In their classic study of the search space of inner- and outer-city residents in Cedar Rapids, Brown and Holmes (1971) found that, with reference to the home from which they had moved, inner-city migrants searched an almost circular area whereas the search of outer-city migrants was sectorally biased. Of course the area from which migrants come may not be the only determinant of the shape of search space. Of crucial importance in a successful search for a house vacancy may be the length of residence; after all, more accurate information available to long-term residents may shorten the length of search (Whitelaw and Gregson 1972; Bible and Brown 1980). This possibility has led Huff (1986) to hypothesize three sorts of relationship between the search pattern of potential migrants and the locational variables influencing it:

1. A constrained choice model which focuses on the relationship between the spatial variation in the number of vacancies in the household's possibility set and the pattern of vacancies visited by the household;
2. An area-based model which highlights the tendency for households to continue searching in a particular community area once they have begun searching there;
3. An anchor point model which focuses on the way in which the locational preference in a household's search behaviour declines with increasing distance from critical points (e.g. prior residence or workplace) in the household activity space.

Despite this, it remains true that information restriction of one sort or another may prevent a household from searching the whole area potentially available to it. For example, a shortage of vacancies may shorten the duration of search (Clark and Smith 1982). Brown (1983) has also indicated that constraints imposed by job-related motives for migration halve the period of search, increase the number of properties viewed per week of search, and increase by fifteen times distances moved by job movers as opposed to non-job movers. In a somewhat different vein, Conway and Graham (1982) argued that high stress levels tend to make the search more purposive and intense and also tend to reduce the time involved. Interestingly, even within an administered (cf. market-oriented) housing system, based on evidence from Glasgow, Kintrea and Clapham (1986) argued that a certain degree of choice and search takes place. Finally, it should be noted that there is overwhelming evidence to suggest that income, ethnic background, and gender have a significant influence upon spatial knowledge and locational choice (Preston 1987).

Despite the broadening of the explanatory basis of migration by the inclusion of the family life cycle, residential evaluation, and spatial searching, the behavioural approach to the study of migration still tends to be demand oriented. It is not surprising, then, that recently an attempt has been made to rectify this weakness and to incorporate into migration analysis consideration of the conditions which control the supply of labour and housing. This has been done on the grounds that these factors affect not only the amount of mobility that can be achieved but also the decision-making context within which migration is undertaken.

Access to housing

In any decision to migrate, whether interregional or intra-urban, access to housing is a vital component. Until recently, the prime focus of those studies attempting to link mobility and housing was upon the concept of *filtering*. Briefly, this concept conceives of the housing market as a ladder up which households move as they adjust the quality and price of their property to their aspirations. As a result of the post-war suburban housing boom which characterized most Western economies, such filtering tends to be outward towards the periphery of cities (although the move to new suburban housing is, of course, only one possible option in 'filtering' since the key elements in the process are the price and quality of dwellings not the age and location). Above all, in the notion of filtering, there is an implicit assumption that housing tenure will be allocated according to ill-defined market competition.

In order to gain some understanding of the way in which markets allocate resources and the way in which this imposes limitations on household choice and movement, the focus of research attention in recent years has shifted to the wider socio-political context within which the supply of housing is determined. This necessitates the study of state policy, financial institutions, builders, real estate agents, and developers (Boddy and Gray 1979). These organizations employ managers and *gatekeepers* who perform a key role in shaping individual and group access to housing stock. They thereby shape the form of housing choice (Pahl 1977). However, despite a good deal of enthusiasm among researchers, the integration of the housing market and its structural constraints with models of residential choice, has so far been limited. A significant contribution has nevertheless been Fischer's (1984) model of 'structural constraint' in which the individual is seen to operate within opportunity fields, thus linking the mobility intentions of households to the opportunities present at any given time.

A glance at the housing market of most Western societies reveals that the degree of access to housing is not uniform for all households even when income differences are taken into account (Ball *et al.* 1988). For instance, in the competition for housing, some individuals have greater power to choose than others, and even those who share the same position in the labour market sometimes differ in their access to housing space. The range of housing choice can, however, be enlarged by financial credit being made available to a greater number of households and by builders contributing more cheap houses for sale. In other words, access to housing and, therefore, migration potential, is a reflection not only of individual constraints but also of the mode of operation of those involved in organizing the housing market.

In Britain, there are three ways in which households can gain access to housing: by credit or capital accumulation; by obtaining a public housing tenancy; and by renting in the private sector. In such a situation households can be seen as competing for limited resources, particularly mortgage capital and subsidized public housing. This results, according to Rex and Moore (1967), in the emergence of classes with different access to these resources (see Table 2.4). Despite considerable controversy over the details of these categorizations (Saunders 1984), the scheme does emphasize that class membership acts as an increasing constraint from the first to the fifth tenure type, with the most fundamental division being between access to mortgage capital and obtaining a public housing tenancy. But, of course, there is no clear cut correspondence between a particular tenure and a particular class, since owner occupation has become widespread and socially dispersed (Karn *et al.* 1985; Forrest 1983).

Access to *owner occupation*, apart from when based on an accumulation of capital, is dependent on mortgage finance. In Britain, building societies provide the largest amount of credit. A major qualifying criterion for access to such credit is security of income, as well as its size; Barbolet (1969), for instance, has shown that for manual and clerical workers earning similar incomes, the latter obtain higher mortgages because of greater job security. Also mortgages are easier to obtain on new premises, except where the applicant is particularly credit worthy. Often this means that those on lower incomes have to resort to insurance or local authority mortgages, with their higher interest rates. They are therefore forced into older, smaller housing. This policy means that owner occupation is less likely to develop in the oldest and most deprived parts of the city, what is often referred to disparagingly as 'the inner city'. This is a tendency which is further accentuated by restrictions on local government lending.

For those unable to attain home ownership, the best alternative is *public tenancy*. In Britain, entering into public housing is dependent on housing 'needs', preference usually being given to large families and those affected by slum clearance, commonly with a residential qualification of about 5 years' duration (Niner 1975). In the allocation of public housing, impressionistic evidence points towards a tendency for certain local authorities to grade their tenants and prospective tenants both on the grounds of cleanliness and on the grounds of regularity in their rent payments. Such a grading results in the 'segregating' of 'bad' tenants into older, poorer housing (Gray 1976). Those households who fail to qualify for a mortgage or a public housing tenancy are of course forced to live in the *private rented sector*, which in 1985 occupied only 13 per cent of all housing in Britain (compared to 54 per cent in 1950). Access to the private rented sector is based on the production of a week's or a month's rent only, yet here again there are constraints. Landlords have considerable influence over the kind of person to whom they are prepared to rent property, and even estate agents, when acting on behalf of a landlord, can discriminate against certain people.

Recent government housing strategies being followed in capitalist welfare societies such as Britain (Forrest and Murie 1988), Canada (Bourne 1986), Australia, and New Zealand (Thorns 1988) have tended to intensify financial and ideological support for ownership at the same time that they have progressively privatized public sector housing. This has resulted in owner occupation no longer being the privilege of the minority élite. In principle, such strategies have therefore reduced the constraints on mobility within the city. So widespread has this process been that Saunders (1984: 203) has claimed that

> social and economic divisions arising out of the ownership of key means of consumption such as housing are now coming to represent a new fault line in British society (and perhaps others too), that privatization of welfare provision is intensifying this cleavage to the point where sectoral alignments in regard to consumption may come to outweigh class alignments in respect of production, and that housing tenure remains the most important single aspect of such alignments because of the accumulative growth of house ownership and the significance of private housing as an expression of personal identity and as a source of ontological security.

In response to this claim, a debate has emerged as to whether occupational class is of primary importance in structuring access to consumption (and the pattern of benefits thereby gained) or whether consumption sectors share certain fundamental and common interests which really do not coincide with occupational class. Housing is not of course the only dimension of consumption cleavages referred to in this debate but it is the most prominent and the one for which fundamental interests can most easily be argued. At the heart of this debate is the possibility that the key role of home ownership is not the fostering of mobility but rather the appropriation of exchange value and thereby wealth accumulation (Murie 1986).

In any evaluation of this claim it should be borne in mind that a growing socio-tenurial polarization has taken place between the 'privileged majority' (who can achieve satisfaction through ownership) and the 'marginalized minority' (those who are left in a dwindling, under-resourced public sector) (Willmott and Murie 1988; Forrest and Murie 1983). However, it must be noted that not all privately owned housing is good housing because a significant minority of such dwellings are materially substandard in one or more respects. Many owners of these dwellings therefore fail to fully obtain the economic benefits normally associated with ownership. For example, cheap, old, inner-city property does not appreciate in market value as rapidly as suburban property (and indeed sometimes fails to keep place with inflation). What appears to be emerging, then, is a cardinal divide between conventional and substandard owner occupation, with the autonomy and self-determination conferred by the former being relatively absent from the latter (Murie 1983; Forrest 1987). Cater and Jones (1989) have incorporated these considerations into an adaptation of Rex and Moore's original model of access to housing (Table 2.4). Briefly, this model emphasizes that the emergence of a new form of 'submerged housing' within the owner occupied sector means that, for many of the 'new' home owners, mobility potential will remain constrained. This corroborates the view of Murie (1986) that the quality of the housing which people own, and the rate of accumulation they experience, will increasingly reflect income and wealth. He argues that the debate over the extent to which consumption cleavages in housing cross-cut occupational class should be tempered by a recognition that we have been looking at a system in transition and that, at the end of the transition, there is likely to be a better fit between occupational class and housing stock than presently exists.

Availability of housing

So far this chapter has been concerned with those institutions that control the consumption and exchange of housing. Other institutions control the type of housing that is available; these include landowners, builders and developers, and politicians and planners, all of whose actions are highly interdependent and often constrain the range of housing choices. For example, development control policies devised by politicians and operated locally by planners, indirectly influence the actions of landowners and builders.

It is appropriate to conclude this chapter therefore by illustrating how some of these housing institutions have influenced residential mobility by reference to two examples. First, the phenomenon of 'gentrification', which involves the buying up and modernization of working-class housing by or for middle-class households (Williams 1978). This is a process which became apparent in the inner parts of London during the early 1960s, and more particularly after the Housing Act 1969, as a result of the availability of standard improvement grants, together with local authority expenditure on environmental improvements. Since there was no check on owners selling their improved housing, these grants could add both extra financial incentive to the activities of property companies and the promise of capital gains to private owners. A change in the law in the Housing Act 1974 reduced opportunities for speculation by limiting the availability of improvement grants but this has not completely removed speculation. Essentially, 'gentrification' has been initiated by the middle classes and several institutions involved in housing, within the context of the government's overall housing policy. Although it has achieved the worthwhile objective of improving existing housing, it has done so at the expense of those least able to compete.

The second example involves homeless households (Thorns 1989). While there was a 25 per cent increase in owner occupation in Britain during the 1980s, there was, on the other hand, a sharp rise in the number of homeless households. The largest number was in the South-east though proportionately the increase was greatest in the North. Many commentators attribute this situation directly to changes in government policy, particularly a cut in the housing investment programme. For example, the reduction in the council house building programme and the shift in the responsibility for low-income housing provision to non-profit-making housing associations, meant that in 1992 there were only 6000 new council houses compared to a housing association contribution of 23 500 dwellings. According to the Director of Shelter, in Britain there is realistically a shortfall of some 1 million homes: 'for years we have been saying a crisis exists and for years it has been allowed to get worse. We are going to have a homelessness problem in this country the likes of which has not been seen outside New York' (McKechnie 1989).

In the analysis of housing availability, a recurring issue is the definition of *housing need*. For example, homelessness in Britain is usually defined by reference to the number of people registered under the Homeless Persons Act, but it is well established that these figures understate the reality of the situation. Therefore, many researchers claim that the 'hidden' number of homeless can best be identified by a more qualitative approach. An illustration of this procedure is exemplified in the adaptation by Forrest and Murie (1987) and Mooney (1991) of *residence history analysis* (Pryor 1979). In the context of housing needs and mobility, this entails a biographical method of life history analysis focused on the compilation of records of an individual's changing housing circumstances; for example, note is taken of the factors influencing changes in housing needs and the problems encountered when attempting to change a housing situation to match changing housing needs. In other words, housing histories provide the researcher with access to real life experiences of people as they negotiate their way through the housing system.

Thorns (1985) has taken these ideas a stage further

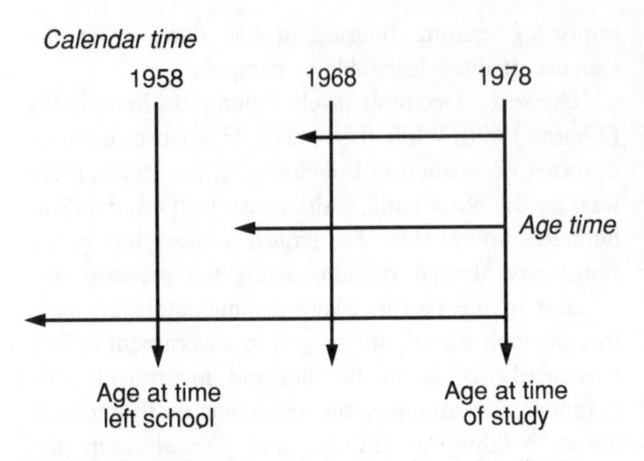

8.9 Residential history analysis: time dimensions (Source: Thorns 1985: 834)

by incorporating both structure and individual dimensions within a time–space framework. As Fig. 8.9 illustrates, a distinction is made between age time (family life cycle) and calendar time (the date individuals move), thus emphasizing the significance of the economic and social context of the mobility process. The blending of structure, process, and historical context allows both the institutional organizations of the housing market (which set the arena for residential adjustment), and the patterns of individual choices, to be seen more clearly. This can be illustrated with reference to interregional migration in Britain during the 1980s.

Migration decision-making in Britain during the last decade cannot really be understood without reference to restructuring of the British economy, the recession, the continued vitality of the South-east region, the deregulation of the financial market, and the changing role of government in the housing market. During the early 1980s, the labour market was providing a clear signal in terms of employment prospects for people to move from North to South, leading a politician (Norman Tebbit) to state, 'I grew up in the 'thirties with an unemployed father. He did not riot – he got on his bike and looked for work and he kept on looking until he found it.' But the message from the housing market has been almost exactly the opposite of that from the labour market.

Since the Second World War, property ownership has been encouraged in Britain by the almost universal belief that housing represented one of the safest and best performing forms of investment. The growth of owner occupation was initially stimulated by tax relief on mortgage repayments for home ownership and, during the 1980s, it has been boosted further by a number of policy changes. The liberalization of the financial system has led to a greater variety of institutions investing in housing and thus making loans available to a greater number of households. An additional stimulus was given by the government's decision to sell council houses to tenants at below market price. Traditionally, owner occupiers as a whole represent the most mobile sector of the population and so, with the spread of owner occupation across a wider spectrum of households, it was thought that constraints on residential mobility would be attenuated. However, compared with the 1970s, the level of interregional migration actually fell by over two-thirds in the 1980s.

One of the effects of the 1979–80 recession, the deepest since the 1930s, was to discourage workers from moving to where work was more likely to be found. The fact that all regions experienced high unemployment appears to have had a generally discouraging effect on people's willingness to consider moving to another region. For example, the existence of significant levels of unemployment even in the South in the early 1980s suggested to potential migrants that those jobs that were available would soon by snapped up by the local pool of unemployed labour.

The prospect of moving to the South became an even more daunting if not impossible one for northerners as a result of a clear and widening house price differential, compared with earlier periods. Part of this was, of course, a result of the self-contained economic success of the South-east, compared with the North, and the consequent 'ripple effect' of high house prices spreading out to adjacent regions. For potential migrants, unless the job that they could move to was very highly paid, the extra cost of the mortgage would appear to represent an insurmountable barrier to mobility. The choice confronting a potential migrant is therefore between paying a much higher mortgage and trading down in the housing market by moving to a smaller property. Those already living in small and cheap property probably do not have even this choice. Moreover, for the former council tenant, owner occupation does not necessarily mean greater mobility potential. The process of privatization takes time, involving both a time limit to the resale of the property without penalty and a desire by the former tenants to distinguish their property from those houses still under council control prior to selling. Predictably, former council properties also become more saleable as the proportion of houses on an estate which have been taken into private ownership increases. This too takes time.

The level of immobility for those remaining in the council sector continues to be significant. Since a council house is something (1) which has to be queued for, and (2) which still has substantial advantages over private renting, it is not going to be given up easily. Given that the demand for council accommodation invariably exceeds its supply, local councils are reluctant to allow outsiders to jump the queue. Councils do run local schemes of property exchange and, at the regional level, a National Mobility Scheme has been instituted, but so far such initiatives have merely scratched at the surface of the mobility problem.

Another feature of government housing reform in the 1980s was the revival of the private rented sector. The philosophy behind the encouragement of the private sector is that, by increasing housing supply and choice in those parts of the country where work is more available, a significant barrier to mobility will be removed. But a question has to be asked as to who are the potential tenants. Owner occupation remains attractive through taxation subsidies and increasing house prices and, unless public housing rents are lifted to reflect truly their market value, council tenancies will continue to be in great demand and sought after in preference to private tenancy. In other words, the best hope for the private rented sector will not be in some sudden switch by existing owner occupiers and council tenants to private rented accommodation, but in new households taking up that form of tenure, probably those households which would formerly have gone for council housing but who in future will be less able to do so.

Conclusion

Over two decades ago, Mangalam (1968: 16) complained that 'migration is practically a virgin area for those who want to study the phenomenon from a behavioural point of view'. From the evidence reviewed in this chapter, it can be safely concluded that the last decade has seen a proliferation of attempts to conceptualize the migration process within a decision-making framework. However, much of this research has been criticized for its tendency to view decisions within an unconstrained context which many consider only applicable to the wealthy groups in Western society. 'If we continue to place our faith in analytical studies of the individual without regard to the societal relations which define the context within which relocation adjustments are made, we are liable to design programmes and policies whose outcomes fail to live up to expectations' (Clark and Moore 1980: 18). During the past decade, therefore, behavioural research has emphasized the point that migration takes place within the context not only of imperfect knowledge but also considerable constraint. These constraints reflect the structured nature of society and manifest themselves locally in terms of differential access to employment and housing. Differences in the availability of, and access to, these resources result from the interaction of several institutions and organizations, which operate within their own spatial and social biases to provide a highly structured framework for household decision-making. However, despite considerable research on the nature and operation of these institutions involved in the provision of employment and housing, so far little attention has been given to how they directly influence the decisions made by individual households as to when and where to move.

9 Shops and Shopping

Despite the fact that shopping accounts, on average, for less than 1 per cent of an individual's time (Chapin and Brail 1969), and despite the fact that retailing comprises only a tiny proportion of all land use, the study of consumer behaviour has generated a vast literature, both in business (e.g. Foxall 1980) and in geography (e.g. Shepherd and Thomas 1980). As a result, 'consumer spatial behaviour' has become the focus of a good deal of geographical research. Basically, this term refers to both the processes by which shoppers choose which shops and shopping centres to visit, and their actual travel patterns. Early studies of consumer spatial behaviour by geographers focused on the relationship between consumer travel and central place theory (Berry 1967). Application of the theoretical ideal of rational economic behaviour to the journey to shop led to the proposition that consumers *minimize travel* and therefore patronize the nearest shopping centre offering whatever goods are required. Put simply, much of this early work followed simple central place theory in assuming that shopping centres on the same hierarchical level are equally attractive to consumers with the result that there is no reason for a shopper to patronize other than the nearest one (Halperin 1988).

This proposition of course ignores the fact that rational economic human beings are sensitive to price and so may choose to patronize centres other than the nearest one if the price of the goods on offer is suitably low. It is, moreover, a proposition that is greatly at odds with real-world behaviour. Although it is sometimes difficult to distil the general from the particular (and therefore to differentiate between the study of 'spatial behaviour' and the examination of 'behaviour in space' (see Chapter 1)), studies in a number of places have shown that only a proportion of consumers minimize travel between their place of residence and the shopping centres which they visit (see Thomas 1976: 37) and that many consumers do not use the central place hierarchy as 'rationally' as they should (Johnston and Rimmer 1967). There are several reasons for this: there is evidence that some consumers like to vary their shopping behaviour from time to time and therefore occasionally change the shopping centres that they visit; factors like quality, service, and friendliness can outweigh economic considerations; and a great deal of shopping is undertaken on multi-purpose trips, or in such a way as to fit in with the management of a consumer's other travel commitments.

This is not of course to say that no consumers minimize travel. Some most certainly do. Indeed, in terms of the paths that pedestrians follow *within* shopping centres, Garling and Garling (1988) found that up to two-thirds of consumers minimize the distance that purchases have to be carried. Rather the point is that researchers working on shopping behaviour quickly realized that an understanding of shopping patterns necessitates an understanding of why individuals behave as they do. Simple models of economic rationality do not provide accurate descriptions of real-world behaviour despite recent attempts to extend their scope (e.g. Bacon 1984). More recently, it has come to be recognized that such models also have inherent logical inconsistencies which render them an inadequate basis for an understanding of real-world behaviour (see Barnes 1988).

The major weakness with the proposition of distance minimization is that it is deterministic: it reduces the complexity of the real world to only one variable (distance) which consumers treat in a specified way (avoid). In reality, of course, a decision as to which shopping centre to visit is often reached only after a trade-off between considerations like size, attractiveness, and distance. The nature of this trade-off has been studied in the approach usually referred to as *spatial interaction theory* (see Chapter 2). Basically, this approach takes the view that account can be taken of the variability of individual shopping behaviour by examining such behaviour *in the aggregate*. The focus, in other words, is on overall shopping patterns. What the individual shopper does is not of concern providing the overall pattern is understood. Early work in this vein was often very deterministic. It is exemplified by Reilly's (1931) law of retail gravitation which sought to pinpoint such things as the 'breakpoints' between the trade areas of different shopping centres, using the equation

$$D_B = \frac{D_{AB}}{1 + \sqrt{(P_A/P_B)}}$$

where D_B is the distance to the breakpoint from the smaller town B, D_{AB} the distance between towns A and B, and P_A and P_B the populations of towns A and B respectively. Clearly, this sort of equation still assumes that all shoppers behave in an identical manner (i.e. they minimize distance).

More realistic portrayals of behaviour are to be found in probabilistic gravity models (see Shepherd and Thomas 1980). One of the first and best known of these models is the one developed by Huff (1963) and based on floor space and travel time:

$$P(C_{ij}) = \frac{\dfrac{S_j}{T_{ij}\lambda}}{\sum_{j=1}^{n} \left[\dfrac{S_j}{T_{ij}\lambda}\right]}$$

where $P(C_{ij})$ is the probability of a consumer at i going to j, S_j the square footage of space devoted to a particular class of good at j, T_{ij} the travel time, and λ a parameter to be estimated empirically to reflect the effect of travel time on various kinds of shopping trips. There are several problems with this type of approach. First, it does not do justice to the host of different factors that influence how a consumer evaluates a shopping centre (e.g. quality of goods and services, perceived attractiveness, ease of parking, price, friendliness). Second, there are very great problems involved in calculating parameter values for both distance (travel time) and attractiveness (square footage) (Openshaw 1973). Third, the approach lacks any sound theoretical underpinning (Jensen-Butler 1972) and violates certain technical aspects of neo-classical economic theory on which it is supposedly based (Hubbard and Thompson 1981). Above all, the approach tends to produce descriptions of behaviour as it occurs in specific situations and it is not always clear whether the findings are applicable beyond the specific situation. Specifically, the value of λ is calibrated in each application of the model in such a way as to provide a fit with observed data and there is no way of knowing whether the same value would apply in a different retailing context (e.g. a different country). In short, then, there are serious weaknesses with aggregate-level gravity models to the extent their use tends to be context specific.

One attempt to model consumer behaviour in a way that enables consumer choice to be generalized beyond a specific situation is the *revealed space preference* approach developed by Rushton (1969a, 1976). This approach proposes that consumers rank all alternative shopping centres on a scale of preferredness, that consumers patronize the most preferred centre, and that overt behaviour can therefore be predicted from a knowledge of preferences. Rushton believed that preferences were capable of measurement by means of non-metric multidimensional scaling of pairwise comparison data. In other words, shoppers could be presented, in a survey, with pairs of shopping centres and asked to specify which member of the pair they preferred. From the results of such surveys, average preference structures (indifference curves) could be calculated. Rushton introduced generality to his model by discussing 'locational types' (various combinations of size and distance such as 'small, nearby centres' or 'large distant centres') rather than specific shopping complexes. Computationally, the procedure is complex and it is not surprising to find that most studies have simplified the problem and considered only the two variables of size/attractiveness and distance. Figure 9.1 shows what the approach does: pairwise comparison of

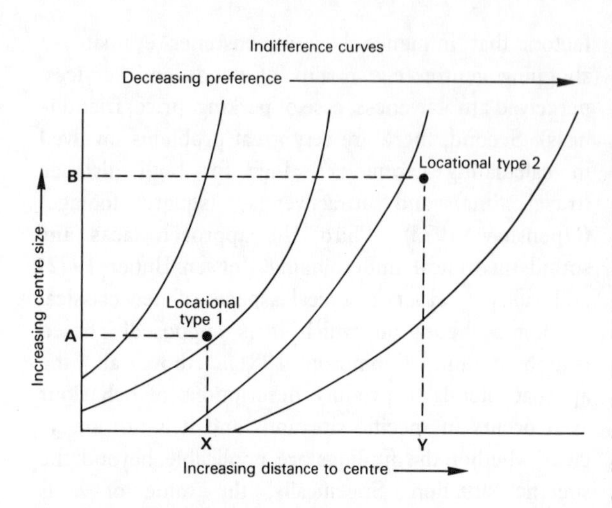

9.1 Revealed space preference

a large number of locational types produces indifference curves such that it is possible to see that a small centre (A) at a short distance (*X*) (locational type 1) is preferred to (and therefore used more than) a large centre (B) at a great distance (*Y*) (locational type 2).

Rushton's model was developed for inter-urban travel but has been adapted to *intra*-urban consumer behaviour by Timmermans (1979) in a paper that defined locational types on the basis of number of employees (cf. size) and 'reasonable travel time' (cf. distance). In other words, Timmermans, unlike Rushton, made the distance limit endogenous to the model. That is to say, instead of specifying distance constraints in the definition of locational types, he permitted respondents to define for themselves what they considered to be a 'reasonable' amount of travelling time. Despite this refinement, the revealed space preference approach is open to many criticisms (Pirie 1976): locational types tend to be defined on only two variables (some measure of size and distance) in order to keep the number of alternatives to manageable proportions for inclusion in the time-consuming paired-comparison methodology; there is no real evidence that consumers use locational types in their evaluation of shopping centres; the approach can only really be applied to known and experienced situations and hence the findings cannot be extrapolated to the hypothetical situations encountered in planning; indifference curves represent only the average preference of the respondent group and so are of little use in predicting individual behaviour; and, perhaps more fundamentally, distance should not be regarded as a dimension of choice so much as a *constraint* on choice in so far as whether or not a shopper goes to a distant centre is more likely determined by time, cost, and mobility constraints rather than by an evaluation of the preferredness of different centres (MacLennan and Williams 1979). In short, then, the revealed space preference approach suffers from the weakness that it is very much an aggregate level approach in that it relies on the specification of overall indifference curves and overall evaluations of locational types. Thus, although Rushton recognized the variability of individual shopping behaviour, the revealed space preference approach glosses over that variability in favour of summarizing overall choice patterns.

The defence against these criticisms has been spirited. Timmermans (1981), for example, has attempted to show how preferences can be transferred from one context to another (e.g. there might be an enduring preference among shoppers for big, new planned shopping malls). He has also suggested that size and accessibility are, in any case, the aspects of shopping centres that individuals consider most often (Timmermans *et al.* 1982). It remains true none the less that the revealed space preference approach fails to take stock of influences on consumers other than those related to the shopping centre. It does not, for instance, explicitly consider the characteristics of the *shopper*. Yet the importance of such characteristics has been recognized for some time. Huff (1960), for example, expressed some of these influences schematically (Fig. 9.2) and emphasized geographical location, social differentiation, and personal mobility as key characteristics. This work was of course purely schematic and no attempt was made to quantify the influence of the various characteristics. Despite the fact that Huff's work is now over three decades old, subsequent research has not materially added to the variables listed in his diagram, although geographical location was considered in rather more detail by Cadwallader (1975) who emphasized the salience of three variables in consumer decisions as to which shopping centre to visit: the attractiveness of stores (as

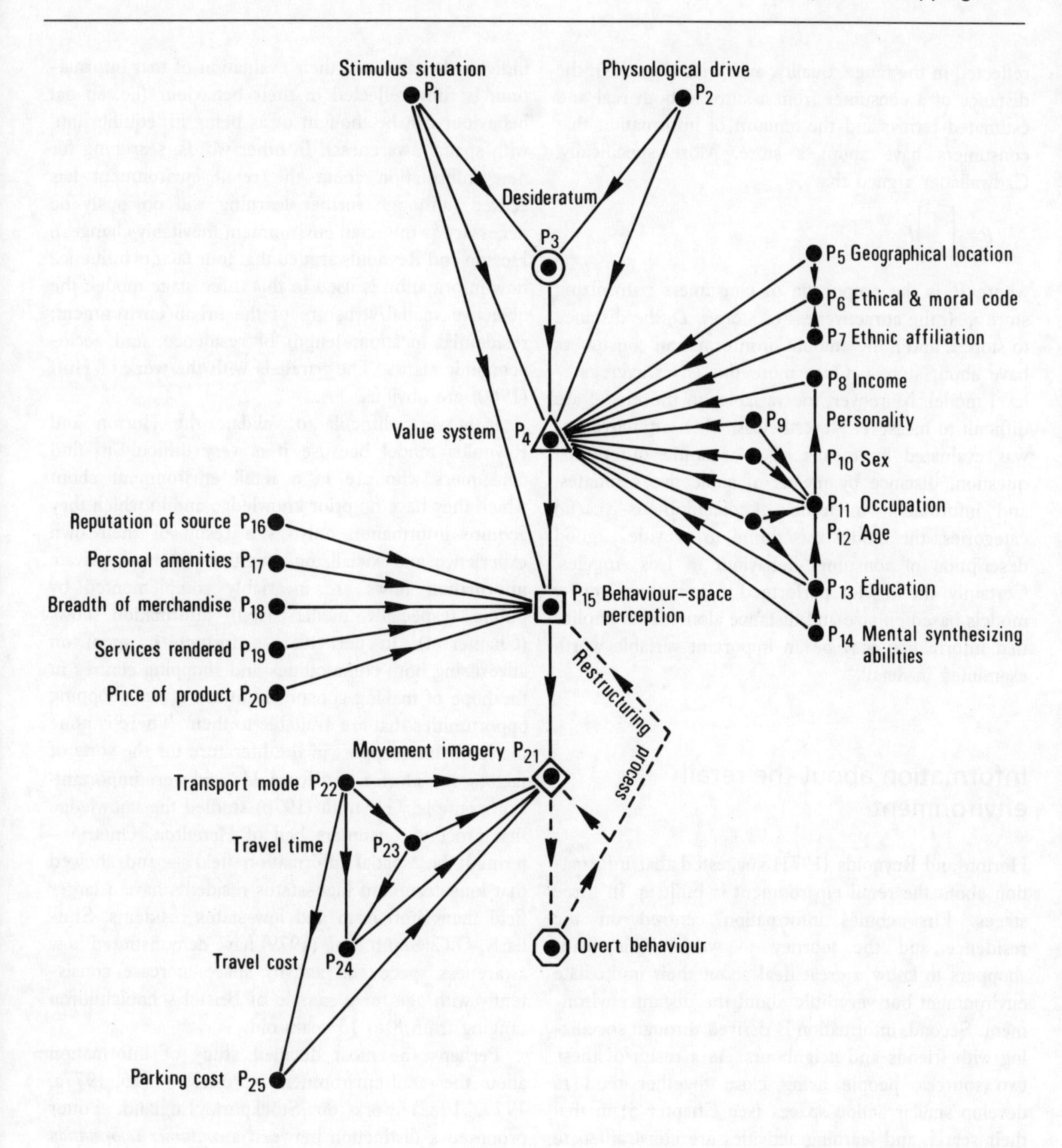

9.2 Influences on consumer behaviour (Source: Huff 1960: 165)

reflected in the range, quality, and price of goods); the distance of a consumer from a store in both real and estimated terms; and the amount of information that consumers have about a store. More specifically, Cadwallader argued that

$$P_i = \left[\frac{A_i}{D_i}\right] I_i$$

where P_i is the proportion of consumers patronizing store i, A_i the attractiveness of store i, D_i the distance to store i, and I_i the amount of information consumers have about store i. Once more this is an aggregate-level model. Moreover, the variables in the model are difficult to measure. Nevertheless, when attractiveness was evaluated by means of a semantic differential question, distance by means of subjective estimates, and information in terms of dichotomous yes/no categories, the model was found to provide a good description of consumer behaviour in Los Angeles. Certainly the model performed better than gravity models based on size and distance alone. This implies that information may be an important variable worth examining in detail.

Information about the retail environment

Horton and Reynolds (1971) suggested that information about the retail environment is built up in three stages. First comes information centred on the residence and the journey to work. This leads shoppers to know a great deal about their immediate environment but very little about the distant environment. Second, information is derived through socializing with friends and neighbours. As a result of these two sources, people living close together tend to develop similar action spaces (see Chapter 5) in that their search and learning activities are constrained to approximately the same area. This process is of course helped by residential differentiation whereby similar sorts of people come to be grouped together within cities (see Chapter 8). The third stage in the build-up of information occurs when shopping has become a matter of routine. A state of 'spatial equilibrium' can be said to exist to the extent that the information that individuals have, and their evaluation of that information, is fully reflected in their behaviour (i.e. spatial behaviour can be thought of as being in 'equilibrium' with spatial awareness). In other words, searching for new information about the retail environment has ceased (although further learning will obviously be necessary as the retail environment inevitably changes). Horton and Reynolds argued that four factors influence how information is used in this three stage model: the objective spatial structure of the urban environment; residential location; length of residence; and socio-economic status. The parallels with the work of Huff (1960) are obvious.

It is very difficult to validate the Horton and Reynolds model because it is very difficult to find consumers who are in a retail environment about which they have no prior knowledge and in which they acquire information only as a result of their own experience and socializing activities. After all, private information flows are invariably complemented by public (especially media-based) information flows (Chapter 4). In particular, a fortune is spent on advertising both commodities and shopping centres in the hope of making consumers aware of the shopping opportunities that are available to them. There is none the less some support in the literature for the sorts of factors that Horton and Reynolds said were important. For example, G. Smith (1976) studied the knowledge that grocery consumers had of Hamilton, Ontario – termed the 'spatial information field' – and showed that long-term and high-status residents have a larger field than short-term and low-status residents. Similarly, G.C. Smith *et al.* (1979) have demonstrated how awareness space and activity space increase consistently with age for a sample of Bristol schoolchildren ranging from 8 to 15 years old.

Perhaps the most detailed study of information about the retail environment is Potter's (1976, 1977a, 1979, 1982) work on Stockport, England. Potter proposed a distinction between a *consumer information field* and a *usage field*. These concepts are analogous to action space and activity space (see Chapter 5). The former refers to the shopping centres about which the consumer holds information and the latter refers to the centres that are actually visited. Potter suggested that both the information field and the usage field occur as wedge-shaped sectors centred on the place of residence

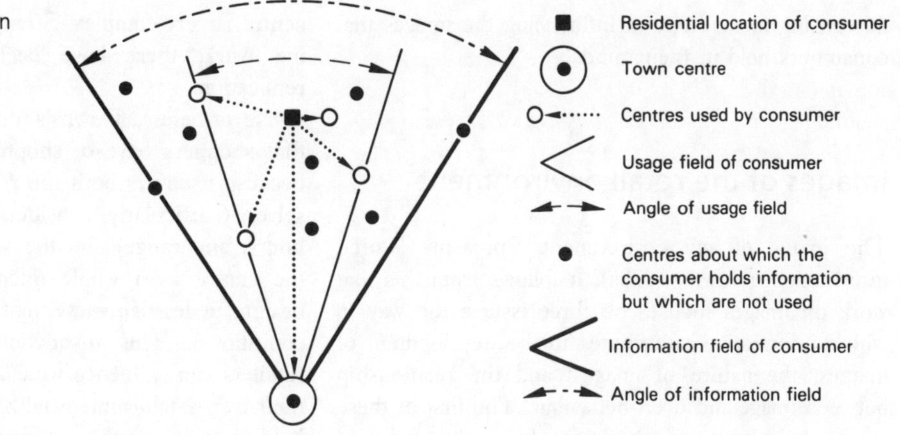

9.3 Consumer information and usage fields (Source: Potter 1979: 21)

and focused on the central business district in such a way that the usage field has a narrower angle than the information field (Fig. 9.3). Diagrammatic representation of these fields for a large number of consumers showed that the angle of the information field increased as social status increased. In the case of the highest-status groups, the angle of the information field was commonly greater than 180°, indicating that the shoppers in question were familiar with more than half of the city. Similarly age and family size were important to the extent that the over sixties and the families with five or more children had a more limited information field than the rest of the population. This has been corroborated in G.C. Smith's (1989) finding that elderly consumers tend to have more limited and more localized images of the retail environment than younger consumers, possibly because of the constraints imposed by decreasing physical competency, diminished mobility, and decreased disposable income. In Potter's work there was also a directional bias in shopping behaviour in that two downtown centres (towards the central business district) were visited for each uptown centre (away from the central business district).

The same tendency for information to be biased to the area around the home and for behaviour to be directionally biased towards the city centre was noted by Hanson (1977) in a study of consumers in Uppsala. This work also demonstrated that individuals had more information about the location of stores than about their interior characteristics, thereby emphasizing the importance of cognitive mapping in providing shoppers with a frame of reference on which to 'hang' information that they acquire about the facilities on offer at particular locations..

A major determinant of the amount of information available to consumers may be the degree of *search* that has been undertaken. For example, Hudson (1975) has taken the view that, as search proceeds, consumers shift attention from simple factual knowledge to what they consider to be positive features of shopping centres (e.g. price, service). Of course, searching is by no means a simple process and market researchers have differentiated different types of search behaviour. Bucklin (1967), for example, distinguished between (1) 'full search' involving visits to many stores, (2) 'directed search' where some comparisons are warranted, especially in relation to price, but full search is not worthwhile, and (3) 'casual search' where information is gained in a non-purposeful manner. Moreover, the type of searching undertaken can be influenced by the type of good required: a consumer may treat the purchase of a costly item as an exercise in problem-solving that warrants extensive searching but may adopt 'routinized response behaviour' in relation to more familiar items, particularly those items whose purchase can be undertaken on multi-purpose (i.e. partially non-retail) trips (Howard and Sheth 1969). And, of course, information about the retail environment is not something that is lying around waiting to be gathered. In many cases it is something actively promoted by retailers through advertising in the hope of influencing the images that consumers hold in their minds.

Images of the retail environment

The image of any environment represents 'stored information' (Downs 1970a). It follows from this that work on images focuses on three issues: the way in which information contributes to the development of images; the nature of images; and the relationship between image and overt behaviour. The first of these issues has been largely ignored in the field of consumer spatial behaviour, possibly because the link between information and image is extremely complex and encompasses both private information (gained largely from experience) and public information (gained largely as a result of exposure to advertising). Cadwallader (1978) has shown, for instance, that the information surfaces and preference surfaces of Los Angeles consumers are convoluted and only very weakly related to each other. It is not surprising, therefore, that behavioural geographers have directed most of their attention to the nature of the image of the retail environment and to its relationship to shopping behaviour.

The pioneering work on the nature of the image of shopping centres was produced by Downs (1970b). This study was concerned with the perception of the internal characteristics of the downtown shopping centre of Broadmead in Bristol. Downs hypothesized that there would be nine elements to the image of the centre: price, structure and design, ease of internal movement and parking, visual appearance, reputation, range of shops, quality of service, shopping hours, and atmosphere. Four semantic differential questions for each of these nine elements were presented to a sample of women. The wealth of data that was generated was reduced by subjecting the responses to a factor analysis which identified eight components of the image of the centre. Of the original nine elements, reputation and atmosphere did not prove to be significant. In contrast, ease of internal movement and parking was separated into two separate components: internal pedestrian circulation, and traffic conditions. In short, Downs showed that the image of a shopping centre is very complex. Strangely, for such a pioneering work, there have been no real attempts at replication.

Nevertheless, Downs's delimitation of the image that shoppers have of shopping centres is important because it covers both the retail establishment themselves (particularly considerations of service, price, hours, and range) and the structure and function of the centre as a whole (incorporating structure and design, pedestrian movement, appearance, and traffic conditions). This distinction is important because retailers can influence to a large extent the image of their own establishment (although the image that they hold may be rather different from that held by consumers, Bloomestein *et al.* 1980) but they can do little about the image of the centre as a whole except indirectly through local chambers of commerce or through management in the case of planned shopping centres that are operated as a unit. Some types of shopping centre may in fact have poor images and this poor image may well influence consumer behaviour. For example, Potter (1977b: 352) has drawn attention to the 'poor imageability' of arterial shopping centres that results from 'their specialized functional character, linear morphology, fabric discontinuity and low qualitative tone' relative to nuclear shopping centres. In a similar vein, Walmsley (1972b) has shown how consumers travel further to patronize a planned centre than an unplanned centre. The demise of central business district shopping areas might also relate to the poor image that some city centres have in the eyes of shoppers, often for non-retail reasons (e.g. crime, congestion, dilapidation).

It is one thing to suggest that particular types of shopping centres have favourable or unfavourable images but it is altogether a different problem to explain how the image of the retail environment as a whole may influence the way consumers choose between competing centres. Human geography has in fact been weak in showing the links between cognition and behaviour. Hanson (1984), for example, has shown that, in Uppsala, Sweden, shoppers with information about more locations within the city travel more frequently and make more complex, multi-stop trips than do shoppers with limited information, and yet the overall areal extent of the activity systems of the highly mobile was approximately the same as that of the less mobile. Likewise, at an aggregate level, it is possible to show that consumer spatial behaviour is related more closely to a cognitive map than to a 'real' map (Mackay *et al.* 1975), but it is difficult to translate this into predictions about individual behaviour because of the very formidable methodological problems that surround the definition and measurement of images (Hudson 1974).

One way of overcoming these difficulties may be to adopt the repertory grid methodology approach to personal construct theory (Chapter 5). Certainly, use of personal construct theory helped Hudson (1976a) to design an interesting model of how consumers decide which shop to patronize, albeit in a model that is, like much theorizing about shopping behaviour, conceptually appealing rather than empirically proven. A three-stage process is involved. First, consumers build up knowledge and preferences over time. That is to say, consumers learn about shops and develop constructs (e.g. cheap–expensive, good range–poor range) for evaluating these to the point where there emerges the notion of an 'ideal shop'. Second, specific needs arise and this leads to an evaluation of a subset of the shops known to the individual, possibly using a subset of the evaluative constructs. In other words, the basic image of the environment is transformed to give one relevant to the particular need. Third, an actual choice is made of one shop from within this simplified image, the shop selected being that which is most similar to the 'ideal shop'. Although Hudson's study of student shopping behaviour in Bristol lent some support to the model, its general applicability is limited because it focuses on very simple situations (the choice of a single shop) and because it has not been related closely to work on consumer learning and decision-making. Moreover, as with many attempts to understand shopping behaviour, little or no emphasis is given to consideration of the constraints under which shoppers operate.

Nevertheless, the idea of studying personal constructs is a sound one that has the potential to contribute significantly to the understanding of shopping behaviour. Timmermans *et al.* (1982), for example, have used personal constructs to discover how shoppers in Eindhoven, Netherlands, differentiated between shopping centres that functionally appeared quite similar. What they found was that parking facilities, proximity to home, perceived atmosphere, and variety of products were the critical concerns to shoppers. And, at a somewhat different scale, Opacic and Potter (1986) have used personal constructs successfully to study how high- and low-income shoppers evaluate individual store types (e.g. superettes, supermarkets, hypermarkets), showing among other things that low-income consumers place a higher premium on price and distance than do high-income consumers.

Consumer learning and decision-making

Shopping is a form of recursive behaviour in that it is repeated at relatively short intervals. It can be presumed therefore that the genuine decision-making of a consumer when first placed in a new environment becomes routinized as behaviour is repeated. In other words, behaviour changes over time as learning occurs. The end result of this process is sometimes a form of repetitive behaviour focused on loyalty to particular stores. Some authorities have suggested that such store loyalty is a negative attribute, forced upon shoppers through constraints on time, money, and transport (Charlton 1973). In contrast, Dunn and Wrigley (1984) have argued that certain retail chains in Britain (e.g. Sainsburys, Marks and Spencers) have been remarkably successful in fostering store loyalty especially among high-income households where loyalty has traditionally been low, at least in comparison to families on low income, with young children, and only one partner in the paid workforce. There is, in other words, some evidence that loyalty results from choice based on learning. According to Hudson (1970), four factors can influence this learning process: the nature of the retail environment; the nature of individual 'behaviour spaces' (a term approximating what might otherwise be called 'action space'); variations in individual evaluation of elements within 'behaviour spaces'; and random fluctuations in behaviour. Moreover, learning itself can be conceptualized as either a change in behaviour or a change in cognitive structure. Hudson saw in this distinction a parallel with the difference between stimulus–response learning theories and cognitive learning theories (see Chapter 4). In Hudson's view, the former are based largely on need reduction and derived from experimental psychology (e.g. experiments where rats, motivated by a need for food, find their way around mazes and have 'correct' choices 'reinforced' through rewards). They are therefore of little value in the study of consumer behaviour. By way of contrast cognitive theories are much more realistic in that they emphasize knowledge of 'behaviour space' and allow for changing expectations of behavioural outcomes. They therefore focus on how individuals come to know the overall retail environment in which they live.

Golledge has probably done more than anyone else to popularize geographical studies of learning (Golledge 1967, 1969; Golledge and Brown 1967). He argued that models of 'place loyalty', whereby individuals patronize a particular centre, represent only one of a number of solutions produced by the learning process. It is just as easy to see consumer behaviour as a market-sharing activity that is amenable to mathematical modelling. Specifically, Golledge drew attention to four types of learning model: 'concept identification' models simulate the trial and error process whereby consumers move from initial search through to routinized behaviour; 'paired associate' models use scaling to describe how consumers select between alternative shopping centres; 'interactance-process' models focus on the establishment of probabilities in competitive choice situations; and 'avoidance conditioning' models employ linear difference equations to account for the rewards and penalties involved in using different shopping centres. Each of these types of model was developed largely from laboratory experiments. Their application to the real world is therefore problematical despite Golledge's enthusiasm.

Burnett (1973), for example, argued against models that simply emphasize changing behavioural probabilities and in favour of a greater emphasis on the attributes that are used to differentiate between alternative shops and shopping centres. She also maintained that the attributes that consumers consider cannot be specified a priori because they are subjective and can only be revealed by direct interrogation. In this context she was able to show, in a study of comparison goods shopping (e.g. clothing) for two samples of women in Sydney, that consumers in the same sample use the same dimensions to discriminate between alternative shopping centres (although the synthetic dimensions produced by multidimensional scaling were, as always, complex and did not correspond closely to the objective retail environment). She also demonstrated that verifiable mathematical relationships exist between the probability of choosing each alternative centre and the decision-maker's subjective assessment of that centre in terms of the dimensions used to discriminate between centres. Furthermore, a difference emerged between the two samples and, because these groups differed in terms of their length of residence, this difference was interpreted in terms of learning.

Burnett's paper brought stern criticism from Louviere (1973) in a commentary that highlights many of the problems encountered in the modelling of consumer learning. To begin with, Louviere argued that Burnett failed to take account of existing theory in psychology and that the Markov approach, which Burnett had advocated, is conceptually inadequate in that it assumes a static environment whereas the retail environment is always changing. He went on to criticize the view that the dimensions used to discriminate between shops can be revealed a posteriori by multidimensional scaling. He also contended that learning cannot be inferred from cross-sectional data (comparison of two points in time) because there is no evidence that shopping centre choice settles down to a stable pattern over time. His point of view was expressed clearly when he claimed that 'models of human information processing, learning, choice, and the like, are similar to equivalent models in the physical sciences: they are "as if" models. Their builders use them to describe the behaviour of people "as if" these people (subjects) obeyed the rules expressed by the mathematics' (Louviere 1973: 322). Clearly, Louviere's view is that consumers do not behave in the way specified by the mathematics. Undeterred, Burnett (1977) proposed a linear learning model for household grocery shopping in Uppsala and used Bayesian probability theory to derive predictions about successive destination choices. This study represents an attempt to develop testable statements about heterogeneous population aggregates from assumptions about individual behaviour. In other words, an attempt is made to integrate the micro- and macro-scales of inquiry.

A similar motivation led Wrigley (1980) to advocate a 'stationary purchasing behaviour model' of shopping activity. This type of model focuses on *how* consumers behave, not *why* they behave as they do. Its emphasis is therefore on the description of behaviour rather than explanation. It allows for individual variability in purchasing while coping with the fact that the overall aggregate purchasing pattern may be unchanged. It does this by assuming that purchase frequencies follow a Poisson distribution (i.e. that purchases are independent random events), that overall purchasing patterns follow a gamma distribution, and that the resultant distribution of purchases at any point in time follows a negative binomial distribution. The model was originally devised for application to product purchasing but was applied successfully by Wrigley (1980) to shopping centre choice in Bradford. Unfortunately, however, the model copes with only one retail outlet at a time and Wrigley's suggestion that it be generalized to produce a stochastic model of multi-outlet purchasing behaviour has not yet been followed, largely because little is known about how consumers evaluate multidimensional alternatives.

Concern about the failure of 'as if' models to get to grips with what motivates individual shoppers has encouraged many researchers to question precisely how consumers compare and contrast retail outlets. In this connection, it is important to note that shopping centres vary in many ways – size, price, quality, layout, service, atmosphere, accessibility, facilities – and it may be that consumers, in making comparisons, evaluate different centres on different dimensions. Thus consumers may not be able to order competing centres into a transitive preference structure since the same centre is frequently not the most preferred on all dimensions. Indeed, there is some evidence for this (Root, 1975). In contrast, Walmsley (1977) has shown that 88 per cent of women in an area of Sydney could order major suburban centres (with more than sixty shops) into transitive preferences and that 67 per cent of the sample had overt behaviour that was congruent with their preferences. Of course, simple observation of transitive preferences leaves unanswered the question of how those preferences were arrived at or, in other words, how the many dimensions of each centre were combined to give an overall preference score.

The answer to this question may lie in some sort of *conjoint analysis* (Golledge and Timmermans 1990). Schuler (1979) used this technique to model supermarket choice in Bloomington in a paper which suggested that consumers decide what attributes of a store are important, weight those attributes, calculate a summary score for each store based on the weighted attributes, and then maximize utility by patronizing the store with the highest score. This approach has been given some theoretical backing by Timmermans (1980) who has demonstrated that conjoint analysis is basically a method of measuring the joint effect of two or more independent variables (e.g. attributes of shopping centres) on the ordering of a dependent variable (e.g. a preference ranking of shopping centres). To be specific, conjoint analysis is based on the assumption that it is possible to measure the relative contributions of two or more independent variables, even though their individual effects may not be measurable in any direct sense. It does this by (1) asking individuals (shoppers) to rank order choice alternatives (shopping centres) in terms of their overall utility, and then (2) using a priori specified composition rules (which describe in mathematical terms the way in which individuals combine part-worth utilities to arrive at an overall utility measure) to estimate the contribution of different attributes. In other words, conjoint analysis provides a mathematical model in which part-worth utilities for various attributes of multi-attribute alternatives can be estimated from subjects' preference orderings of a set of factorially designed alternatives (Timmermans *et al.* 1984: 40).

According to Timmermans (1984b), there are several composition rules whereby evaluations of attributes are combined. Most studies have assumed an additive rule (sometimes with the weighting of individual attributes) which is compensatory in that a low value on one attribute is made up for by a high value on some other attribute. Timmermans suggested, however, that such 'trade-offs' might not be universal and that the best fit with his experimental data was achieved by a multiplicative composition rule which is non-compensatory in that a low value on one attribute tends to be compounded into a low overall utility. The corollary to this is that varying opinions as to the salience of attributes are likely to be reflected in varied behaviour patterns. The study of consumer decision making is therefore an important field of research and much attention has been directed in recent years to exploring non-compensatory composition rules (Timmermans 1984b), to treating distance as other than a monotonically decreasing utility function (thereby recognizing the effect of distance as a constraint) (Timmermans 1983), and to extending traditional choice theory to more realistic situations (Timmermans and Golledge 1990).

Variability in overt behaviour

A large number of factors have been found to influence consumer spatial behaviour: income, gender, education, age, occupation, ethnic affiliation, geographical location, mobility, culture, attitude, and personality (Opacic and Potter 1986; Williams 1981; Shepherd and Thomas 1980; Gayler 1980; Lloyd and Jennings 1978; Kassarjian 1971; Ray 1967) (see Fig. 9.2). As a result overt behaviour patterns are very varied and generalizations hard to make. It does seem, however, that consumers often divide their purchases of particular commodities between a number of shopping centres. Hanson and Huff (1988), for instance, found that shopping records for 149 people over 35 days revealed that each person exhibited more than one 'typical' daily behaviour pattern. On the face of it this appears to be sub-optimal behaviour because it seems to violate the principle of distance minimization and the idea of transitive preferences. In practice, however, such behaviour may simply be a rational response to the constraints under which individual shoppers find themselves operating. It might also be a rational response to either the varying utility offered by shopping centres or to the costs in both time and money involved in searching for information about centres (Hay and Johnston 1979).

Because search is costly, it tends to be constrained in both distance and directional terms to a relatively small geographical area. For example, other things being equal, consumers are more likely to visit centres that lie between their home and the central business district than centres that lie in the direction away from the town centre (Lee 1962; Potter 1977b). This tendency is probably related to the way in which shops are concentrated in central city areas which are also the hub of transport networks. It is also related to distance perception. Thompson (1963) was the first researcher to point out that subjective distance differed significantly from objective distance and, since his work appeared, several people have demonstrated that distances along familiar and preferred routes tend to be underestimated relative to distances along unfamiliar and disliked routes, both at the urban scale (Pacione 1975) and the scale of individual streets (Meyer 1977). Of course it would be wrong to overemphasize the importance of distance perception because its influence on travel behaviour is still unclear. Some of the differences that have been observed in journeys to shop may in fact be more apparent than real. For example, the apparent tendency for consumers to travel further for high-order goods (e.g. good quality clothing) than for low-order goods (e.g. weekly groceries) may be no more than an artefact of the denser 'packing' of low-order goods outlets (e.g. corner stores) because there is evidence that the amount of excess travel (i.e. travel over and above that which is necessary to buy a good) is very similar for different commodities (Walmsley 1975).

Observation of variability in overt behaviour has led some researchers to propose that there are different types of consumer. One of the first to do this was Stone (1954) who identified four sorts of consumers: price-conscious individuals for whom economic considerations are paramount; individuals who are bound by habit and who emphasize personal contacts; individuals who adopt an 'ethical' position such as supporting 'the little retailer'; and apathetic individuals who do not care which shop they visit. What this typology does, in other words, is highlight the social and psychological influences acting upon consumers. This point was taken up by Brooker-Gross (1981) who argued that 'outshopping' (travel *between* towns for shopping) should be interpreted as a social group activity. A more explicitly behavioural typology of consumers was developed by Williams (1979). He proposed that individuals could be differentiated according to mode of travel (walk, bus, car), frequency of trip (frequent, infrequent), and size of centre visited (small, medium, large), and he went on to identify seven types of consumer in Birmingham (e.g. those who walked frequently to small centres comprised 28 per cent of the sample, those who drove seldom to large centres comprised 25 per cent of the sample).

The idea behind the development of typologies of consumers is that the classificatory technique on which it is based provides a way of linking micro- and macro-scale studies, that is to say a way of relating descriptions of individual behaviour to descriptions of the behaviour of aggregate populations. Critics of the revealed space preference approach have shown that it is impossible to work from the macro-scale to the micro-scale and to make assertions about individuals on the basis of groups. Typologies represent an attempt to go the other way, from the micro-scale to the macro-scale, by grouping together individuals whose behaviour is similar. Of course, in such a move, care must be taken to guard against laying too much emphasis on *consumer sovereignty*. Implicit in many studies has been the idea that consumers make free choices and that they go to whatever shopping centres they wish. In reality, of course, decisions by consumers are very highly constrained. This is most obvious in the case of women. Women undertake the majority of shopping trips and yet their gender role renders them relatively immobile for a variety of reasons:

1. Family role-playing often denies women access to cars;
2. Gender-related tasks restrict the time that women have available for access to shops;
3. The conditions under which they are often forced to travel (e.g. with young children) lessens the willingness of many women to travel.

In many instances, then, overt behaviour may reflect the dictates of transport availability and time just as much as it reflects a rational evaluation of the centres 'on offer'. It is questionable therefore whether geographical studies of consumer behaviour will make great strides until these constraints are taken into account. Basically the constraints are of two sorts: those associated with the environment (poor shops, restricted trading hours, poor transport) and those associated with the consumer (lack of mobility, poverty, age, all of which impact differentially on different groups) (Shepherd and Thomas 1980: 47–8). In this sense, therefore, it is incumbent upon students of consumer behaviour to look at developments within the retail environment as well as the constraints acting upon consumers. For example, the manner in which retailers influence the nature and amount of information available to consumers has not yet received the attention it deserves. As a result geographical studies of consumer behaviour have been slow to take stock of the changes that are under way in retailing generally (see Davies and Kirby 1980). It is appropriate therefore to consider these changes in some detail.

The changing retail scene

Change is, in many senses, the norm in the retail scene (Brown 1987). Fundamental changes are taking place in the management, technology, organization, and environment of retailing to the extent that Dawson (1979) has drawn a parallel with the Industrial Revolution. Just as the Industrial Revolution underpinned social changes in the nineteenth century, so a marketing revolution is having a major impact on society in the late twentieth century. In Dawson's words, the hypermarket is as different from the corner store as the woollen mill was from the weaver's cottage. Despite the significance of these changes, very little geographical research has been directed at understanding of the behaviour of some of the principal 'actors' in the changes, namely retailers (Halperin *et al.* 1983).

The most notable change in the retail scene in Western society in the last 40 years has been the suburbanization of retail activity (Wrigley 1988; Kellerman 1985; Potter 1982; Guy 1980). The reasons for this suburbanization have been very well summarized by Lord (1980). Basically, there are four contributing factors: the changing nature of downtown areas; demographic changes within the city generally; competition within the retail industry; and changes in the nature of society generally.

Downtown areas in Western cities, particularly in North America, are often characterized by traffic and parking problems. Moreover, many retail premises are cramped and inadequate, largely on account of their age. Urban renewal obviously can combat this ageing process but the net result is often a higher-priced upmarket retail scene that effectively denies access to some of the less affluent groups in society.

Where renewal does not take place, downtown areas can be perceived as unpleasant and perhaps even unsafe, if they are linked with crime and vandalism as well as with physical blight (Kivell and Shaw 1980).

The relative decline in importance of central business districts owes much to the suburban sprawl of population, itself a function of increased personal mobility. Some cities have developed at extremely low densities: Sydney, Australia, for example, has a population of approximately 3 million spread over 1200 km^2. The effect of increased car ownership and suburban sprawl has not only been to put people at increasing distances from the city centre but also to free them from reliance on city centre oriented public transport systems (Dawson 1979). In this sense, it is important to note that the real cost of travel has often decreased over the years. In Australia, for example, the real cost of fuel fell from 1950 to 1977 and is still below the level of the 1960s (Walmsley and Weinand 1991). Alongside population growth in the suburbs, there has of course been a decline in inner-city population levels. To some extent this has come about because of expansion of commercial activity into previously residential areas. To some extent, too, it is due to the removal of old and sometimes inadequate housing stock. The net effect in all cases has been to reduce the total spending power of the inner suburbs (despite recent trends to gentrification, see Chapter 8), and thus lessen the buying power of the immediate catchment area of the central business district (Thorpe 1983).

Problems for central city retailers have been exacerbated by increased competition within the retail industry generally. This has taken many forms. At a simple level, there has often been a disproportionate increase in the expense of city centre trading, often associated with increases in local rates (Thorpe 1983). One response to this has been to substitute capital for labour in an attempt to increase productivity and efficiency. Such changes are, however, only one part of the story. Other changes are perhaps more far-reaching. One of the most obvious of these changes has been the advent of chain stores. Also important is franchising whereby independent stores survive by affiliating with a wholesaler who provides advice on store organization, layout, accounting and stock control (Cardew and Simons 1982). 'Own brand' retailing has also increased in prominence in recent years, representing an attempt on the part of many retailers to cut out the wholesaler from the distribution network (Wrigley 1989). Another change has been the emergence of 'scrambled merchandising' as long-established conventions on lines of trade have been abandoned (Dawson 1979). Grocery stores, for example, now commonly stock household equipment (e.g. light bulbs), clothing (e.g. socks), and garden equipment (e.g. fertilizer). In some instances these stores compete with bigger ones by offering longer trading hours, particularly in the case of petrol stations that have diversified into drive-in convenience stores (see Maher and Mercer 1984). Paradoxically, this trend towards scrambled merchandising has appeared alongside an emphasis on market segmentation and perhaps even 'niche marketing' as specialist stores have emerged catering for a particular clientele (e.g. do-it-yourself shops, ethnic stores).

By far and away the most important change in retail competition results, however, from the development of planned shopping centres or malls (Dawson and Lord 1985). The first recognizably planned shopping centre was built at Roland Park in Baltimore in 1907 and the first suburban planned centre was developed at the Country Club Plaza in Kansas City in 1923 (Davies 1976). Planned centres therefore have a long history. However, it is only in recent years that they have come to occupy a distinct niche in the retail hierarchy (Walmsley 1974c), possibly because competition between rival retail corporations, in association with retail property developers, has led to the development of new sites in an attempt to maintain or increase market share (Cardew and Simons 1982). Despite their number, however, many planned centres have a sameness about them that was well captured in Relph's (1987: 180) caricature:

> the entrances should be well marked on the outside but on the inside should be as inconspicuous as is possible under fire escape regulations, a multitude of mirrors should remind us of our shabbiness, and corridors should have sight lines of no more than 200 feet and be no more than 20 feet wide. Lighting and sound levels are controlled to keep people alert, benches and other furnishings are placed at the centres of

walkways so that shoppers are forced close to the always open store entrances, and the seats should not be too comfortable (flat wooden surfaces are common) so that shoppers do not waste good buying time sitting down.

Planned shopping centres themselves face competition from other, more recent retail developments, notably superstores, hypermarkets, and retail warehouses. Superstores usually comprise up to 4000 m^2 of selling space and provide a wide range of household goods, clothing, and food (Davies and Sparks 1989; see also Thorpe 1991). Hypermarkets contain upwards of 5000 m^2 of retail space and are free-standing car-oriented stores (Hallsworth 1989; Potter 1982). Retail warehouses first began to appear in the late 1970s. They provide major points of direct sale to the public (see P. Jones 1982). Such has been the popularity of this type of endeavour that it has contributed to a new phenomenon on the British retail scene, the retail park. These parks are largely unplanned (i.e. non-integrated) agglomerations of superstores, hypermarkets, and retail warehouses, often on industrial estates or in Enterprise Zones (Thomas and Bromley 1987). Indeed, it is something of a paradox that retailing, which is the most location-sensitive of all land uses, should choose to locate in Enterprise Zones which have been supposedly selected because they are so unattractive that major incentives are felt necessary to attract private sector investment (Schiller 1986).

In many ways, the changes that are occurring in retailing are supported, and perhaps even facilitated, by changes in the nature of society generally. For example, there has been a marked increase in the last 40 years in the number of females in the paid labour force (Dawson 1979). The significance of this lies in the fact that it has stimulated changes in shopping patterns. Although gender roles within households have been remarkably resilient to change and women still do most of the shopping, pressure on time for women in the paid labour force has undoubtedly altered shopping behaviour. Similarly, improved storage facilities (especially the widespread use of freezers) has altered the frequency of shopping. The 'cashless society' (credit cards and debit cards) and EFTPOS (electronic funds transfer at point of sale) have freed shoppers from the need to tie shopping in with banking activities (Beaumont 1989). And a rise of consumerism has made it harder for the unscrupulous retailer to survive (Dawson 1979).

Technological innovations have already caused many changes in shopping and in the behaviour of shoppers. More changes undoubtedly lie ahead. One possibility is the widespread adoption of 'teleshopping' based on cable television and perhaps home computers. Unfortunately, however, there is currently much speculation and an ill-informed debate as to the likely impact and extent of teleshopping. Davies and Edyvean (1984), for example, have pointed out that, historically, major innovations in retailing have required long periods of gestation before widespread adoption. Self-service, to give but one illustration, languished for years before becoming widely accepted. It may even be that the effect of new technology will be to entrench the position of the present retailing giants who are, after all, in the best position to take advantage of changes (A. Taylor 1984).

In the speculation about new technology, it is very easy to lose sight of the fact that the nature of shopping is perhaps already changing in fundamental ways. For example, the distinction between shopping and other aspects of life (leisure, health care, education) is perhaps breaking down with the emergence of vast retail–leisure–service complexes. The most notable of these is the West Edmonton Mall (Hopkins 1990). This comprises a novel mixture of retail, commercial, recreational, and entertainment facilities on an unprecedented scale (almost 500 000 m^2, over 600 stores and services, almost 3 ha of water park, and an 18-hole miniature golf course). The Mall incorporates Miami Beach, Versailles, Park Lane, and the Grand Canyon. This pastiche of what are in effect 'icons' serves to foster the view of retailing as *spectacle*. The placelessness that has come to characterize many planned shopping centres has given way in Edmonton to 'elsewhereness' as shoppers are invited and enticed to escape to different places.

Innovations such as the West Edmonton Mall throw into question the social function of consumption in Western society. They therefore invite reflection on changes in society generally. Jackson (1989: 5), for example, has asked whether, in Britain, the age of yuppie culture, urban heritage, rural nostalgia, football hooliganism, and inner-city rioting is a response to national economic decline or a sign of the growing confidence of the 'consumption classes' and of the increasing alienation of the impoverished and despairing 'underclass'. Perhaps developments like that at Edmonton are a sign of the times in that they cater for the 'consumption classes'. Perhaps West Edmonton Mall is not unique except in scale. Perhaps smaller shopping centres are coming to rely on spectacle. Relph (1987), for instance, has described two other examples of spectacle-based shopping: St Andrew's Village (Toronto) is a shopping mall that attempts to create the charm of a century-old village thereby harnessing the 'back-to-roots' sentiment; and The Barnyard (Monterey) is an upmarket plaza consisting of replicas of eight Californian barns-cum-stores selling to the internationally affluent in a lavishly landscaped arena tended by janitors wearing blue overalls and straw boaters.

It may well be then that affluent consumers are making new demands and thereby changing the nature of the retail complexes within which those demands are satisfied. Features such as Disneyland have been around for a long time (see Zukin 1990). Perhaps the emergence of 'retailing as spectacle' will see such elements of the cultural landscape become much more common and more varied in nature. In any event, such developments promise interesting times for the shopper and challenging times for geographers concerned to understand shopping behaviour. There has long been a tension in the study of consumer behaviour between macro-scale, aggregate approaches that simply describe overall shopping patterns and micro-scale approaches focused on the individual and geared to explaining why behaviour takes the form that it does. Likewise, there has been a tension between the study of consumers as if they are sovereign decision-makers and recognition of the very real constraints under which many shoppers operate. The rise to prominence of such phenomena of cultural capital as the West Edmonton Mall perhaps suggests that neither the macro- nor the micro-scale approach is sufficient in itself. Instead researchers perhaps need to look at the meaning that different 'spectacles' have for shoppers and the way in which that meaning is contested and negotiated (Jackson 1989). In this way, a more penetrating understanding of constraints might be achieved. In summary, then, there is still a great deal to be learnt about what goes on in the minds of both retailers and shoppers.

10 Leisure and Recreation

In recent years all developed economies have experienced a significant growth in leisure-time activities. Part of this can be related to the rapid increase in most people's disposable incomes, mobility levels, and available discretionary time (Patmore 1970; Patmore and Collins 1980). Part also derives from the fact that, in the Western world, the last two decades have seen significant increases in the numbers of unemployed and, therefore, in the size of the group with enforced leisure time. The term 'leisure' encompasses activities in which individuals may indulge of their own free will either to rest, amuse themselves, to add to their knowledge and improve their skills disinterestedly and to increase their voluntary participation in the life of the community after discharging their professional, family, and social duties (Appleton 1974: 63).

Among commentators, there has often been a tendency to exaggerate the mount of leisure available to, and sought by, individuals. For example, faced by the choice of more leisure time or more income, a good many British workers still choose more income from overtime, a second job, or do-it-yourself activities which are hardly leisure (Parker 1983: xi). Nevertheless, there is no denying that, overall, the amount of leisure has increased and that this trend of increasing leisure time can be related to the progression of society from pre-industrial and feudal status towards industrial and eventually post-industrial status (Roberts 1978). As illustrated in Table 10.1, this progression involves an increasing separation of work and leisure in time and space, so making leisure the vacuum between work periods. In the nineteenth century the distinction between work and leisure was the preserve of the wealthy but today it has filtered down to all income groups (Bailey 1978; Lowerson and Myerscough 1977). This point is explored in Fig. 10.1 which suggests that the amount of leisure time available to an individual might, in the future, be related to that individual's intelligence (and thus to her or his position in a labour force that puts a premium on qualifications). Thus what we are perhaps witnessing in Western society is an inversion of the 'traditional' arrangement whereby leisure was the privilege of the 'upper class' who were not obligated to work, and the

Table 10.1 The uses of time

	Fully committed	Highly committed	Leisure
Sleeping	Essential sleep	—	Relaxing
Exercise	Health	—	Sport
Eating	Eating	—	Dining
Shopping	Essential	Optional	—
Work	Primary	Overtime	—
Housework	Cooking	Repairs	Gardening
Education	Schooling	Further education	—
Culture	—	—	Reading, TV, radio, hobbies and passive play
Travel	To work/school	—	Working/driving for pleasure

Source: After Cosgrove and Jackson (1972)

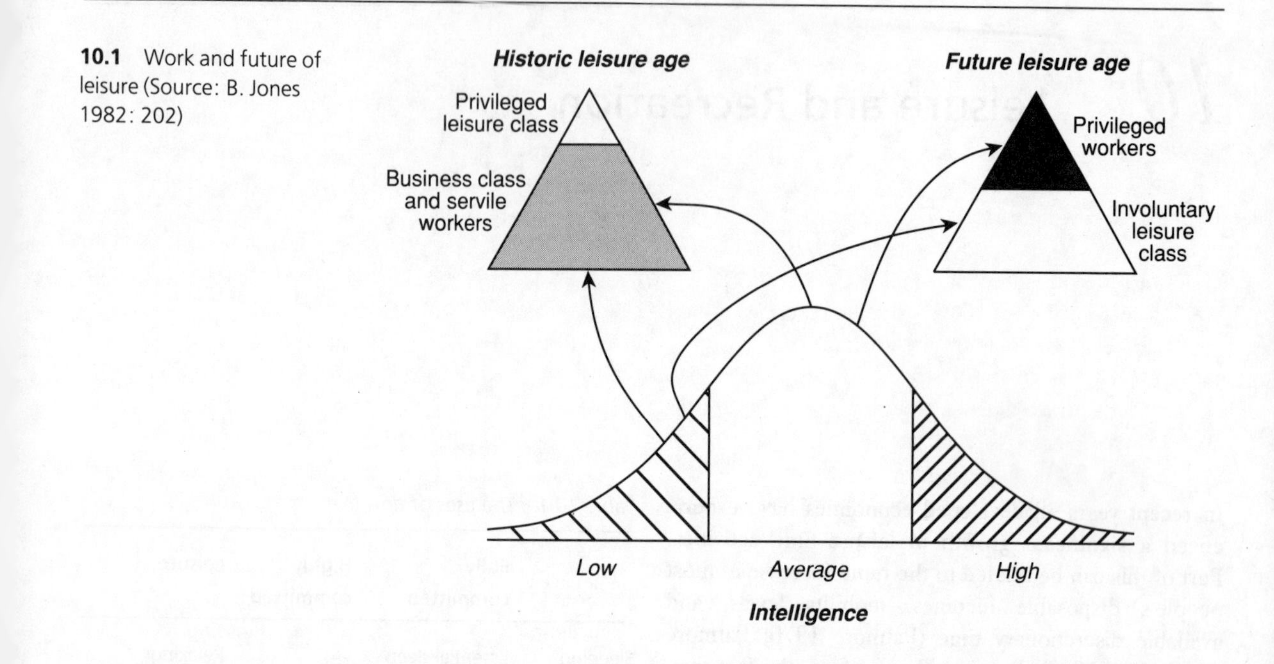

10.1 Work and future of leisure (Source: B. Jones 1982: 202)

emergence of a situation where true leisure will be the preserve of those educated and trained individuals who hold down well-paid jobs (see B. Jones 1982). In other words, there might emerge a difference between the type of leisure enjoyed by those in gainful employment and the type enjoyed by those marginalized in the labour force. This is important because, for a long time, leisure has been regarded as one of the good things in life. The overwhelming experience of the majority of the population in the industrial era was the appropriation of time by work; leisure was something 'won in the interstices, something precious and insufficient' (Seabrook 1988: 1). And, like most good things, leisure has never been shared equally among all members of society.

However, the 1960s and early 1970s seemed to herald change. As Glyptis (1989) points out, the buoyant climate of full employment, shorter working hours, longer holidays, earlier retirement, rising incomes, and improved mobility seemed to promise greater leisure for all with a result that leisure was the new measure of quality of life. 'An age of leisure, rumoured to be on the near horizon, seemed both attractive and achievable' (Glyptis 1989: xi). All that altered with structural changes to the international economy from the mid-1970s onwards. Labour-saving technology, deflation, and monetarist policies all contributed to the 'collapse of work' (Jenkins and Sherman 1979). As a result, the three groups today with the most leisure are the rich, the retired, and the redundant (Seabrook 1988: 2). The real winners in the future might be those for whom the distinction between work and leisure is blurred, those who are able to say 'I'm only doing what I would be doing anyway, and getting paid for it.' Prominent among such groups are likely to be media people, designers, planners, consultants, entertainers, researchers, and communicators of all types (Seabrook 1988: 4).

Speculation about what the future might hold is interesting but it should not cause us to lose sight of the fact that leisure time has increased for many people over the course of their lifetime and that, at the same time, a whole commercial entertainment business has developed (Kando 1975). Often this commercialism has manifested itself in the phenomenon of tourism. Thus, according to Burns (1973: 46), 'social life outside the work situation . . . has been created afresh, in forms which are themselves the creation of industrialization, which derive from it and which contribute to its development, growth and further articulation'. From this perspective, the growth of leisure is something that can be seen as part and parcel of the capitalist mode of production. As Aronowitz (1982) has pointed out, leisure is important to the reproduction of labour and the relations of production in so far as its provision reduces the likelihood of workers attempting to change their conditions of work due to a lack of time for play. It is thus in the interests of business to provide recreational facilities and encourage leisure-time activities. This tendency for capitalism to cater for leisure and recreation can take many forms. One manifestation is the marketing of particular items of hardware which, in itself, has become a major industry. These items may range from the expensive (e.g. recreational vehicles) to the inexpensive (e.g. cheap forms of leisure wear) and even the fatuous (e.g. hula-hoops), and the marketing may be segmented by class, race, or gender, but in all cases leisure is 'commodified' (Kirby 1985). And not only is business involved. The state too takes a part: in an abstract way, national teams can bolster ideological integrity; and in more concrete ways, the state can provide recreational facilities such as playing fields, often to the point where the 'professionalization' of leisure services accounts for a substantial share of the expenditure of local authorities (Kirby 1985).

In view of these trends, it is not surprising that the provision and use of leisure facilities have become the focus of a good deal of research. At least four different perspectives have been adopted in this work:

1. Certain activities have been deemed to be inherently pleasurable thereby contributing to leisure;
2. Leisure has been viewed from a residual perspective as that which is left over after work has been completed;
3. Leisure has been seen as that component of discretionary time that is given over to pleasure (a variant of (2) that recognises that not all non-work activity is pleasant);
4. Attempts have been made to devise a comprehensive and holistic definition of leisure that takes into account notions of work, psychological needs, and cultural values (Kirby 1985).

On top of this, some researchers have viewed leisure from a political economy perspective (see Aronowitz 1982). This variety of approaches is perhaps not surprising given the increasing awareness of the importance of leisure in modern life. After all, leisure is now deemed vital for the maintenance of a reasonable quality of life (Hookway and Davidson 1970). Allen and Beattie (1984), for example, found that resident satisfaction with leisure is a good predictor of overall community satisfaction. In short, leisure signifies 'play' and, as such, as thinkers from Plato to Sartre have observed, it represents the peak of human freedom and dignity (Csikszentmihalyi 1975: xi). Predictably, given its importance to human well-being, researchers from several disciplines, including geography (Pigram 1983; Smith 1983; Simmons 1975; Lavery 1974; Cosgrove and Jackson 1972; Patmore 1970), sociology (Parker 1976; Smith *et al.* 1973), and economics (Vickerman 1975; Clawson and Knetsch 1966), as well as multidisciplinary researchers (Chubb and Chubb 1981; Dunn 1980; Mercer 1977, 1981; Johnson 1977), have investigated the nature of leisure-time behaviour and thereby provided the basis for policy initiatives and land-use management strategies (Jubenville *et al.* 1987; Torkildsen 1986; Kaplan 1975; Appleton 1974). There is also an enormous and growing body of research explicitly concerned with tourism (Gunn 1988; Pearce 1987; Murphy 1985; Mathieson and Wall 1982; Pearce 1982a) and with particular activities such as sport (Bale 1988). It is important therefore to look at how people use their leisure time and at the distinction between leisure and recreation.

Leisure-time activities

According to Young and Willmott (1973), any attempt to describe and measure leisure activities is like attempting to grab a jellyfish with bare hands because leisure has different meanings in different societies and to different people. Indeed in some instances it is difficult to demarcate leisure from work and other forms of human behaviour. Two examples serve to make the point: first, the growing popularity of 'hobby' farming among the middle class is essentially a leisure activity, yet farming is a means of livelihood for millions and a source of income even for hobby farmers; and second, many painters or writers are able to earn income from what they consider to be non-work activities. In other words, for any individual,

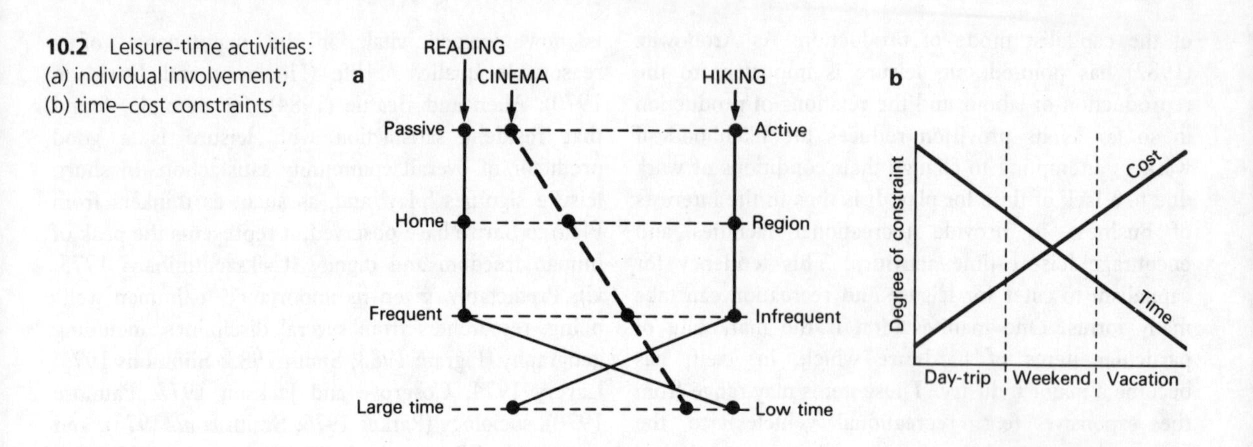

10.2 Leisure-time activities: (a) individual involvement; (b) time–cost constraints

depending on circumstances, the same activity can be considered as an essential or leisure act. As a result, the distinction between obligatory activity and discretionary activity is blurred. This has led Clawson and Knetsch (1966: 6) to suggest that the distinguishing characteristic of leisure and recreation is not the activity itself but the attitude with which it is undertaken. Therefore, leisure can be conceived singly, or in combination, either as a period of time, as a certain activity, or an attitude of mind (Tinsley and Kass 1979).

Once a pursuit has been defined as a leisure activity it then becomes necessary to distinguish the nature of that activity and its significance to the individual (Pierce 1980). Broadly, a leisure activity can be classified, on the basis of individual involvement, into *passive* and *active*, and on the basis of whether it takes place in the home, the local neighbourhood or a distant region. A further complication is provided by the frequency of involvement in a particular activity and the length of time that the pursuit may involve, whether it be hours, days, or weeks (Wilman 1980). Figure 10.2(a) illustrates three very different leisure-time activities: an individual who reads every evening for a couple of hours at home, an individual who spends an evening at a city centre cinema once a month, and an individual who goes hiking once a year in a distant mountainous region for 3 weeks. Likely participation in such alternatives is, of course, constrained by the availability of leisure opportunities within a time–cost framework (Fig. 10.2(b)). The illustration in Fig. 10.2 is inevitably an extremely simple one. In reality individuals may read, go to the cinema, and hike (as well as indulge in a great many other leisure-time activities) and may be constrained by factors other than the availability of facilities and time–cost. Indeed it is quite wrong to interpret discretionary travel as being relatively free from constraints because the same information, social, internal, and institutional constraints apply as with travel behaviour generally (see Desbarats 1983) (see also Chapter 5).

Despite the development of a vast commercial entertainment industry, the focus for most leisure time activities in the Western world is still the home which may provide the venue for something like three-quarters of all leisure-time pursuits (Mercer 1980). The single most popular leisure-time activity is in fact watching television (Kirby 1985). Although frequently overlooked by researchers who seem to have a predilection for studying travel patterns, the focus of leisure activity on the home is important. Indeed it provides support for Wingo's (1964: 138) view that 'the popularity of the low density suburban dwelling probably reflects the surging demands for more private forms of recreation among middle-class, child-oriented families, suggesting that metropolitan scatteration is as much a recreation as a housing phenomenon'. It is important therefore to look more closely at the phenomenon of recreation.

Recreation

Leisure-time activities that involve travel beyond the home are often thought of as *recreation*. Such activities encompass both passive pursuits (like pleasure driving) and active pursuits (like mountain walking) and since the 1950s a great deal of research has looked at the planning, provision, use, and impact of recreation facilities (Coppock and Duffield 1975; Mercer 1981).

On the supply side, extensive investment in recreation facilities has been undertaken by both the public and private sectors on both sides of the Atlantic (Vickerman 1975). Much research, particularly in the United States, has therefore been concerned with the cost-effectiveness of investments and the economic benefits accruing from recreation (Clawson 1972). Marketing has also been identified in the United States as an important way of promoting recreational facilities to the public (La Page 1979). Similar work in the United Kingdom, by contrast, has been largely confined to the related area of tourism (Sillitoe 1969). Inventories of the number, size, and location of recreation sites, and studies of managerial and planning problems associated with recreational pressures and land-use conflicts, have also been common in the literature (Fischer *et al.* 1974; Ontario Research Council on Leisure 1977). Increasingly the importance of careful planning and management has been stressed (Bannon 1981) together with the need to bring users and resources together in such a way as to prevent the despoilation of resources while maximizing the satisfaction of recreationists (Bosselman 1978; Brockman and Morrison 1979).

One of the most innovative planning concepts to have emerged has been the *recreation opportunity spectrum*. This spectrum describes the range of recreational experiences that could be demanded by potential users in the event that a full array of recreational opportunity *settings* was available. In reality, of course, no environment is likely to provide a full array of opportunities and the role of the planner is often to decide what is viable and acceptable. Basically, the idea is that planners attempt to provide a range of recreation opportunity settings, each such setting being a combination of the biological, physical, managerial, and social conditions that give a place value for recreational purposes. In this sense, a recreational opportunity setting encompasses qualities provided by nature (e.g. topography, scenery, vegetation), qualities associated with the activity of recreation (these being influenced by the level and type of use made of the setting), and conditions provided by management (e.g. roads, facilities, regulations, charges). By combining variations of these qualities and conditions, the planners and managers of an area can provide a variety of opportunities to the receationist (Clark and Stankey 1979).

Paralleling this work on the recreation opportunity spectrum, a good deal of research has been done on the carrying capacity of resources. Often this work has attempted to identify a threshold beyond which further use of a particular setting is considered undesirable. Initially this threshold was defined using physical, economic, or ecological criteria (see Owens 1984; Burton 1971) but, of late, psychological thresholds have been introduced into the evaluation (Becker 1978). Indeed, it is now widely recognized that any attempt to measure carrying capacity must consider tangible resource limits, the tolerance of the host population, the satisfaction of visitors, and the rate of change (to guard against the perception that the rate of change is excessive) (Getz 1983). Generally, a greater number of users leads to a reduction in the satisfaction which individuals derive from the recreational experience, although the degree of sensitivity to, or tolerance of, crowding varies between different groups of recreationists (Cesano 1980). For example, canoeing in a wilderness area can have a low tolerance threshold compared to motorboating at the seaside. These ideas on carrying capacity, and others, have led to the *limits of acceptable change* system for recreation planning (Stankey *et al.* 1985). In simple terms this approach argues that there is a level of use or carrying capacity beyond which the impacts become unacceptable and it is up to environmental managers to monitor change in order to identify the point at which both the resources themselves and the perceived quality of the recreational experience begin to deteriorate (Long *et al.* 1990).

On the demand side of recreation studies, research emphasis has been largely on observed usage patterns at different facilities within a given time–space framework (Davidson 1970; Outdoor Recreation Resources Review Commission 1962). This has involved sophisticated surveys, designed to identify participation rates in various activities, the trends in participation, and the factors influencing recreation behaviour. At an early stage it was found, for example, that age, education, and income were major factors influencing recreational activity (Rodgers 1969; Dottavio *et al.* 1980) and special attention has been given to the effect of transport availability and distance in determining recreational travel (Baxter 1979; Smith and Kopp 1980; Cooper 1981). For example, Mercer (1971) found in Melbourne that when trips by outer suburban residents to extra-metropolitan recreation sites are plotted in the aggregate, then the evidence shows a directional bias in trip-making (i.e. individuals tended to stay within their own 'sector' of the city) which in turn suggests a wedge-shaped mental map. Similarly, in Britain, Harrison (1983) has shown that recreation sites on the urban fringe attract visitors from a very localized catchment area. These types of survey have often shown remarkable levels of recreational activity. For example, in New South Wales, which had at the time a population of less than 6 million, there were, in 1979–80, no fewer than 68 million nights spent by individuals at locations more than 40 km from their home. This recreation involved 16 million trips (NSW State Pollution Control Commission 1978). By their very nature, of course, such surveys involve large statistical aggregates and focus on what might be called 'average' behaviour.

The information collected in large-scale surveys has been useful for managing and marketing recreational facilities, and provides some indication of the significance of tourism in a national economy. However, many researchers have been sceptical as to the relevance of user surveys since 'most recreation research to date cannot stand the question, "so what?"' (Brown *et al.* 1973: 16). All too often *ad hoc* studies have proliferated without researchers having any particular intention of making a contribution to the development of a testable theory (Owens 1984: 173). Research has therefore tended to be non-cumulative in the sense that findings from one particular situation often cannot be extrapolated to other situations. This is particularly the case with studies of tourism. Indeed there are signs that the growth in studies of leisure, recreation, and tourism has been too rapid. Not only have idiographic studies been predominant but the approaches used have been fragmented and specialized with the result that the overall subject area has very little coherence (Boniface and Cooper 1987: xi). Critics have suggested that, in order to rectify these weaknesses, research should consider the significance of recreational activities for the well-being of the individual and the small group as well as the 'broader social contexts of the role of recreation in competing among alternative uses of resources' (Brown *et al.* 1973: 17).

Thus there has been a shift in recent years to a greater emphasis on the social and psychological aspects of recreational behaviour and activity participation (Ulrich and Addoms 1981). In the particular case of tourism, Canter (1982) has suggested that coherence in research and progress in understanding human behaviour can be achieved by focusing on a series of questions: What aspirations do potential tourists have for their recreation activities? What possibilities are different classes of tourist aware of? Once a decision is made regarding behaviour, what processes come into play in evaluating it? What is noticed and ignored in the tourist experience? And what criteria are used for evaluating experience? These questions would seem to be equally relevant to an understanding of leisure activities and recreation behaviour generally.

The stimulus for much of this behaviourally oriented work on recreation has been derived from wilderness studies in the United States, which have identified such experiential notions as 'wilderness as a locale for sport and play' and 'wilderness as a sanctuary' (Mercer 1976). Likewise recognition of varying sensitivity to crowding and wilderness intrusions has emphasized the need for more flexible management in order to maintain satisfaction for all groups of users. No fewer than three sets of factors have been found to influence attitudes to crowding:

1. Individual characteristics such as class, life style, experience and expectations;
2. Objective site characteristics (e.g. physical conditions, carrying capacity);
3. Visitor perceptions of the social environment in terms of the type of people present and their behaviour (Westover and Collins 1987).

In other words, group membership might be critical in recreational experience. Precisely what comprises a

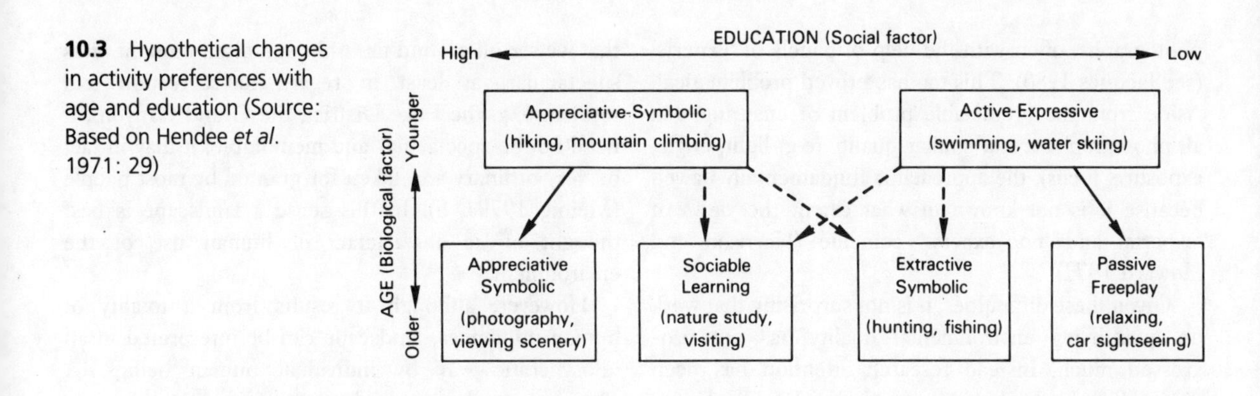

10.3 Hypothetical changes in activity preferences with age and education (Source: Based on Hendee *et al.* 1971: 29)

group of users is of course difficult to define (Womble and Studebaker 1981). One interesting approach to grouping recreationists is that developed by Hendee *et al.* (1971) based on the perceived similarities in the underlying meanings of activities to participants (Fig. 10.3). Such a classification, based on an experiential emphasis, allows researchers to examine psychological, sociological, or environmental/situational effects on each grouping, and it has been employed as an initial framework in several contexts which have generally revealed that different localities can provide different experiences for recreationists, even though sometimes the same activity may be involved (Cool 1979; Warburton 1981).

In a similar vein the United States National Advisory Council on Regional Recreation Planning has used information on recreational dimensions and related it to the planning process by advocating the classification of people into 'user-groups' based on the values – physical, emotional, aesthetic, educational, social, and intellectual – which people place on different leisure and recreational experience. In this sense research is increasingly recognizing the fact that people can be differentiated according to their overall life styles. Mitchell (1983), for instance, has distinguished survivor, sustainer, belonger, emulator, achiever, 'I-am-me', experiential, socially conscious, and integrated life styles. However, despite this shift towards an analysis of the recreational experience as a key component of recreational behaviour, the vast majority of studies approach behaviour in the aggregate (simply describing who does what) and therefore contribute little to an understanding of the underlying motives involved in the decision to recreate. One of the few areas where a significant attempt has been made to study behaviour from an experiential perpective is in relation to landscape appreciation.

Landscape appreciation

Landscape is important to a great many forms of recreation. Most outdoor recreation for example is reliant upon either the direct availability of resources (e.g. water surfaces for fishing and boating) or the provision of a context within which the recreation activity can be enjoyed (e.g. sightseeing, painting). Similarly, the existence of what is deemed attractive landscape is critical to the well-being of many tourist regions. It is not surprising then that a good deal of attention has been directed by both academic researchers and planners to the issue of landscape appreciation.

The stimulus for much of this work came from post-war British planning legislation which required local authorities to evaluate landscape quality as a basis for environmental protection and enhancement (Blacksell and Gilg 1975). This requirement spawned two types of study (Deardon 1980). On the one hand, there were attempts to quantify landscape quality by allocating points for the presence/intensity of designated features (see Fines 1968). This approach has attracted stern criticism, not least because of the subjectivity involved (Brancher 1969). The alternative approach has therefore been widely adopted. This has involved using photographs as surrogates for actual visits to landscapes, and then assessing the photographs, often with the help of panels of 'experts' (see Jacques 1980). This too has proved problematical. Aside from the predictable problem of ensuring that all photographs are of similar quality (e.g. light, angle, exposure, focus), the approach is fundamentally flawed because it is not known to what extent the views of 'experts' and 'non-experts' coincide (Harrison and Howard 1972).

Given these difficulties, it is not surprising that work on quantifying environmental quality has not progressed much. Instead research attention has been directed to landscape *interpretation* and to landscape *aesthetics* (Punter 1982). Work on landscape interpretation has tried to identify what gives some landscapes their value (Gold and Burgess 1982). This is no easy task. As a result researchers are far from being able to say what constitutes a good view or a beautiful landscape, nor have they found a satisfactory means by which to articulate the nature and significance of the intimate attachments that people develop for landscape (Burgess and Gold 1982: 3). Prominent among the landscapes that appear to be both valuable and attractive are those that seem to 'capture' what might be thought of as bygone times. These are presumably appealing because they foster a sense of escape from the 'here and now' (Lowenthal 1982). Recreationists can visit such landscapes and derive pleasure from their visits. Whether or not such landscapes are also aesthetically pleasing is an entirely different question. Basically there are three elements whereby the aesthetic quality of a landscape can be assessed: unity (do the components of the landscape cohere into a single harmonious unit?); vividness (does the landscape have features that make it distinct and striking?); and variety (does the landscape have enough variety to engage human interest without so much complexity as to make it chaotic?) (Litton 1972).

Clearly these are profound questions and they can only be fully answered by adopting a humanistic approach to landscape appreciation (see Chapter 5). In fact landscape interpretation is one of the fields of geographical inquiry where a humanistic approach has been most prominent. Landscape is, however, an ambiguous term. It usually implies nature, scenery, environment, and place, and yet it is synonymous with none of these (see Cosgrove 1985a, b). Rather landscape is best thought of as that continuous surface that we see all around us. Although often thought of as spectacular, at least in regard to recreation and tourism (e.g. the Lake District, the Grand Canyon), it need not be spectacular and memorable; it may in fact be very ordinary and taken for granted by most people (Meinig 1979a, b). In this sense a landscape is best thought of as an artefact of human use of the environment.

However, although it results from a totality of human influences, landscape can be interpreted in an idiosyncratic way by individual human beings. A particular landscape might mean different things to different men and women. For example, landscapes can be imbued with symbolic value which, in turn, can generate strong attachments. Likewise, how a landscape is interpreted can reflect the values and attitudes of the individuals concerned. Thus a capitalist may interpret a landscape in monetary terms, an artist in aesthetic terms, a scientist in ecological terms, and a social activitist in terms of disorder and injustice (Meinig 1979c). And, of course, a recreationist may interpret a landscape in terms of the extent to which it gratifies a desire for pleasurable experience. Despite these differences most landscapes do have some intersubjectively shared meaning. In large measure this intersubjective meaning derives from the cultural nature of landscapes and in this context P.F. Lewis (1979) has sugggested a series of axioms to guide the study and understanding of landscapes:

1. Landscapes are clues to culture in that they yield evidence as to the type of people that occupy them.
2. All elements in a landscape reflect culture and in this sense all are equally important.
3. Common landscapes that are often taken for granted are difficult to study by conventional academic means and therefore demand new approaches such as those offered by a humanistic perspective (see Chapter 5).
4. Consideration of the meaning of a landscape demands consideration of the history of that landscape.
5. Elements of a cultural landscape make little sense unless they are studied within their geographical context.
6. Because most cultural landscapes are intimately related to the physical environment, interpretation of landscape depends on a knowledge of that physical landscape.

7. Almost all objects in a landscape convey some sort of message, although not necessarily in any obvious way.

The development of axioms for landscape interpretation rests on the assumption that all people experience their surroundings in a similar way and that their response to their environment is mediated by the culture and society to which they belong. Shared images and shared experiences result, in other words, from shared social situations. In other words, there exists a socialization process with regard to landscape. Clearly this extends to recreation. The consensus on what is deemed appropriate holiday-making has been a strong influence on recreation travel ever since Thomas Cook organized his first trip from Leicester to Loughborough on 7 July 1851.

Although it may seem plausible to hypothesize that socialization influences the development of attitudes to landscape, this is by no means the only view that has been suggested. Some writers have preferred to interpret the human response to landscape as reflecting innate properties of the human body and the human mind. Tuan (1979), for example, has argued that images of landscape are potentially infinite but that they tend to have a 'family likeness' that results from a common principle of organization. Thus the upright carriage of the human body lends itself to simple geometrical concepts like 'top', 'bottom', 'back', and 'front'. Likewise the development and increasing sophistication of children's minds result in landscapes being seen in an increasingly complex manner. Appleton (1975) has gone even further and put forward the proposition that the human response to landscape is mainly a function of atavistic mechanisms that derive from the time when humans had to use the landscape in an immediate sense for survival (Cosgrove 1978). This proposition finds expression in what is known as *habitat theory*. According to this theory, 'aesthetic satisfaction, experienced in the contemplation of landscape, stems from the immediate perception of landscape features which, in their shapes, colours, spatial arrangements and other visible attributes, act as sign–stimuli indicative of environmental conditions favourable to survival' (Appleton 1975: 2). The spontaneous, aesthetic response of humans to a landscape is, in other words, conditioned by the survival mechanisms developed by our primitive ancestors.

This type of reasoning has been refined even more in *prospect–refuge theory* which holds that the landscapes which appear most satisfying to humans are those that provide an ability to see (prospect) without being seen (refuge) (Appleton 1975). Use of this theory has allowed different landscapes to be compared. It has also provided a basis for the evaluation of landscape painting, poetry, and literature (Gold 1980: 121). For example, both prospect and refuge can be characterized symbolically: the former by light, open areas (e.g. sea-shores) and the latter by dark, impenetrable areas (e.g. woodland). In short, prospect–refuge theory argues that the sensitivity which humans experience in regard to a landscape, whether it be a painting or the real world, hinges on whether the landscape in question provides them with an ability to see without being seen and that this, in turn, derives from behavioural mechanisms which are innate (Appleton 1975). This means that although landscape appreciation can be trained, it can never escape these innate mechanisms. Therefore, according to prospect–refuge theory, there is a limit to the ability of humans to replace natural surroundings with artificial ones without inhibiting the process of aesthetic enjoyment (Appleton 1975). Predictably, this theory has come in for criticism (Cosgrove 1978), not least because it rests on an assumption that many archaeologists would now challenge, namely that primitive humans were necessarily hunters. It also flies in the face of the observation that most recreationists seem to be able to find enjoyment in a wide variety of landscapes. Moreover, the attraction that beaches hold for recreationists can presumably be explained without resort to prospect–refuge theory.

The number of landscapes that impinge upon any single human being directly is obviously limited. Yet the vast majority of humans are aware of many landscapes that are beyond their direct experience. This is particularly so with recreationists. They encounter these landscapes through the medium of literature, film, television, and advertising generally. A 'literary' perspective on landscape interpretation has, however, been generally neglected in geography because the evidence it provides is not in a quantitative form (i.e. it has not fitted the dominant paradigm of the last quarter of a century, see Chapter 1) and because the experiences described in such 'literary' sources have rarely thrown much light on the sorts of rational economic behaviour with which much of geography has been preoccupied for so long (Tuan 1976a: 271). Nevertheless, literary sources can be of value to the geographer in several ways: they provide clues as to how people value the environments about them by capturing the ambience of a setting (Salter and Lloyd 1976); they can articulate experience and can transpose inchoate feelings into comprehensible forms (Tuan 1976a: 263); they can reveal something of environmental perceptions and preferences; and, above all, the way that literature can balance the subjective and the objective may provide a more generally applicable model of geographic synthesis (Tuan 1978a).

Any consideration of literary sources of course raises the issue of mass culture. This has been investigated in relation to landscape by Ley and Olds (1988) who (1) questioned the meaning of such public 'spectacles' as the world fair held in Vancouver in 1986 and (2) considered the thesis that such fairs are an instrument of hegemonic power whereby the culture industry imposes meanings on to a depoliticized mass audience. The hegemonic view of such spectacles was found to be an oversimplification in that it represents the consciousness of the masses as being monolithic, unproblematic, passive, and without the potential for resistance. In blunt terms, the hegemonic view of spectacles locates mass culture ultimately in economic relations, thereby overlooking the influence of factors such as race, religion, gender, and life style. The view of Ley and Olds (1988: 195) is that 'popular culture is turbulent and multilayered, heterogenous and actively negotiated'. It is not therefore dictated to by the coalitions of business corporations, politicians, designers, and artists who blend the spectacle, fantasy, and entertainment that is a world fair. Nevertheless there is no denying that such fairs are a source of attraction for millions of people. They are, for instance, a prominent focus of tourist activity. It is important therefore to understand the role of tourism in contemporary society.

Tourism

The distinction between leisure, recreation, and tourism is very blurred. A distinction between leisure and recreation can be sustained to the extent that the former, for most people, can be viewed as the time not given over to work whereas recreation can be viewed as the activities undertaken in that discretionary time (Fig. 10.4). The distinction between recreation and tourism is more problematical. The main thing differentiating the two is the location at which the activity is undertaken. To be specific, if an activity demands a stay overnight away from home base then that activity tends to be termed 'tourism' rather than 'recreation'. Clearly, this is a rather unsatisfactory distinction. Temporary dislocation is a poor basis for defining what has become in recent years a fundamentally important human activity and an economic undertaking that in many countries accounts for in excess of 5 per cent of gross domestic product. Tourism *can* be defined as the temporary movement of people to destinations outside their normal place of work and residence, together with both the activities undertaken during their stay at those destinations and the facilities created to cater for visitors (see Mathieson and Wall 1982; Burkardt and Medlik 1981), but to do so is to take a rather simple view of the phenomenon in question. Tourism is much more than mere dislocation. In one sense it is a significant means whereby modern people assess their world, defining their own sense of identity in the process (Jakle 1985: xi). In a practical sense, it is a major generator of employment. In 1988, for example, the World Tourism Organization estimated that spending on domestic and international trips to places more than 40 km from home amounted to $US 2 *trillion*; in the USA alone there were 1.24 *billion* person-trips extending more than 160 km (Guy *et al.* 1990).

Given the importance that attaches to tourism, it is not surprising that a great deal of effort has gone into trying to understand tourist behaviour (see Pearce 1982a, 1988). This research has shown that tourism is a complex phenomenon. For instance, tourist travel patterns have been found to be influenced by at least the following variables: the size of the population centres in question; economic prosperity; the amount of spare time; demographic characteristics, especially

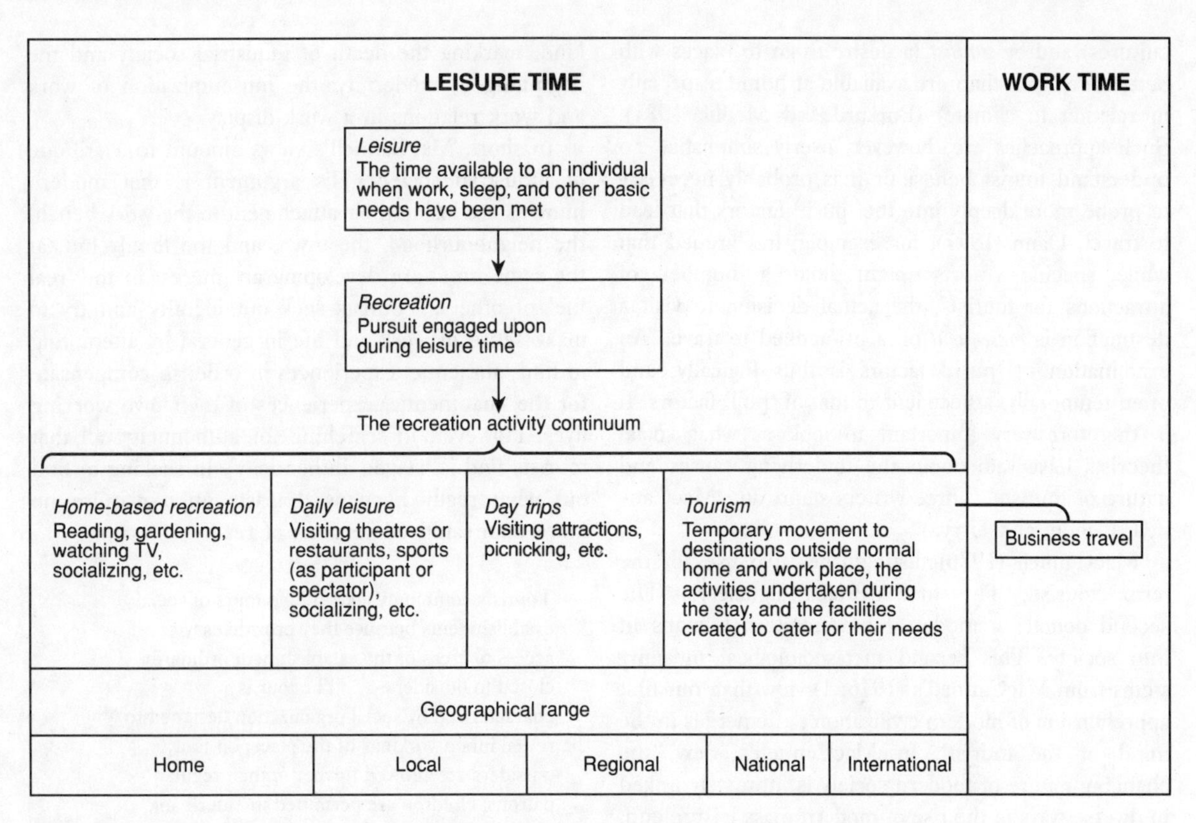

10.4 The definition of leisure, recreation, and tourism (Source: Boniface and Cooper 1987: 2)

age; motivations and preferences; constraints on travel; transport facilities; fares and other travel costs; exchange rates; and promotional activities (Leiper 1990). Moreover there have been at least three different approaches to the study of tourism. Some studies have adopted an idiographic perspective and focused on the uniqueness of the attributes (e.g. scenery) on offer at a particular place. Others have taken an organizational perspective and looked at management, promotion, and capacity. And some have focused on the cognitive images that tourists have of the areas they visit (Lew 1987). Additionally, a good deal of attention has been paid in recent years to tourism *marketing* (Holloway and Plant 1988; Jefferson and Lickorish 1988). However, the marketing of tourism is not an easy task. To begin with there is a tension between the promotion of places (e.g. resorts) and the promotion of facilities (e.g. hotels, attractions) available at particular places. Business enterprises obviously have a vested interest in promoting themselves but this can leave a vacuum where no authority takes responsibility for marketing places (Ashworth and Goodall 1990). And ultimately, of course, marketing can only be successful if it is based on a sound understanding of tourist behaviour.

Herein lies a dilemma: it is not at all clear what motivates tourists. Tourists typically assert that they want 'peace, a change and something totally different from everyday life' and yet they are found 'in crowds, in concrete holiday towns, in artificial tourist landscapes, in a Disneyland atmosphere, and in situations that are sometimes more complex, more restless and less free than the situation at home' (Krippendorf 1987: 30). Some simple claims have been made as to why tourists behave as they do. For example, some writers have argued that tourists are motivated by *wanderlust* (a desire to change the known for the unknown and to see different places, peoples, and cultures) and by *sunlust* (a desire to go to places with better amenities than are available at home, especially in relation to climate) (Burkard and Medlik 1981). Such approaches are, however, overly simplistic. To understand tourist behaviour it is probably necessary to probe more deeply into the 'push' factors that lead to travel. Dann (1977), for example, has argued that while specific resorts might hold a number of attractions for tourists, the actual decision to visit a destination is *consequent* on a prior need to travel. An examination of 'push' factors is thus logically, and often temporally, antecedent to that of 'pull' factors. It is therefore very important to look at what social theorists have said about the underlying causes and nature of tourism. Three writers stand out: MacCannell, Cohen, and Urry.

MacCannell (1976) distinguishes two uses of the term 'tourist'. The first denotes sightseers. The second denotes a model of how modern humans fit into society. This second metasociological meaning stems from MacCannell's (1976: 1) view that 'our first apprehension of modern civilization . . . emerges in the mind of the tourist'. In MacCannell's view, the changing nature of modern society is intimately linked in diverse ways to the rise of modern mass leisure and, especially, to the growth of international tourism. Tourism, in this sense, is the key to understanding how people cope with contemporary society. The study of tourism represents 'an ethnography of modernity' (MacCannell 1976: 4). At the heart of MacCannell's thinking is the idea that leisure is displacing work from the centre of modern social arrangements. The tourist attractions of today are analogous to the religious symbolism of primitive peoples. MacCannell (1976: 6) argues that work is ceasing to have meaning: 'It is only by making a fetish of the work of others, by transforming it into "amusement" ("do-it-yourself"), a spectacle (Grand Coulee), or an attraction (the guided tours of the Ford Motor Company), that modern workers on vacation can apprehend work as part of a meaningful totality.' He argues that industrial society bound workers to jobs but, because of the specialization and fragmentation of tasks, failed to provide them with any sense of identity. Instead workers had to seek identity in off-the-job activities. Among these diversions, according to MacCannell (1976: 36), is to be found a cultural production of a curious and special kind, marking the death of industrial society and the beginning of modernity: the museumization of work and work relations in a work display.

In short, MacCannell's views amount to a critique of industrial society. His argument is that modern humans are losing their attachment to the work bench, the neighbourhood, the town, and the family but, at the same time, are developing an interest in the 'real life' of others. Tourists seek out identity and try to make sense of work and life in general by attempting to find 'authentic' experiences in order to compensate for the 'inauthentic' experiences of their own working lives. However, in searching for authenticity, all that tourists find is 'staged authenticity'. In seeking to find out what really happens, tourists often acquire no more than sanitized versions of reality.

> Tourists commonly take guided tours of social establishments because they provide easy access to areas of the establishment ordinarily closed to outsiders. . . . The tour is characterized by social organization designed to reveal inner workings of the place; on tour, outsiders are allowed further *in* than regular patrons; children are permitted to enter bank vaults to see a million dollars, allowed to touch cows' udders, etc. At the same time, there is a staged quality to the proceedings that lends to them an aura of superficiality, albeit a superficiality not always perceived as such by the tourist, who is usually forgiving in these matters (MacCannell 1976: 98).

There is, in other words, an element of false consciousness in the fact that what is taken to be authentic is actually staged. MacCannell's (1973) view is that tourist settings are arranged so as to give the impression that the visitor is being given an intimate glimpse of life 'backstage' when, in reality, what is on offer is a 'false backstage' that has been created simply to satisfy the tourist demand. Instead of intimacy, closeness, and meaning, tourists get only inauthenticity and staging. 'The touristic experience that comes out of the tourist setting is based on inauthenticity, and as such it is superficial when compared with careful study; it is morally inferior to mere experience' (MacCannell 1973: 599). In MacCannell's view, presenting tourists with a 'false backstage' is insidious and dangerous even if it can at times protect a minority group from overly close attention, as in the case of the Amish staging tourist activities to protect their privacy (Buck 1978).

In short, what MacCannell is attempting is nothing less than a theory of modern life that seeks to explain how humans engage a world that is too big and too complex for them to know it entirely and intimately. In some respects the argument is not new. Boorstin (1964) lampooned 'pseudo-events' in his denigration of tourism a decade earlier (Cohen 1988). Nevertheless, MacCannell's work is much cited. Moreover it has been extended somewhat by Pearce and Moscardo (1986) to include not just the authenticity of the setting but also the authenticity of the people. After all, it is the relationship between the tourist and host that determines authenticity, not the arena. Despite this modification, there are problems with MacCannell's approach. For one thing there is a tendency for authenticity to be treated as a 'thing-like social fact' when in reality the concept is much more problematical and able to be understood only when the taken-for-granted world of the individual concerned is called into question. Authenticity in this sense is not a property of a setting so much as something that emerges in a relationship (Turner and Manning 1988). And, on top of this weakness, there remains the issue of the extent to which tourists are aware of inauthenticity. This is a point taken up by Cohen.

Cohen (1979a) devised a typology of authenticity in tourist settings depending on whether the scene is real (not manipulated by hosts or the tourist establishment to create a false impression) or staged, and whether or not tourists are aware of the staging (Fig. 10.5). The typology covers:

1. Authentic settings perceived as such;
2. Settings where tourists are unaware of the staging;
3. Cases where tourists are suspicious, possibly because of a prior bad experience where they have been taken in by events purporting to be authentic;
4. Contrived situations that are recognized as such.

It follows from Fig. 10.5 that the stereotype of a tourist as a 'slightly funny, quaintly dressed, camera-toting foreigner, ignorant, passive, shallow and gullible' (Cohen 1974: 527) is a grossly inaccurate caricature. There are in fact many different types of tourist and,

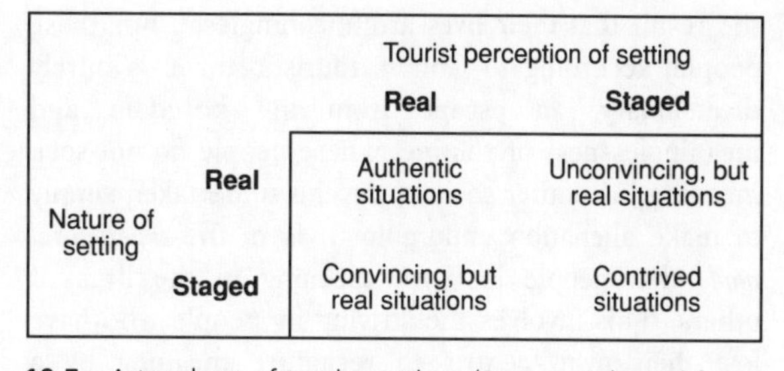

		Tourist perception of setting	
		Real	Staged
Nature of setting	Real	Authentic situations	Unconvincing, but real situations
	Staged	Convincing, but real situations	Contrived situations

10.5 A typology of tourist settings (Source: Adapted from Cohen 1979a)

according to Cohen (1979b), the best way to understand them is through a *phenomenological interpretation* of their behaviour. The central argument of Cohen's work is that there exist phenomenologically distinct types of tourist experience depending on what sort of 'centre' an individual tourist adheres to. In Cohen's view, every society possesses a 'centre' which is the charismatic nexus of its supreme and ultimate moral values. Much of structural–functionalist sociology has adopted a similar view. Indeed from the functionalist perspective, the tension that inevitably arises when individuals suppress their own wishes and conform to society's ultimate values requires an antidote and that antidote is often found in the form of recreation and leisure, with tourism being the modern recreational activity *par excellence*. Cohen sees this view of modern humans as conformists who indulge in tourism as a temporary getaway from their 'centre' as far too simple. Instead he argues that modern society is characterized by multiple 'centres' (political, religious, cultural) with the result that there are likely to be different modes of touristic experience.

Cohen in fact identifies five modes of experience. In the *recreational mode* the tourist trip is used simply to restore physical and mental powers and well-being. There is no deep commitment to travel as a means of self-realization and the tourist often appears to be taken in by blatantly contrived displays. The goal of the tourist in this mode is entertainment not authenticity. The trip is merely a pressure valve in modern life. This is therefore the mode of experience that best approximates the structural–functionalist view of recreation. The *diversionary mode* of experience is very different. Cohen argues that some people are alienated from the 'centre' of their society and culture with the result that their lives are 'meaningless'. For these people, according to Cohen, touristic travel is purely diversionary, an escape from the boredom and meaninglessness of routine. These people do not seek meaning in another society; travel is undertaken simply to make alienation endurable. It is in the *experiential mode* that people look for meaning in the lives of others. This involves the striving by people who have lost their own 'centre' to recapture meaning by a vicarious (and often essentially aesthetic) experience of authenticity in the lives of others. However, a gulf remains between these tourists and the people they observe. The tourist is not converted to a new way of life. This is rather different from the *experimental mode* of tourist experience. In this case people who no longer adhere to the 'centre' of their own society try out alternative ways of life in a quest for meaning. Tourists in the experimental mode sample different ways of life in what is essentially a search for self. In some cases, the search may become a way of life in itself. The 'drifter' is perhaps a case in point. Finally, in the *existential mode* of experience, tourists settle on a spiritual 'centre' external to their native society and culture. This is analogous to religious conversion and is perhaps exemplified by the individual who 'drops out' to find fulfilment in closeness to nature (Cohen 1979b).

A third, contrasting, metatheoretical viewpoint on tourism is that of Urry (1990a). Urry's (1990b: 25–6) argument is that

> consumption in the case of many tourist services is a rather complex and inchoate process . . . the minimal characteristic of tourist activity is the fact that we look at, or gaze upon, particular objects, such as piers, towers, old buildings, artistic objects, food, countryside, and so on. The actual purchases in tourism (the hotel bed, the meal, the ticket etc.) are often incidental to the gaze, which may be no more than a momentary view.

In other words, the 'gaze' is vitally important. According to Urry, a large proportion of the population of modern society engages in 'gazing', commonly at landscapes that separate them from everyday and routine experiences. Furthermore, the gaze is often captured and objectified in photographs, films, postcards, and models. Consequently, a large number of tourist professionals are employed in creating new objects of the tourist gaze. They therefore develop a series of signs as to what is worth gazing upon. In fact tourists are in many senses semioticians who make sense of the signs around them. 'The denigrators of tourists are upset by the proliferation of tacky representations – postcards, ashtrays, little silk pillows, ugly painted plates – and fail to grasp the essential semiotic functions of these markers' (Culler 1981: 133). As Boorstin (1964) and others have pointed out, such souvenirs act as signs and markers. In fact tourists are often more interested in what is 'Japanesy' than in what is Japanese (see Culler 1981).

Urry's views are of course not dissimilar from those of MacCannell and Cohen. They do, however, have implications that the others do not. To begin with, Urry's emphasis is very much on *positional goods*. In such cases consumption takes on a social as well as an individual aspect. In other words, the satisfaction that individuals get from goods and services depends in large measure not only on their own consumption but on the consumption patterns of others (see Hirsch 1978). The thing about positional goods is of course that they involve a zero-sum game; if someone consumes more, then someone else must consume less. Unlike material goods, the supply of positional goods cannot be increased. The consumption of positional goods therefore says a lot about social relationships. Many holiday destinations are 'consumed', not because they are intrinsically superior, but because they convey taste and superior status (Urry 1990b). In short, some tourist attractions are status symbols. To have 'gazed' upon a particular feature can be a measure of social standing.

In all this theorizing about tourism, it is important to remember that the nature of tourism might be changing. Urry (1988), for instance, has pointed out that mass holiday-making might have been the quintessential form of tourism in the industrial era but that the future is likely to be characterized by the 'post-tourist' (really the 'post-mass-tourist'). Post-tourists are people who do not have to leave home to gaze on attractions (because these are available on television and video), and they are people who realize that tourism is a game where there are no authentic experiences. Perhaps tourism in the postmodern world will be different. It may well be a world of spectacle where nothing is new, where the distinction between high and low culture and between taste and tastelessness disappears, where social groups are less bound by mass norms, and where heterogeneity and variety are valued (Urry 1988). It will certainly be a world where recreation is important and where an understanding of recreational behaviour is needed. It is important therefore to complement metatheoretical perspectives on leisure, recreation, and tourism with an appreciation of behaviour at the micro-scale of the individual decision-making unit.

Decision-making and images

According to Warburton (1981), recreational behaviour can be conceptualized as a system of five interrelated components: antecedent conditions; user aspirations; intervening variables; user satisfaction; and real benefits. Essentially, the idea is that the level of benefit will, with experience, either heighten or dampen the urge for recreational activity within the context of an individual's needs and opportunities. In other words, decisions are made about whether or not to recreate, and where to go, on the basis of opportunity (geographical access, transport availability, time), knowledge, favourable social milieu (family, peer support), receptiveness (willingness to enter into an activity), perceptions, and needs (Iso-Ahela 1980; Brandenburg *et al.* 1982).

A similar interpretation is provided by Aldskŏgius's (1977) decision-making model of recreational-trip behaviour in Sweden, which distinguishes, on the one hand, the underlying motives of the individual from, on the other, the recreational resource endowment of an area, within the context of constraints such as an individual's mobility potential, 'real' income, and time availability (Fig. 10.6). The central core of the model is the individual's cognitive image of recreational opportunities, within which information is integrated through a process of selection, transformation, and evaluation. In this information-processing phase, an important part is played by the individual's social environment in general, and by socio-economic and household characteristics in particular. Two factors are of specific interest. First, there is the residential history of the individual. This is likely to have influenced the diversity of recreational experience and familiarity with different environments, as well as contributing to preferences and attitudes with respect to different activities and resources. Second, it seems reasonable to assume that the greater the length of residence of an individual in an area, the greater will be that individual's knowledge of available recreational opportunities. A similar view has been expressed by Stabler (1990) and Kent (1990) who introduced the idea of 'opportunity sets' to signify the tourist attractions which are known to individuals and meet their needs within a framework of constraints. In this sense an opportunity set is the group of locations that results when the push (motivation) and pull (attraction) forces acting on an individual have been taken into account.

Despite the lead of Aldskŏgius, very little work has been done on the cognitive images held by recreationists. Mercer (1971) has suggested that outer suburban residents in Melbourne might have wedge-shaped mental maps but there have been very few attempts to explore images directly. Exceptions are to be seen in Pearce's (1977) study of first-time visitors to Oxford and his study of the mental route maps held by tourists in north Queensland (Pearce 1981). Both these studies showed learning takes place with experience (see also Guy *et al.* 1990). The cognitive images of recreationists are, however, relatively primitive, as shown by Fig. 10.7 which presents the Lynch-type maps held by (a) visitors to and (b) residents of the coastal resort of Coffs Harbour, Australia (see Jenkins and Walmsley 1992; Walmsley and Jenkins 1992a). Clearly, visiting recreationists have limited local knowledge and very simple images. The sketch-mapping approach to cognitive imagery of course reveals nothing about *why* recreationists travel to particular places. Such detail is perhaps better obtained through the use of multi-dimensional scaling (see Goodrich 1978) or repertory grid analysis (see Embacher and Buttle 1989). Both these techniques permit the researcher to elicit the recreationist's views on the key characteristics of an area and on how and why one place is differentiated from another. They therefore go some way to identifying what it is that is taken into account in the definition of preferences.

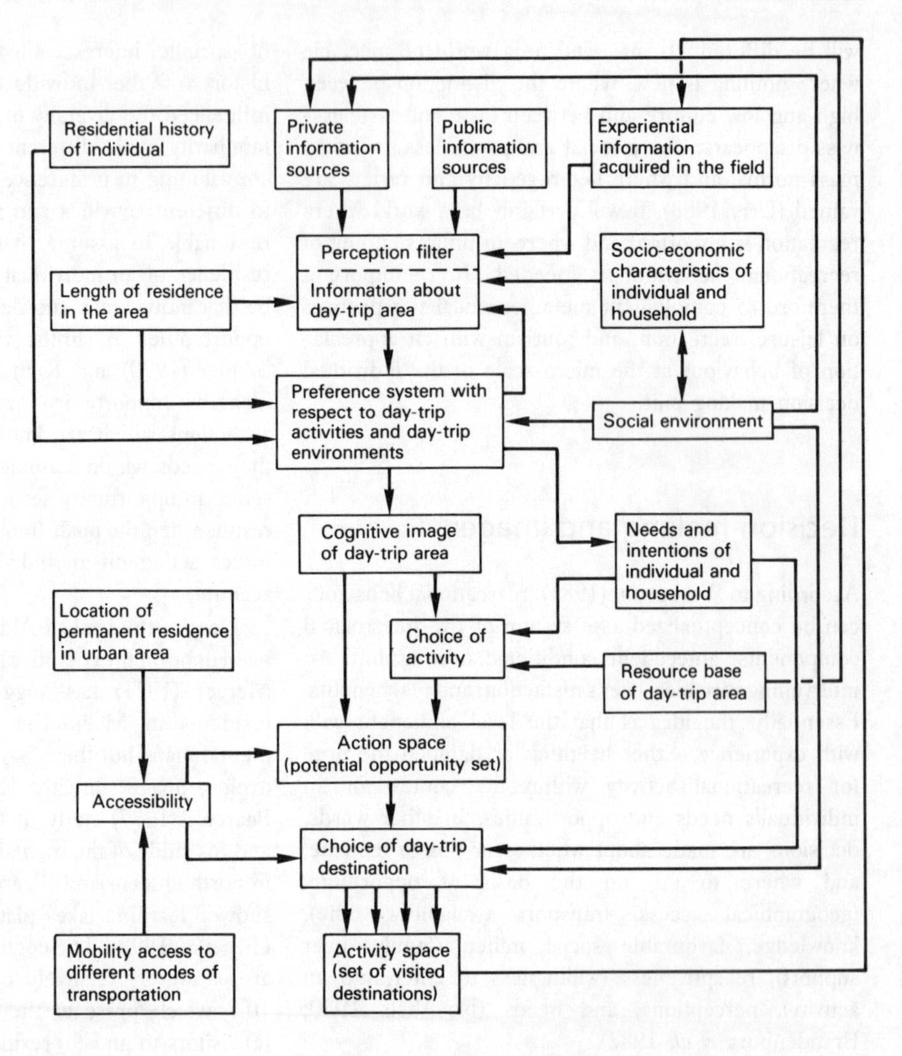

10.6 A conceptual model of recreational behaviour (Source: Aldskǒgius 1977: 165)

Motivation and preference

In any consideration of why individuals differ in their choice of recreational activity some attention must be paid to the underlying motives for recreation in general (Crandall 1980). In this context it is important to realize that individuals have a series of needs which they strive to fulfil. These needs can be thought of in terms of a hierarchy and only when low-order physiological needs (e.g. shelter) have been met will higher-order needs such as recreation play a significant role in an individual's life style (Maslow 1954; Glyptis 1981). Such higher-order needs can be characterized by two opposite forces – tension-seeking and tension-reducing – between which a balance is struck. For example, recreation can serve both to ameliorate tension and also at times actually to increase it (Iso-Ahela 1980). However, it is extremely difficult for a researcher to identify whether an activity is fulfilling a tension-reducing or tension-seeking function. This is partly because needs and tensions are socially and culturally, as well as personally, defined, with the result that the researcher needs to consider the personality, experience, aspirations, and work situation of the individual recreationist (Iso-Ahela 1982). Perhaps the importance of tension-seeking and

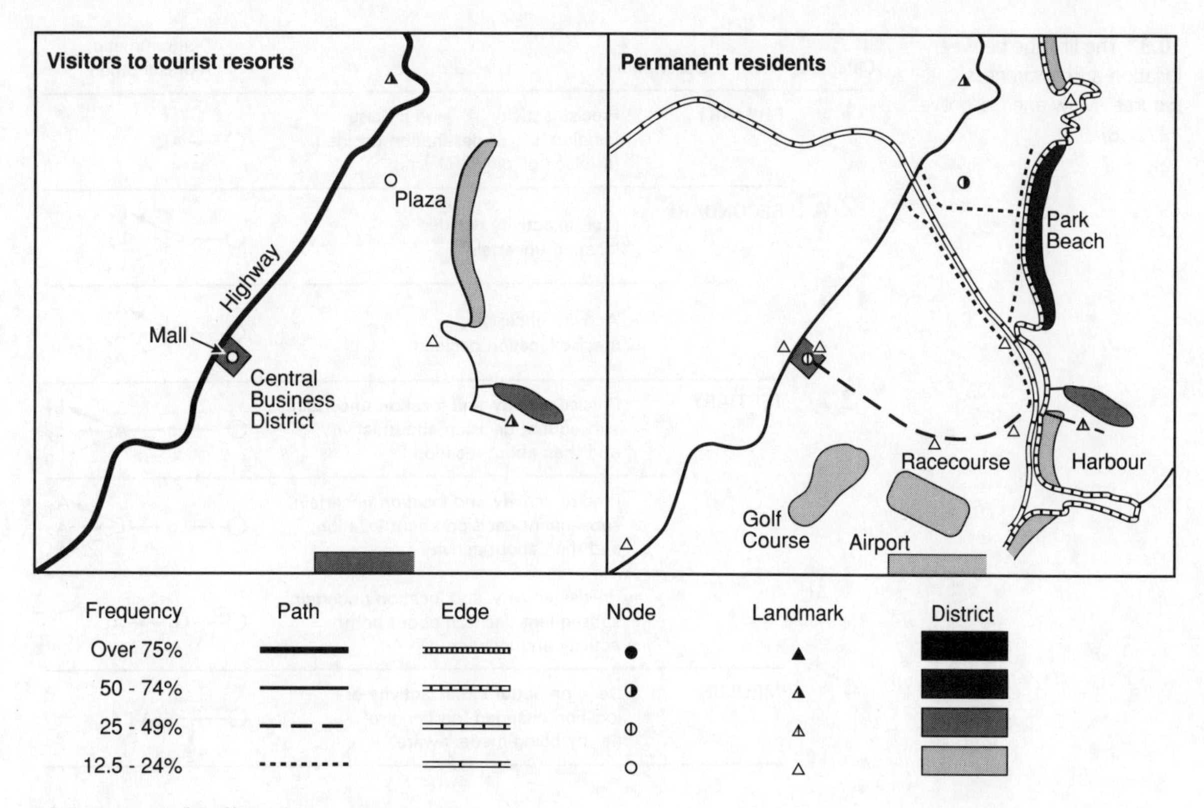

10.7 Images of Coffs Harbour, Australia

tension-reducing behaviour in recreation was most vividly expressed in Plog's (1973) claim that tourists can be arranged, in terms of underlying personality and motivation, on a spectrum from *allocentric* (adventure-seeking, preferring the exotic) to *psychocentric* personality types (safety-seeking, preferring the familiar). In Plog's view, a person's score on this personality scale has a big bearing on where they go for recreation and what they do. Although Smith (1990) has questioned the validity of Plog's distinction, it remains one that is widely cited (see Plog 1990; Walmsley and Jenkins 1991).

Tension-seeking and tension-reducing perspectives underlie many general theories that have been advanced to explain choice behaviour. For example, *compensatory theory* hypothesizes that, whenever a person has the chance to avoid regular routine, that person will tend to choose a directly opposite activity. In the context of recreation this means that people will adopt a leisure style that is opposite to their occupational life style. Evidence for this has come from Hendee (1969) who has illustrated that urban dwellers tend to develop an appreciative rather than a utilitarian attitude towards nature while rural dwellers are more inclined to seek out the 'busy and exciting' activities of the towns and cities. In contrast, *familiarity theory* claims that individuals seek out recreational activities which are not markedly different from their everyday occupational life style. Again Hendee (1969) found support for this to the extent that farmers are often involved in such activities as hunting and fishing. Very little research, however, has assessed the relative significance of these contradictory theories, no doubt due to the inherent difficulties involved in determining what is 'opposite' or 'familiar'. It may be that familiarity with an activity breeds specialization whereby behaviour becomes reinforced (Schreyer and Beaulieu 1986). What begins as compensation may thereby end

10.8 The linkage between location and awareness (Source: Maw and Cosgrove 1972: 8)

Category		Description		Diagrammatic representation
1		PRIMARY	Precise activity 'A' and precise location 'L' of destination decided (at place of origin 'O')	O → AL
2	A	SECONDARY	Precise activity decided, location uncertain	O — A → L_1 / L_2 / L_3
	L		Activity uncertain, precise location decided	O — L → A_1 / A_2 / A_3
3	A	TERTIARY	Precise activity and location uncertain, subsequent decision about activity, and then about location	O — Lg — A → L_1 / L_2 / L_3
	L		Precise activity and location uncertain, subsequent decision about location, and then about activity	O — Lg — L → A_1 / A_2 / A_3
	AL		Precise activity and location uncertain, subsequent decision about both activity and location	O — Lg → AL
4		IMPULSE	Decision about either activity or location changed 'on impulse' i.e. by being made 'aware'	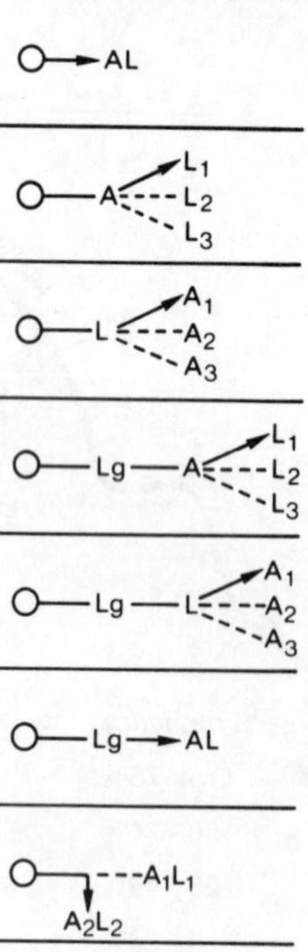 O - - - A_1L_1 / ↓ A_2L_2

up as familiar. It may even be that recreational experiences for the majority of people are both familiar and compensatory. After all, such a situation fits experimental evidence that individuals require a continually varied sensory input. Given this conflict, it is not surprising that a good deal of research on based activities in the recreational preference structure of urban residents, and argued that the relationship between preference and behaviour was not really significant because, for the majority of the respondents, a number of constraints precluded their preferences from becoming operative. Given the rather tenuous link between preferences and behaviour, it is perhaps best to view actual recreational behaviour patterns as arising from a decision-making process that is characterized by uncertainty, limited awareness, and limited opportunity (Chase and Cheek 1979). Such a perspective is central to Maw and Cosgrove's (1972) model of recreational choice (Fig. 10.8). At the scale of the individual day-trip this model suggests that quite often the choice of activity and location is not based activities in the recreational preference structure of urban residents, and argued that the relationship between preference and behaviour was not really significant because, for the majority of the respondents, a number of constraints precluded their preferences from becoming operative. Given the rather tenuous link between preferences and behaviour, it is perhaps best to view actual recreational behaviour patterns as arising from a decision-making process that is characterized by uncertainty, limited awareness, and limited opportunity (Chase and Cheek 1979). Such a perspective is central to Maw and Cosgrove's (1972) model of recreational choice (Fig. 10.8). At the scale of the individual day-trip this model suggests that quite often the choice of activity and location is not clearly planned when the trip begins (as in behavioural categories 2, 3 and 4 of Fig. 10.8). As a result recreational behaviour needs to be seen as a continually unfolding learning process rather than an event or series of events isolated in time and space (Murphy and Rosenblood 1974). This is especially

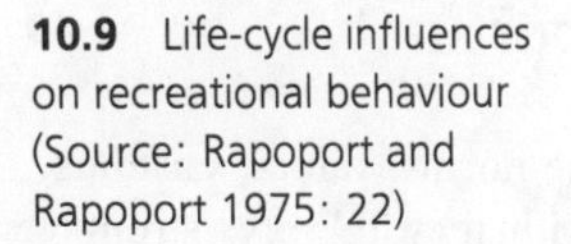
10.9 Life-cycle influences on recreational behaviour (Source: Rapoport and Rapoport 1975: 22)

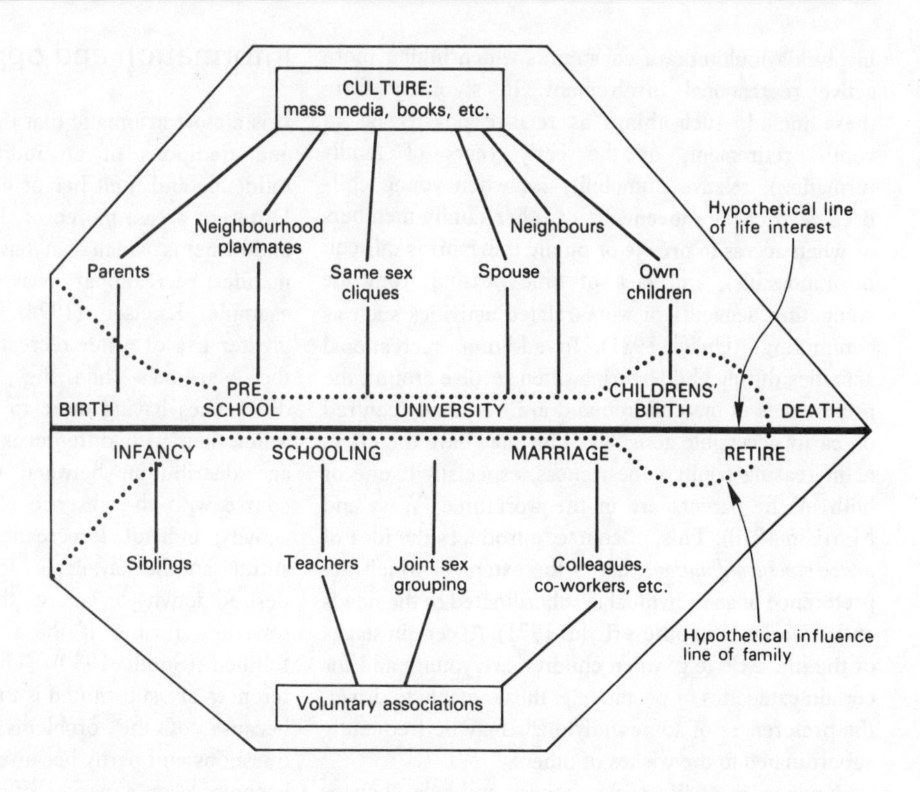

true of an individual or a family moving to an entirely new residential location; in such cases recreational trip-making behaviour starts off as a more or less random search process (possibly with a bias to attractive landscapes) but later (sometimes a matter of years) settles down to a habit phase (Elson 1976). Learning is not, however, a simple function of personal experience because recreational behaviour patterns are at all times influenced by the social context in which they occur (Mercer 1976). As a result any attempt to explain recreational preferences needs to go beyond personality and gender, financial and class background, and time–space factors to consider the role of family life-cycle, residential history, and information availability (Witt and Bishop 1970).

Personal community theory hypothesizes that individual recreational preferences are influenced by those around them (especially relatives, friends, and peer groups) and that a preferred activity developed in childhood is often maintained into adulthood. However, it is evident even to the casual observer that different groups influence individuals at different stages in their life cycle; for example, during infancy parents are dominant whereas during schooldays teachers and school friends become important. These developmental life events have been incorporated by Rapoport and Rapoport (1975) into a simplified model of lifetime influences on recreational choice (Fig. 10.9). The model highlights the changing influences acting upon an individual throughout a lifetime and the way in which different roles – dependent child, college student, breadwinner – are adopted. Each stage brings with it drives towards, and constraints on, different types of recreational behaviour.

Of course, at several stages of the life cycle an individual may be subject to a number of different roles with which are associated quite different social influences and recreational pursuits (Preston and Taylor 1981). In fact a constellation of roles can result in a diversity of recreational pursuits being carried on at the same stage of the life cycle as, for example, when business executives pursue some activities with their families, some with friends, and some with business associates. However, each life-cycle stage also involves a multitude of constraints which inhibit more active recreational involvement. In summary form these include such things as relative poverty (as in youth, retirement, or the early years of family formation), relative immobility (as when young children restrict the movements of other family members or when access to private or public transport is difficult or impossible), and lack of time (arising from the competing demands of work-related activities such as commuting) (Unkel 1981). In addition, recreational activities during child-rearing often revolve around the preferences of the children and are frequently centred on easily accessible activities. This may have the effect of increasing family cohesiveness, especially if one or both of the parents are in the workforce (West and Merriam 1970). This, of course, introduces the idea of *preference subordination*, that is the extent to which the preference of an individual is subordinated to the needs and preferences of others (Eyles 1971). At certain stages of the life cycle (e.g. when children are young) and for certain categories of people (e.g. those unable to drive), the preferences of some individuals may be frequently subordinated to the wishes of others.

Knowledge of life-cycle changes and role changes is crucial in any attempt to determine the preference structure of individual recreational behaviour, not least because it draws attention to the difference between real and professed motives for involvement in many recreational pursuits. Such motives are often highly complex and obscure. It may be, for example, that many motives are so deep-seated as not to be uncovered in the standard questionnaire schedule. This is perhaps why the metatheoretical perspectives of MacCannell (1976), Cohen (1979a), and Urry (1990a) have attracted so much attention. It might also explain why apparently different 'reasons' may be given by similar people for participation in the same activities (Dann 1981). There is, in other words, a great need to consider the *meaning* of recreational pursuits for individuals, particularly in view of Kelly's (1974: 192) observation that 'the same activity may have differing social meanings and role relations at different times'. This, in turn, means that it might be profitable to explore the hypothesis that leisure is more identity-seeking for youth, more role-related for parents, more work and community-related in later years, and morc interpersonal or solitary in retirement.

Information and opportunity

It is almost axiomatic that the norms, values, customs, and traditions of an individual will very strongly influence and limit her or his range of leisure choice. Different areas, moreover, have different recreational endowments which can have a direct impact on the manifest recreational behaviour of the population. For example, Knetsch (1974) found that people make greater use of water recreation facilities per capita in the Maritimes than they do in the Prairies, the differences having more to do with the availability of water than with differences in income, education, or age distribution between the two populations. Of course with the absence of a particular recreational activity, individual recreationists have either to substitute another available activity for the unavailable desired activity or enlarge their recreational 'space' by travelling further if the recreational need is to be fulfilled (Smith 1980). The degree to which preferences are substituted is difficult to determine, partly because of the problems of formulating suitable questions and partly because the substitute activity can become, over a period of time, the desired activity. Enlargement of recreational awareness space is more readily assessed. However, use of this enlarged space is not automatic but depends on the mobility potential of the individual (Harrison and Stabler 1981).

Overall, though, an individual's action space plays a crucial role in determining recreational choices; for example, the recreational universe of slum-dwellers differs markedly from that of the middle-class cosmopolites, as demonstrated by Orleans's (1973) mental mapping exercise. Further, the residential history of an individual is an immensely important factor influencing recreational preferences and needs. An individual who has lived in a number of different places has probably had the opportunity to engage in a variety of recreational pursuits, is therefore aware of different recreational possibilities, and, inevitably, becomes sensitized to differences between the environments with which contact is made. Despite this, many individuals have recreation patterns that become habituated with regard to favoured activities and location. For these individuals a change in behaviour only comes about as a consequence of life-cycle changes and/or changes in the availability of information.

An individual's choice of recreational pursuit is influenced markedly by the information available about likely recreational opportunities. All the various forms of media, as well as friends and relatives, provide details on places to visit, activities to undertake, and societies to join. Moreover a number of specialized magazines are devoted to providing information related to hundreds of individual recreational pursuits. Advertising, too, has become strongly associated with leisure and recreation since it not only encourages people to participate in certain activities but also frequently uses a variety of recreational settings as a backdrop for its general promotional activities. However, the effect of this information on recreational behaviour has yet to be analysed in any detail (see Phelps 1986). The one exception is the tourist brochure, the importance of which is starting to be appreciated (Dilley 1986, Adams 1984).

Whatever the source of an individual's information, recreation and tourism were generally regarded until relatively recently as beneficial both for individual well-being and for those organizations involved in providing recreational facilities. For example, great emphasis has been placed by planners on the regional multiplier effect and the way in which investment in tourism is supposed to trickle through to the local community (Archer 1973; Smith and Wilde 1977). However, of late, some attention has been drawn to issues of environmental quality and social justice (Cushman and Hamilton-Smith 1980). For the most part, this attention has focused on land use pressures and land management policies (Cheng 1980; Turner 1981; Bosselman 1978), and only rarely has there been any appreciation of the social costs of tourism in advanced Western economies (Birrell and Silverwood 1981; Pizam 1978). This is perhaps changing. One of the earliest authors to sound a cautionary note in respect to the side-effects of tourism was Young (1973) who drew attention to the regional and local disbenefits of tourist development. More recently, Walmsley *et al.* (1981) have shown how increasing tourism is related to increasing crime rates on the north coast of New South Wales. Given these problems it may well be that the response of host communities to recreationists will follow a predictable path whereby an initial increase in favourability in the early days gives way to increasing criticism as the resultant problems become apparent (Long *et al.* 1990).

Looking into the future, the advancement of the microchip technology will almost certainly result in people working less and retiring earlier. As was pointed out at the start of this chapter, this may mean a dramatic increase in leisure time (Seltzer and Wilson 1980). If there is no work to be had then people must find identity and purpose in a work substitute. Leisure seems to be the obvious answer (Glyptis 1989). The study of leisure, recreation, and tourism is therefore likely to become an increasingly important part of human geography.

11 Belonging and Well-being

The significance of *place* in day-to-day living and in the quality of urban life has been of concern to geographers and other social scientists for a long time. For example, it formed the basis of much of the work of the Chicago ecologists (Park *et al.* 1925) (see Chapter 2). Within the context of ecological theory, such studies provided detailed descriptions of life within the city's social areas, emphasizing their personality, way of life, and community structure. Much of the focus was upon ethnic or 'deprived' areas within the 'downtown' parts of the city (an emphasis which persists in more recent research such as the proliferation of studies of the 'black ghetto', see Jackson and Smith 1984). Essentially, the way of life of these communities was interpreted in terms of the *local milieux* which provided the basis for community and the circumstances for deprivation.

However, by focusing on downtown communities, the Chicago ecologists tended to ignore other urban areas which differed in terms of social structure, way of life, and territoriality. As a result, when later researchers considered these previously ignored communities, the ecological approach was found wanting and the concept of a 'natural area' or community became less tenable. In its place, attention switched to the impact of urbanization and industrialization in altering community organization. This impact is best appreciated in the context of a continuum model: at one end is a *Gemeinschaft* society, that is a territorially related community that is socially homogeneous and bound together by a tightly knit pattern of primary relationships; and, at the other end, is a *Gesellschaft* society where individuals participate in an impersonal way in various specialized institutions such that their relationships with other people are often compartmentalized, formal, and role-directed (Tönnies 1887). Cities are prominent in *Gesellschaft* societies and in such places 'people become members of groups larger than neighbourhoods, and merely reside in residential areas in contrast to living in rural or village neighbourhoods as was true in the past' (Isaacs 1948: 18). The shift from *Gemeinschaft* to *Gesellschaft* is not of course uniform and some urban neighbourhoods, small towns, and villages have retained some of the characteristics of the former at a time when society as a whole exhibits the features of *Gesellschaft* (Suttles 1972; G.J. Lewis 1979).

In the 1960s, Webber (1963, 1964a, b) enlarged upon the shift from *Gemeinschaft* to *Gesellschaft*. He viewed the decline of community that was characteristic of much advanced Western society as being the result of a set of technological, sociological, and psychological developments. For instance, the advent of the railway, the motor car, mass media, and mass culture has led to a lessening of the salience of place and therefore a weakening of the links between communities and milieux. According to Laslett (1968), in the 'world we have lost', everything physical was on a human scale and everything temporal was tied to the human life span, resulting in all communities being highly localized in their structure and activities (G.J. Lewis 1979). In contrast, the 'world we have gained' is distinguished by a widening of geographical horizons, a fact which has led Webber (1964a: 116) to suggest that nowadays 'it is interaction, not place, that is the essence of the city and city life'. For example, Webber suggested that, for professional and managerial groups, communities might be geographically far-flung but nevertheless close-knit, intimate, and held together by shared interests and values. In contrast, for the working class, community organization might be still territorially coterminous with neighbourhood. What Webber argued was that 'non-place' communities would eventually permeate all segments of society (as a result of increasing affluence and increasing mobility) with the result that the environmental knowledge of the majority of the population would broaden in scope and deepen in understanding.

Despite Webber's critique of the significance of place in urban living, planners and politicians continued for many years to conceive of locality as a significant parameter in economic and social development. Even the argument that many so-called 'areal' problems are caused by the structure of society rather than by the local environment, failed to deflect the popularity of area-based programmes in many countries (Lawless 1979). This is partly because not all researchers accepted Webber's argument. Indeed, during the 1980s there has been a distinct revival of interest in the role of the local environment in the economic and social life of both urban and rural residents. Three recent trends in social and behavioural research provide a basis for this renewed interest.

First, growing dissatisfaction with structuralist approaches, coupled with postmodernist thinking on how society reproduces itself (Soja 1989), has made place a major focus of interest. For example, it has been shown that, even in a highly mobile society, there are large sections of the population (such as the elderly, the poor, and the young) who are still highly place-bound. Likewise, social heterogeneity can inhibit travel within cities; few people, for example, like to travel through areas where they feel uncomfortable or unwelcome. Nowhere is the importance of place better seen that in childhood where the immediate locality, both in a physical and social sense, is of utmost significance in the development of the individual (see Chapter 4). Moreover, even mobile individuals retain a strong feeling for their homeland and birthplaces, as exemplified by the growing number of 'return' migrants in North America and Western Europe (Lewis 1982). Similarly, much of the recent boom in locality-based interest groups, whether opposing new housing developments or motorway construction, or demanding improved social services or recreational facilities, is a reflection of middle-class desires for place identity (Proshansky 1978). At a larger scale, the growth in regional consciousness, whether it be the Welsh or the Scots, and the emergence of environmental consciousness, demonstrated in architectural and countryside conservation (Shoard 1980), is further evidence of the feeling that people have for place. Some have gone as far as to claim that greater spatial mobility may, in any case, create a reciprocal need for a firm home base, thereby strengthening place attachment (Norberg-Schultz 1971). In this way, the local area can serve as a repository of social traditions which individuals cherish and wish to protect, and as a haven in an uncertain environment. What this illustrates is the continuing significance of place in the everyday life of all urban residents and the view that 'mobility will never destroy the importance of locality' (Pahl 1968: 48).

Second, with the advent of socially relevant geography in the 1970s, researchers for the first time addressed the issue of how individuals in socially deprived milieux overcome their disadvantage. Until that time research in social geography was overwhelmingly concerned with the description of the aggregate features of city structure. The significance of locality in the quality of everyday life has been cogently argued by Faulkner (1978) in his adaptation of Maslow's (1954) hierarchy of *needs* (Table 11.1). According to Faulkner, each level of an individual's needs is met by the attributes of the local environment in which the individual resides. Clearly, this suggests that the nature and form of that locality have a significant influence on the quality of life.

Third, even among the neo-Marxists there has been a re-evaluation of the significance of *locale* in so-called post-structuralist thinking, and a questioning of universalist interpretations of spatial organization and change. According to Swyngedouw (1989: 31), this greater preoccupation of Marxists with fragmentation and uniqueness is 'imbedded in and associated with the contradictory dynamics of increasingly footloose and mobile capital seeking out profitable locations amidst a highly disjointed and fragmented mosaic of uneven development in which competitive places try to secure a lucrative development niche'. As part of this

Table 11.1 A typology of urban needs

Need category	Description	Attributes of the urban environment associated with the satisfaction of needs (examples)
1. Physiological	Provision of food, shelter, and health care	Retailing/wholesaling systems distributing food, clothing and health supplies Health care clinics and hospitals Essential services (water, sewerage, power) Dwellings
2. Safety–security	Protection from physical harm and intruders Privacy and absence of over-crowding Protection of property	Fire and police protection services Road safety Absence of noxious environ-mental elements (pollutants) Residential areas that ensure privacy
3. Affection–belonging	Harmonious relationships with other members of the community Identification with and acceptance of groups within the community	Facilities for community organ-izations (meeting places) Physical layout of neighbour-hood such that co-operative and harmonious inter-family relationships are fostered Physical identity of the neighbourhood
4. Esteem	Status and recognition by others in the community	Opportunities of home owner-ship Prestige of neighbourhood
5. Self-actualization	Role relationship *vis-à-vis* others Realization of one's potential Creativity/self-expression	Built environment that facilitates creativity and self-expression Employment opportunities and community organizations that enable the use and develop-ment of skills
6. Cognitive/aesthetic	Provision of educational experience, intellectual stimu-lation and experiences Aesthetically appealing events and phenomena	Educational and cultural facilities Recreational facilities Aesthetically appealing built and natural environment

Source: Faulkner (1978: 62)

overall shift of focus and orientation, there has been a growing interest in the study of *localities*, particularly highlighting their economic restructuring and social recomposition (Cooke 1989a).

Belonging to place

In the context of a mobile society, places may be regarded as 'foci where we experience the meaningful events of our existence ... [and] ... points of departure from which we orient ourselves and take possession of the environment' (Norberg-Schultz 1971: 919). They are not therefore analogous to places in the animal world (Dyson-Hudson and Smith 1978) but rather can be viewed, in Chombart de Lauwe's (1975) terms, as involving a hierarchy of social spaces each reflecting increasingly temporary use of the urban environment (Fig. 11.1). Therefore, to know that a part of a city is occupied by a particular group, does not allow prediction of how any individual or group will react to any changes which might affect their neighbourhood. This sort of understanding can only be achieved if attention is directed to the ways in which the urban environment is 'known' by its residents and has 'meaning' to them (Krampen 1981). Knowing and meaning are, in other words, crucial mediating factors in the interaction between individuals and the built environment (McGill and Korn 1982). Put simply, places are more than mere objects in a physical sense since they are also things that are given meaning by individuals and, hence, their reality is both socially constructed and socially contingent (see Chapter 5). It follows from this that 'place' is likely to have a variety of different realities, each reflecting different concerns, needs, and cultures.

The phenomenologists have taken this view a stage further by suggesting that, within the so-called *life-world*, there is no conceptual distinction between object and subject since every object has its purpose and every place its meaning (Seamon 1979). Such a world is, of course, anthropocentric because it is constructed around the individual and the group and their particular concerns. In reality, the relationship between people and the environment is *reciprocal* in that individuals and groups not only create a sense of place but are also themselves influenced by the places they inherit and frequent. According to Park (1916: 579), the city is, on the one hand 'a product of nature and particularly human nature' and, on the other, 'in making the city, man [*sic*] has remade himself' (Park 1929: 3). This reciprocal relationship between people and place has been conceptualized by psychologists as a *behaviour setting*. According to Barker (1963: 19), 'behaviour settings and their inhabitants are mutually, causally related. Settings have plans for their inhabitants' behaviour, and inputs are activated within the limits of the setting's control systems to produce the planned behaviour.'

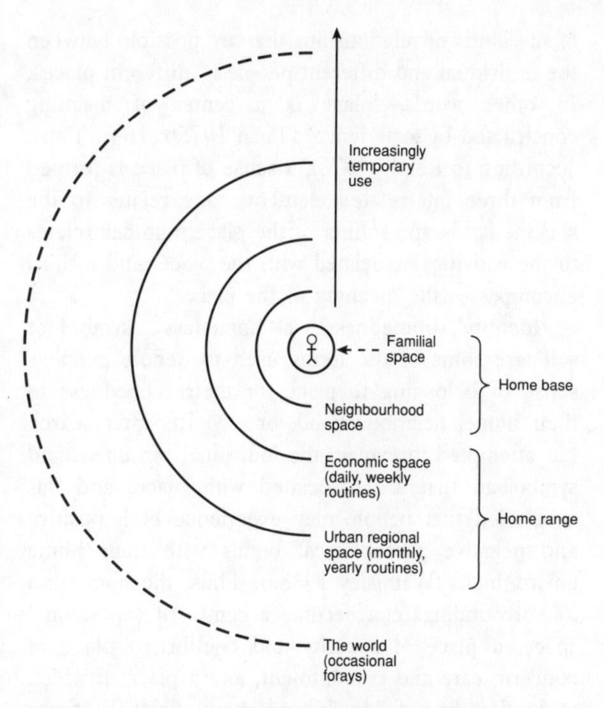

11.1 A hierarchy of social spaces (Source: Porteous 1977: 92)

Space is, therefore, 'irrevocably humanized, being both a mirror and moulder of human purposes' (Ley 1983: 143). McDermott (1975: 21) suggests that the city is perceived in terms of embodied neighbourhoods, where boundaries are the boundaries of typical experience. In short, the city is an experiential entity. Thus the image of the city held by an individual is overwhelmingly a relational one, a spatial consequence of the kinds of relationships that are possible between the individual and different people in different places. In other words, 'place is a center of meaning constructed by experience' (Tuan 1975b: 164). Thus, according to Relph (1976), a sense of place is formed from three interrelated elements: one relates to the specific landscape setting of the place, another relates to the activities associated with the place, and a third encompasses the meaning of the place.

'Identity', 'rootedness', 'at homeness', 'symbol of self' are some of the terms used to denote people's sense of belonging to place, or their relatedness to their home, neighbourhood, or city. Recent research has attempted to explain the individual sentiment and symbolism that are associated with place, and has concluded that people may experience both positive and negative psychological bonds with their home environment (Walmsley 1988a). Thus, the home and its surroundings can become a centre of experiential space, a place of security and comfort, a place of concern, care and commitment, and a place in which to develop the individual's conception of self (Buttimer and Seamon 1980; Duncan 1982; Cooper 1974). Proshansky (1978: 155), for instance, has succinctly summarized the factors influencing a person's *place identity* by emphasizing that such identity results from a complex pattern of conscious and unconscious ideas, beliefs, preferences, feelings, values, goals, behavioural tendencies, and skills relevant to the particular environment in question. The evidence used to illustrate psychological bonds with place is strong and includes:

1. People's expression of grief and the fragmentation of self-identity that can result when people are forced to change residence (Brett 1980);
2. People's refusal to move from their homes despite the possibility that a major disaster or natural hazard might recur (Burton *et al.* 1978);
3. Expressions of strong, positive emotional ties with the places people call 'home' even though their dwelling and neighbourhood conditions might impose considerable hardships and might objectively be thought of as poor (Schorr 1970).

Several disciplines have been involved in the study of people's bonds with their home and its surroundings. For example, geographers have adopted a phenomenological approach to the *rootedness* that binds individuals to home places (Tuan 1977; Buttimer and Seamon 1980), psychologists and sociologists have analysed community attachments (Fried and Gleicher 1961), and interdisciplinary investigations have focused on the 'home place' as a contributory factor in the development and expression of self-identity (Cooper 1974; Duncan 1982). All of these lines of inquiry illustrate common threads in the development of bonds between people and home places. In particular, they emphasize that these bonds develop through long-term residence in, and involvement with, a particular home place. Through purposeful daily activities in a home, place becomes imbued with positive effect and thereby distinguished from its surroundings. The home environment becomes, therefore, a unique place of familiar, known, and predictable activities, people, and physical elements. In other words, it serves as a focal point in an individual's experiential space. Such psychological bonds with places are most often unconscious, or taken-for-granted, experiences involving a sense of belonging, comfort, ease, and security. Furthermore, people not only come to know and have feelings about these places, they also come to know and have feelings about themselves through knowledge of the places they frequent. The home setting of daily activities is, therefore, one of the means by which people develop, reaffirm, and change their self-conceptions (their beliefs and evaluations of their own unique self, their integration in a social group, and the role they play in the broader society). Thus, through habitual, focused, and satisfying involvement in a neighbourhood, the tangible home setting becomes an enduring symbol of self, a symbol of the continuity of one's experience, and a symbol of that which is significant and valued by the resident.

It is well established that the psychological bond that develops between people and place is not simply a matter of satisfaction with the surroundings of the home. For example, in their investigation of residents' reactions to displacement resulting from urban renewal, Fried (1963) and Fried and Gleicher (1961) found that while most residents were satisfied with living in the West End of Boston, only some residents grieved for their lost homes when forced to move. This strong affective grief reaction could not be explained solely by the break with positively evaluated people and places; rather it resulted from the fragmentation of the place and group identity of the residents in question. Other researchers have found that, while people may express satisfaction with their home environment, they may not have a commitment to maintain their residence in that locality (Tuan 1980). In short, the psychological bond that humans develop with places involves not only satisfaction with the home environment but also ideas, feelings, and behavioural dispositions that relate the identity of a person to that of a place.

Within Western society there has been considerable concern for some time over a growing rupture between people and places that often leads to the loss of a place-bound way of life, often as a result of an increasingly mobile way of life. Arguments about the 'rootlessness' of contemporary society focus not only on high rates of mobility but also on the 'placeless' character of new residential environments. Commercialization and standardization (especially in design and building practices) are deemed by some to be the cause of the destruction of the uniqueness of residential neighbourhoods. However, the evidence to support such assertions and a general concern about the effects of mobility is far from convincing. Rather, it would appear that residential mobility can actually reinforce a sense of place. For instance, migratory movements within the city tend to be short distance and usually within the same neighbourhood or to a neighbourhood with similar characteristics. Likewise, in migratory moves to another city, most households tend to choose residential neighbourhoods that are as similar as possible to their previous neighbourhood. When similar residential areas are not available, people often invest considerable psychic and physical effort, as well as material resources, in an attempt to alter the new residential setting so that it resembles the home of the past. For example, immigrants to the United States have often transferred the physical, social, political, and economic order of their mother country to the new community. Thus, Little Italys, Chinatowns, and other ethnic enclaves in many North American cities serve as evidence of attempts to make a new place into 'a home'.

Even among the homeless, 'rootedness' can form a significant component in their daily life. For example, in a study of homeless women in Skid Row, Los Angeles, Rowe and Wolch (1990) suggested that, despite the transiency of the homeless, the composition of both peer and 'homed' networks (as well as places where social interaction occurred) can in the short term be relatively stable (Fig. 11.2). Nevertheless, for these homeless people, social network relationships appeared to replace the role of locationally fixed places in daily routines, especially in creating time–space continuity and providing material, emotional, and logistical support. This replacement is, however, often not entirely fulfilling because 'the preeminence of short-term needs and a devalued locale can lead to an altered assessment of life plans and priorities, and a transformed sense of self' (Rowe and Wolch 1990: 191). In this context, supportive homeless social networks are particularly vital to the restoration of a positive and valued personal identity.

Nearly all of the studies interested in the psychological bond between people and place have been overwhelmingly concerned with the home and its immediate surroundings. A major exception to this has been Proshansky's (1978) attempt to determine how city life can produce an 'urban' type of personality. This work therefore raises the possibility that an individual's psychological bonds with places may function *transpatially*; that is, they may transcend a relationship tied to one specific place. Proshansky (1978) proposed that an individual's place identity is influenced by her or his unique environmental experiences, as well as by experiences common to all people living in particular kinds of physical settings.

Feldman (1990) extended this line of argument by suggesting that such a conception of an *urban place-identity* could be applied to the whole of the urban system. By sustaining a relationship with a *type* of settlement, residentially mobile people are able to maintain order and predictability in their residential experience. Feldman argued that this psychological phenomenon of *settlement identity* (cf. place identity) provides the information and the skills necessary for individuals to interact successfully with new residential environments, as well as allowing for the transference of symbolic and affective ties and the maintenance of some of the properties of home places that give support to, and reaffirmation of, people's self-identity. In a study of different settlement types – city, suburb, small town, and country/mountain areas – in the

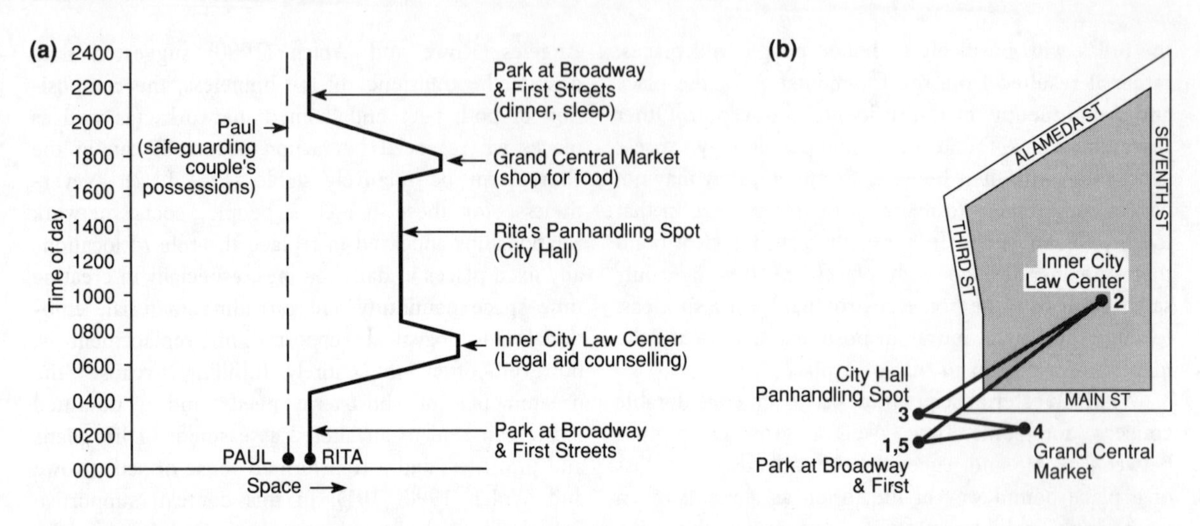

11.2 Homeless social networks in Los Angeles: (a) time–space prism; (b) a typical daily path (Source: Rowe and Wolch 1990: 191)

Denver region, Feldman concluded that, for those respondents who identified themselves with a particular type of settlement, the majority had lived in that type of settlement in the past, were presently residing in that type of settlement, and expressed intentions to maintain their residence in that type of settlement in the future. On the other hand, those whose psychological bonds did not conform with the type of settlement in which they lived tended to be characterized by a variety of constraints (including limited resources, constraints in the housing market, and personal compromises). This group tended to value the type of settlement they lived in less favourably than those respondents who lived in a consonant situation. Interestingly, there was also a group who indicated that they did not identify themselves with any type of settlement. From this kind of evidence, it appears that

> as we grow up, we develop a generalizable system of environmental categories, concepts, and relationships which form our coding system for the city – our personal urban model. When we encounter a new city we match each new experience against our general expectations; events are 'placed', never-before-seen buildings are identified as belonging to a particular class of building, functional and social patterns are inferred (Appleyard 1973: 110).

There is also plenty of evidence to suggest that individuals can have a sense of place without necessarily feeling 'at home' in that locale. In the view of Norberg-Schultz (1980), places can develop a character, or spirit which, despite the turnover of people and events, transcends the behaviour of those living in that place. According to Ley (1983: 145–64), some places within the city are viewed as being of a high *status* irrespective of whether or not the residents identify with such places; on the other hand, slums, ghettos, and large public housing estates are often seen as places of *stigma* to many. To the young, suburbs and villages are frequently conceived of as places of *ennui* or *boredom*; in contrast, the city centre may be a place of *stimulus*. Other parts of the city, particularly inner-city neighbourhoods, are sometimes thought of as places of *stress*, as illustrated by Ley (1974) in a study of 'Monroe' in Philadelphia (Fig. 11.3). Such stressful environments contrast with the feeling of *security* engendered by well-established suburbs or villages (Cohen 1982). The character of a place is not, however, immutable. In the process of change, whether it be the 'gentrification' of inner suburbs or the turnover of village population, there is inevitably a

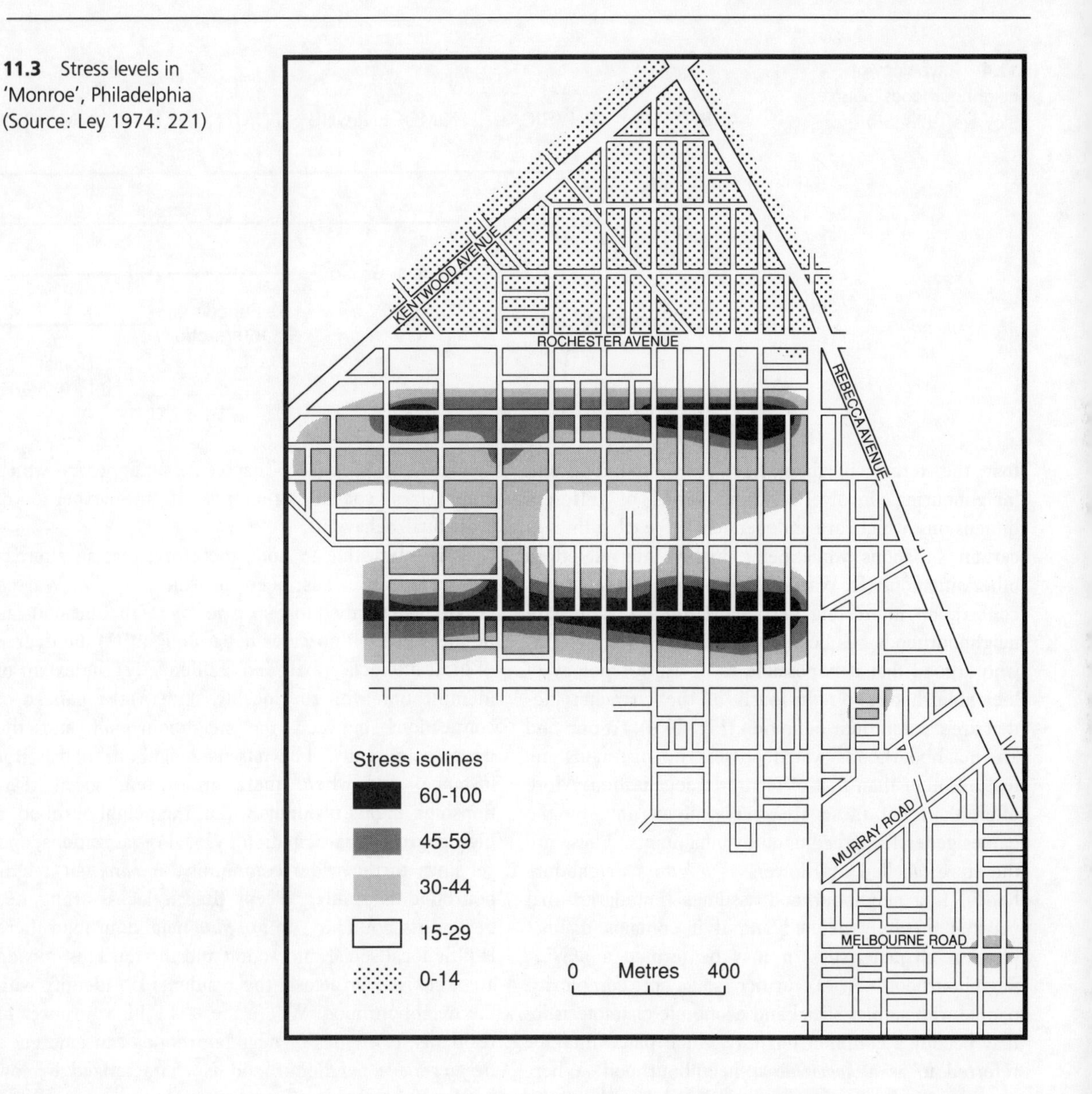

11.3 Stress levels in 'Monroe', Philadelphia (Source: Ley 1974: 221)

change in the sense of place that people experience. Pred (1984) has discussed such change at length within the context of societal power relations and shown how the changing character of a place is inextricably bound up with, and reproduced by, the behaviour of the individuals and organizations that operate within such specific locales. In such places, nature, culture, and individual consciousness are fused together (see Fig. 5.10).

The local neighbourhood

Of the various places, or social spaces, discussed in the previous section, the neighbourhood is the one which has attracted most attention from researchers, planners, and the media. However, despite the detail and diversity of the literature on neighbourhoods, there is little agreement as to a neighbourhood's true nature and definition (Keller 1968; Altman and Wandersman 1989). Much of the confusion results from the fact

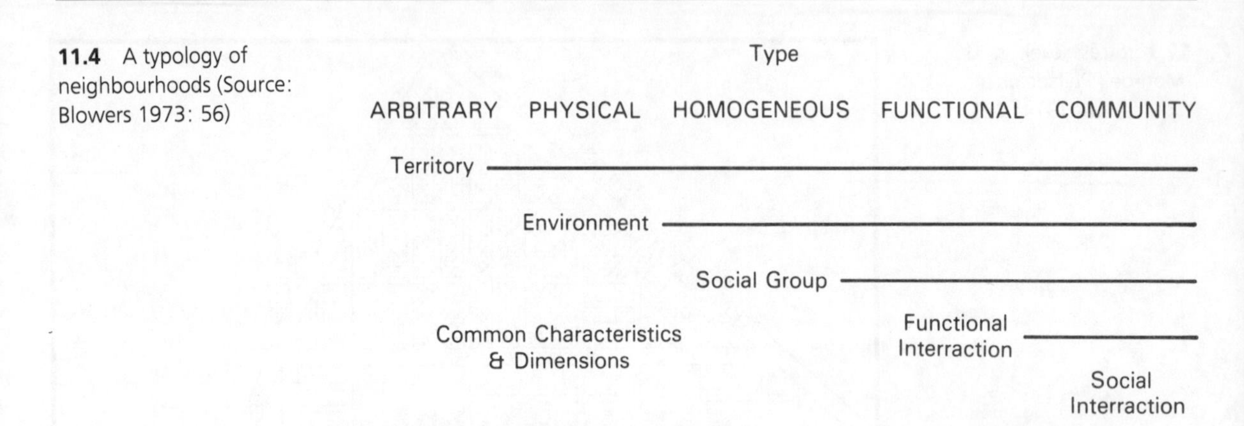

11.4 A typology of neighbourhoods (Source: Blowers 1973: 56)

that the terms 'neighbourhood', 'community', and 'neighbouring' involve a whole series of different dimensions which are independent of each other in certain situations while being highly interrelated in other situations (R. Warren 1978; Lyon 1987). In this context, an interesting perspective on the nature of neighbourhoods has been provided by Blowers (1973) who argued that it is possible to devise a typology of 'neighbourhoods' on the basis of the characteristics that are used in their definition (Fig. 11.4). At one end are neighbourhoods which possess no distinguishing feature other than that of territorial localization. More often than not these areas are known only by the names generally agreed upon by inhabitants. These are therefore labelled by Blowers as *arbitrary* neighbourhoods. If a neighbourhood assumes boundaries that may be precisely defined and if it contains distinct physical characteristics, it may be termed a *physical* neighbourhood. At a further stage of complexity, people with similar social and economic characteristics may occupy a common territory. Such places may be referred to as a *homogeneous* neighbourhood. Where considerable interaction is engendered (e.g. by shops or other community facilities), then a *functional* neighbourhood may emerge. Finally, there are neighbourhoods where primary relationships may be developed to form a localized *community*.

Clearly, this typology attempts to incorporate the likely role of place in urban living, and Blowers even went so far as to suggest that, for different individuals and groups, all, or some, of these different types of neighbourhood are of significance. Despite this, the typology remains, like social area analyses and factorial ecologies (see Chapter 2), an approach which emphasizes spatial structures at the expense of individual behaviour.

It is interesting to note, therefore, that an alternative perspective has been provided by D. Warren (1978) who claimed that six types of neighbourhood can be identified within cities depending on (1) the degree of local interaction between residents, (2) the extent of identification with the locality, and (3) the pattern of connections between the neighbourhood and the larger community. The *integral* neighbourhood is, for instance, one where there are strong social relationships among residents within the neighbourhood, a high degree of participation in local organizations, and yet links to the wider community. A *parochial* neighbourhood is similar except that it lacks strong ties beyond its boundary. In a *diffuse* neighbourhood there is little local social interaction and the linkages outside are weak; nevertheless, the residents do identify with the neighbourhood. Where there is a high turnover of residents, two types of neighbourhoods can emerge: a *stepping-stone* neighbourhood is characterized by low levels of local identification but quite strong local social interaction and quite strong ties beyond the neighbourhood; and a *transitory* neighbourhood emerges where there is an absence of social interaction and identification but strong outside linkages. Finally, according to Warren, there is the *anomic* neighbourhood which lacks participation and identification, both with the local community and beyond. Such a neighbourhood may be described as a completely disorganized and atomized residential area. Despite its intrinsic appeal, Warren's formulation does have a number of weaknesses. For example, it does not consider the specific environmental qualities of localities and the ways in which these affect people's relationships with each other and with the rest of the neighbourhood. In addition, it does not stress sufficiently the issue of diversity. It therefore leaves the impression that neighbourhoods are largely culturally and socially homogeneous, with the possible exception of the transitory neighbourhoods.

Despite such typologies, the significance of neighbourhood in everyday life is unclear. Several sociologists have questioned its relevance in determining behaviour (Pahl 1966) while others have been highly critical of its political overtones, notably the normative role sometimes ascribed to neighbourhoods in socialization processes of one sort or another. These differences of opinion might well be regarded as purely academic were it not for the fact that planners and architects commonly use the neighbourhood concept as a basis for the development of the modern city. Although the nature of the application varies a good deal, most planning uses of the neighbourhood concept derive from Perry's (1929) original six principles:

1. The neighbourhood unit should contain between 3000 and 6000 people, in order that one elementary (primary) school can be sustained.
2. The neighbourhood should have distinct boundaries.
3. There should be a system of parks and recreation spaces.
4. The school, and other services, should be located at the centre of the neighbourhood.
5. The shopping district should be located on the periphery so that (1) several neighbourhoods may be served by such facilities, and (2) traffic will be able to avoid the residential areas.
6. The structure of neighbourhood streets should be proportional to their traffic load.

Inevitably the indiscriminate application of these principles by planners in North America and Western Europe during the post-war era has led to considerable criticism, particularly of the ethics and utility of the neighbourhood concept. For example, Isaacs (1948: 20) saw the neighbourhood unit acting 'as an instrument for implementing segregation of racial and cultural groups', while Herbert (1963–64: 171) argued that the concept was based on a misconception of what constituted a primary group since basically 'people are not contained, or constrained in their behaviour by the planner's imposition of a territory based community'.

Despite definitional difficulties, as well as continuing controversy over its use for planning purposes, urban residents readily recognize the significance of neighbourhoods in their daily lives. Neighbourhoods are therefore very important in the lives of city dwellers. For example, neighbourhoods provide a means of translating social distance into geographical distance, thus keeping like with like in different parts of the city. Planners also use neighbourhoods as a convenient geographical scale for the provision of goods and services such as schools and shops (Hallman 1984). In addition, neighbourhoods give identity to what might otherwise be anonymous residential suburbs, as well as forming 'a territorial group the members of which meet on common ground for spontaneous and organized social contacts' (Glass 1948: 124). The resultant sense of belonging can contribute significantly to self-identity and well-being. Of course, neighbourhoods can also serve as arenas for personal development and for the socialization of children to the point where they accept existing norms of behaviour. In other words, despite increased mobility on the part of the population generally, there are contexts within the city where there is considerable attachment to the local neighbourhood. This is particularly true of those areas where a common bond of deprivation can lead to community action (Ley and Cybriwsky 1974; Cox and McCarthy 1980). However, there is considerable research which suggests that neighbourhood feeling is so multi-faceted that the deprivation argument for local bonding has to be tempered with other evidence. For example, in a comparative study of two neighbourhoods in Leicester, one deprived and the other prestigious, Lewis (1981) found that when compared in terms of knowledge, overt behaviour, and satisfaction, it was only the latter which distinguished prestigious from poor. Moreover, the basis of the high degree of satisfaction with the more prestigious area was not as a place to make friends or rear children but rather as a place to live. Conversely, the widespread dissatisfaction with the deprived neighbourhood stemmed from deficiencies in its housing stock and from a general lack of amenities

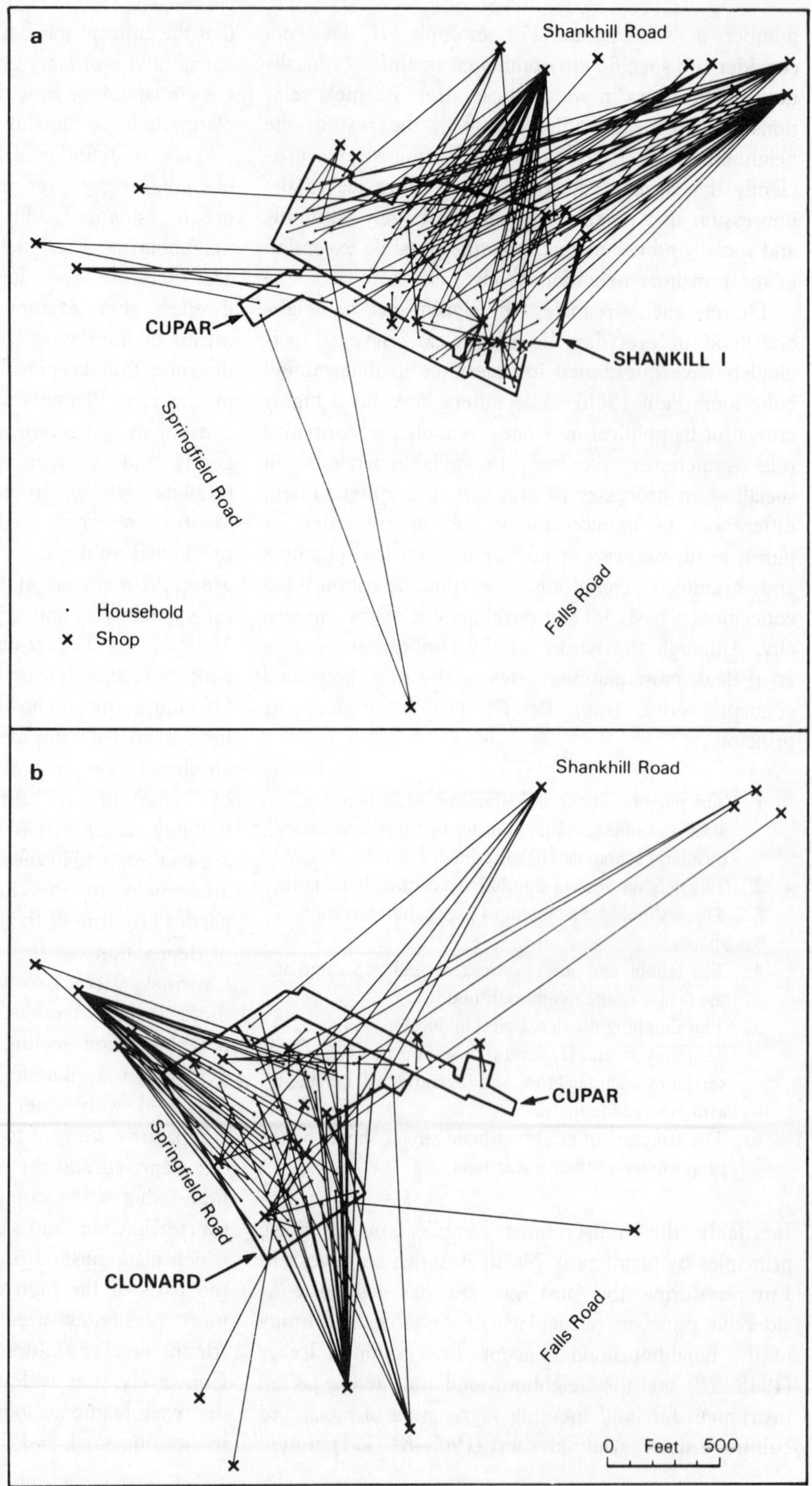

11.5 Activity patterns for the purchase of groceries in Belfast: (a) Protestants; (b) Roman Catholics (Source: Boal 1969: 41, 42)

rather than from any failings of neighbourliness. This confirms Ley's (1974) finding, in his study of the black inner suburb of 'Monroe' (Philadelphia), that two-thirds of the residents would not object to the area being razed and redeveloped.

In addition to serving as residential units, neighbourhoods can also serve as a repository of social traditions or as a haven in an uncertain world. In such cases there emerge what have been described as *defended spaces*, where ethnic, religious, or socio-economic groups cluster together and actively discourage the intrusion of outsiders (Newman 1972; Mawby 1977). According to Suttles (1972: 21), such neighbourhoods may be identified by the 'residential group which seals itself off through the efforts of the delinquent gang, by restrictive covenant, by sharp boundaries, or by forbidding reputation'. In other words, these neighbourhoods can develop not only in the inner city but also in more prestigious districts, as exemplified by the so-called 'gilded ghettos' of exclusive housing (Fainstein and Martin 1978). The means by which a neighbourhood can be defended is clearly demonstrated in Firey's (1945) classic study of Beacon Hill in Boston where restrictive covenants and a neighbourhood association successfully defended a prestigious residential area from encroachment by commercial interests for 150 years. The isolation of such 'defended' neighbourhoods is vividly illustrated in a series of papers by Boal (1969) on the activity patterns of the Catholic and Protestant population of Belfast where, even before the recent troubles, activity segregation was so marked that the main criterion in mobility appeared to be the desire to avoid crossing alien territory (Fig. 11.5).

In short, the available evidence suggests that the neighbourhood concept, despite its weaknesses, remains significant in urban life. Indeed, much of the controversy surrounding its use appears to result from terminological confusion, differing assumptions, and a lack of comparability in research findings. As Stein (1960), a major proponent of neighbourhood theory, has suggested, this controversy and confusion can only be resolved after a more rigorous analysis of human behaviour at a local level. Despite the difficulties inherent in such an approach, there are at least two ways of ascertaining the presence of neighbourhoods within cities: assessment of the degree of localization of individual activities, and examination of the degree of agreement in the identification and demarcation of local residential environments.

Neighbouring and the community

Neighbouring, or the localization of social interaction as a basis of neighbourhood identification, has been a focus of considerable research, particularly by sociologists. Much of this research has, however, been weakened by problems of measurement and definition (Irving 1978). Nevertheless, following Bott's (1957) lead, some success has been achieved in conceptualizing social interaction in terms of a network (Mitchell 1969; Duncan and Duncan 1976; Wellman and Crump 1978). Basically, *social network analysis* identifies the structure of social interaction by treating persons as points and relationships as connecting lines (Granovetter 1976), thus allowing the researcher to 'map out the complex reality of the interpersonal worlds surrounding specific individuals' (Smith 1978: 108). This procedure has the advantage of not being confined to any level of analysis. It can, for example, be applied at a variety of geographical scales from the street to the city. It also measures the connectedness, centrality, proximity, and range of social interaction (Irving 1977). In the application of the technique a distinction is usually drawn between primary relationships (such as those involving kinsfolk and friends) and more purposive secondary relationships, including those within voluntary associations (*expressive interaction*) and those associated with work, trade unions, political parties, and pressure groups (*instrumental interaction*) (Irving and Davidson 1973).

A central feature of any discussion of neighbouring is the concept of *community*. Despite considerable confusion as to its meaning (and, according to Hillery (1955), there exist over ninety definitions), Nisbet (1970: 47) has stated explicitly that community 'encompasses all forms of relationship which are characterized by a high degree of personal intimacy, emotional depth, moral commitment, social cohesion, and continuity in time. Community is founded on man [*sic*] conceived in his wholeness.' Bell and Newby (1976) have taken this definition a stage further by distinguishing between *community* and *communion.* They argued that a community involves an implicit sense of belonging in a taken-for-granted situation, so that any social area with agreed boundaries can constitute a community. In contrast, communion rests upon a form of human association which refers to effective bonds, thus implying an active and involved group rather than one which is passive and apathetic.

Despite this distinction, geographers and sociologists have over the years tended to confound the question of community with that of local ties. The result of this, according to Wellman (1979: 1202), is that 'the fundamentally structural community question has often been transmuted into a search for local solidarity, rather than a search for functioning primary ties, wherever located and however solidary'. Therefore, because empirical research has often (but not always) found that locally based ties are weak and that local sentiments and solidarity are scarce phenomena, it is frequently concluded that urban life leads to a 'decline of community'. In other words, when not found in the neighbourhood, community is thought not to exist. However, as pointed out by Wellman and Leighton (1979: 366), such a perspective omits the wider interaction and networks of urban residents: 'the identification of neighbourhood as a container for communal ties assumes the *a priori* organizing power of space. This is spatial determinism. Even if we grant that space–time costs encourage some relationships to be local, it does not necessarily follow that all communal ties are organized into solidary neighbourhood communities'. In other words, local interaction is a part of wider interactions and, therefore, a failure to consider both types of interaction simultaneously leads to false assumptions as to the importance of local ties and incorrect portrayals of city life.

In a review of the diversity of the community literature, Wellman and Leighton (1979) suggest that the concept of 'community' has been interpreted in at least three ways: as *community lost*, as *community saved*, and as *community liberated.* By far the most widespread perspective is that of *community lost* (Stacey 1969; Stein 1960). Briefly, this view argues that *Gemeinschaft* or pre-industrial society has been replaced by a more impersonal and alienating *Gesellschaft* society, making city people lonely and isolated. In other words, as Fischer (1982: 9) claims, 'Americans generally believe that urban life destroys, or at least distorts, personal relations: that urbanites, as a consequence of where they live, are estranged from kin, friendless, and lonely . . .'. Certainly those studies which have searched for strong and integrated networks within the confines of urban neighbourhoods have reported negative results (Kasarda and Janowitz 1974). The opposing view is, of course, that these findings cannot be taken as proof of the 'community lost' hypothesis. What is argued by advocates of this opposing view is that the personal networks of city dwellers are extensive and intimate, and only a small minority of people are isolated and lonely (Wellman 1979; Fischer 1982). Despite these differing interpretations, it is undoubtedly true that the locally based part of an urbanite's networks is often sparsely knit and made up of relations that are ranked relatively lowly in terms of importance.

The idea of *community saved* is rather different in that it maintains that locally based communities can survive in urban surroundings, serving as important sources for social ties and support (Hunter 1974). Nowhere are these communities more evident than in the so-called 'urban villages' (Gans 1962), located in the inner city, and distinguished by class, as in Bethnal Green, London (Young and Willmott 1962), by life style, as in Greenwich Village, New York (Jacobs 1961), and by ethnicity, as in the former Italian districts in Boston (Fried and Gleicher 1961; Gans 1962). In a review of these studies, Schorr (1963: 42–3) argued that the meaning of community for the residents of such areas was 'security, warmth and a sense of belonging. . . . These are the things he [*sic*] prizes about the space around him, his possession of all of it, his being enclosed by it, its familiarity, its manageability, and its intimacy'. However, by restricting the studies to community-like activities and to social ties within one specific area, researchers seldom asked questions as to the relative importance of this social involvement, compared to the inhabitants' ties and activities located outside the local area. The result has therefore often been a tendency to overestimate both the scope and the importance of local communities.

The *community liberated* approach shares the assumptions of the other two approaches concerning the personal and societal importance of functioning, informal primary ties, but at the same time maintains that personal relations nowadays are liberated from the confines of the neighbourhood. It has often been shown, for example, that a typical urbanite is a member of several networks; some of these may be tightly knit and of a primary-group type, whereas others may be large, sparsely knit and with loose boundaries. From this perspective, networks based on neighbourhoods are seen as but one of several forms of social involvement in which people can take part. Moreover, such networks are usually relatively weak and non-persistent (Fischer 1977, 1982; Schiefloe 1990). Fischer (1977) has pursued this line of argument further and has argued strongly that an extensive pattern of social involvement is a result of rational action and free choice, based on individual strategies for optimizing the quality of social involvement. In his view, the limitation of 'community' within certain physical boundaries is a reflection of limitations on the choice of personal relations available to individuals. Only where people have little choice will their sense of community be a local one.

Of particular interest to geographers is the role of distance and space in stimulating or retarding social interaction (Smith and Smith 1978). Early work suggested that friendship patterns within housing projects appear to be governed by 'the mere physical arrangement of the houses' (Festinger *et al.* 1950: 10) but more recent research has shown that propinquity can initiate friendship and maintain less intense relationships but is not a sufficient basis of itself for creating intense or deeper relationships (Darke and Darke 1969). Likewise, length of residence is an important factor in increasing the circle of acquaintanceship, if not the intensity of relationship (Freudenberg 1986). In addition, certain stages in the life cycle are marked by an increase in dependence upon the local community for social interaction (Carey and Mapes 1971). For example, propinquity is crucial for the development of social relationships among mothers with young children (Pred and Palm 1978) and among the elderly (Rowles 1978). For more mobile sections of society, a mixture of social distance and communality of values is probably the overriding determinant in friendship formation (Gans 1962). Certainly evidence for this is to be seen in the highly localized 'urban village' social relationships of the working classes in Western cities (Connell 1973). Of course, such 'villages' reflect not only limited spatial mobility but also the continuing significance of the extended family in working-class social life (Young and Willmott 1962). There are therefore predictable differences between various sections of working-class communities. For instance, among the diverse residents of the St Paul's district of Bristol, Richmond (1973) found that the Asians and the West Indians had greater levels of 'local' interaction than the Irish and the native population. Obviously, 'defensive' and 'cultural' forces contributed to this greater degree of localization in social relationships.

In contrast, a number of early studies of the middle classes suggested that their 'local' friendship networks were more loosely knit since their greater social skills, greater diversity of interests, and greater social mobility resulted in their maintaining relationships over wider geographical areas (Stein 1960). Subsequent investigation, however, has revealed the need to revise this view because it has become clear that, even though lacking the feelings of mutuality common among inner-city residents, middle-class suburbanites are also involved in high levels of local social interaction (Willmott and Young 1960). In particular, Muller (1976) suggested that, in American cities, social cohesion results from two complementary trends: first, constrained by income and life cycle, suburbs emerge that are based on *life style*; and second, a tendency to withdraw into a 'defended' enclave is sometimes apparent. Although these trends are less significant in European suburbs where housing is more diverse, social cohesion is often as intense as in American suburbs, thereby suggesting the operation of forces which characterize all suburbanites: these forces relate to the homogeneity, self-selection, eagerness to make friends, and physical isolation from previous social contacts of suburbanites (Hunter 1989).

Class differences in the localization of social contact can only really be uncovered through comparative study. In one such study Athanasiou and Yoshioka (1973) found that most individuals, irrespective of class, tended to have a high proportion of their more intense relationships in the immediate vicinity of their home. Support for this claim was also forthcoming from a study of neighbour relations in Leicester, where 69 per cent of the respondents' 'best' friends resided within a half-mile of their home address

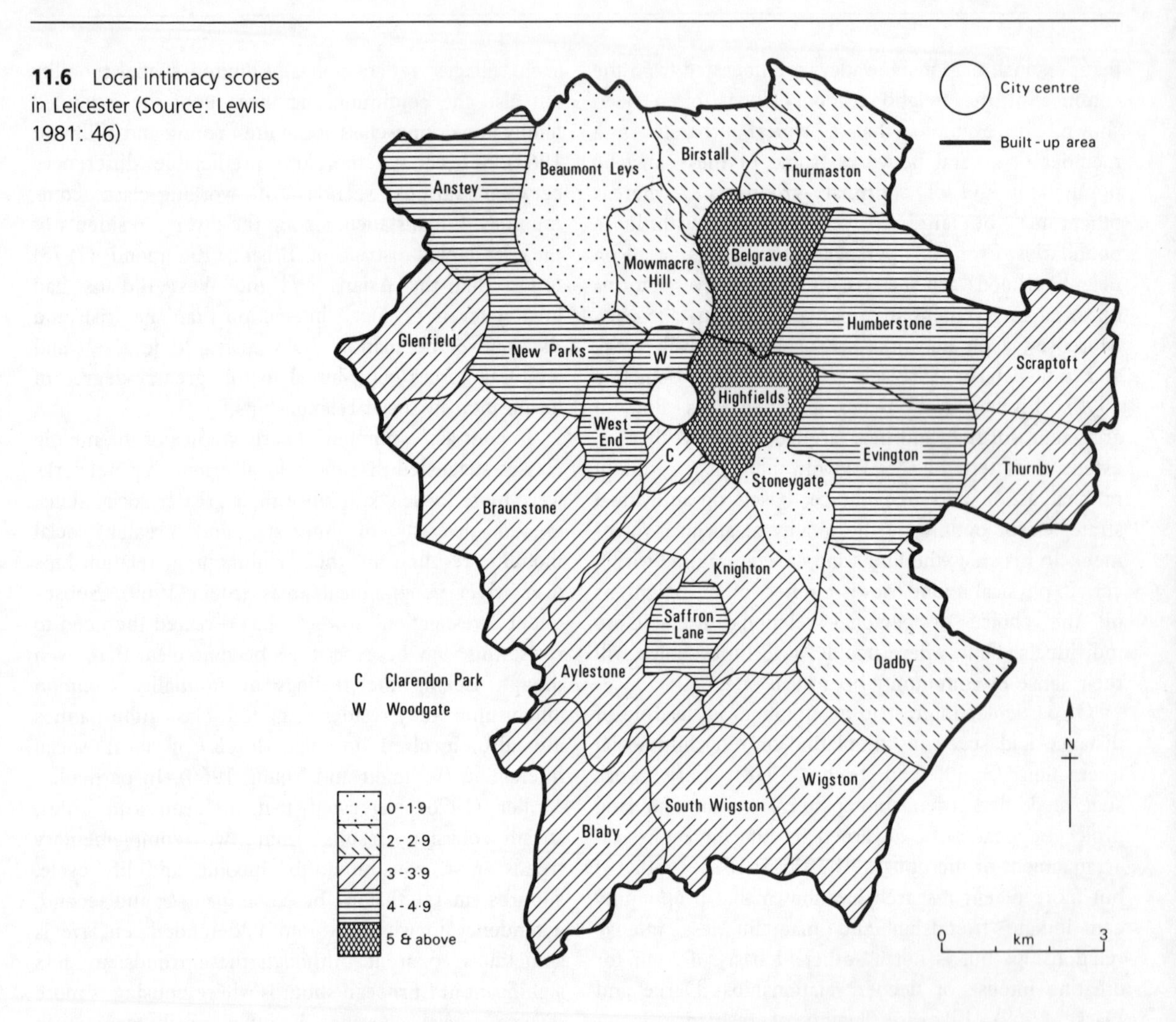

11.6 Local intimacy scores in Leicester (Source: Lewis 1981: 46)

(Lewis 1981). Social class, life cycle, and length of residence were not significant discriminating variables in this pattern. However, when less intense relationships (such as knowing, naming, and speaking to neighbours) were analysed, it was found that low-income areas around the city centre scored highly whereas the wealthier south-east sector and post-war suburbia scored lowly (Fig. 11.6). This suggests that, contrary to the popular view, the 'close' friendships of the wealthy classes are highly localized and it is their less intense relationships which are widely spaced. This, in turn, lends credence to the argument that suburban relationships are neither more nor less superficial than those found in inner-city areas (Baldassare and Fischer 1975). In other words, to dichotomize urban residents into *localites* (with restricted 'urban realms') and *cosmopolites* (for whom distance is elastic) is much too simplistic and naïve. A great amount of research in different neighbourhoods will in fact be required before it will be possible to 'write the equation which specifies the probability that A and B are friends based on the distance between A and B and similarity of A and B' (Athanasiou and Yoshioka 1973: 63).

Fischer (1977, 1981) has argued that network relations in the neighbourhood are of a residual character, and that such relations exist primarily because they are easily available, in the sense that they carry small costs as far as time and distance are concerned. It has also been argued that no relations are established or maintained if the level of reward attached to them is deemed too low, and that the most involved respondents in an area tend to be those with few sources of extra-neighbourhood ties (Schiefloe 1990). In other words, the motivation to engage in local networks varies, depending among other things upon people's alternative sources for social relations and network involvement (Fischer 1982). In this context, Schiefloe (1990) has argued that people's motivation for local interaction is limited, both in quality and quantity, because of a desire to protect privacy from unwanted intrusions, to budget their time, and to refrain from relations where they can be subject to burdensome obligations. This is achieved by *control of accessibility*.

Lastly, it is important to point out that analyses of neighbouring and community have been bedevilled by many researchers taking an *ideal type* as their starting point (Bell and Newby 1976; Kamenka 1982). This ideal type is an urban neighbourhood where all inhabitants have a position in a local *Gemeinschaft* society, where the network includes everybody, and where relations are strong and primary. This ideal type, which is of course a theoretical construction, is often taken as a normative ideal, stating how a 'real community' ought to be. This way of thinking has resulted in the tendency to consider the local community as the natural social habitat for people. When studies demonstrate that such is not the case, at least in urban surroundings, this is taken as proof that a sense of community has been lost. However, this ideal type is itself both theoretically and empirically questionable. Even in small and isolated rural communities, it has been observed that networks are widespread and dense, and differ in strength and quality in ways that are not directly related to distance (Berry *et al.* 1990; G.J. Lewis 1979). In other words, relations in such areas are no different from those in urban neighbourhoods. Theoretically, therefore, the ideal type is misleading since its two basic components – primary-group relationships and locality-based *Gemeinschaft* – are incompatible: a primary group can only consist of a few people while an urban neighbourhood, by definition, is the home for a far greater number.

The perceived neighbourhood

The considerable degree of localization of social activities which exists within cities might be expected to yield some agreement in the identification of the extent of neighbourhoods (Rivlin 1989; Carp and Carp 1982). A major contribution on this theme was the Royal Commission on Local Government in England's (1969) survey of what 2000 residents thought was the most appropriate size for local government units. Over 80 per cent of the respondents claimed to possess some feelings of attachment to a 'home' community area, and this tended to increase with length of residence. Several other studies have also revealed considerable agreement in the extent of 'perceived' neighbourhoods (Unger and Wandersman 1985; Aitken and Prosser 1990). For example, Eyles (1968) confirmed the significance of home, location, and physical landmarks in his determination of the geographical extent of the 'perceived' Highgate village, while Herbert (1975) emphasized the compactness of working-class neighbourhoods in Adamsdown, Cardiff (Fig. 11.7). However, in a study of the working-class terraced housing district of Selly Oak, Birmingham, Spencer (1973) argued that the size of the neighbourhood varied according to the method employed in its delimitation. Notably, a graphic method produced much larger 'neighbourhood' maps than a verbal method because respondents tended to draw a boundary around an area which often included housing and shops which had little significance to them. Only when the 'right' questions were asked did respondents delimit neighbourhoods with the size and form that reflected spatial behaviour (Taylor *et al.* 1984). This variability serves to emphasize the fact that the conception of neighbourhood held by individuals may be very different from what is officially designated as a neighbourhood (Lee 1976).

One of the pioneering studies of the difference between 'perceived' and 'real' neighbourhoods is to be found in Lee's (1968) examination of how 219 middle-class Cambridge housewives mapped the extent of their neighbourhood and recorded details of friendship, club membership, and shopping activity in the immediate locality. Although each *socio-spatial schema* was unique, there was a tendency towards 'some norm formation and shared social and spatial

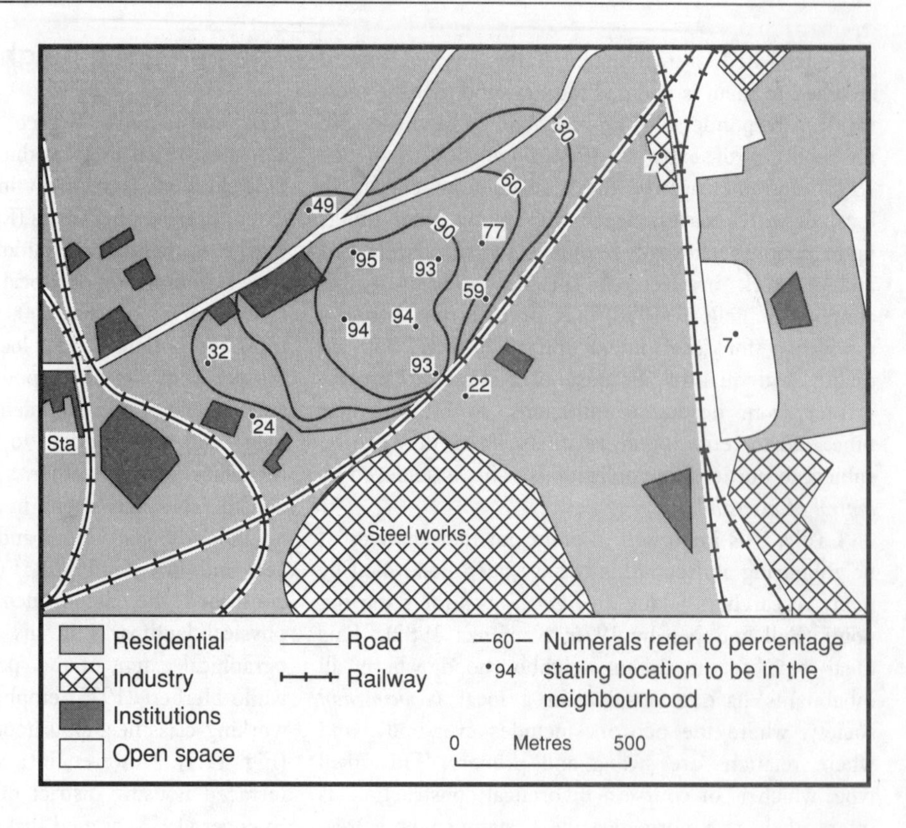

11.7 Perceived neighbourhoods in Adamsdown, Cardiff (Source: Herbert 1975: 472)

experiences' which could be generalized into a threefold typology of neighbourhood schemata (Lee 1968: 246). The first, the *social acquaintance neighbourhood*, was a small area delineated by social interaction, with everybody knowing everybody else, whether they had anything to do with each other or not. In contrast, the boundaries of the *homogeneous neighbourhood* were set by the quality of housing and the kinds of families who resided in them. In this context the most persuasive social relationship was one of mutual awareness rather than overt interaction. The third neighbourhood was generally much larger than the other two and it contained a balanced range of amenities and a heterogeneity of population and house types. In many ways, it corresponded closely to the planner's conception of a neighbourhood and hence was termed the *unit neighbourhood*.

On the basis of the Cambridge maps, Lee developed the argument that a neighbourhood is an amalgam of physical and social factors. By expressing the content of neighbourhood maps (number of houses, shops, and amenity buildings) relative to the content of the locality within a half-mile radius from the respondent's house, it was possible to devise a measure (termed a *neighbourhood quotient*) by which neighbourhood composition and size could be correlated with a number of independent variables (Fig. 11.8). In this way positive relationships were demonstrated between the neighbourhood quotient and social class, age, length of residence, working locally, friendship pattern, and active involvement within the locality. For example, a woman married to a wage earner working locally tended to have a large and physically varied neighbourhood and high social involvement, while native-born, long-term residents had relatively small neighbourhoods since they relied for their social relationships on networks of kin spread throughout the city.

Considerable support is available for Lee's findings. The threefold typology of neighbourhoods found in Cambridge closely parallels the three local groupings – patriarchal, domestic, and parish echelons – distin-

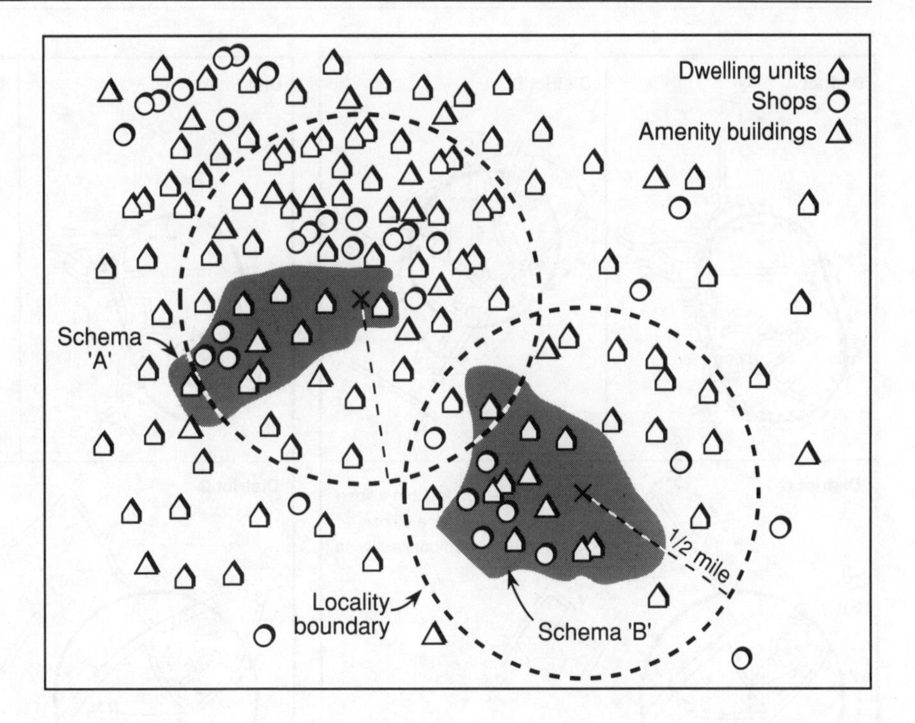

11.8 The neighbourhood quotient. Schema A illustrates an average Nh.Q. subject from a high-density locality and schema B illustrates a high Nh.Q. subject from a low-density locality (Source: Lee 1968: 257)

guished by Bardet (1961) in a previous study of sixty rural and urban places in Europe and Africa. Similarly, Golledge and Zannaras (1973: 85) concluded their study of suburban neighbourhoods in Columbus, Ohio, with the observation that physical and social space overlap so much 'that the two are very closely linked in the mind of the urban dweller'. Again class, length of residence, club membership, local friendships, and shopping activities were the major factors which differentiated those with large neighbourhoods from those with small. Everitt (1976) has also revealed a broad congruence between the perceived home area of a sample of residents in Los Angeles and their movements to work place, friends' residences, and clubs and, in an interesting reanalysis of the Royal Commission data, Berry and Kasarda (1977) reaffirmed that community sentiment, including neighbourhood perception, is influenced by the degree of local participation, and by length of residence. Even when such factors as population size, density, socioeconomic status, and life cycle were held constant, this relationship held true. In short, activity patterns help to determine the extent of perceived neighbourhoods, as was clearly demonstrated in Everitt and Cadwallader's (1981) examination of how women had greater 'perceived' neighbourhoods than men on account of greater local involvement.

There are, however, a number of local variations that confuse any general rule of thumb concerning the geographical extent of neighbourhoods. For example, Herbert and Raine (1976) questioned the residents of three streets in Cardiff as to their definition of neighbourhood. In two of the areas, there was a strong coincidence between social interaction measures and neighbourhood extent but not in the third (Fig. 11.9). In other words, the perceived neighbourhood, identified by the concept of 'home area', not only varied in extent but also was inconsistently related to the measures of social interaction. In area A, the fit between the perceptual and interaction measures of neighbourhood was close while in area C it was minimal. In area C, the composite ellipse was merely a statistical average with very little relation to any one measure of neighbourhood. Clearly this suggests that for some urban residents the concept of neighbourhood is a meaningful one while for others it is of little significance.

Moreover, just as neighbourhoods can vary in

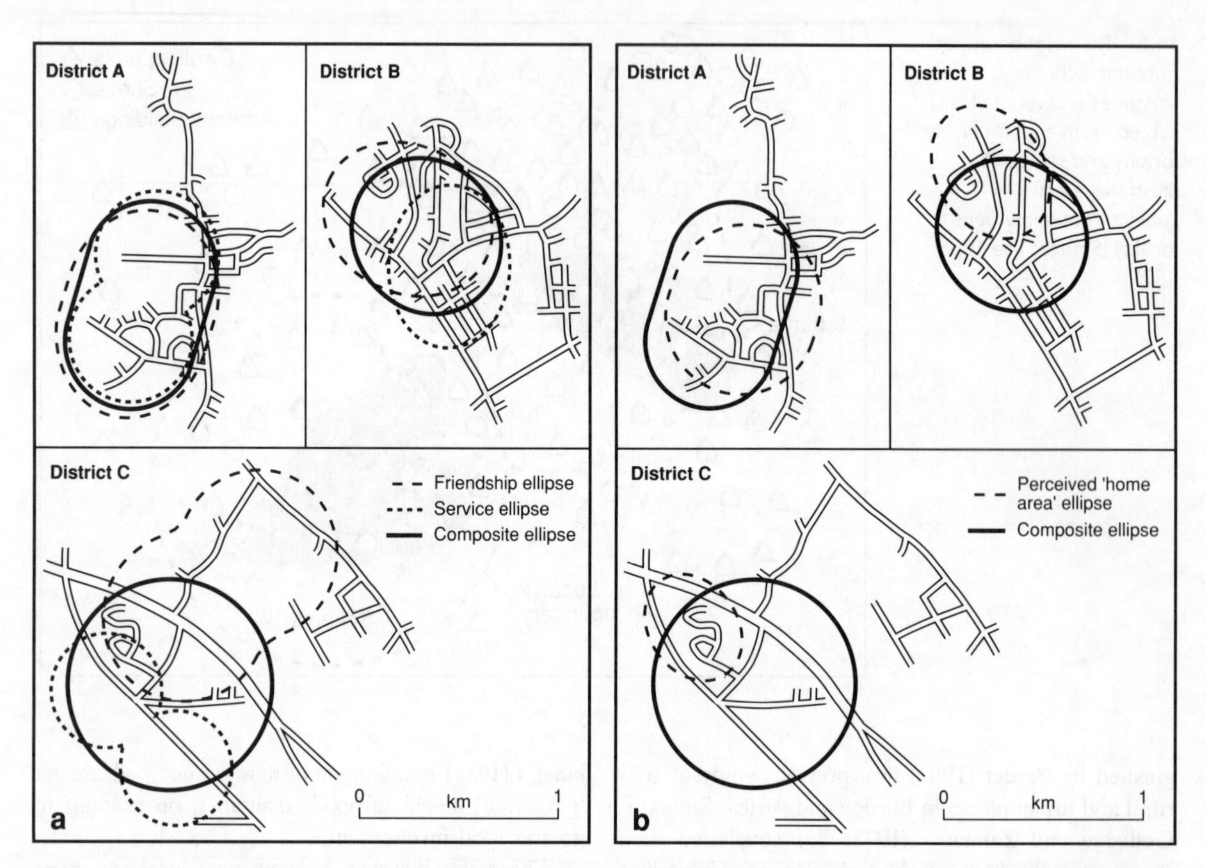

11.9 Definitions of neighbourhood in Cardiff by (a) activity patterns and (b) by perceptions (Source: Herbert and Raine 1976: 328, 329)

extent, so they can vary over time. In some parts of the city, the built and social environment of the residents can be very stable while in other parts the environment may experience rapid change (Schwirian 1983; Beauregard 1990). Such change can be highly disruptive to the residents, resulting in significant changes in their activities, satisfaction, and place identity (Fig. 11.10). For example, in a study of two neighbourhoods in San Diego, one stable and the other changing rapidly, Aitken (1990) concluded that the reaction to change related not only to its proximity and magnitude but also to the type of residents involved and their degree of attachment to the neighbourhood. For example, nearby changes that were perceived as totally unacceptable elicited little or no response in a more remote location. Moreover, the obtrusiveness of a change was mediated by its type rather than by its magnitude. It would appear, then, that a deeper understanding of how people cope with neighbourhood change is necessary if we are to improve the livability of cities. In particular, 'an understanding of different people's ability to cope with change would enable planners and designers to anticipate the cost of change to "vulnerable" populations such as the elderly or the immobile' (Aitken 1990: 266).

Social well-being

Within the context of the contemporary social and economic restructuring of the city, there is a growing interest in the influence of a sense of belonging (or a lack of it) upon social well-being (Townsend 1979).

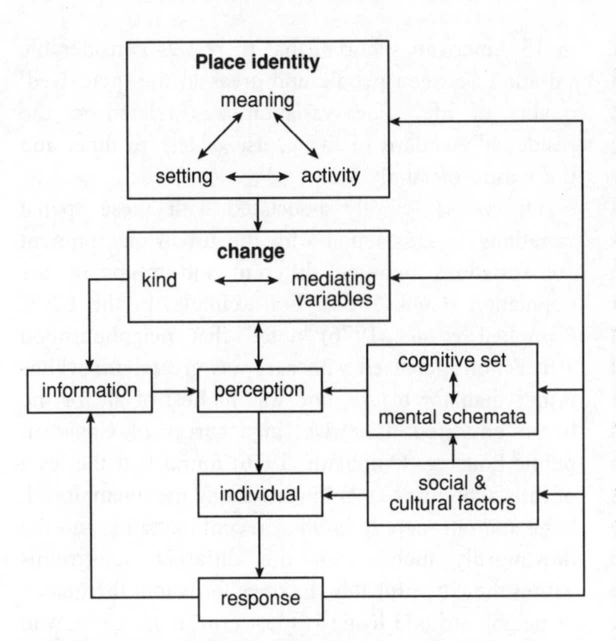

11.10 The impact of person–environment change on perception and place identity (Source: Aitken 1990: 250)

Although this perspective has been of significance in psychology for some time, it has only recently been incorporated into geographical studies of human welfare (Ley 1983; Walmsley 1988a). In other words, geographers have come to realize that both neo-classical economics and neo-Marxism provide only a partial explanation of the human condition; for example, these essentially reductionist perspectives tell us little of how places provide opportunities for human fulfilment (or a lack of opportunities) and, at the same time, how they influence the attitudes and behaviour of residents.

In any assessment of the influence of urban places on the quality of life, three competing perspectives are usually considered. The first is the *deterministic* viewpoint which, following Wirth (1938), argues that 'the concentration of a large and heterogeneous population . . . eventually leads to the weakening of interpersonal ties, primary social structures, and normative consensus. It does so largely for two reasons: the immediate psychological impact of the urban scene . . . and the complex structure differentiation generated by dynamic density' (Fischer 1975: 132). In short, this perspective holds that urban living leads to overstimulation, withdrawal, aberration in behaviour, and low levels of well-being. The second perspective is the *compositional* one which emphasizes the role of the cohesion and the intimacy of distinctive social worlds based on ethnicity, kinship, neighbourhood, occupation, or life style. The argument here is that cohesiveness enhances well-being. Finally, there is the *sub-cultural* perspective which is similar to the compositional except that it argues that the social worlds, or sub-cultures, will be intensified by conflict and competition in urban life. Over time, then, new sub-cultures will be spawned which may have different values and different behaviours from earlier ones. Belonging to a sub-culture, particularly one identified with a certain place, can have a beneficial impact on well-being.

In Chapter 2 it was argued that the concept of well-being involves both the attributes of the population and the conditions of the environment in which they live, thus making it rather problematical to define (Dahmann 1985). Therefore, census-based *objective indicators* only provide a partial description since they fail to describe the social and psychological aspects of the concept of well-being (Campbell 1981; Pacione 1982b). Social and psychological well-being can only be studied by means of a social survey. It is not surprising, therefore, that despite the problems of asking the 'right' type of question, this form of analysis has become increasingly popular in recent times (Lewis and Lyon 1986), particularly when it became 'apparent that there was often a low correlation between objective and subjective indicators, thereby suggesting that the former alone were an inadequate measure of well-being' (Walmsley 1988a: 124).

Inevitably, *subjective indicators* focus largely on the assessment of the satisfaction residents feel for the city and neighbourhood within which they live. Although many believe that finance, employment, health, and friendship are the prime bases for a satisfactory life, there is nevertheless growing evidence to suggest that neighbourhood satisfaction cannot be ignored (Marans and Rodgers 1975). In this regard, a major contribution to the understanding of neighbourhood satisfaction was Campbell *et al.*'s (1976) study of 2164 adults in 40 states in the USA. This found that 15 per cent were dissatisfied with their housing and a few – one in 10 – were dissatisfied with the neighbourhood within which they lived. However, probably of greater significance than such findings is the fact that 'perceived' well-being varies from place to place within cities. For example, in a study of Dundee, Scotland, the highest level of satisfaction was found in inner-city areas, new owner-occupied suburbs, and districts of older public housing, while most dissatisfaction was found in privately rented inner-city areas and in several of the suburban public housing estates (Knox and Maclaren 1978). In the latter case, problems of accessibility to work, friends, relatives, shops, and schooling were major contributors to dissatisfaction. In contrast, in Toronto, central neighbourhoods retained high levels of livability for all groups (Michelson 1977): groups of residents on above average income in single family housing in both the suburbs and the central city expressed neighbourhood satisfaction in excess of 95 per cent of cases; for those in high-rise residences, the level of satisfaction was only 10–15 per cent lower.

On a national basis, inner-city/suburban contrasts in perceived well-being should not be overestimated since there is evidence to suggest that greater contrasts exist *within* these two residential areas than *between* them. For example, in a series of studies of Boston's West End in the 1960s (an area shown by objective indicators to have a low quality of life), over 75 per cent of the residents expressed positive feelings towards the place where they lived (Fried and Gleicher 1961; Gans 1962). In other words, a cohesive ethnic community had created such an elaborate social support system that any urban renewal would cause a marked sense of uprootedness and thus a deterioration in the quality of life for many of the residents. In contrast, in inner Philadelphia, such a feeling was absent because of the area's social problems, notably delinquency and street gangs, drug addiction, crime, and drunkenness (see Ley 1974). In this area, the majority of the residents would prefer to leave altogether and would offer no resistance to demolition. In other words, these findings suggest that it is the social environment rather than the physical environment that the residents regard as most critical to their quality of life. Although the preferred places of residence of the inner-city residents in these studies were often suburban neighbourhoods, Zehner (1972) concluded, in a detailed study of over 6000 residents in 15 American suburbs, that there was considerable variation between people and areas in the 'perceived' quality of life. This variation was related to the residents' standard of living, use of leisure time, and the nature of family life.

Of course, closely associated with these spatial variations in satisfaction with the urban environment are variations between different sub-groups of the population (Cook 1988). For example, in the USA, Campbell *et al.* (1976) noted that neighbourhood satisfaction increased with age, was greater for whites rather than for blacks, and was highest of all for the better educated. Likewise, in a survey of Canadian public housing, Onibokun (1976) found that the level of satisfaction was at its lowest among the unemployed, large and one-parent families, recent migrants, and the downwardly mobile. Of the different sub-groups within the city, probably the ones for whom the quality of neighbourhood living is most crucial are those who are tied to their locality. Significantly, these immobile groups – the young, elderly, poor, and carless – tend quite often to evaluate their neighbourhood much more favourably than estimates using objective indicators would predict. This is particularly the case when the individual in question is unable to improve the situation. According to Carp (1975), this is because individuals are inclined to deny that their living conditions are all that bad so as reduce anxiety and feelings of inadequacy. 'In short, an ego-defense mechanism may influence individuals in their assessment of their quality of life' (Walmsley 1988a: 126).

The contextual setting of the neighbourhood can, over time, lead to a convergence of residents' attitudes and opinions, thus creating a kind of neighbourhood ethos (Guest and Lee 1983; Pacione 1990a). In other words, people start to think alike. This adaptation to a prevailing neighbourhood point of view is the result of a desire to conform to peer group pressure, within the confines of local social interaction (Jeffries and Dobos 1984). The significance of this process in determining behaviour has been highlighted in three particular situations. First, it has been shown that local neighbourhoods can have a marked influence on educational achievement (Moulden and Bradford 1984). In particular, there is a tendency for attitudes to education to conform to the prevailing neighbourhood view irrespective of socio-economic status. Robson (1969) has described this process most systematically in a study of educational attitudes in Sunderland, England. He concluded 'that no matter what the area, the attitudes of individual families were more similar to those around them than to those of their objective social area' (Robson 1969: 244). In unravelling this neighbourhood effect, it was found that a key factor was the level of social interaction in each residential area. For example, in working-class neighbourhoods, strong local integration led to attitudes reflecting class norms; in contrast, where families were isolated from their neighbours, higher than expected educational aspirations were evident. In other words, only social isolation provided immunity from the neighbourhood effect.

Second, there is plenty of evidence to suggest that political attitudes within cities are influenced by the neighbourhood context (Walmsley and Lewis 1984: 145–51). Over 20 years ago, Cox (1969) illustrated the effect of neighbourhood on political party affiliations in both the USA and Britain. In particular, it was shown that, in Columbus, Ohio, neighbourhood networks were of greater influence than national or regional ones in voters' choice of political party affiliation. Of the sub-groups investigated, it was the recent newcomers, and those of moderate political affiliation, that were the most likely to be affected by these forces.

The third situation in which a neighbourhood point of view influences individual attitudes and opinions involves attitudes towards crime, juvenile delinquency, and punishment (Davidson 1981; S.J. Smith 1986). In a major study in Cardiff, Herbert (1976) attempted to analyse the 'perceived' environment of delinquent and non-delinquent neighbourhoods (Fig. 11.11). With class and stage in life cycle held constant, the residents of the former were found to be much less disposed towards reporting petty theft and damage to public property than were the residents of non-delinquent areas. Similarly, parents in non-delinquent areas were much more inclined to administer sanctions in the home when it came to dealing with misbehaviour. In other words, localized values and life styles acted independently of the ecological structure of the city to encourage distinctive behavioural patterns.

Such contextual analysis increasingly forms a significant part of attempts to understand the motivation of offenders and of attempts to understand the distribution of crime. According to Brantingham and Brantingham (1975), crime should be thought of in terms of a space–time convergence of likely offenders, suitable targets, and the absence of protection. S.J. Smith (1987) has gone a stage further by arguing that it is necessary to determine not only the images held by the offenders of the opportunities for crime within the urban environment but also the images held by victims. In this context, Lee (1981) has suggested that a fear of crime is often translated into a fear of places and Ley (1974) has illustrated how perceived unsafe places in Philadelphia are carefully avoided. S.J. Smith (1986) has also claimed that the fear of crime is heightened among those who are on the 'margins' of society, either in economic or social terms.

However, there is plenty of evidence to suggest that real and perceived levels of vulnerability often coincide and those who live in vulnerable areas frequently respond by exhibiting a strong security consciousness (Lewis and Maxfield 1980). Despite this, it must be admitted that wealthier people generally invest more in security and so ensure lower vulnerability (Herbert 1982). Neighbourhood attributes are therefore implicated in perceptions of neighbourhood safety. This was a point made by Baba and Austin (1989) in a study of the fear of crime in Oklahoma City; although differences in perception between victims and non-victims play an important role in determining the perceived level of neighbourhood safety, evaluation of neighbourhood attributes was found to be a more important determinant of perceived neighbourhood safety. This finding implies that while victimization may be an important factor affecting the perception of residents, many other variables associated with neighbourhood life play an important role in the social and psychological assessment of personal safety (Loo 1986).

Of the various contextual variables that influence behaviour, probably one of the most contentious is that of population density and crowding. Early research on the crowding of animals showed that high densities produce severe stress, often resulting in physiological changes that threaten reproduction and thus group survival. In contrast, studies of humans and crowding have revealed far more complex relationships (Merry 1989). High densities of people do not always produce a sense of crowding, stress responses, withdrawal, or

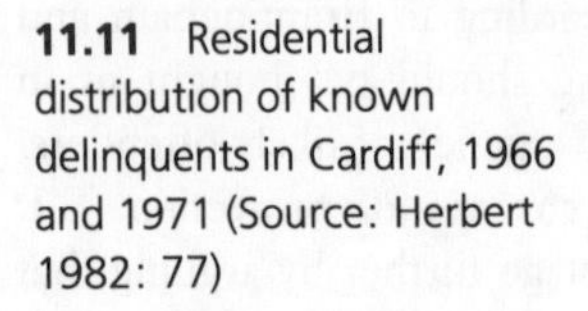

11.11 Residential distribution of known delinquents in Cardiff, 1966 and 1971 (Source: Herbert 1982: 77)

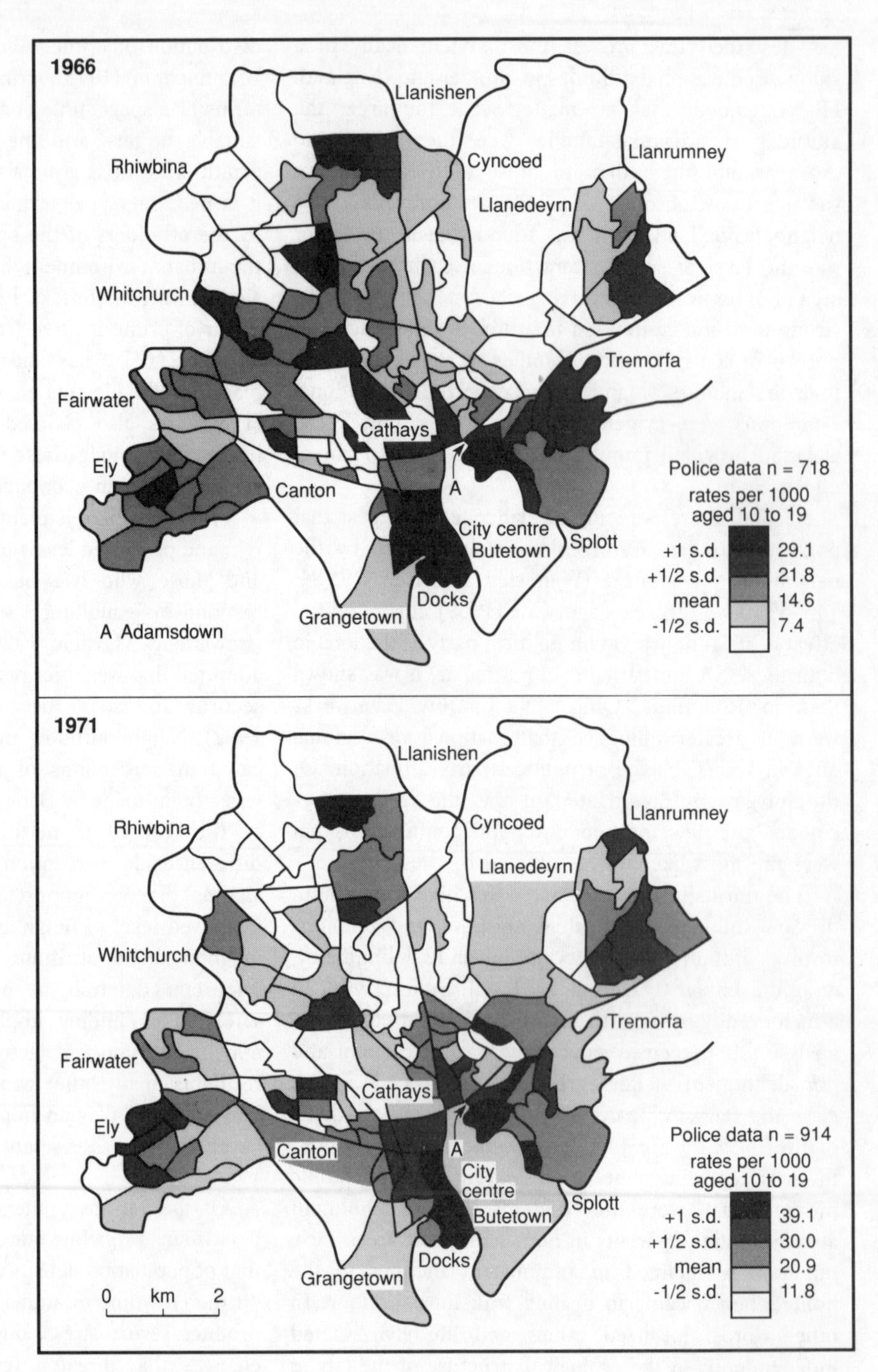

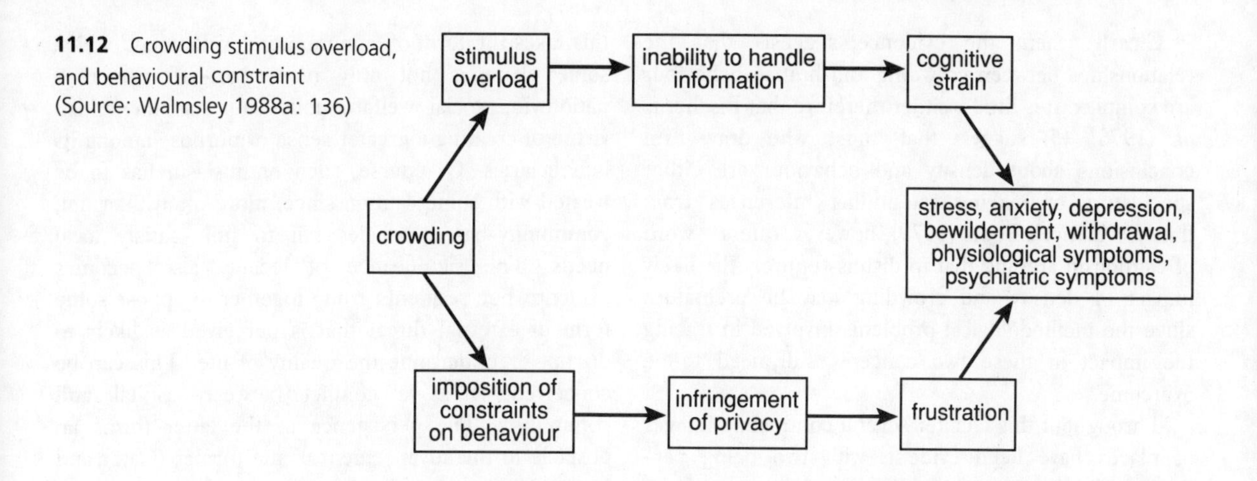

11.12 Crowding stimulus overload, and behavioural constraint (Source: Walmsley 1988a: 136)

the aggressiveness that is observed in animals. Density seems to be a necessary but not a sufficient condition for the experience of crowding (Fig. 11.12). The distinction between density and crowding is therefore an important one. As Baum (1987: 46) points out:

> Density refers to physical conditions associated with numbers of people in given amounts of space. Crowding on the other hand, is an experience – the outcome of appraisal of physical conditions, situational variables, personal characteristics, and coping assets. Under some conditions and for some people, a given level of density in a setting will lead to crowding, while in other conditions or for other people, it may not. In much the same way as people evaluate and interpret stressors, crowding is an outcome of evaluation of settings. Density is only one of several aspects of settings that appear to determine the outcomes of these appraisals.

In short, density can be thought of as a physical condition while crowding is an experiential condition.

More generally, high population densities and crowding have both been cited as the cause of 'physical effects (mass starvation, famine, environmental pollution, slums, disease and physiological breakdown), social effects (deterioration of education and social systems, increased crime, riots, war, economic stress and increased centralized control), and interpersonal and psychological effects (mental illness, increased drug addiction and alcoholism, family disorganization, reduced freedom and decreased quality of life)' (Boots 1979: 13). There is, however, some disagreement as to the evidence for links between crowding and social pathology (see Schmitt 1966; Levy and Herzog 1974). In a detailed study of objective and perceived crowding (the former being virtually a measure of density) among over 800 family members in Toronto, Booth (1976: 100) revealed that, for physical and mental health, family relations, community life, and political activity, 'crowded conditions seldom have any consequence, and even when they do the effects are very modest'. In other words, as Baum's (1987) extensive review of the literature suggests, the effect of density and its relationship to feelings of being crowded, varies significantly among individuals and situations. In some situations density engenders a feeling of crowding, but in others it does not; some individuals can tolerate more crowding than others, and crowding in some settings is more stressful than in others. In an attempt to unravel the complex relationship between density and human behaviour, Freedman (1975) developed a density/intensity hypothesis which claimed that density increases the intensity of an individual's feelings about a situation, no matter whether they are pleasant or unpleasant. There are, of course, significant differences in response to density according to sex, experience, age, and individual psychological characteristics. In other words, there are important cultural and social factors which have a bearing upon whether density equates with crowding.

Clearly, then, the evidence suggests that the relationships between crowding and human behaviour are complex. It is little wonder therefore that Fischer *et al.* (1975: 45) suggest that 'those who draw firm conclusions about density and behaviour are either speculatory or making astounding inferences from flimsy evidence'. Boots (1979), however, offers a word of caution in arguing that to dismiss entirely the likely impact of density and crowding may be premature since the methodological problems involved in tracing the impact of these two concepts still need to be overcome.

Throughout this chapter several contrasting images of places have been evident, with two being particularly dominant: first, places can be seen as a retreat and a haven from the activity of urban living; and second, places can serve as a kind of enclosure or control mechanism for some people. In the constant sifting and sorting of the city's built and social environments, these images constantly reassert themselves and play a significant role in the restructuring of urban space and the way in which that space is interpreted by city residents (Varady 1986). So persistent is the significance of 'community' and 'neighbourhood' in urban affairs that these two ideas are now increasingly incorporated into policy. Usually this takes the form of 'community development' which some claim is not only more cost-effective than nationwide social welfare policy but has the added virtue of creating a greater sense of purpose among its beneficiaries. Of course, such enthusiasm has to be treated with some caution since, more often than not, community-based policies fail to fully satisfy local needs. The significance of locality also becomes evident when residents group together to oppose some form of external threat that is perceived as likely to disrupt or undermine the quality of life. This can be conceived of as a conflict between capital and community. The persistence of the latter forms an obstacle to the advancement of the former (Cater and Jones 1989). In other words, in many instances developments within the city 'transform the spatial form of the city, substantially at labour's expense, and in a manner so spatially and temporally concentrated as to leave localized fractions of labour without a glimmer of doubt that development is impacting adversely upon them' (Cox 1981: 432). In such cases, recourse to community involvement may be the only path open to residents. Local involvement will therefore continue to be important in the lives of city dwellers.

References

Abler, R., Falk, T. (1981) Public information services and the changing role of distance in human affairs, *Economic Geography* **57**, 10–22.

Abu Lughod, J.L. (1969) Testing the theory of social area analysis: the ecology of Cairo, Egypt, *American Sociological Review* **34**, 198–212.

Ackoff, R.L., Emery, F.E. (1972) *On Purposeful Systems*, Tavistock, London.

Adams, J.S. (1969) Directional bias in intra-urban migration, *Economic Geography* **45**, 302–23.

Adams, J.S., Gilder, K.A. (1976) Household location and intra-urban migration. In: Herbert, D.T., Johnston, R.J. (eds) *Social Areas in Cities*, Vol. 1 *Spatial Processes and Form*, Wiley, London, pp.159–92.

Adams, K.M. (1984) Come to Tana Toraja, 'land of heavenly kings': travel agents as brokers in ethnicity, *Annals of Tourism Research* **11**, 469–85.

Agnew, J.A., Duncan, J.S. (1981) The transfer of ideas into Anglo–American human geography, *Progress in Human Geography* **5**, 42–57.

Agnew, J.A., Duncan, J.S. (1989) Introduction. In: Agnew, J.A., Duncan, J.S. (eds) *The Power of Place: bringing together geographical and sociological imaginations*, Unwin Hyman, Boston, pp. 1–8.

Aitken, S.C. (1987) Households moving within the rental sector: mental schemata and search spaces, *Environment and Planning A* **19**, 369–83.

Aitken, S.C. (1990) Local evaluations of neighbourhood change, *Annals of the Association of American Geographers* **80**, 247–67.

Aitken, S.C. (1991a) Person–environment theories in contemporary perceptual and behavioural geography I: personality, attitudinal and spatial choice theories, *Progress in Human Geography* **15**, 179–93.

Aitken, S.C. (1991b) A transactional geography of the image–event: the films of Scottish director, Bill Forsyth, *Transactions of the Institute of British Geographers* **16**, 105–18.

Aitken, S.C., Bjorklund, E.M. (1988) Transactional and transformational theories in behavioural geography, *Professional Geographer* **40**, 54–64.

Aitken, S.C., Prosser, R. (1990) Residents' spatial knowledge of neighbourhood continuity and form, *Geographical Analysis* **22**, 301–25.

Aitken, S.C., Sell, J.L. (1989) People and paradigms. In: Hardie, G., Moore, R., Sanoff, H. (eds) *Changing Paradigms*, Omnipress, Madison, pp. 68–74.

Aitken, S.C., Cutter, S.L., Foote, K.E., Sell, J.L. (1989) Environmental perception and behavioural geography. In: Wilmott, G., Gaile, G. (eds) *Geography in America*, Merrill, Columbus, pp. 218–38.

Ajo, R. (1953) *Contributions to 'Social Physics': a programme sketch with special regard to national planning*, Lund Publications in Geography Series B No. 4, Lund.

Ajzen, I., Fishbein, M. (1980) *Understanding Attitudes and Predicting Social Behaviour*, Prentice-Hall, Englewood Cliffs.

Aldskȯgius, H. (1977) A conceptual framework and a Swedish case study of recreational behaviour and environmental cognition, *Economic Geography* **53**, 163–83.

Alexander, I. (1979) *Office Location and Public Policy*, Longman, London.

Allen, J. (1988) Towards a postindustrial economy? In Allen, J., Massey, D. (eds) *Restructuring Britain: the economy in question*, Sage, London, pp. 91–135.

Allen, J., Massey, D. (eds) (1988) *Restructuring Britain: the economy in question*, Sage, London.

Allen, L.R., Beattie, R.J. (1984) The role of leisure as an indicator of overall satisfaction with community life, *Journal of Leisure Research* **16**, 99–109.

Althusser, L. (1969) *For Marx*, Penguin, Harmondsworth.

Altman, I. (1975) *The Environment and Social Behaviour*, Brooks–Cole, Monterey.

Altman, I., Christensen, K. (eds) (1990) *Environment and Behaviour Studies: emergence of intellectual traditions*, Plenum, New York.

Altman, I., Rogoff, N. (1987) World views in psychology: trait, interactional, organismic and transactional perspectives. In: Stokols, D., Altman, I. (eds) *Handbook of Environment Psychology*, Wiley, New York, pp. 1–40.

Altman, I., Wandersman, A. (eds) (1987) *Neighbourhoods and Community*, Plenum, New York.

Altman, I., Wandersman, A. (eds) (1989) *Neighbourhood and Community Environments*, Plenum Press, New York.

Altman, I., Werner, C.M. (eds) (1985) *Home Environments*, Plenum, New York.

Altman, I., Zube, E.H. (eds) (1989) *Public Places and Spaces*, Plenum, New York.

Alwan, A.J., Parisi, D.G. (1974) *Quantitative Methods for Decision-making*, Eurospan, New York.

Amedeo, D., Golledge, R.G. (1975) *An Introduction to Scientific Reasoning in Geography*, Wiley, New York.

Amin, A., Goddard, J. (eds) (1986) *Technological Change, Industrial Restructuring and Regional Development*, Allen & Unwin, London.

Amin, A., Robins, K. (1990) The re-emergence of regional economies, *Environment and Planning D: Society and Space* **8**, 7–34.

Anderson, J. (1971) Space-time budgets and activity studies in urban geography and planning, *Environment and Planning* **3**, 353–68.

Anderson, K.J. (1988) Cultural hegemony and the race-definition process in Chinatown, Vancouver: 1890–1980, *Environment and Planning D: Society and Space* **6**, 127–49.

Angel, D.P. (1990) New firm formation in the semiconductor industry: elements of a flexible manufacturing system, *Regional Studies* **24**, 211–21.

Anselin, L., Madden, M. (eds) (1990) *New Directions in Regional Analysis*, Belhaven Press, London.

Antes, J.R., McBride, R.B., Collins, J.D. (1988) The effects of a new city traffic route on the cognitive maps of its residents, *Environment and Behaviour* **20**, 75–91.

Appleton, I. (ed.) (1974) *Leisure Research and Policy*, Scottish Academic Press, Edinburgh.

Appleton, J. (1975) *The Experience of Landscape*, Wiley, London.

Appleyard, D. (1969) City designers and the pluralistic city. In: Rodwin L. (ed.) *Planning Urban Growth and Regional Development*, MIT Press, Cambridge, Mass., pp.422–52.

Appleyard, D. (1970) Styles and methods of structuring a city, *Environment and Behaviour* **2**, 100–17.

Appleyard, D. (1973) Notes on urban perception and knowledge. In: Downs R.M., Stea D. (eds) *Image and Environment*, Aldine, Chicago, pp.109–14.

Appleyard, D. (1978) The major published works of Kevin Lynch: an appraisal, *Town Planning Review* **49**, 551–7.

Appleyard, D., Lynch, K., Meyer, J. (1964) *The View from the Road*, MIT Press, Cambridge, Mass.

Archer, B. (1973) *The Impact of Domestic Tourism*, Occasional Papers in Economics No. 2, University of North Wales, Bangor.

Ardrey, R. (1967) *The Territorial Imperative*, Fontana, London.

Aronowitz, S. (1982) On the theorization of leisure. In: Forrest, R., Henderson T., Williams, P. (eds) *Urban Political Economy and Social Theory*, Gower, Farnborough, pp.144–59.

Ashworth, G., Goodall, B. (eds) (1990) *Marketing Tourism Places*, Routledge, London.

Athanasiou, R., Yoshioka, G.A. (1973) The spatial character of friendship formation, *Environment and Behaviour* **5**, 43–65.

Atkinson, J. (1985) The changing corporation. In: Clutterbuck, D. (ed.) *New Patterns of Work*, Gower, Aldershot, pp. 13–35.

Baba, J., Austin, D.M. (1989) Neighbourhood environmental satisfaction, victimization and social participation as determinants of perceived neighbourhood safety, *Environment and Behaviour* **21**, 763–80.

Bacon, R.W. (1984) *Consumer Spatial Behaviour*, Clarendon Press, Oxford.

Badcock, B. (1984) *Unfairly Structured Cities*, Blackwell, Oxford.

Bade, F.J. (1983) Locational behaviour and the mobility of firms in West Germany, *Urban Studies* **20**, 279–97.

Bailey, P. (1978) *Leisure and Class in Victorian Society*, Methuen, London.

Bailly, A.S., Greer-Wootten, B. (1983) Behavioural geography in Francophone countries, *Progress in Human Geography* **7**, 344–56.

Baldassare, M., Fischer, C.S. (1975) Suburban life: powerlessness and need for affiliation, *Urban Affairs Quarterly* **10**, 314–26.

Bale, J. (1988) The place of 'place' in cultural studies of sports, *Progress in Human Geography* **12**, 507–24.

Ball, M., Harloe, M., Martens, M. (1988) *Housing and Social Change in Europe and the USA*, Routledge, London.

Bannister, D., Fransella, F. (1971) *Inquiring Man: the theory of personal constructs*, Penguin, Harmondsworth.

Bannon, J.J. (1981) *Problem Solving in Recreation and Parks*, Prentice-Hall, Englewood Cliffs.

Barbolet, R.H. (1969) *Housing Classes and the Socio-ecological System*, University Working Paper No. 4, Centre for Environmental Studies, London.

Bardet, G. (1961) Social topography: an analytic–synthetic understanding of the urban texture. In: Theodorson, G.A. (ed.) *Studies in Human Ecology*, Row, Peterson, Evanston, Ill., pp. 370–83.

Barker, R.G. (1963) On the nature of the environment, *Journal of Social Issues* **19**, 17–38.

Barker, R.G. (1968) *Ecological Psychology: concepts and methods for studying the environment and behaviour*, Stanford University Press, Stanford.

Barnes, T.J. (1988) Rationality and relativism in economic geography: an interpretive review of the *homo economicus* assumption, *Progress in Human Geography* **12**, 473–96.

Barnes, T.J. (1989) Place, space, and theories of economic value: contextualism and essentialism in economic geography, *Transactions of the Institute of British Geographers* **14**, 299–316.

Barnes, T., Curry, M. (1988) Time and narrative in economic geography, *Environment and Planning A* **20**, 141–9.

Barnett, R.R., Levaggi, R., Smith, P. (1990) An assessment of the regional impact of the introduction of a community charge (or poll tax) in England, *Regional Studies* **24**, 289–97.

Barr, B.M., Fairbairn, K.J. (1978) Linkages and manufacturer's perception of spatial economic opportunity. In: Hamilton, F.E.I. (ed.) *Contemporary Industrialization: spatial analysis and regional development*, Longman, London, pp. 122–43.

Barrett, F.A. (1973) *Residential Search Behaviour: a study of intra-urban relocation in Toronto*, Geographical Monographs No. **1**, York University, Toronto.

Bartlett, F.C. (1932) *Remembering*, Cambridge University Press, Cambridge.

Bartlett, P.G. (1982) *Agricultural Decision Making*, Academic Press, New York.

Bassett, K., Short, J.R. (1980) *Housing and Residential Structure*, Routledge & Kegan Paul, London.

Batey, P.W.J., Rose, A.Z. (1990) Extended input–output models: progress and potential, *International Science Review* **13**, 27–42.

Batty, M. (1970a) An activity allocation model for the Nottingham–Derbyshire subregion, *Regional Studies* **4**, 307–32.

Batty, M. (1970b) Models and projections of the space economy: a sub-regional study in Northwest England, *Town Planning Review* **41**, 121–48.

Batty, M. (1976) Reilly's challenge: new laws of retail gravitation which define systems of central places, *Environment and Planning A* **10**, 185–219.

Baum, A. (1987) Crowding. In: Stokols, D., Altman, I. (eds) *Handbook of Environmental Psychology*, Wiley, New York, pp. 86–118.

Baxter, M.J. (1979) The interpretation of the distance and attractiveness components in models of recreation trips, *Geographical Analysis* **11**, 311–15.

Beard, J.B., Ragheb, M.G. (1980) Measuring leisure satisfaction, *Journal of Leisure Research* **12**, 20–33.

Beaumont, J.R. (1989) Towards an integrated information system for retail management, *Environment and Planning A* **21**, 299–310.

Beauregard, R.A. (1989) Between modernity and postmodernity: the ambiguous position of US planning, *Environment and Planning D: Society and Space* **7**, 381–95.

Beauregard, R.A. (1990) Trajectories of neighbourhood change: the case of gentrification, *Environment and Planning A* **22**, 855–74.

Beck, R.J., Wood, D. (1976) Cognitive transformation of information from urban geographic fields to mental maps, *Environment and Behaviour* **8**, 199–238.

Becker, R.H. (1978) Social carrying capacity and user satisfaction: an experimental function, *Leisure Sciences* **1**, 241–58.

Bell, C., Newby, H. (1976) Community, communion and community action: the social services of the new urban politics. In: Herbert, D.T., Johnston R.J. (eds) *Social Areas in Cities*, Vol. 2, Wiley, London, pp. 189–207.

Bell, D. (1973) *The Coming of Post-Industrial Society*, Heinemann, London.

Benson, B.L., Faminow, M.D. (1990) Geographic price interdependence and the extent of economic markets, *Economic Geography* **66**, 47–66.

Berkman, H.G. (1965) The game theory of land use determination, *Land Economics* **41**, 11–19.

Berry, B.J.L. (1967) *Geography of Market Centres and Retail Distribution*, Prentice-Hall, Englewood Cliffs.

Berry, B.J.L., Kasarda, J.D. (1977) *Contemporary Urban Ecology*, Macmillan, New York.

Berry, E.H., Krannich, R., Greder, T. (1990) A longitudinal analysis of neighbouring in rapidly changing rural places, *Journal of Rural Studies* **6**, 175–86.

Bhaskar, R. (1975) *A Realist Theory of Science*, Harvester, Brighton.

Bible, D.S., Brown, L.A. (1980) A spatial view of intra-urban migration search behaviour, *Socio-Economic Planning Sciences* **14**, 19–23.

Billinge, M. (1983) The Mandarin dialect: an essay on style in contemporary geographical writing, *Transactions of the Institute of British Geographers* **8**, 400–20.

Birch, B.P. (1981) Wessex, Hardy and the nature novelists, *Transactions of the Institute of British Geographers* **6**, 348–58.

Birrell, R., Silverwood, R. (1981) The social costs of environmental deterioration: the case of the Victorian coastline. In: Mercer, D. (ed.) *Outdoor Recreation: Australian perspectives*, Sorrett, Melbourne, pp. 118–24.

Blacksell, M., Gilg, A. (1975) Landscape evaluation in practice – the case of south-east Devon, *Transactions of the Institute of British Geographers* **66**, 135–40.

Blades, M. (1989) Children's ability to learn about the environment from direct experience and from spatial representations, *Children's Environment Quarterly* **6**, 5–14.

Blaikie, P. (1978) The theory of the spatial diffusion of innovations: a spacious cul-de-sac, *Progress in Human Geography* **2**, 268–95.

Blaikie, P.M., Brookfield, H.C. (1987) *Land Degradation and Society*, Methuen, London.

Blake, R.B., Monton, J.S., Bidwell, A.C. (1962) The managerial grid, *Advanced Managements Office Executive* **1**, 10–19.

Blakemore, M. (1981) From way-finding to map-making: the spatial information fields of aboriginal peoples, *Progress in Human Geography* **5**, 1–24.

Blaut, J.M. (1977) Two views of diffusion, *Annals of the Association of American Geographers* **67**, 343–9.

Blaut, J.M. (1991) Natural mapping, *Transactions of the Institute of British Geographers* **16**, 55–74.

Blaut, J.M., Stea, D. (1971) Studies of geographic learning, *Annals of the Association of American Geographers* **61**, 387–93.

Bloomestein, H., Niskamp, P., Van Veenendaad, W. (1980) Shopping perceptions and preferences: a multi-dimensional attractiveness analysis of consumer and entrepreneurial attitudes, *Economic Geography* **56**, 155–74.

Blouet, B.W., Lawson, M.P. (eds) (1975) *Images of the Plains: the role of human value in settlement*, University of Nebraska Press, Lincoln.

Blowers, A. (1973) The neighbourhood: exploration of a concept. In: *The City as a Social System*, Open University, Milton Keynes, pp. 49–90.

Boal, F.W. (1969) Territoriality on the Shankill–Falls divide, Belfast, *Irish Geography* **6**, 30–50.

Boal, F.W. (1976) Ethnic residential segregation. In: Herbert, D.T., Johnston, R.J. (eds) *Social Areas in Cities*, Vol. 1, Wiley, Chichester, pp. 41–79.

Boal, F.W., Livingstone, D.N. (eds) (1989) *The Behavioural Environment*, Routledge, London.

Boddy, M., Gray, F. (1979) Filtering theory, housing policy and the legitimation of inequality, *Policy and Politics* **7**, 39–54.

Boehm, M., Ihlandfeldt, K.R. (1986) Residential mobility and neighbourhood quality, *Journal of Regional Science* **26**, 411–24.

Boniface, B.G., Cooper, C.P. (1987) *The Geography of Travel and Tourism*, Heinemann, Oxford.

Boorstin, D.J. (1964) *The Image: a guide to pseudo-events in America*, Harper & Row, New York.

Booth, A. (1976) *Human Crowding and its Consequences*, Praeger, New York.

Boots, B.N. (1979) Population density, crowding and human behaviour, *Progress in Human Geography* **3**, 13–64.

Bosselman, F.P. (1978) *In the Wake of the Tourist: managing special places in eight countries*, Conservation Foundation, Washington.

Bott, E. (1957) *Family and Social Networks*, Tavistock, London.

Boulding, K.E. (1956) *The Image: knowledge in life and society*, University of Michigan Press, Ann Arbor.

Boulding, K.E. (1959) National images and international systems, *Journal of Conflict Resolution* **3**, 120–31.

Boulding, K.E. (1973) Foreword. In: Downs, R.M., Stea, D. (eds) *Image and Environment*, Aldine, Chicago, pp. vii–xi.

Bourne, L. (1986) Recent housing policy issues in Canada, *Housing Studies* **1**, 122–6.

Bowden, M.J. (1975) Desert wheat belt, plains corn belt: environmental cognition and behaviour of settlers in the plains margin, 1850–99. In: Blouet, B.W., Lawson, M.P. (eds) *Images of the Plains*, University of Nebraska Press, Lincoln, pp. 189–202.

Bowden, M.J., Lowenthal, D. (eds) (1975) *Geographies of the Mind*, Oxford University Press, London.

Boyce, D.E. (1988) Renaissance of large-scale models, *Papers of the Regional Science Association* **65**, 1–10.

Boyle, M.J., Robinson, M.E. (1979) Cognitive mapping and understanding. In: Herbert, D.T., Johnston, R.J. (eds) *Geography and the Urban Environment: progress in research and application*, Vol. II, Wiley, London, pp. 59–82.

Brancher, D.M. (1969) Critique of K.D. Fines, *Regional Studies* **3**, 91–2.

Brandenburg, J., Greiner, W., Hamilton-Smith, E., Scholten, H., Senior, R. *et al.* (1982) A conceptual model of how people adopt recreation activities, *Leisure Studies* **1**, 263–76.

Brantingham, P.L., Brantingham, J. (1975) Residential burglary and urban form, *Urban Studies* **12**, 273–84.

Braverman, H. (1974) *Labour and Monopoly Capital: the degradation of work in the twentieth century*, Monthly Review Press, New York.

Brett, J.M. (1980) The effect of job transfer on employees and their families. In: Cooper, C.L., Payne, R. (eds) *Current Concerns in Occupational Stress*, Wiley, New York, pp. 1–19.

Briggs, R. (1973) Urban cognitive distance. In: Downs, R.M., Stea, D. (eds) *Image and Environment*, Aldine, Chicago, pp. 361–88.

Briggs, R. (1976) Methodologies for the measurement of cognitive distance. In: Moore, G.T., Golledge, R.G. (eds) *Environmental Knowing*, Dowden, Hutchinson & Ross, Stroudsburg, pp. 325–34.

Britton, J.N.H. (1974) Environmental adaptation of industrial plants: service linkages, locational environment and organization. In: Hamilton, F.E.I. (ed.) *Spatial Perspectives on Industrial Organization and Decision-making*, Wiley, London, pp. 363–90.

Britton, S. (1990) The role of services in production, *Progress in Human Geography* **14**, 529–46.

Broadbent, G. (1976) A plain man's guide to the theory of signs in architecture, *Architectural Design* **47**, 474–82.

Brockman, C.F., Morrison, L.G. (eds) (1979) *Recreational Use of Wild Lands*, McGraw-Hill, New York.

Brooker-Gross, S.R. (1981) Shopping behaviour in two sets of shopping destinations: an interactionist interpretation of outshopping, *Tijdschift voor Economische en Sociale Geografie* **72**, 28–34.

Brooker-Gross, S.R. (1983) Spatial aspects of newsworthiness, *Geografiska Annaler* **65B**, 1–9.

Brookfield, H.C. (1969) On the environment as perceived, *Progress in Geography* **1**, 51–80.

Brookfield, H.C. (1989) The behavioural environment: how, what for, and whose?. In: Boal, F.W., Livingstone, D.N. (eds) *The Behavioural Environment*, Routledge, London, pp. 311–28.

Brower, S.N. (1980) Territory in urban settings. In: Altman, I., Rapoport, A., Wohlwill, J.F. (eds) *Human Behaviour and*

Environment – Advances in Theory and Research, Vol. 4 *Environment and Culture*, Plenum, New York, pp. 179–207.

Brown, J.M. (1983) The structure of motives for moving: a multidimensional model of residential mobility, *Environment and Planning A* **15**, 1531–44.

Brown, L.A. (1982) *Innovation Diffusion*: *a new perspective*, Methuen, London.

Brown, L.A., Cox, K.R. (1971) Empirical regularities in the diffusion of innovation, *Annals of the Association of American Geographers* **61**, 551–9.

Brown, L.A., Holmes, J. (1971) Search behaviour in an intra–urban migration context: a spatial perspective, *Environment and Planning* **3**, 307–26.

Brown, L.A., Lawson, V.A. (1985) Migration in third world settings, uneven development, and conventional modelling: a case study of Costa Rica, *Annals of the Association of American Geographers* **75**, 29–47.

Brown, L.A., Moore, E.G. (1970) The intra-urban migration process: a perspective, *Geografiska Annaler* **52B, 1–13.**

Brown, M.A. (1981) Behavioural approaches to the geographic study of innovation diffusion: problems and prospects. In: Cox, K.R., Golledge, R.G. (eds) *Behavioural Problems in Geography Revisited*, Methuen, London, pp. 123–44.

Brown, M.A., Broadway, M.J. (1981) The cognitive maps of adolescents: confusion about inter-town distances, *Professional Geographer* **33**, 315–25.

Brown, P.J., Dyer, A., Walley, R.S. (1973) Recreation research – so what? *Journal of Leisure Research* **5**, 16–24.

Brown, R.V., Kahr, A.S., Petersen, C. (1974) *Decision Analysis for the Manager*, Holt, Rinehart & Winston, New York.

Brown, S. (1987) Institutional change in retailing: a geographical interpretation, *Progress in Human Geography* **11**, 181–206.

Brummell, A.C. (1979) A model of intra-urban mobility, *Economic Geography* **55**, 338–52.

Bruner, J.S., Oliver, R.R., Greenwood, P.M. (eds) (1966) *Studies in Cognitive Growth*, Wiley, New York.

Brunswik, E. (1956) *Perception and the Representative Design of Psychological Experiments*, UCLA Press, Los Angeles.

Bryant, C.G.A., Jary, D. (eds) (1991) *Giddens' Theory of Structuration*: *a critical appreciation*, Routledge, London.

Buck, R.C. (1978) Boundary maintenance revisited: tourist experience in an Old Order Amish community, *Rural Sociology* **43**, 221–34.

Bucklin, L.P. (1967) The concept of mass in intra-urban shopping, *Journal of Marketing* **31**, 37–42.

Bull, P.J. (1985) Intra-urban industrial geography. In: Pacione, M. (ed.) *Progress in Industrial Geography*, Croom Helm, London, pp. 82–110.

Bulmer, M. (1984) *The Chicago School of Sociology*, University of Chicago Press, Chicago.

Bunting, T.E., Guelke, L. (1979) Behavioural and perception geography: a critical appraisal, *Annals of the Association of American Geographers* **69**, 448–62.

Burgess, E.W. (ed.) (1920) *The Urban Community*, University of Chicago Press, Chicago.

Burgess, J.A. (1982) Selling places: environmental images for the executive, *Regional Studies* **16**, 1–17.

Burgess, J. (1985) News from nowhere: the press, the riots and the myth of the inner city. In: Burgess, J. Gold, J.R. (eds) *Geography, the Media and Popular Culture*, Croom Helm London, pp. 192–228.

Burgess, J. (1990) The production and consumption of environmental meanings in the mass media: a research agenda for the 1990s, *Transactions of the Institute of British Geographers* **15**, 139–61.

Burgess, J., Gold, J. (1982) On the significance of valued environments. In: Gold, J., Burgess, J. (eds) *Valued Environments*, Allen & Unwin, London, pp. 1–9.

Burgess, J., Wood, P. (1988) Decoding Docklands: place advertising and decision-making strategies of the small firm. In: Eyles, J., Smith, D.M. (eds) *Qualitative Methods in Human Geography*, Polity Press, Cambridge, pp. 94–117.

Burgess, J., Limb, M., Harrison, C.M. (1988a) Exploring environmental values through the medium of small groups: 1. Theory and practice, *Environment and Planning A* **20**, 309–26.

Burgess, J., Limb, M., Harrison, C.M. (1988b) Exploring environmental values through the medium of small groups: 2. Illustrations of a group at work, *Environment and Planning A* **20**, 457–76.

Burkardt, A.J., Medlik, S. (1981) *Tourism*: *past, present, and future* (2nd edn), Heinemann, London.

Burnett, P. (1973) The dimensions of alternatives in spatial choice processes, *Geographical Analysis* **5**, 181–204.

Burnett, P. (1976) Behavioural geography and the philosophy of mind. In: Golledge, R.G., Rushton, G. (eds) *Spatial Choice and Spatial Behaviour*, Ohio State University Press, Columbus, pp. 23–48.

Burnett, P. (1977) Tests of a linear learning model of destination choice: applications to shopping travel by heterogeneous population groups, *Geografiska Annaler* **59B, 95–108.**

Burnett, P. (1978a) Time cognition and urban travel behaviour, *Geografiska Annaler* **60B, 107–15.**

Burnett, P. (1978b) Markovian models of movement within urban spatial structures, *Geographical Analysis* **10**, 142–53.

Burnett, P. (1980) Spatial constraints-oriented modelling as an alternative approach to movement, microeconomic theory, and urban policy, *Urban Geography* **1**, 53–67.

Burnett, P. (1981) Theoretical advances in modelling economic and social behaviours: applications to geographical policy–oriented models, *Economic Geography* **57**, 291–303.

Burns, T. (1973) Leisure in industrial society. In: Smith, M.A., Parker, S., Smith, C.S. (eds) *Leisure and Society in Britain*, Allen Lane, London, pp. 40–55.

Burton, I. (1963) The quantitative revolution and theoretical geography, *Canadian Geographer* **7**, 151–62.

Burton, I., Kater, R.W., White, G.F. (1978) *The Environment as Hazard*, Oxford University Press, New York.

Burton, T.L. (1971) *Experiments in Recreation Research*, Allen & Unwin, London.

Butler, E.W., Charin, F.S., Hemmens, G.L., Kaiser, E.J., Steeman, M.A. (1969) *Moving Behaviour and Residential Choice: a national survey*, National Co-operative Research Program Report No. 81, Washington.

Buttimer, A. (1969) Social space in interdisciplinary perspective, *Geographical Review* **59**, 417–26.

Buttimer, A. (1974) *Values in Geography*, Association of American Geographers Commission on College Geography, Resource Paper No. 24, Washington.

Buttimer, A. (1976) Grasping the dynamism of lifeworld, *Annals of the Association of American Geographers* **66**, 277–92.

Buttimer, A. (1978) Charism and context: the challenge of La Géographie Humaine. In: Ley, D., Samuels, M.S. (eds) *Humanistic Geography*: *prospects and problems*, Maaroufa Press, Chicago, pp. 58–76.

Buttimer, A. (1979) Erewhon or nowhere land. In: Gale, S., Olsson, G. (eds) *Philosophy in Geography*, Reidel, Dordrecht, pp. 9–37.

Buttimer, A. (1990) Geography, humanism, and global concern, *Annals of the Association of American Geographers* **80**, 1–33.

Buttimer, A., Seamon, D. (eds) (1980) *The Human Experience of Space and Place*, Croom Helm, London.

Byrne, R.W. (1979) Memory for urban geography, *Quarterly Journal of Experimental Psychology* **31**, 147–54.

Cadwallader, M. (1975) A behavioural model of consumer spatial decision making, *Economic Geography* **51**, 339–49.

Cadwallader, M. (1976) Cognitive distance in intraurban space. In: Moore, G.T., Golledge, R.G. (eds) *Environmental Knowing*, Dowden, Hutchinson, & Ross, Stroudsburg, pp. 16–24.

Cadwallader, M. (1977) Frame dependency in cognitive maps: an analysis using directional statistics, *Geographical Analysis* **9**, 284–92.

Cadwallader, M. (1978) Urban information and preference surfaces: their patterns, structures, and interrelationships, *Geografiska Annaler* **60B**, 97–106.

Cadwallader, M. (1979) Problems in cognitive distance: implications for cognitive mapping, *Environment and Behaviour* **11**, 559–76.

Cadwallader, M. (1981) Towards a cognitive gravity model: the case of consumer spatial behaviour, *Regional Studies* **15**, 275–84.

Cadwallader, M. (1988) Urban geography and social theory, *Urban Geography* **9**, 227–52.

Cadwallader, M. (1989a) A conceptual framework for analysing migration behaviour in the developed world, *Progress in Human Geography* **13**, 494–511.

Cadwallader, M. (1989b) A synthesis of macro and micro approaches to explaining migration: evidence from inter-state migration in the United States, *Geografiska Annaler* **71B**, 85–94.

Campbell, A. (1981) *The Sense of Well-being in America*, McGraw–Hill: New York.

Campbell, A., Converse, P.E., Rodgers, W.L. (1976) *The Quality of American Life*, Russell Sage Foundation, New York.

Canter, D.V. (1977) *The Psychology of Place*, Architectural Press, London.

Canter, D. (1982) Psychology and tourism management, *Tourism Management* **3**, 193–5.

Canter, D., Kenny, C. (1975) The spatial environment. In: Canter, D., Stringer, P. (eds) *Environmental Interaction*, Surrey University Press, London, pp. 127–63.

Canter, D.V., Tagg, S.K. (1975) Distance estimation in cities, *Environment and Behaviour* **7**, 59–80.

Cardew, R.V., Simons, P.L. (1982) Retailing in Sydney. In: Cardew, R.V., Langdale, J.V., Rich, O.C. (eds) *Why Cities Change*, Allen & Unwin, Sydney, pp. 151–64.

Carey, H.C. (1858) *Principles of Social Science*, Lippincott, Philadelphia.

Carey, L., Mapes, R. (1971) *The Sociology of Planning*: *a study of social activity on new housing estates*, Batsford, London.

Carlstein, T., Thrift, N.J. (1978) Afterword: towards a time–space structured approach to society and environment. In: Calstein, T., Parker, O., Thrift, N.J. (eds) *Timing Space and Spacing Time*, Vol. 2 *Human Activity and Time Geography*, Arnold, London, pp. 235–63.

Carlstein, T., Parker, D., Thrift, N. (eds) (1978a) *Timing Space and Spacing Time*, Vol. 1 *Making Sense of Time*, Arnold, London.

Carlstein, T., Parker, D., Thrift, N. (eds) (1978b) *Timing Space and Spacing Time*, Vol.2 *Human Activity and Time Geography*, Arnold, London.

Carlstein, T., Parker, D., Thrift, N. (eds) (1978c) *Timing Space and Spacing Time*, Vol. 3 *Time and Regional Dynamics*, Arnold, London.

Carp, F.M. (1975) Ego-defense or cognitive consistency effects of environmental evaluations, *Journal of Gerontology* **30**, 707–11.

Carp, F.M., Carp, A. (1982) Perceived environmental quality of neighbourhoods, *Journal of Environmental Psychology* **2**, 295–312.

Carr, M. (1983) A contribution to the review and critique of behavioural industrial location theory, *Progress in Human Geography* **7**, 386–401.

Castells, M. (1977) *The Urban Question*, Arnold, London.

Castells, M. (1978) *City, Class and Power*, Macmillan, London.

Castles, F.G., Murray, D.J., Potter, D.C. (1977) *Decisions, Organizations and Society*, Penguin, Harmondsworth.

Cater, J., Jones, T. (1989) *Social Geography*: *an introduction to contemporary issues*, Longman, Arnold.

Cebula, R.J. (1980) *The Determinants of Human Migration*, Lexington Books, Lexington.

Cesano, F.J. (1980) Congestion and the valuation of recreation benefits, *Land Economics* **56**, 329–38.

Chapin, F.S. (1968) Activity systems and urban structure: a working schema, *Journal of the American Institute of Planners* **34**, 12–18.

Chapin, F.S. (1974) *Human Activity Patterns in the City: things people do in time and space*, Wiley, New York.

Chapin, F.S. (1978) Human time allocation in the city. In: Carlstein, T. *et al.* (eds) *Timing Space and Spacing Time*, Vol. 2 *Human Activity and Time Geography*, Arnold, London, pp. 13–26.

Chapin, F.S., Brail, R.K. (1969) Human activity systems in the metropolitan United States, *Environment and Behaviour* **1**, 107–30.

Chapin, F.S., Hightower, H.C. (1965) Household activity patterns and land use, *Journal of the American Institute of Planners* **31**, 222–31.

Chapman, G.P. (1974) Perception and regulation: a case study of farmers in Bihar, *Transactions of the Institute of British Geographers* **62**, 71–93.

Chapman, K., Walker, D. (1987) *Industrial Location*, Basil Blackwell, Oxford.

Charlton, P. (1973) A review of shop loyalty, *Journal of the Marketing Research Society* **15**, 25–41.

Chase, D.R., Cheek, N.H. (1979) Activity preferences and participation: conclusions from a factor analytic study, *Journal of Leisure Research* **11**, 92–101.

Cheng, J.R. (1980) Tourism: how much is too much? lessons for Canmore from Banff, *Canadian Geographer* **24**, 72–80.

Chombart de Lauwe, P.H. (1975) *Des Hommes et des Villes*, Payot, Paris.

Chomsky, N. (1965) *Aspects of the Theory of Syntax*, MIT Press, Cambridge, Mass.

Chouinard, V., Fincher, R. (1983) A critique of 'Structural Marxism and human geography', *Annals of the Association of American Geographers* **73**, 137–46.

Christopherson, S. (1989) Flexibility in the US service economy and the emerging spatial division of labour, *Transactions of the Institute of British Geographers* **14**, 131–43.

Chubb, M., Chubb, H.R. (1981) *One Third of Our Time? An Introduction to Recreation Behaviour and Resources*, Wiley, New York.

Clark, D., Unwin, K.I. (1981) Telecommunications and travel: potential impact in rural areas, *Regional Studies* **15**, 47–56.

Clark, G.L. (1988) Time, events, and places: reflections on economic analysis, *Environment and Planning A* **20**, 187–94.

Clark, G.L. (1990) Unethical secrets, lies and legal retaliation in the context of corporate restructuring in the United States, *Transactions of the Institute of British Geographers* **15**, 403–20.

Clark, R., Stankey, G. (1979) *The Recreation Opportunity Spectrum*, US Department of Agriculture General Technical Report PNW98, Seattle.

Clark, W.A.V. (1982a) Recent research on migration and mobility: a review and interpretation, *Progress in Planning* **18**, 1–56.

Clark, W.A.V. (1982b) *Modelling Housing Market Search*, Croom Helm, London.

Clark, W.A.V. (1986a) *Human Migration*, Sage, Beverley Hills.

Clark, W.A.V. (1986b) Residential segregation in American cities: a review and interpretation, *Population Research and Policy Review* **5**, 95–127.

Clark, W.A.V. (1988) Understanding residential segregation in American cities: interpreting the evidence – a reply to Glaster, *Population Research and Policy Review* **8**, 193–7.

Clark, W.A.V., Cadwallader, M. (1973) Locational stress and residential mobility, *Environment and Behaviour* **5**, 29–41.

Clark, W.A.V., Flowerdew, R. (1982) A review of search models and their application to search in the housing market. In: Clark, W.A.V. (ed.) *Modelling Housing Market Search*, Croom Helm, London, pp. 4–29.

Clark, W.A.V., Moore, E.G. (eds) (1980) *Residential Mobility and Public Policy*, Sage, New York.

Clark, W.A., Onaka, J.L. (1983) Life cycle and housing adjustment as explanations of residential mobility, *Urban Studies* **20**, 47–57.

Clark, W.A.V., Smith, T.R. (1979) Modelling information uses in a spatial context, *Annals of the Association of American Geographers* **69**, 575–88.

Clark, W.A.V., Smith, T.R. (1982) Housing market search behaviour and expected utility theory: 2 The process of search, *Environment and Planning A* **14**, 717–38.

Clark, W.A.V., Smith, T.R. (1985) Production system models of search behaviour: a comparison of behaviour in computer–simulated and real-world environments, *Environment and Planning A* **17**, 555–68.

Clark, W.A.V., White, K. (1990) Modelling elderly mobility, *Environment and Planning A* **22**, 909–24.

Clark, W.A.V., Huff, J.D., Burt, J.E. (1979) Calibrating a model of the decision to move, *Environment and Planning A* **11**, 689–704.

Clarke, I. (1985) *The Spatial Organization of Multinational Corporations*, St Martin's Press, New York.

Clarke, J.I. (1989) A quarter of a millennium of world population movement, 1780–2030s, *Espace, Populations, Sociétés* **89**, 295–304.

Clarke, M., Longley, P., Williams, H. (1989) Microanalysis and simulation of housing careers: subsidy and accumulation in the UK housing market, *Papers of the Regional Science Association* **66**, 105–22.

Clawson, M. (1972) *Methods of Measuring the Demand for and Use of Outdoor Recreation*, Resources for the Future, Washington.

Clawson, M., Knetsch, J.L. (1966) *Economics of Outdoor Recreation*, Johns Hopkins University Press, Baltimore.

Cliff, A.D., Ord, J.K. (1973) *Spatial Autocorrelation*, Pion, London.

Cohen, A.P. (1982) *Belonging: identity and social organization in British rural cultures*, Manchester University Press, Manchester.

Cohen, E. (1974) Who is a tourist? *Sociological Review* **22**, 527–55.

Cohen, E. (1979a) Rethinking the sociology of tourism, *Annals of Tourism Research* **6**, 18–35.

Cohen, E. (1979b) A phenomenology of tourist experiences, *Sociology* **13**, 179–201.

Cohen, E. (1988) Traditions in the qualitative sociology of tourism, *Annals of Tourism Research* **15**, 29–46.

Cohen, I.J. (1989) *Structuration Theory*, Macmillan, Basingstoke.

Cohen, R., McManus, J., Fox, D., Kastelnin, C. (1973) *Psych City: a simulated community*, Pergamon, Oxford.

Collingwood, R.G. (1956) *The Idea of History*, Oxford University Press, London.

Connell, J. (1973) Social networks in urban society. In: *Social Patterns in Cities*, Institute of British Geographers Special Publication No. **5**, London, pp. 41–52.

Conway, D., Graham, M.M. (1982) Court-ordered bussing and housing searches, *Environment and Behaviour* **14**, 45–71.

Cook, C. (1988) Components of neighbourhood satisfaction: responses from urban and suburban single-parent women, *Environment and Behaviour* **20**, 115–49.

Cooke, P. (1986) The changing urban and regional system in the United Kingdom, *Regional Studies* **20**, 243–51.

Cooke, P. (ed.) (1989a) *Localities: the changing face of urban Britain*, Unwin Hyman, London.

Cooke, P. (1989b) Locality, economic restructuring and world development. In: Cooke, P. (ed.) *Localities: the changing face of urban Britain*, Unwin Hyman, London, pp. 1–44.

Cooke, P. (1990) *Back to the Future*, Unwin Hyman, London.

Cool, S.F.M. (1979) Recreation activity packages at water-based resources. In: Van Doren, C.S., Priddle, G.B., Lewis, J.E. (eds) *Land and Leisure*, Methuen, London, pp. 123–41.

Cooper, C. (1974) The house as a symbol of the self. In: Lang, J., Burnette, C., Moleski, W., Vachon, O. (eds) *Designing for Human Behaviour*, Dowden, Hutchinson & Ross, Stroudsburg, pp. 130–46.

Cooper, C.P. (1981) Spatial and temporal patterns of tourist behaviour, *Regional Studies* **15**, 359–71.

Coppock, J.T., Duffield, B. (1975) *Recreation in the Countryside*, Macmillan, London.

Cornish, M., Jackson, C., Ursell, G., Walker, R. (1977) Regional culture and identity in industrialized societies: a critical comment, *Regional Studies* **11**, 113–16.

Cosgrove, D. (1978) Place, landscape, and the dialectics of cultural geography, *Canadian Geographer* **22**, 66–72.

Cosgrove, D. (1985a) *Social Formation and Symbolic Landscape*, Barnes & Noble, Totowa, NJ.

Cosgrove, D. (1985b) Prospect, perspective and the evolution of the landscape idea, *Transactions of the Institute of British Geographers* **10**, 45–62.

Cosgrove, D. (1989a) A terrain of metaphor: cultural geography 1988–9, *Progress in Human Geography* **13**, 566–75.

Cosgrove, D. (1989b) Geography is everywhere: culture and symbolism in the human landscape. In: Gregory, D., Walford, R. (eds) *Horizons in Human Geography*, Macmillan, London, pp. 118–25.

Cosgrove, D., Daniels, S. (eds) (1988) *The Iconography of Landscape*, Cambridge University Press, Cambridge.

Cosgrove, I., Jackson, R. (1972) *The Geography of Recreation and Leisure*, Hutchinson, London.

Couclelis, H. (1983) Some second thoughts about theory in the social sciences, *Geographical Analysis* **15**, 28–33.

Couclelis, H. (1986) A theoretical framework for alternative models of spatial decision and behaviour, *Annals of the Association of American Geographers* **76**, 95–113.

Couclelis, H., Golledge, R. (1983) Analytic research, positivism, and behavioural geography, *Annals of the Association of American Geographers* **73**, 331–9.

Couclelis, H., Golledge, R.G., Gale, N., Tobler, W. (1987) Exploring the anchor-point hypothesis of spatial cognition, *Journal of Environmental Psychology* **7**, 99–122.

Coupe, R.T., Morgan, B.S. (1981) Toward a fuller understanding of residential mobility: a case study in Northampton, *Environment and Planning A* **13**, 201–16.

Cowen, H., Livingstone, I., McNab, A., Harrison, S., Howes, L. *et al.* (1989) Cheltenham – affluence amid recession. In: Cooke, P. (ed.) *Localities*, Unwin Hyman, London, pp. 86–128.

Cox, K.R. (1969) The voting decision in a spatial context, *Progress in Geography* **1**, 81–117.

Cox, K.R. (1981) Capitalism and conflict around the communal living space. In: Dear, M., Scott, M.B. (eds) *Urbanization and Urban Planning in Capitalist Society*, Methuen, London, pp. 116–29.

Cox, K.R., Golledge, R.G. (eds) (1981) *Behavioural Problems in Geography Revisited*, Methuen, London.

Cox, K.R., McCarthy, J.J. (1980) Neighbourhood activism in the American city: behavioural relationships and evaluation, *Urban Geography* **1**, 22–38.

Cox, K.R., McCarthy, J.J., Nartowitz, F. (1979) The cognitive organization of the North American city: empirical evidence, *Environment and Planning A* **11**, 327–34.

Craik, K.H. (1968) The comprehension of the everyday physical environment, *Journal of the American Institute of Planners* **34**, 29–37.

Craik, K.H. (1970) Environmental psychology, *New Directions in Psychology* **4**, 1–121.

Craik, K.H. (1973) Environmental psychology, *Annual Review of Psychology* **24**, 403–22.

Craik, K.H. (1977) Multiple scientific paradigms in environmental psychology, *International Journal of Psychology* **12**, 147–57.
Crandall, R. (1980) Motivations for leisure, *Journal of Leisure Research* **12**, 45–54.
Csikszentmihalyi, M. (1975) *Beyond Boredom and Anxiety: the experience of play in work and games*, Jossey-Bass, San Francisco.
Cullen, G. (1961) *Townscape*, Architectural Press, London.
Cullen, I.G. (1976) Human geography, regional science, and the study of individual behaviour, *Environment and Planning A* **8**, 397–409.
Cullen, I.G. (1978) The treatment of time in the explanation of spatial behaviour. In: Carlstein, T., Parker, D., Thrift, N.J. (eds) *Timing Space and Spacing Time*, Vol. 2 *Human Activity and Time Geography*, Arnold, London, pp. 27–38.
Cullen, I.G., Godson, V. (1975) Urban networks: the study of activity patterns, *Progress in Planning* **4**, 5–96.
Culler, J. (1981) Semiotics of tourism, *American Journal of Semiotics* **1**, 127–40.
Cunningham, C., Jones, M. (1991) Girls and boys come out to play: play, gender and urban planning, *Landscape Australia* **2**, 257–63.
Curry, M. (1982) The idealist dispute in Anglo–American geography, *Professional Geographer* **26**, 37–50.
Curry, M.R. (1991) Postmodernism, language, and the strains of modernism, *Annals of the Association of American Geographers* **81**, 210–28.
Cushman, G., Hamilton-Smith, E. (1980) Equity issues in urban recreational services. In: Mercer, D., Hamilton-Smith, E. (eds) *Recreation Planning and Social Change in Urban Australia*, Sorrett, Melbourne, pp. 167–78.
Cyert, R.M., March, J.G. (1963) *A Behavioural Theory of the Firm*, Prentice-Hall, Englewood Cliffs.

Dahmann, D.C. (1985) Assessments of neighbourhood quality in metropolitan America, *Urban Affairs Quarterly* **20**, 511–35.
Daniels, P.W. (1985) Service industries: some new directions. In: Pacione, M. (ed.) *Progress in Industrial Geography*, Croom Helm, London, pp. 111–41.
Daniels, P.W. (1986) The geography of services, *Progress in Human Geography* **10**, 436–44.
Daniels, P.W. (1987) The geography of services, *Progress in Human Geography* **11**, 433–47.
Daniels, P.W., Warnes, A.M. (1980) *Movement in Cities: spatial perspectives on urban transport and travel*, Methuen, London.
Dann, G.M.S. (1977) Anomie, ego-enhancement and tourism, *Annals of Tourism Research* **4**, 184–94.
Dann, G.M.S. (1981) Tourist motivation: an appraisal, *Annals of Tourism Research* **8**, 187–219.
Darke, J., Darke, R. (1969) *Physical and social factors in neighbourhood relations*, Centre for Environmental Studies Working Paper No. 41, London.
Davidson, J. (1970) *Outdoor Recreation Surveys: design and use of questionnaires for site surveys*, Countryside Commission, London.
Davidson, R.N. (1981) *Crime and Environment*, Croom Helm, London.
Davies, K., Sparks, L. (1989) The development of superstore retailing in Great Britain 1960–1986: results from a new database, *Transactions of the Institute of British Geographers* **14**, 74–89.
Davies, R.L. (1976) *Marketing Geography*, Retailing & Planning Associates, Corbridge.
Davies, R.L., Edyvean, D.J. (1984) The development of teleshopping, *The Planner* **70**(8), 8–10.
Davies, R.L., Kirby, D.A. (1980) Retail organization. In: Dawson, J.A. (ed.) *Retail Geography*, Croom Helm, London, pp. 156–92.
Davies, W.K., Lewis, G.J. (1973) The urban dimensions of Leicester, England. Institute of British Geographers, Special Publications No. 5, *Social Patterns in Cities*, 71–86.
Dawson, J. (1979) Retail trends in the EEC. In: Davies, R.L. (ed.) *Retail Planning in the European Community*, Saxon House, Farnborough, pp. 21–49.
Dawson, J.A., Lord, J.D. (eds) (1985) *Shopping Centre Development: policies and prospects*, Croom Helm, London.
Day, R.A. (1976) Urban distance cognition: review and contribution, *Australian Geographer* **13**, 193–200.
Dear, M.J. (1986) Postmodernism and planning, *Environment and Planning D: Society and Space* **4**, 367–84.
Dear, M.J. (1988) The postmodern challenge: reconstructing human geography, *Transactions of the Institute of British Geographers* **13**, 262–74.
Dear, M.J., Moos, A.I. (1986) Structuration theory in urban analysis. 2. empirical application, *Environment and Planning A* **18**, 351–74.
Deardon, P. (1980) Landscape assessment: the last decade, *Canadian Geographer* **24**, 316–25.
De Jong, G.F., Fawcett, J.T. (1979) Motivations and migration: an assessment and value expectancy research model. In: *Workshop in Microlevel Approaches to Migration Decisions*, East-West Center, Honolulu, pp. 48–64.
De Jonge, D. (1962) Images of urban areas: their structure and psychological foundations, *Journal of the American Institute of Planners* **28**, 266–76.
De Saussure, F. (1966) *Course in General Linguistics*, McGraw–Hill, New York.
Desbarats, J. (1983) Spatial choice and constraints on behaviour, *Annals of the Association of American Geographers* **73**, 340–57.
Deurloo, M.C., Clark, W.A.V., Dieleman, F.M. (1988) Generalized log-linear models of housing choice, *Environment and Planning A* **20**, 55–69.
Dicken, P. (1977) A note on location theory and the large business enterprise, *Area* **9**, 138–43.

Dicken, P. (1988) The changing geography of Japanese direct foreign investment in manufacturing industry: a global perspective, *Environment and Planning A* **20**, 633–53.

Dilley, R.S. (1986) Tourist brochures and tourist images, *Canadian Geographer* **30**, 59–65.

Doherty, S., Gale, N., Pellegrino, J.W., Golledge, R. (1989) Children's versus adult's knowledge of places and distances in a familiar neighbourhood environment, *Children's Environment Quarterly* **6**, 65–71.

Doob, L.W. (1978) Time: cultural and social anthropological aspects. In: Carlstein, T., Parker, O., Thrift, N.J. (eds) *Timing Space and Spacing Time*, Vol. 1 *Making Sense of Time*, Arnold, London, pp. 56–65.

Dottavio, F.D., O'Leary, J.T., Koth, B.A. (1980) The social group variable in recreational studies, *Journal of Leisure Research* **12**, 357–67.

Downs, R.M. (1970a) Geographic space perception: past approaches and future prospects, *Progress in Geography* **2**, 65–108.

Downs, R.M. (1970b) The cognitive structure of an urban shopping centre, *Environment and Behaviour* **2**, 13–39.

Downs, R.M. (1976) Personal constructions of personal construct theory. In: Moore, G.T., Golledge, R.G. (eds) *Environmental Knowing*, Dowden, Hutchinson & Ross, Stroudsburg, pp. 72–87.

Downs, R.M. (1979) Critical appraisal or determined philosophical skepticism? *Annals of the Association of American Geographers* **69**, 468–71.

Downs, R.M. (1981) Cognitive mapping: a thematic analysis. In: Cox, K.R., Golledge, R.G. (eds) *Behavioural Problems in Geography Revisited*, Methuen, London, pp. 95–122.

Downs, R.M., Meyer, J.T. (1978) Geography and the mind: an exploration of perceptual geography, *American Behavioural Scientist* **22**, 59–77.

Downs, R.M., Stea, D. (1973a) Cognitive maps and spatial behaviours: processes and products. In: Downs, R.M., Stea, D. (eds) *Image and Environment*, Aldine, Chicago, pp. 8–26.

Downs, R.M., Stea, D. (1973b) Cognitive representations. In: Downs, R.M., Stea, D. (eds) *Image and Environment*, Aldine, Chicago, pp. 79–87.

Downs, R.M., Stea, D. (1973c) Preface. In: Downs, R.M., Stea, D. (eds) *Image and Environment*, Aldine, Chicago, pp. xiii–xviii.

Downs, R.M., Stea, D. (1977) *Maps in Minds: reflections on cognitive mapping*, Harper & Row: New York.

Downs, R.M., Liben, L.S., Daggs, D.G. (1988) On education and geographers: the role of cognitive developmental theory in geographic education, *Annals of the Association of American Geographers* **78**, 680–700.

Duncan, J.S. (ed.) (1982) *Housing and Identity*, Croom Helm, London.

Duncan, J. (1987) Review of urban imagery: urban semiotics, *Urban Geography* **8**, 473–83.

Duncan, J.S., Duncan, N.G. (1976) Housing as presentation of self and the structure of social networks. In: Moore, G.T., Golledge, R.G. (eds) *Environmental Knowing*, Dowden, Hutchinson & Ross, Stroudsburg, pp. 247–53.

Duncan, J., Duncan N. (1988) (Re)reading the landscape, *Environment and Planning D: Society and Space* **6**, 117–26.

Duncan, J., Ley, D. (1982) Structural Marxism and human geography: a critical assessment, *Annals of the Association of American Geographers* **72**, 30–59.

Duncan, O.D., Duncan, B. (1955) A methodological analysis of segregation indexes, *American Sociological Review* **20**, 210–17.

Dunham, H.W. (1937) The ecology of functional psychoses in Chicago, *American Sociological Review* **2**, 467–79.

Dunn, D.R. (1980) Urban recreation research: an overview, *Leisure Sciences* **3**, 25–58.

Dunn, R., Wrigley, N. (1984) Store loyalty for grocery products: an empirical study, *Area* **16**, 307–14.

Dyck, I. (1990) Space, time, and renegotiating motherhood: an exploration of the domestic workplace, *Environment and Planning D: Society and Space* **8**, 459–83.

Dyson-Hudson, R., Smith, E.A. (1978) Human territoriality: an ecological reassessment, *American Anthropologist* **80**, 21–41.

Edney, J.J. (1976a) Human territories: comment of function properties, *Environment and Behaviour* **8**, 31–48.

Edney, J.J. (1976b) The psychological role of property rights in human behaviour, *Environment and Planning* **8**, 811–22.

Edwards, L.E. (1983) Towards a process model of office location decision making, *Environment and Planning A* **15**, 1327–42.

Edwards, W., Tversky, A. (1967) *Decision Making*, Penguin, Harmondsworth.

Eibl-Eibesfeldt, I. (1970) *Ethology: the biology of behaviour*, Holt, New York.

Ekinsmyth, C. (1988) The urban environment: the learning process of new residents, unpublished Ph.D thesis, University of Leicester.

Ekman, G., Bratfisch, O. (1965) Subjective distance and emotional involvement: a psychological mechanism, *Acta Psychologica* **24**, 430–7.

Elson, M.J. (1976) Activity spaces and recreation trip behaviour, *Town Planning Review* **47**, 241–55.

Embacher, J., Buttle, R. (1989) A repertory grid analysis of Austria's image as a summer vacation destination, *Journal of Travel Research* **27**, 3–7.

Entrikin, J.N. (1976) Contemporary humanism in geography, *Annals of the Association of American Geographers* **66**, 615–32.

Entrikin, J.N. (1990) *The Betweenness of Place: towards a geography of modernity*, Macmillan, London.

Ericksen, R.H. (1975) *The Effects of Perceived Place Attributes on Cognition of Urban Distance*, University of Iowa, Department of Geography Discussion Paper No. 23.

Esser, A.H. (ed.) (1971) *Behaviour and Environment: the use of space by animals and man*, Plenum, New York.

Evans, A.W. (1973) *The Economics of Residential Location*, Macmillan, London.

Evans, G.W., Skorpanich, M.A., Garling, T., Bryant, K.J., Bresolin, B. (1984) The effects of pathway configuration, landmarks and stress on environmental cognition, *Journal of Environmental Psychology* **4**, 323–35.

Evans-Pritchard, E.E. (1940) *The Nuer*, Clarendon Press, Oxford.

Everitt, J.C. (1976) Community and propinquity in a city, *Annals of the Association of American Geographers* **66**, 104–16.

Everitt, J.C., Cadwallader, M.T. (1981) Husband–wife role variation as a factor in home area definition, *Geografiska Annaler* **63B**, 23–34.

Eyles, J. (1968) *The inhabitants' images of Highgate Village, London*, London School of Economics, Department of Geography Graduate Discussion Paper **15**.

Eyles, J. (1971) Pouring new sentiments into old theories: how else can we look at behavioural patterns? *Area* **3**, 242–50.

Eyles, J. (1985) *Senses of Place*, Silverbrook Press, Warrington.

Eyles, J. (1988) Interpreting the geographical world: qualitative approaches in geographical research. In: Eyles, J., Smith, D.M. (eds) *Qualitative Methods in Human Geography*, Polity Press, Cambridge, pp. 1–16.

Eyles, J. (1989) The geography of everyday life. In: Gregory, D., Walford, R. (eds) *Horizons in Human Geography*, Macmillan, London, pp. 102–17.

Eyles, J., Donovan, J. (1986) Making sense of sickness and care: an ethnography of health in a West Midlands town, *Transactions of the Institute of British Geographers* **11**, 415–27.

Eyles, J., Peace, W. (1990) Signs and symbols in Hamilton: an iconology of a steeltown, *Geografiska Annaler* **72B**, 73–88.

Eyles, J., Smith, D.M. (eds) (1988) *Qualitative Methods in Human Geography*, Polity Press, Cambridge.

Fainstein, M.J., Martin, M. (1978) Support for community control among local urban elites, *Urban Affairs Quarterly* **13**, 443–68.

Falk, T., Abler, R. (1980) Intercommunications, distance, and geographical theory, *Geografiska Annaler* **62B**, 59–67.

Faris, R.E.L., Dunham, H.W. (1939) *Mental Disorders in Urban Areas*, University of Chicago Press, Chicago.

Faulkner, H.W. (1978) Locational stress on Sydney's metropolitan fringe, unpublished Ph.D thesis, Australian National University, Canberra.

Featherstone, M. (ed.) (1988) *Postmodernism*, Sage, New York.

Feldman, R.M. (1990) Settlement identity: psychological bonds with home places in a mobile society, *Environment and Behaviour* **22**, 183–229.

Festinger, L., Schacter, S., Bach, K. (1950) *Social Pressures in Informal Groups*, Harper, New York.

Fincher, R. (1983) The inconsistency of eclecticism, *Environment and Planning A* **15**, 607–22.

Fines, K.D. (1968) Landcape evaluation: a research project in East Sussex, *Regional Studies* **2**, 41–55.

Firey, W. (1945) Sentiment and symbolism as ecological variables, *American Sociological Review* **10**, 140–8.

Fischer, C.S. (1975) Towards a subcultural theory of urbanism, *American Journal of Sociology* **80**, 131–41.

Fischer, C. (1977) *Networks and Places: social relations in the urban setting*, Free Press, New York.

Fischer, C. (1981) The public and private worlds of city life, *American Sociological Review* **46**, 396–416.

Fischer, C. (1982) *To Dwell among Friends: personal networks in town and city*, University of Chicago Press, Chicago.

Fischer, C. (1984) *The Urban Experience*, Sage, Beverly Hills.

Fischer, C., Baldassare, M., Ofshe, R.J. (1975) Crowding studies and urban life: a critical review, *Journal of the American Institute of Planners* **31**, 406–18.

Fischer, D.W., Lewis, J.E., Priddle, G.B. (eds) (1974) *Land and Leisure: concepts and methods in outdoor recreation*, Maaroufa Press, Chicago.

Fischer, M.M., Nijkamp, P. (1985) Developments in explanatory discrete spatial data and choice analysis, *Progress in Human Geography* **9**, 515–51.

Fishbein, H.D. (1976) An epigenetic approach to learning theory: a commentary. In: Moore, G.T., Golledge, R.G. (eds) *Environmental Knowing*, Dowden, Hutchinson & Ross, Stroudsburg, pp. 131–6.

Florence, P.S. (1964) *Economics and Sociology and Industry: a realistic analysis of development*, Watts, London.

Flowerdew, R. (ed.) (1982) *Institutions and Geographical Patterns*, Croom Helm, London.

Ford, R.G., Smith, G.C. (1990) Household life cycle change in the urban council housing sector 1971–81, *Environment and Planning A* **22**, 53–67.

Forer, P.C., Kivell, H. (1981) Space-time budgets, public transport, and spatial choice, *Environment and Planning A* **13**, 497–509.

Forrest, R. (1983) The meaning of home ownership, *Environment and Planning D: Space and Society* **1**, 205–16.

Forrest, R. (1987) Spatial mobility, tenure mobility and emerging social divisions in the UK housing market, *Environment and Planning A* **19**, 1611–30.

Forrest, R., Murie, A. (1983) Residualization and council housing: aspects of the changing social relations of housing tenure, *Journal of Social Policy* **12**, 453–68.

Forrest, R., Murie, A. (1987) The affluent home owner: labour market position and the shaping of housing histories, *Sociological Review* **35**, 370–403.

Forrest, R., Murie, A. (1988) *Selling the Welfare State*, Routledge, London.

Fotheringham, A.S. *et al.* (1989) Diffusion-limited aggregation and the (fractal) nature of urban growth, *Papers of the Regional Science Association* **67**, 55–69.

Foxall, G.R. (1980) *Consumer Behaviour: a practical guide*, Croom Helm, London.

Franck, K.A. (1984) Exercizing the ghost of physical determinism, *Environment and Behaviour* **16**, 411–36.

Fransella, F., Bannister, D. (1977) *A Manual for Repertory Grid Technique*, Academic Press, London.

Fredriksson, C.G., Lindmark, L.G. (1979) From firms to systems of firms: a study of interregional dependence in a dynamic society. In: Hamilton, F.E.I., Linge, G.J.R. (eds) *Spatial Analysis, Industry and the Industrial Environment: progress in research and applications*, Vol. 1 *Industrial Systems*, Wiley, Chichester, pp. 155–86.

Freedman, J.L. (1975) *Crowding and Behaviour*, Freeman, San Francisco.

Freudenberg, W.R. (1986) The density of acquaintanceship: an overlooked variable in community research? *American Journal of Sociology* **92**, 27–63.

Frey, A. (1981) The geographer's burden of office, *Progress in Human Geography* **5**, 131–9.

Frey, W.H. (1985) Mover destination selectivity and the changing suburbanization of metropolitan whites and blacks, *Demography* **22**, 223–43.

Fried, M. (1963) Grieving for a lost home. In: Duhl, L. (ed.) *The Urban Condition*, Basic Books, New York, pp. 151–71.

Fried, M., Gleicher, P. (1961) Some sources of residential satisfaction in an urban slum, *Journal of the American Institute of Planners* **27**, 305–15.

Fröbel, F., Henricks, J., Kreye, O. (1980) *The New International Division of Labour*, Cambridge University Press, Cambridge.

Gale, N., Golledge, R.G., Pellegrino, J., Doherty, S. (1990) The acquisition and integration of route knowledge in a familiar neighbourhood, *Journal of Environmental Psychology* **10**, 3–25.

Galster, G. (1988) Residential segregation in American cities: a contrary view, *Population Research and Policy Review* **7**, 93–112.

Gans, H.J. (1962) *The Urban Villagers: groups and class in the life of Italian-Americans*, Free Press, New York.

Gans, H. (1979) *Deciding What's News*, Pantheon Books, New York.

Garling, T., Garling, A. (1988) Distance minimization in downtown pedestrian shopping, *Environment and Planning A* **20**, 547–54.

Gayler, H.J. (1980) Social class and consumer spatial behaviour: some aspects of variation in shopping patterns in metropolitan Vancouver, Canada, *Transactions of the Institute of British Geographers* **5**, 427–45.

Geertz, C. (1980) *Negara: the theatre state in nineteenth century Bali*, Princeton University Press, Princeton.

Gershuny, J.I. (1978) *After Industrial Society? Emerging Self-service Economy*, Macmillan, London.

Gertler, M.S. (1988a) Some problems of time in economic geography, *Environment and Planning A* **20**, 151–64.

Gertler, M.S. (1988b) The limits to flexibility: comments on the post-Fordist vision of production and its geography, *Transactions of the Institute of British Geographers* **13**, 419–32.

Getis, A., Boots, B.N. (1971) Spatial behaviour: rats and man, *Professional Geographer* **23**, 11–14.

Getz, D. (1983) Capacity to absorb tourism: concepts and implications for strategic planning, *Annals of Tourism Research* **10**, 239–63.

Gibson, J.J. (1979) *The Ecological Approach to Visual Perception*, Houghton Mifflin, Boston.

Giddens, A. (1976) *New Rules of Sociological Method*, Hutchinson, London.

Giddens, A. (1979) *Central Problems in Social Theory*, University of California Press, Berkeley.

Giddens, A. (1981) *A Critique of Contemporary Historical Materialism*, Macmillan, London.

Giddens, A. (1984) *The Constitution of Society*, University of California Press, Berkeley.

Giddens, A. (1987) *Social Theory and Modern Society*, Basil Blackwell, Oxford.

Giggs, J.A. (1973) The distribution of schizophrenics in Nottingham, *Transactions of the Institute of British Geographers* **51**, 55–76.

Glacken, C.J. (1967) *Traces on the Rhodian Shore*, University of California Press, Berkeley.

Glass, R. (1948) *The Social Background of a Plan: a study of Middlesborough*, Kegan Paul, London.

Glyptis, S.A. (1981) Leisure life styles, *Regional Studies* **15**, 311–26.

Glyptis, S. (1989) *Leisure and Unemployment*, Open University Press, Milton Keynes.

Goddard, J.B. (1973) *Office Linkages and Location*, Pergamon, Oxford.

Goddard, J.B. (1975a) *Office Location in Urban and Regional Development*, Oxford University Press, London.

Goddard, J.B. (1975b) Organizational information flows and the urban system. In: Massey, D., Morrison, W.I. (eds) *Industrial Location: alternative frameworks*, Centre for Environmental Studies Conference Paper **15**, London, pp. 87–135.

Goddard, J.B. (1978) The location of non-manufacturing activities within manufacturing industries. In: Hamilton, F.E.I. (ed.) *Contemporary Industrialization: spatial analysis and regional development*, Longman, London, pp. 62–85.

Godelier, M. (1972) Structure and contradiction in capital. In: Blackburn, R. (ed.) *Ideology in Social Science*, Fontana, London, pp. 334–68.

Godkin, M.A. (1980) Identity and place: clinical applications based on notions of rootedness and uprootedness. In: Buttimer, A., Seamon, D. (eds) *The Human Experience of Space and Place*, Croom Helm, London, pp. 73–85.

Goetze, R., Colton, K.W. (1980) The dynamics of neighbourhood: a fresh approach to understanding housing and neighbourhood change, *Journal of the American Planning Association* **46**, 184–94.

Gold, J.R. (1980) *An Introduction to Behavioural Geography*, Oxford University Press, London.

Gold, J.R. (1982) Territoriality and human spatial behaviour, *Progress in Human Geography* **6**, 44–67.

Gold, J., Burgess, J. (eds) (1982) *Valued Environments*, Allen & Unwin, Oxford.

Gold, J.R., Goodey, B. (1984) Behavioural and perceptual geography, *Progress in Human Geography* **7**, 578–86.

Gold, J.R., Goodey, B. (1989) Environmental perception: the relationship with age, *Progress in Human Geography* **13**, 99–106.

Goldschreider, F.K., Da Vanzo, J. (1985) Living arrangements and the transition to adulthood, *Demography* **22**, 545–63.

Goldstein, S. (1976) Facets of redistribution: research challenges and opportunities, *Demography* **13**, 34–43.

Golledge, R.G. (1967) Conceptualizing the market decision process, *Journal of Regional Science* **7**, 239–58.

Golledge, R.G. (1969) The geographical relevance of some learning theories. In: Cox, K.R., Golledge, R.G. (eds) *Behavioural Problems in Geography: a symposium*, NorthWestern University Studies in Geography No. **17**, Evanston, pp. 101–45.

Golledge, R.G. (1976) Methods and methodological issues in environmental cognition research. In: Moore, G.T., Golledge, R.G. (eds) *Environmental Knowing*, Dowden, Hutchinson & Ross, Stroudsburg, pp. 300–14.

Golledge, R.G. (1978) Learning about urban environments. In: Carlstein, T., Parker, D., Thrift, N.J. (eds) *Timing Space and Spacing Time*, Vol. 1 *Making Sense of Time*, Arnold, London, pp. 76–98.

Golledge, R.G. (1979) Reality, process, and the dialectical relation between man and environment. In: Gale, S., Olsson, G. (eds) *Philosophy in Geography*, Reidel, Dordrecht, pp. 109–20.

Golledge, R.G. (1981) Misconceptions, misinterpretations, and misrepresentations of behavioural approaches in human geography, *Environment and Planning A* **13**, 1325–44.

Golledge, R.G. (1982) Substantive and methodological aspects of the interface between geography and psychology. In: Golledge, R.G., Rayner, J. (eds) *Proximity and Preference: problems in the multidimensional analysis of large data sets*, University of Minnesota Press, Minneapolis, pp. x–xix.

Golledge, R.G. (1983) Models of man, points of view, and theory in social science, *Geographical Analysis* **15**, 57–60.

Golledge, R.G. (1985) Teaching behavioural geography, *Journal of Geography in Higher Education* **9**, 111–27.

Golledge, R.G. (1988) Science and humanism in geography: multiple languages in multiple realities. In: Golledge, R.G., Conclens, H., Gould, P. (eds) *A Ground for Common Search*, The Geographical Press, Santa Barbara, pp. 63–71.

Golledge, R.G., Brown, L.A. (1967) Search, learning and the market decision process, *Geografiska Annaler* **49B**, 116–24.

Golledge, R.G., Couclelis, H. (1984) Positivist philosophy and research on human spatial behaviour. In: Saarinen, T.F., Seamon, D., Sell, J.L. (eds) *Environmental Perception and Behaviour*, Research Paper No. 209, Department of Geography, University of Chicago, pp. 179–90.

Golledge, R.G., Hubert, L.J. (1982) Some comments on non–Euclidean mental maps, *Environment and Planning A* **14**, 107–18.

Golledge, R.G., Rushton, G. (1976) Introduction. In: Golledge, R.G., Rushton, G. (eds) *Spatial Choice and Spatial Behaviour*, Ohio State University Press, Columbus, pp. 1–2.

Golledge, R.G., Rushton, G. (1984) A review of analytic behavioural research in geography. In: Herbert, D.T., Johnston, R.J. (eds) *Geography and the Urban Environment*, Vol. 6, pp. 1–43.

Golledge, R.G., Spector, A.N. (1978) Comprehending the urban environment: theory and practice, *Geographical Analysis* **10**, 125–52.

Golledge, R.G., Stimson, R.J. (1987) *Analytical Behavioural Geography*, Croom Helm, London.

Golledge, R.G., Timmermans, H. (1988) Introduction. In: Golledge, R.G., Timmermans, H. (eds) *Behavioural Modelling in Geography and Planning*, Croom Helm, London, pp. xix–xxix.

Golledge, R.G., Timmermans, H. (1990) Applications of behavioural research on spatial problems I: cognition, *Progress in Human Geography* **14**, 57–99.

Golledge, R.G., Zannaras, G. (1973) Cognitive approaches to the analysis of human spatial behaviour. In: Ittelson, W.H. (ed.) *Environment and Cognition*, Seminar Press, New York, pp. 59–94.

Golledge, R.G., Briggs, R., Demko, D. (1969) The configuration of distances in intraurban space, *Proceedings of the Association of American Geographers* **1**, 60–5.

Golledge, R.G., Brown, L.A., Williamson, F. (1972) Behavioural approaches in geography: an overview, *Australian Geographer* **12**, 59–79.

Golledge, R.G., Rayner, J.N., Parnicky, J.J. (1983) The spatial competence of selected mentally retarded populations. In: Pick, H., Acredolo, L. (eds) *Spatial Orientation and Spatial Representation*, Plenum, New York, pp. 79–100.

Golledge, R.G., Smith, T.R., Pellingrino, J.W., Doherty, S., Marshall, S.P. (1985) A conceptual model and empirical analysis of children's acquisition of spatial knowledge, *Journal of Environmental Psychology* **5**, 125–52.

Goodchild, B. (1974) Class differences in environmental perception: an exploratory study, *Urban Studies* **11**, 157–69.

Goodey, B., Gold, J.R. (1987) Environmental perception: the relationship with urban design, *Progress in Human Geography* **11**, 126–33.

Goodrich, J.N. (1978) A new approach to image analysis through multidimensional scaling, *Journal of Travel Research* **16**, 3–7.

Gopal, S., Smith, T.R. (1990) Human way-finding in an urban environment: a performance analysis of a computational process model, *Environment and Planning A* **22**, 169–91.

Gould, P.R. (1963) Man against his environment: a game theoretic framework, *Annals of the Association of American Geographers* **53**, 290–7.

Gould, P. (1969) Methodological developments since the fifties, *Progress in Geography* **1**, 1–49.

Gould, P.R. (1972) Pedagogic review: entropy in urban and regional modelling, *Annals of the Association of American Geographers* **62**, 689–700.

Gould, P. (1973a) On mental maps. In: Downs, R.M., Stea, D. (eds) *Image and Environment*, Aldine, Chicago, pp. 182–220.

Gould, P. (1973b) The black boxes of Jonkoping: spatial information and preference. In: Downs, R.M., Stea, D. (eds) *Image and Environment*, Aldine, Chicago, pp. 235–45.

Gould, P. (1975) Acquiring spatial information, *Economic Geography* **51**, 87–99.

Gould, P., Lafond, N. (1979) Mental maps and information surfaces in Quebec and Ontario, *Cahiers de Géographie de Québec* **23**, 371–98.

Gould, P., White, R. (1968) The mental maps of British school leavers, *Regional Studies* **2**, 161–82.

Gould, P., White, R. (1974) *Mental Maps*, Penguin, Harmondsworth.

Gouldner, A. (1980) *The Two Marxisms*, Seabury, New York.

Graham, E. (1976) What is a mental map? *Area* **8**, 259–62.

Graham, J., Gibson, K., Horvath, R., Shakons, D.M. (1988) Restructuring in US manufacturing: the decline of monopoly capitalism, *Annals of the Association of American Geographers* **78**, 473–90.

Granovetter, M.S. (1976) Network sampling: some first steps, *American Journal of Sociology* **81**, 1287–303.

Graves, T.D. (1966) Alternative models for the study of urban migration, *Human Organization* **25**, 295–9.

Gray, F. (1976) Selections and allocation in council housing, *Transactions of the Institute of British Geographers* **1**, 34–46.

Green, A.E. (1986) The likelihood of becoming and remaining unemployed in Great Britain, 1984, *Transactions of the Institute of British Geographers* **11**, 37–56.

Green, D.H. (1977) Industrialists' information levels and regional incentives, *Regional Studies* **11**, 7–18.

Gregory, D. (1978) *Ideology, Science and Human Geography*, Hutchinson, London.

Gregory, D. (1980) The ideology of control: systems theory and geography, *Tijdschrift voor Economische en Sociale Geografie* **71**, 327–42.

Gregory, D. (1981) Human agency and human geography, *Translations of the Institute of British Geographers* **6**, 1–18.

Gregory, D. (1989a) The crisis of modernity? Human geography and critical social theory. In: Peet, R., Thrift, N.J. (eds) *New Models in Geography*, Vol. 2, Unwin Hyman, London, pp. 348–85.

Gregory, D. (1989b) Areal differentiation and post-modern human geography. In: Gregory, D., Walford, R. (eds) *Horizons in Human Geography*, Macmillan, London, pp. 67–96.

Gregory, D., Urry, J. (eds) (1985) *Social Relations and Spatial Structures*, Macmillan, London.

Gregson, N. (1986) On duality and dualism: the case of structuration and time geography, *Progress in Human Geography* **10**, 184–205.

Gregson, N. (1987) Structuration theory: some thoughts on the possibilities for empirical research, *Environment and Planning D: Society and Space* **5**, 73–91.

Guelke, L. (1974) An idealist alternative in human geography, *Annals of the Association of American Geographers* **64**, 193–202.

Guelke, L. (1977) The role of laws in human geography, *Progress in Human Geography* **1**, 376–86.

Guelke, L. (1981) Idealism. In: Harvey, M.E., Holly, B.P. (eds) *Themes in Geographic Thought*, Croom Helm, London, pp. 133–47.

Guelke, L. (1985) *Geography and Humanistic Knowledge*, University of Waterloo Lectures in Geography No. 2, Waterloo.

Guelke, L. (1989) Forms of life, history, and mind: an idealist proposal for integrating perception and behaviour in human geography. In: Boal, F.W., Livingstone, D.N. (eds) *The Behavioural Environment*, Routledge, London, pp. 289–310.

Guest, A.M., Lee, B.A. (1983) The social organization of local areas, *Urban Affairs Quarterly* **19**, 217–40.

Gulliver, F.P. (1908) Orientation of maps, *Journal of Geography* **7**, 55–8.

Gunn, C.A. (1988) *Vacationscape: designing for tourist regions* (2nd edn), Van Nostrand Reinhold, New York.

Guy, B.S., Curtis, W.W., Crotts, J.C. (1990) Environmental learning of first-time travellers, *Annals of Tourism Research* **17**, 419–31.

Guy, C.M. (1980) *Retail Location and Retail Planning in Britain*, Gower, Aldershot.

Habermas, J. (1972) *Knowledge and Human Interests*, Heinemann, London.

Hägerstrand, T. (1952) *The Propagation of Innovation Waves*, Lund Studies in Geography, Series B No. 4.

Hägerstrand, T. (1957) Migration and area. In: Hannbergerg, D., Hägerstrand, T., Deving, B.O. (eds) *Migration in Sweden*, Lund Studies in Geography 13B, pp. 27–158.

Hägerstrand, T. (1970) What about people in regional science? *Papers and Proceedings of the Regional Science Association* **24**, 7–21.

Hägerstrand, T. (1975) Time, space and human conditions. In: Karlqvist, A., Lundqvist, L., Snickers, F. (eds) *Dynamic Allocation of Urban Space*, Saxon House, Farnborough, pp. 3–14.

Hägerstrand, T. (1978) Survival and arena. In: Carlstein, T., Parker, D., Thrift, N.J. (eds) *Timing Space and Spacing Time*, Vol. 2 *Human Activity and Time Geography*, Arnold, London, pp. 122–45.

Hägerstrand, T. (1984) Presence and absence: a look at conceptual choices and bodily necessities, *Regional Studies* **18**, 373–9.

Hagey, M.J., Malecki, E.J. (1986) Linkages in high technology industries: a Florida case study, *Environment and Planning A* **18**, 1477–98.

Haggett, P. (1965) *Locational Analysis in Human Geography*, Arnold, London.

Hakanson, L. (1979) Towards a theory of location and corporate growth. In: Hamilton, F.E.I., Linge, G.J.R. (eds) *Spatial Analysis, Industry and the Industrial Environment*: *progress in research and applications*, Vol. 1 *Industrial Systems*, Wiley, Chichester, pp. 115–38.

Hall, D.R. (1982) Valued environments and the planning process: community consciousness and urban structure. In: Gold, J.R., Burgess, J. (eds) *Valued Environments*, Allen & Unwin, London, pp. 172–88.

Hall, E.T. (1959) *The Silent Language*, Doubleday, New York.

Hall, E.T. (1966) *The Hidden Dimension*, Doubleday, New York.

Hall, P., Markusen, A. (eds) (1985) *Silicon Landscapes*, Allen & Unwin, Winchester, Mass.

Hall, P., Breheny, M. McQuaid, R., Hart, O. (1987) *Western Sunrise*: *the genesis and growth of Britain's high tech corridor*, Allen & Unwin, London.

Hallman, H.W. (1984) *Neighbourhoods: their place in urban life*, Sage, London.

Hallsworth, A.G. (1989) *The Human Impact of Hypermarkets and Superstores*, Gower, Aldershot.

Halperin, W.C. (1988) Current topics in behavioural modelling of consumer choice. In: Golledge, R.G., Timmermans, H. (eds) *Behavioural Modelling in Geography and Planning*, Croom Helm, London, pp. 1–26.

Halperin, W.C., Gale, N., Golledge, R.G. (1983) Exploring entrepreneurial cognitions of retail environments, *Economic Geography* **59**, 3–15.

Hamilton, F.E.I. (1974) A view of spatial behaviour, industrial organizations and decision-making. In: Hamilton, F.E.I. (ed.) *Spatial Perspectives on Industrial Organization and Decision Making*, Wiley, London, pp. 3–43.

Hamilton, F.E.I. (1978) The changing milieu of spatial industrial research. In: Hamilton, F.E.I. (ed.) *Contemporary Industrialization*: *spatial analysis and regional development*, Longman, London, pp. 1–19.

Hamilton, F.E.I., Linge, G.J.R. (1979) Industrial systems. In: Hamilton, F.E.I., Linge, G.J.R. (eds) *Spatial Analysis, Industry and the Industrial Environment*: *progress in research and applications*, Vol. 1 *Industrial Systems*, Wiley, Chichester, pp. 1–23.

Hamnett, C. (1979) Area-based explanations: a critical appraisal. In: Herbert, D.T., Smith, D.M. (eds) *Social Problems in the City*, Oxford University Press, London, pp. 244–60.

Hamnett, C. (1984) Gentrification and residential location theory:a review and assessment. In: Herbert, D.T., Johnston, R.J. (eds) *Geography and the Urban Environment*, Vol. 6, Wiley, New York, pp. 283–319.

Hannerz, U. (1980) *Exploring the City*, Columbia University Press, New York.

Hanson, S. (1976) Spatial variation in the cognitive levels of urban residents. In: Golledge, R.G., Rushton, G. (eds) *Spatial Choice and Spatial Behaviour*, Ohio State University Press, Columbus, pp. 157–88.

Hanson, S. (1977) Measuring the cognitive levels of urban residents, *Geografiska Annaler* **59B**, 67–81.

Hanson, S. (1980) Spatial diversification and multipurpose trips: implications for choice theory, *Geographical Analysis* **12**, 245–57.

Hanson, S. (1984) Environmental cognition and travel behaviour. In: Herbert, D.T., Johnston, R.J. (eds) *Geography and the Urban Environment*: *progress in research and applications*, Vol. 6, Wiley, Chichester, pp. 95–125.

Hanson, S., Hanson, P. (1980) Gender and urban activity patterns in Uppsala, Sweden, *Geographical Review* **70**, 291–9.

Hanson, S., Huff, J. (1988) Repetition and day-to-day variability in individual travel patterns: implications for classification. In: Golledge, R.G., Timmermans, H. (eds) *Behavioural Modelling in Geography and Planning*, Croom Helm, London, pp. 368–98.

Harries, K.D. (1980) *Crime and the Urban Environment*, Thomas, Springfield.

Harris, L. (1988) The UK economy at a crossroads. In: Allen, J., Massey, D. (eds) *Restructuring Britain*: *the economy in question*, Sage, London, pp. 8–44.

Harris, R. (1984) Residential segregation and class formation in the capitalist city: a review and directions for research, *Progress in Human Geography* **8**, 26–49.

Harrison, A.J.M., Stabler, M.J. (1981) An analysis of journeys for canal based recreation, *Regional Studies* **15**, 345–58.

Harrison, C. (1983) Countryside recreation and London's urban fringe, *Transactions of the Institute of British Geographers* **8**, 295–313.

Harrison, E.F. (1975) *Managerial Decision-making Process*, Houghton Mifflin, Boston.

Harrison, J., Howard, W.A. (1972) The role of meaning in the urban image, *Environment and Behaviour* **4**, 389–411.

Harrison, J., Sarre, P. (1971) Personal construct theory in the measurement of environmental images: problems and methods, *Environment and Behaviour* **3**, 351–74.

Harrison, J.A., Sarre, P. (1975) Personal construct theory in the measurement of environmental images, *Environment and Behaviour* **7**, 3–58.

Harrison, R.T., Livingstone, D.N. (1979) There and back again – towards a critique of idealist human geography, *Area* **11**, 75–9.

Harrison, R.T., Livingstone, D.N. (1982) Understanding in geography: structuring the subjective. In: Herbert, D.T., Johnston, R.J. (eds) *Geography and the Urban Environment*, Vol. 5, Wiley, Chichester, pp. 1–39.

Hart, P.W.E. (1980) Problems and potentialities of the behavioural approach to agricultural location, *Geografiska Annaler* **62B**, 99–107.

Hart, R.A., Moore, G.T. (1973) The development of spatial cognition: a review. In: Downs, R.M., Stea, D. (eds) *Image and Environment*, Aldine, Chicago, pp. 246–88.

Harvey, D. (1969a) *Explanation in Geography*, Arnold, London.

Harvey, D. (1969b) Conceptual and measurement problems in the cognitive–behavioural approach to location theory. In: Cox, K.R., Golledge, R.G. (eds) *Behavioural Problems in Geography: a symposium*, Northwestern University Studies in Geography No. 17, Evanston, pp. 35–68.

Harvey, D. (1973) *Social Justice and the City*, Arnold, London.

Harvey, D. (1974) Class monopoly rent, finance capital and urban revolution, *Regional Studies* **8**, 239–55.

Harvey, D. (1975) Class structure in a capitalist society and the theory of residential differentiation. In: Peel, R., Chisholm, M., Haggett, P. (eds) *Processes in Physical and Human Geography*, Heinemann, London, pp. 354–72.

Harvey, D. (1982) *The Limits to Capital*, Blackwell, Oxford.

Harvey, D. (1985a) *The Urbanization of Capital*, Blackwell, Oxford.

Harvey, D. (1985b) *Consciousness and the Urban Experience*, Blackwell, Oxford.

Harvey, D. (1987a) Flexible accumulation through urbanization: reflections on 'post-modernism' in the American city, *Antipode* **19**, 260–86.

Harvey, D. (1987b) Three myths in search of a reality in urban studies, *Environment and Planning D: Society and Space* **5**, 367–76.

Harvey, D. (1989) *The Condition of Postmodernity*, Basil Blackwell, Oxford.

Harvey, D. (1990) Between space and time: reflections on the geographical imagination, *Annals of the Association of American Geographers* **80**, 418–34.

Harvey, D., Chatterjee, L. (1974) Absolute rent and the structuring of space by governmental and financial institutions, *Antipode* **6**, 22–36.

Hawley, A.H. (1950) *Human Ecology: a theory of community structure*, Ronald Press, New York.

Hay, A.M., Johnston, R.J. (1979) Search and the choice of shopping centre: two models of variability in destination selection, *Environment and Planning A* **11**, 791–804.

Hayter, R., Watts, H.D. (1983) The geography of enterprise: a reappraisal, *Progress in Human Geography* **7**, 157–81.

Hebb, D.O. (1949) *The Organization of Behaviour*, Wiley, New York.

Hebb, D.O. (1966) *A Textbook of Psychology*, Saunders, Philadelphia.

Heft, H. (1988a) Review essay: the development of Gibson's ecological approach to perception, *Journal of Environmental Psychology* **8**, 325–34.

Heft, H. (1988b) The vicissitudes of ecological phenomena in environment–behaviour research: on the failure to replicate the 'angularity effect', *Environment and Behaviour* **20**, 92–9.

Held, D., Thompson, J. (eds) (1989) *Social Theory of Modern Societies: Anthony Giddens and his critics*, Cambridge University Press, Cambridge.

Hendee, J.C. (1969) Rural–urban differences reflected/outdoor recreation participation, *Journal of Leisure Research* **1**, 333–41.

Hendee, J.C., Gale, R.P., Catton, W.R. (1971) Typology of outdoor recreation activity preferences, *Journal of Environmental Education* **3**, 28–34.

Henscher, D., Johnson, L. (1981) *Applied Discrete Choice Modelling*, Croom Helm, London.

Herbert, D.T. (1972) *Urban Geography: a social perspective*, David Charles, Newton Abbot.

Herbert, D.T. (1973) The residential mobility process: some empirical observation, *Area* **5**, 44–8.

Herbert, D.T. (1975) Urban neighbourhood and social geographical research. In: Phillips, A.D.M., Turton, B.J. (eds) *Environment, Man and Economic Change*, Longman, London, pp. 459–78.

Herbert, D.T. (1976) The study of delinquency areas: a social geographical approach, *Transactions of the Institute of British Geographers* **1**, 472–92.

Herbert, D.T. (1977) Crime, delinquency and the urban environment, *Progress in Human Geography* **1**, 208–39.

Herbert, D.T. (1982) *The Geography of Urban Crime*, Longman, London.

Herbert, D.T., Raine, J.W. (1976) Defining communities within urban areas, *Town Planning Review* **47**, 325–38.

Herbert, D.T., Smith, D.M. (eds) (1979) *Social Problems and the City*, Oxford University Press, London.

Herbert, D.T., Thomas, C.J. (1982) *Urban Geography*, Wiley: Chichester.

Herbert, G. (1963–64) The neighbourhood unit: principles and organic theory, *Sociological Review* **11–12**, 165–213.

Herbst, P.G. (1964) Organizational commitment: a decision model, *Acta Sociologica* **7**, 34–45.

Herlyn, U. (1989) Upgrading and downgrading of the urban area, *Tijdschrift voor Economische en Sociale Geografie* **80**, 97–105.

Hewings, G.J.D., Sonis, M., Jensen, R.C. (1988) Fields of influence of technological change in input–output models, *Papers of the Regional Science Association* **64**, 25–36.

Higgs, G. (1975) An assessment of the action component of action space, *Geographical Analysis* **7**, 35–50.

Hillery, G.A. (1955) Definitions of community: areas of agreement, *Rural Sociology* **20**, 779–91.

Hillery, G.A., Brown, J.S. (1965) Migrational systems of the southern Appalachians: some demographic observations, *Rural Sociology* 30, 3–48.

Hirsch, F. (1978) *Social Limits to Growth*, Routledge & Kegan Paul, London.

Holahan, C.J., Dobrowolny, M.B. (1978) Cognitive and behavioural correlates of the spatial environment: an interactional analysis, *Environment and Behaviour* **10**, 317–34.

Holcomb, B. (1986) Geography and urban women, *Urban Geography* **7**, 448–56.

Holloway, J.C., Plant, R.V. (1988) *Marketing for Tourism*, Pitman, London.

Hookway, A.J., Davidson, J. (1970) *Leisure: problems and prospects for the environment*, Countryside Commission, London.

Hoover, E.M. (1937) *Location Theory and the Shoe and Leather Industries*, Harvard University Press, Cambridge, Mass.

Hopkins, J.S.P. (1990) West Edmonton Mall: landscapes of myths and elsewhereness, *Canadian Geographer* **34**, 2–17.

Horton, F., Reynolds, D.R. (1971) Effects of urban spatial structure on individual behaviour, *Economic Geography* **47**, 36–48.

Hotelling, H. (1929) Stability in competition, *Economic Journal* **39**, 41–57.

Howard, J.A., Sheth, J.N. (1969) *The Theory of Buyer Behaviour*, Wiley, New York.

Hubbard, R., Thompson, A.F. (1981) Preference structures and consumer spatial indifference behaviour: some theoretical problems, *Tijdschift voor Economische en Sociale Geografie* **72**, 35–9.

Huber, G. (1980) *Decision Making*, Scott, Foreman, Glenview, Ill.

Huckfeldt, R.R. (1983) Social contexts, social networks, and urban neighbourhoods: environmental constraint on friendly choice, *American Journal of Sociology* **89**, 651–69.

Hudson, R. (1970) *Personal Construct theory, Learning Theories and Consumer Behaviour*, University of Bristol, Department of Geography Seminar Paper Series A No. 21.

Hudson, R. (1974) Images of the retailing environment: an example of the use of repertory grid methodology, *Environment and Behaviour* **6**, 470–94.

Hudson, R. (1975) Patterns of spatial search, *Transactions of the Institute of British Geographers* **65**, 141–54.

Hudson, R. (1976a) *Environmental Images, Spatial Choice and Consumer Behaviour*, University of Durham, Department of Geography Occasional Paper No.9.

Hudson, R. (1976b) Linking studies of the individual with models of aggregate behaviour: an empirical example, *Transactions of the Institute of British Geographers* **1**, 159–74.

Huff, D.L. (1960) A topographical model of consumer space preferences, *Papers and Proceedings of the Regional Science Association* **6**, 159–73.

Huff, D.L. (1963) A probabilistic analysis of shopping centre trade areas, *Land Economics* **39**, 81–90.

Huff, J.O. (1986) Geographic regularities in residential search behaviour, *Annals of the Association of American Geographers* **76**, 208–27.

Hull, C.L. (1952) *A Behaviour System*, Yale University Press, New Haven.

Humphreys, J.S. (1990) Place learning and spatial cognition: a longitudinal study of urban newcomers, *Tijdschrift voor Economische en Sociale Geografie* **81**, 364–80.

Hunter, A. (1974) *Symbolic Communities: the resistance and change of Chicago's local communities*, University of Chiacgo Press, Chicago.

Hunter, A. (1989) The symbolic ecology of suburbia. In: Altman, I., Wandersman, A. (eds) *Neighbourhood and Community Environments*, Plenum Press, New York, pp. 191–218.

Hurwicz, L. (1970) Optimality criteria for decision making under risk, Cowles Commission Paper, Statistics 350 (mimeo).

Husserl, E. (1931) *Ideas: general introduction to pure phenomenology*, (trans. W.R.B. Gibson), Allen & Unwin, London.

Huttman, E., van Vliet, W. (eds) (1987) *Handbook of Housing and the Built Environment in the US*, Greenwood Press, Westport, Connecticut.

Hwang, S.S., Albrecht, D.E. (1987) Constraints to the fulfillment of residential preferences among Texas homebuyers, *Demography* **24**, 61–76.

Hymer, S. (1976) *The International Operations of National Firms*, MIT Press, Cambridge, Mass.

Irving, H. (1977) Social networks in the modern city, *Social Forces* **55**, 867–80.

Irving, H.W. (1978) Space and environment in interpersonal relations in Herbert, D.T., Johnston, R.J. (eds) *Geography and the Urban Environment*, Vol.1, Chichester, Wiley, pp. 249–84.

Irving, H.W., Davidson, R.N. (1973) A working note on the measurement of social interaction, *Transactions of the Bartlett Society* **9**, 7–19.

Isaacs, R.R. (1948) The neighbourhood theory: an analysis of its adequacy, *Journal of the American Institute of Planners* **14**, 15–23.

Isaacs, R.R. (1948) The neighbourhood theory: an analysis of its adequacy, *Journal of the American Institute of Planners* **14**, 15–23.

Isbell, E.C. (1944) Internal migration in Sweden and intervening opportunities, *American Sociological Review* **9**, 627–39.

Iso-Ahela, S.E. (ed.) (1980) *Social Psychological Perspectives on Leisure and Recreation*, Thomas, Springfield, Ill.

Iso-Ahela, S.E. (1982) Towards a social psychological theory of tourism motivation: a rejoinder, *Annals of Tourism Research* **9**, 256–61.

Ittelson, W.H. (1978) Environmental perception and urban experience, *Environment and Behaviour* **10**, 193–214.

Ittelson, W.H., Proshansky, H.M., Rivlin, L.G., Winkel, G.H. (1974) *An Introduction to Environmental Psychology*, Holt, Rinehart & Winston, New York.

Jackson, E.L. (1981) Response to earthquake hazard: the West Coast of North America, *Environment and Behaviour* **13**, 387–416.

Jackson, L.E., Johnston, R.J. (1972) Structuring the image: investigation of the elements of mental maps, *Environment and Planning* **4**, 415–27.

Jackson, P. (1983) Principles and problems of participant observation, *Geografiska Annaler* **65B**, 39–46.

Jackson, P. (1985) Urban ethnography, *Progress in Human Geography* **9**, 157–76.

Jackson, P. (1987) *Race and Racism*, Allen & Unwin, London.

Jackson, P. (1989) *Maps of Meaning: an introduction to cultural geography*, Unwin Hyman, London.

Jackson, P. (1991) Mapping meanings: a cultural critique of locality studies, *Environment and Plannning A* **23**, 215–28.

Jackson, P., Smith, S.J. (1984) *Exploring Social Geography*, Allen & Unwin, London.

Jackson, R.H. (1978) Mormon perception and settlement, *Annals of the Association of American Geographers* **68**, 317–34.

Jacobs, J. (1961) *The Death and Life of Great American Cities*, Random House, New York.

Jacques, D.L. (1980) Landscape appraisal: the case for a subjective theory, *Journal of Environmental Management* **10**, 107–13.

Jakle, J.A. (1985) *The Tourist*, University of Nebraska Press, Lincoln.

Jakle, J.A., Brunn, S., Roseman, C.C. (1976) *Human Spatial Behaviour: a social geography*, Duxbury, North Scituate.

Jarvis, I., Mann, L. (1977) *Decision Making: a psychological analysis of conflict, choice and commitment*, Free Press, New York.

Jayet, H. (1990) Spatial search processes and spatial interaction, *Environment and Planning A* **22**, 583–99, 719–32.

Jefferson, A., Lickorish, L. (1988) *Marketing Tourism*, Longman, London.

Jeffries, R.W., Dobos, J. (1984) Communication and neighbourhood mobilization, *Urban Affairs Quarterly* **20**, 97–112.

Jencks, C. (1985) *Modern Movements in Architecture* (2nd edn), Penguin, Harmondsworth.

Jenkins, C., Sherman, B. (1979) *The Collapse of Work*, Eyre Methuen, London.

Jenkins, J.M., Walmsley, D.J. (1992) The environmental images of tourists: rationale, methodology, implications. In: *Proceedings of the Joint Institute of Australian Geographers–New Zealand Geographical Society Conference*, NZGS, Auckland, pp. 62–9.

Jensen-Butler, C. (1972) Gravity models as planning tools: a review of theoretical and operational problems, *Geografiska Annaler* **54B**, 68–78.

Jensen-Butler, C. (1981) *A Critique of Behavioural Geography: an epistemological analysis of cognitive mapping and of Hägerstrand's time–space model*, Aarhus University Department of Geography Arbejdsrapport No. 12.

Johnson, C.R. (1977) *Leisure and Recreation: introduction and overview*, Lea & Febinger, Philadelphia.

Johnson, J.R., Roseman, C.C. (1990) Increasing black outmigration from Los Angeles: the role of household dynamics and kinship system, *Annals of the Association of American Geographers* **80**, 205–22.

Johnston, R.J. (1973a) Spatial patterns in suburban evaluations, *Environment and Planning A* **5**, 385–95.

Johnston, R.J. (1973b) The factorial ecology of major New Zealand urban areas: a comparative study, Institute of British Geographers, Special Publications No. 5, *Social Patterns in Cities*, 143–68.

Johnston, R.J. (1979) *Geography and Geographers: Anglo–American human geography since 1945*, Arnold, London.

Johnston, R.J. (1980) On the nature of explanation in human geography, *Transactions of the Institute of British Geographers* **5**, 402–12.

Johnston, R.J. (1983) *Philosophy and Human Geography*, Arnold, London.

Johnston, R.J. (1989) Philosophy, ideology and geography. In: Gregory, D., Walford, R. (eds) *Horizons in Human Geography*, Macmillan, London, pp. 48–66.

Johnston, R.J. (1991) *Geography and Geographers* (4th edn), Arnold, London.

Johnston, R.J., Rimmer, P.J. (1967) A note on consumer behaviour in an urban hierarchy, *Journal of Regional Science* **7**, 161–6.

Johnston, R.J., Haner, J., Hoekreld, G.A. (1990) *Regional Geography: current developments and future prospects*, Routledge, London.

Jones, B. (1982) *Sleepers, Wake!* Oxford University Press, Melbourne.

Jones H. (1986) Evolution of Scottish migration patterns: a social relations of production approach, *Scottish Geographical Magazine* **102**, 151–64.

Jones, P. (1982) The locational policies and geographical expansion of multiple retail companies: a case study of MFI, *Geoforum* **13**, 39–43.

Jordan, T.G. (1978) Perceptual regions in Texas, *Geographical Review* **68**, 293–307.

Jubenville, A., Twight, B.W., Becker, R.H. (1987) *Outdoor Recreation Management*, Venture Publishing Inc., Philadelphia.

Kamenka, E. (ed.) (1982) *Community as a Social Ideal*, Edward Arnold, London.

Kando, T.M. (1975) *Leisure and Popular Culture in Transition*, Mosby, St Louis.

Kaplan, M. (1975) *Leisure: theory and policy*, Wiley, New York.

Kaplan, R. (1987) Validity in environment/behaviour research: some cross-paradigm concerns, *Environment and Behaviour* **19**, 495–500.

Kaplan, R., Kaplan, S. (1989) *The Experience of Nature: psychological perspectives*, Cambridge University Press, Cambridge.

Kaplan, S. (1973) Cognitive maps in perception and thought. In: Downs, R.M., Stea, D. (eds) *Image and Environment*, Aldine, Chicago, pp. 63–78.

Kaplan, S. (1976) Adaptation, structure, and knowledge. In: Moore, G.T., Golledge, R.G. (eds) *Environmental Knowing*, Dowden, Hutchinson & Ross, Stroudsburg, pp. 32–45.

Karn, V., Keheny, J., Williams, P. (1985) *Home Ownership in the Inner City: salvation or despair*, Gower, Aldershot.

Kasarda, J.D., Janowitz, M. (1974) Community attachment in mass society, *American Sociological Review* **39**, 328–39.

Kasmar, J.V. (1970) The development of a usable lexicon of environmental descriptors, *Environment and Behaviour* **2**, 153–69.

Kassarjian, H.H. (1971) Personality and consumer behaviour: a review, *Journal of Marketing Research* **8**, 409–18.

Kates, R.W. (1962) *Hazard and Choice Perception in Flood Plain Management*, University of Chicago, Department of Geography Research Paper No. 78.

Keeble, D, (1976) *Industrial Location and Planning in the United Kingdom*, Methuen, London.

Keeble, D. (1990) Small firms, new firms and uneven regional development in the United Kingdom, *Area* **22**, 234–45.

Keller, S. (1968) *The Urban Neighbourhood: a sociological perspective*, Random House, New York.

Kellerman, A. (1984) Telecommunications and the geography of metropolitan areas, *Progress in Human Geography* **8**, 222–46.

Kellerman, A. (1985) The suburbanization of retail trade: a US nationwide view, *Geoforum* **16**, 15–23.

Kelly, G.A. (1955) *The Psychology of Personal Constucts*, Norton, New York.

Kelly, J.R. (1974) Socialization toward leisure: a developmental approach, *Journal of Leisure Research* **6**, 181–93.

Kendig, H. (1984) Housing careers, life cycle and residential mobility: implications for the housing market, *Urban Studies* **21**, 271–83.

Kent, P. (1990) People, places and priorities: opportunity sets and consumers' holiday choice. In: Ashworth, G., Goodall, B. (eds) *Marketing Tourism Places*, Routledge, London, pp. 42–62.

King, L.J., Golledge, R.G. (1978) *Cities, Space, and Behaviour: the elements of urban geography*, Prentice-Hall, Englewood Cliffs.

Kintrea, K., Clapham, D. (1986) Housing choice and search strategies within an administered housing system, *Environment and Planning A* **18**, 1281–96.

Kirby, A. (1985) Leisure as commodity: the role of the state in leisure provision, *Progress in Human Geography* **9**, 64–84.

Kirk, W. (1952) Historical geography and the concept of the behavioural environment, *Indian Geographical Journal* **25**, 152–60.

Kirk, W. (1963) Problems of geography, *Geography* **48**, 357–71.

Kirk, W. (1978) The road from Mandalay: towards a geographical philosophy, *Transactions of the Institute of British Geographers* **3**, 381–94.

Kivell, P.T., Shaw, G. (1980) The study of retail location. In: Dawson, J.A. (ed.) *Retail Geography*, Wiley, New York, pp. 95–155.

Knetsch, J.L. (1974) *Outdoor Recreation and Water Resources Planning*, American Geophysical Union, Washington.

Knox, P. (1982) *Urban Social Geography: an introduction*, Longman, London.

Knox, P. (1987) The social production of the built environment, *Progress in Human Geography* **11**, 354–77.

Knox, P., MacLaren, A. (1978) Values and perceptions in descriptive approaches to urban social geography. In: Herbert, D.T., Johnston, R.J. (eds) *Geography and the Urban Environment*, Vol. 1, Wiley, London, pp. 197–247.

Kobyashi, A., MacKenzie, S. (eds) (1989) *Remaking Human Geography*, Unwin Hyman, London.

Koestler , A. (1967) *The Ghost in the Machine*, Pan, London.

Koffka, K. (1935) *Principles of Gestalt Psychology*, Harcourt–Brace, New York.

Koroscil, P.M. (1971) The behavioural environmental approach, *Area* **3**, 96–9.

Krampen, M. (1981) *Meaning in the Urban Environment*, Pion, London.

Krippendorf, J. (1987) *The Holiday Makers*, Heinemann, Oxford.

Krupat, E. (1985) *People in Cities: the urban environment and its effects*, Cambridge University Press, Cambridge.

Kuipers, B. (1982) The 'map in the head' metaphor, *Environment and Behaviour* **14**, 202–20.

Kurtz, L. (1984) *Evaluating Chicago Sociology: a guide to the literature with an advanced bibliography*, University of Chicago Press, Chicago.

Kurtzweil, E. (1980) *The Age of Structuralism*, University of Columbia Press, New York.

Ladd, F.C. (1970) Black youths view their environment: neighbourhood maps, *Environment and Behaviour* **2**, 74–99.

Langdale, J.V. (1989) The geography of international business telecommunications: the role of leased networks, *Annals of the Association of American Geographers* **79**, 501–22.

Langer, E.J., Piper, A.T. (1987) The prevention of mindlessness, *Journal of Personality and Social Psychology* **53**, 280–7.

La Page, W.F. (1979) Market analysis for recreation managers.

In: van Doren, C.S., Priddle, G.B., Lewis, J.E. (eds) *Land and Leisure: concepts and methods in outdoor recreation*, Methuen, London, pp. 151–73.

Lash, S. (1990) *Sociology of Postmodernism*, Routledge, London.

Lash, S., Urry, J. (1987) *The End of Organized Capitalism*, Polity Press, Cambridge.

Laslett, P. (1968) *The World We Have Lost*, Methuen, London.

Lavery, P. (ed.) (1974) *Recreational Geography*, David & Charles, Newton Abbot.

Lawless, P. (1979) *Urban Deprivation and Government Initiative*, Faber & Faber, London.

Leach, E.R. (1961) *Pul Eliya – A Village in Ceylon: a study of land tenure and kinship*, Cambridge University Press, Cambridge.

Leach, E.R. (1981) British social anthropology and Lévi–Straussian structuralism. In: Blau, P.M., Merton, R.K. (eds) *Continuities in Structural Enquiry*, Sage, Beverly Hills, pp. 27–50.

Lee, B.A. (1981) The urban unease revisited: perception of local safety and neighbourhood among metropolitan residents, *Social Science Quarterly* **62**, 611–29.

Lee, E.S. (1960) A theory of migration, *Demography* **3**, 47–57.

Lee, R. (1990) Regional geography: between scientific theory, ideology, and practice (or What use is regional geography?). In: Johnston, R.J., Haver, J., Hoerveld, G. (eds) *Regional Geography*, Routledge, London, pp. 103–21.

Lee, T.R. (1962) 'Brennan's Law' of shopping behaviour, *Psychological Reports* **11**, 662.

Lee, T. (1968) Urban neighbourhood as a socio-spatial schema, *Human Relations* **21**, 241–67.

Lee, T.R. (1970) Perceived distance as a function of direction in the city, *Environment and Behaviour* **2**, 40–51.

Lee, T.R. (1976) *Psychology and the Environment*, Methuen, London.

Lee, Y., Schmidt, C.G. (1988) Evolution of urban spatial cognition: patterns of change in Guangzhou, China, *Environment and Planning A* **20**, 339–51.

Leff, H.L., Gordon, L.R. (1979) Environmental cognitive sets: a longitudinal study, *Environment and Behaviour* **11**, 291–327.

Leff, H.L., Gordon, L.R., Ferguson, J.G. (1974) Cognitive set and environmental awareness, *Environment and Behaviour* **6**, 395–447.

Leigh, R., North, D. (1975) A framework for the study of the spatial aspects of acquisition activity in manufacturing industry. In: Massey, D., Morrison, W.I. (eds) *Industrial Location: alternative frameworks*, Centre for Environmental Studies Conference Paper No. **15**, London, pp. 49–78.

Leiper, N. (1990) *Tourism Systems*, Department of Management Studies Occasional Paper **2**, Massey University, Palmerston North, New Zealand.

Leiser, D., Zilbershatz, A. (1989) The Traveller: a computational model of spatial network learning, *Environment and Behaviour* **21**, 435–63.

Lenntorp, B. (1978) A time-geographic simulation model of individual activity programmes. In: Carlstein, T., Parker, D., Thrift, N.J. (eds) *Timing Space and Spacing Time*, Vol. 2 *Human Activity and Time Geography*, Arnold, London, pp. 162–80.

Lentiner, B., Marwitz, M., Narula, S.C. (1988) The optimal market area for a single (store) region: new wine in an old bottle, *Papers of the Regional Science Association* **65**, 167–80.

Lenz-Romeiss, F. (1973) *The City: new town or home town?* (trans. E. Kustner and J.A. Underwood), Pall Mall, London.

Lever, W.F. (1985) Theory and methodology in industrial geography. In: Pacione, M. (ed.) *Progress in Industrial Geography*, Croom Helm, London, pp. 10–39.

Lévi-Strauss, C. (1963) *Structural Anthropology* (trans. Jacobson, C., Schoepf, B.G.), Basic Books, New York.

Lévi-Strauss, C. (1969) *The Elementary Structures of Kinship* (trans. Bell, J.H., von Stumer, J.R., Needham, R.), Beacon Press, Boston.

Lévi-Strauss, C. (1973) *Totemism* (trans. Needham, R.), Penguin, Harmondsworth.

Levy, L., Herzog, A.N. (1974) Effects of population density and crowding on health and social adaptation in the Netherlands, *Journal of Health and Social Behaviour* **15**, 228–40.

Lew, A. (1987) A framework of tourist attraction research, *Annals of Tourism Research* **14**, 533–75.

Lewin, K. (1936) *Principles of Topological Psychology*, McGraw–Hill, New York.

Lewin, K. (1951) *Field Theory in Social Science: selected theoretical papers* (ed. by D. Cartwright), Harper, New York.

Lewis, D.A., Maxfield, M.G. (1980) Fear in the neighbourhoods: an investigation of the impact of crime, *Journal of Research in Crime and Delinquency* **17**, 160–89.

Lewis, G.J. (1979) *Rural Communities: a social geography*, David & Charles, Newton Abbot.

Lewis, G.J. (1981) Urban neighbourhood and community: exploratory studies in Leicester. In: Turnock, D. (ed.) *Leicester Geographical Studies*, Occasional Paper No. 1, Department of Geography, University of Leicester, pp. 37–52.

Lewis, G.J. (1982) *Human Migration: a geographical perspective*, Croom Helm, London.

Lewis, G.J. (1986) Welsh rural communities: retrospect and prospect, *Cambria* **13**, 27–40.

Lewis G.J., Davies, W.K.D. (1974) The social patterning of a British city: the case of Leicester, *Tijdschrift voor Economische en Sociale Geographie* **65**, 194–207.

Lewis, P.F. (1979) Axioms for reading the landscape: some guides to the American scene. In: Meinig, D.W. (ed.) *The Interpretation of Ordinary Landscapes*, Oxford University Press, New York, pp. 11–32.

Lewis, S., Lyon, L. (1986) The quality of community and the quality of life, *Social Spectrum* **6**, 397–410.

Ley, D. (1974) *The Black Inner City as Frontier Outpost: images and behaviour of a Philadephia neighbourhood*, Association of American Geographers Monograph Series No. 7, Washington.

Ley, D. (1977) Social geography and the taken-for-granted world, *Transactions of the Institute of British Geographers* **2**, 498–512.

Ley, D. (1978) Social geography and social action. In: Ley, D., Samuels, M.S. (eds) *Humanistic Geography: prospects and problems*, Maaroufa Press, Chicago, pp. 41–57.

Ley, D. (1980) *Geography without Man: a humanistic critique*, Oxford University, Department of Geography Research Paper No. 24.

Ley, D. (1981) Behavioural geography and the philosophies of meaning. In: Cox, K.R., Golledge, R.G. (eds) *Behavioural Problems in Geography Revisited*, Methuen, London, pp. 209–30.

Ley, D. (1983) *A Social Geography of the City*, Harper & Row, New York.

Ley, D. (1988) Interpretive social research in the inner city. In: Eyles, J. (ed.) *Research in Human Geography*, Basil Blackwell, Oxford, pp. 121–38.

Ley, D. (1989) Fragmentation, coherence and limits to theory. In: Kobyashi, A., MacKenzie, S. (eds) *Remaking Human Geography*, Unwin Hyman, London, pp. 227–44.

Ley, D., Cybriwsky, R. (1974) The spatial ecology of stripped cars, *Environment and Behaviour* **6**, 63–7.

Ley, D., Olds, K. (1988) Landscape as spectacle: world's fairs and the culture of heroic consumption, *Environment and Planning D: Society and Space* **6**, 191–212.

Ley, D., Samuels, M.S. (1978a) Introduction: contexts of modern humanism in geography. In: Ley, D., Samuels, M.S. (eds) *Humanistic Geography: prospects and problems*, Maaroufa Press, Chicago, pp. 1–17.

Ley, D., Samuels, M.S. (1978b) Epistemological orientations. In: Ley, D., Samuels, M.S. (eds) *Humanistic Geography: prospects and problems*, Maaroufa Press, Chicago, pp. 19–21.

Ley, D., Samuels, M.S. (1978c) Methodological implications. In: Ley, D., Samuels, M.S. (eds) *Humanistic Geography: prospects and problems*, Maaroufa Press, Chicago, pp. 121–2.

Lieber, S.R. (1976) A comparison of metric and nonmetric scaling models in preference research. In: Golledge, R.G., Rushton, G. (eds) *Spatial Choice and Spatial Behaviour*, Ohio State University Press, Columbia, pp. 191–210.

Lieber, S.R. (1979) An experimental approach for the migration decision process, *Tijdschrift voor Economische en Sociale Geografie* **70**, 75–85.

Lighter, D.T. (1985) Racial concentration and segregation across US counties, 1950–1980, *Demography* **22**, 603–9.

Linde, M.A.J., Dieleman, F.M., Clark, W.A.V. (1986) Starters in the Dutch housing market, *Tijdschrift voor Economische en Sociale Geografie* **77**, 243–50.

Linge, G.J.R. (1988) Peripheralization and industrial change. In: Linge, G.J.R. (ed.) *Peripheralization and Industrial Change: impacts on nations, regions, firms and people*, Croom Helm, London, pp. 1–21.

Lipsey, R.G. (1975) *An Introduction to Positive Economics* (4th edn), Weidenfeld & Nicolson: London.

Litton, R.B. (1972) Aesthetic dimensions of the landscape. In: Krutilla, J.V. (ed.) *Natural Environments*, Johns Hopkins University Press, Baltimore, pp. 262–91.

Liu, B.C. (1975) *Quality of Life Indicators in United States Metropolitan Areas, 1970*, US Environmental Protection Agency, Washington.

Livingstone, D.N., Harrison R.T. (1981) Immanuel Kant, subjectivism, and human geography: a preliminary investigation, *Transactions of the Institute of British Geographers* **6**, 359–74.

Lloyd, P.E., Dicken, P. (1972) *Location in Space: a theoretical approach to geography*, Harper & Row, London.

Lloyd, R. (1989) Cognitive maps: encoding and decoding information, *Annals of the Association of American Geographers* **79**, 101–24.

Lloyd, R., Heivly, C. (1987) Systematic distortions in urban cognitive maps, *Annals of the Association of American Geographers* **77**, 191–207.

Lloyd, R., Hooper, H. (1991) Urban cognitive maps: computation and structure, *Professional Geographer* **43**, 15–28.

Lloyd, R., Jennings, D. (1978) Shopping behaviour and income: comparisons in an urban environment, *Economic Geography* **54**, 157–68.

Long, P.T., Perdine, R.R., Allen, L. (1990) Rural resident tourism perceptions and attitudes by community level of tourism, *Journal of Travel Research* **28**, 3–9.

Loo, C. (1986) Neighbourhood satisfaction and safety: a study of low-income ethnic areas, *Environment and Behaviour* **18**, 109–31.

Lord, J.D. (1980) Intra-metropolitan area comparisons in retailing productivity: implications for central business district retail firms, *Annals of Regional Science* **14**, 95–105.

Lösch, A. (1954) *The Economics of Location* (trans. W.H. Woglom and W.F. Stolper), Yale University Press, New Haven.

Louviere, J.J. (1973) The dimensions of alternatives in spatial choice processes: a comment, *Geographical Analysis* **5**, 315–25.

Louviere, J.J. (1976) Information-processing theory and functional form in spatial behaviour. In: Golledge, R.G., Rushton, G. (eds) *Spatial Choice and Spatial Behaviour*, Ohio State University Press, Columbus, pp. 211–48.

Louviere, J.J. (1981) A conceptual and analytical framework for understanding spatial and travel choices, *Economic Geography* **57**, 304–14.

Louviere, J., Timmermans, H. (1990a) A review of recent advances in decompositional preference and choice models, *Tijdschrift voor Economische en Sociale Geografie* **81**, 214–24.

Louviere, J., Timmermans, H. (1990b) Using hierarchical

information integration to model consumer responses to possible planning actions: recreation destination choice illustration, *Environment and Planning A* **22**, 291–308.

Lovering, J. (1990) Fordism's unknown successor: a comment on Scott's theory of capital accumulation and the re-emergence of regional economies, *International Journal of Urban and Regional Research* **14**, 159–74.

Lowenthal, D. (1961) Geography, experience, and imagination: towards a geographical epistemology, *Annals of the Association of American Geographers* **51**, 241–60.

Lowenthal, D. (1972) Research in environmental perception and behaviour: perspectives on current problems, *Environment and Behaviour* **4**, 333–42.

Lowenthal, D. (1975) Past time, present place: landscape and memory, *Geographical Review* **65**, 1–36.

Lowenthal, D. (1982) Revisiting valued landscapes. In: Gold, J., Burgess, J. (eds) *Valued Environments*, Allen & Unwin, London, pp. 74–99.

Lowenthal, D., Prince, H.C. (1964) The English landscape, *Geographical Review* **54**, 309–46.

Lowenthal, D., Prince, H.C. (1965) English landscape tastes, *Geographical Review* **55**, 186–222.

Lowerson, J., Myerscough, J. (1977) *Time and Space in Victorian England*, Harvester Press, Hassocks.

Lowrey, R.A. (1973) A method for analysing distance concepts of urban residents. In: Downs, R., Stea, D. (eds) *Image and Environment*, Aldine, Chicago, pp. 338–60.

Lynch, K. (1960) *The Image of the City*, MIT Press, Cambridge, Mass.

Lynch, K. (1972) *What Time is This Place*? MIT Press, Cambridge, Mass.

Lynch, K. (1976) Foreword. In: Moore, G.T., Golledge, R.G. (eds) *Environmental Knowing*, Dowden, Hutchinson & Ross, Stroudsburg, pp. v–viii.

Lyon, L. (1987) *The Community in Urban Society*, Dorsey Press, Chicago.

Lyons, J. (1977) *Chomsky*, Fontana, London.

Mabogunje, A.K. (1970) Systems approach to a theory of rural–urban migration, *Geographical Analysis* **2**, 1–18.

MacCannell, D. (1973) Staged authenticity: arrangements of social space in tourist settings, *American Journal of Sociology* **79**, 589–603.

MacCannell, D. (1976) *The Tourist: a new theory of the leisure class*, Schocken Books, New York.

McDaniel, R. (1975) How information structures influence spatial organization. In: Abler, R., Janeke, D., Philbrick, A., Sommer, J. (eds) *Human Geography in a Shrinking World*, Duxbury, North Scituate, pp. 57–66.

McDermott, P.J., Taylor, M.J. (1976) Attitudes, images and location: the subjective context of decision-making in New Zealand manufacturing, *Economic Geography* **52**, 325–46.

McDermott, P., Taylor, M. (1982) *Industrial Organization and Location*, Cambridge University Press, Cambridge.

McDermott, R. (1975) *Towards an Embodied Map of Urban Neighbourhoods*, Anthropological Society, Washington.

McGee, M.G. (1979) Human spatial abilities: psychometric studies and environmental, genetic, hormonal, and neurological influences, *Psychological Bulletin* **86**, 889–918.

McGill, W., Korn, J.H. (1982) Awareness of an urban environment, *Environment and Behaviour* **14**, 186–201.

McHarg, I. (1969) *Design with Nature*, Natural History Press, New York.

McHugh, K.E. (1984) Explaining migration intentions and destination selection, *Professional Geographer* **36**, 315–25.

Mackay, D.B., Olshausky, R.W., Sentell, G. (1975) Cognitive maps and spatial behaviour of consumers, *Geographical Analysis* **7**, 19–34.

Mackay, J.B. (1976) The effect of spatial stimuli on the estimation of cognitive maps, *Geographical Analysis* **8**, 439–52.

McKay, J., Whitelaw, J.S. (1977) The role of large private and government organizations in generating flows of inter-regional migrants: the case of Australia, *Economic Geography* **53**, 28–44.

McKechnie, G.E. (1977) The environmental response inventory in application, *Environment and Behaviour* **9**, 255–76.

McKechnie, S. (1989) quoted in Utting, D. Homelessness, *Sunday Correspondent*, 12 November, p. 12.

MacLennan, D., Williams, N.J. (1979) Revealed space preference theory – a cautionary note, *Tijdschrift voor Economische en Sociale Geographie* **70**, 307–9.

McNee, R.B. (1960) Towards a more humanistic economic geography: the geography of enterprise, *Tijdschrift voor Economische en Sociale Geographie* **50**, 201–6.

McNee, R.B. (1974) A systems approach to understanding the geographic behaviour of organizations, especially large corporations. In: Hamilton, F.E.I. (ed.) *Spatial Perspectives on Industrial Organization and Decision-making*, Wiley, London, pp. 47–76.

McNeil, E.B. (1970) *The Psychoses*, Prentice-Hall, Englewood Cliffs.

Madden, J.F., Chen Chui, L-I. (1990) The wage effects of residential location and commuting contrasts on employed married women, *Urban Studies* **27**, 353–69.

Magnusson, D. (ed.) (1981) *Towards a Psychology of Situations: an interactional perspective*, Erlbaum Associates, Hillsdale, N.J.

Maher, C.A. (1984) *Residential Mobility within Australian cities*, Australian Bureau of Statistics, Canberra.

Maher, C.A., Mercer, D.C. (1984) Retail services and the role of the state: convenience stores and the Planning Appeals Board in Melbourne, *Environment and Planning A* **16**, 733–48.

Mallory, W.E., Simpson-Housley, P. (eds) (1989) *Geography and Literature: a meeting of the disciplines*, Syracuse University Press, Syracuse.

Malmberg, T. (1980) *Human Territoriality*, Mouton, The Hague.

Mangalam, J.J. (1968) *Human Migration*, University of Kentucky Press, Lexington.

Mann, M. (1973) *Workers on the Move*, Cambridge University Press, London.

Marans, R.W., Rodgers, W. (1975) Towards an understanding of community satisfaction. In: Hawley, A.H., Roch, V.P. (eds) *Metropolitan America in Contemporary Perspective*, Halstead Press, New York, pp. 89–101.

March, J.G., Simon, H.A. (1958) *Organizations*, Wiley, New York.

Marchand, B. (1972) Information theory and geography, *Geographical Analysis* **4**, 234–58.

Marshall, J.N. (1979) Ownership, organization and industrial linkage: a case study in the Northern Region of England, *Regional Studies* **13**, 531–57.

Marshall, J.N. (1982) Organizational theory and industrial location, *Environment and Planning A* **14**, 1667–83.

Mårtensson, A. (1977) Childhood interaction and temporal organization, *Economic Geography* **53**, 99–125.

Martin, A.E. (1967) Environment, housing and health, *Urban Studies* **4**, 1–21.

Martin, J.E. (1981) Location theory and spatial analysis, *Progress in Human Geography* **5**, 258–62.

Maslow, A. (1954) *Motivation and Personality*, Harper & Row, New York.

Mason, C.M., Harrison, R.T. (1985) The geography of small firms in the UK, *Progress in Human Geography* **9**, 1–37.

Massam, B.H., Bouchard, D. (1976) A comparison of observed and hypothetical choice behaviour, *Environment and Behaviour* **8**, 367–74.

Massey, D. (1975a) Is the behavioural approach really an alternative?. In: Massey, D., Morrison, W.I. (eds) *Industrial Location: alternative frameworks*, Centre for Environmental Studies Conference Paper No. **15**, London, pp. 79–86.

Massey, D. (1975b) Behavioural research, *Area* **7**, 201–3.

Massey, D. (1979) A critical evaluation of industrial-location theory. In: Hamilton, F.E.I., Linge, G.J.R. (eds) *Spatial Analysis, Industry and the Industrial Envirorment: progress in research and applications*, Vol. 1 *Industrial Systems*, Wiley, Chichester, pp. 57–72.

Massey, D. (1984) *Spatial Divisions of Labour: social structures and the geography of production*, Methuen, New York.

Massey, D. (1988) What's happening to UK manufacturing? in Allen, J., Massey, D. (eds) *Restructuring Britain: the economy in question*, Sage, London, pp. 46–90.

Massey, D. (1991) Flexible sexism, *Environment and Planning D: Society and Space* **9**, 31–57.

Massey, D., Meegan, R.A. (1979) The geography of industrial reorganization: the spatial effects of restructuring of the electrical engineering sector under the Industrial Reorganization Corporation, *Progress in Planning* **10**, 155–237.

Massey, D., Meegan, R. (1982) *The Anatomy of Job Loss*, Methuen, London.

Mathieson, A., Wall, G. (1982) *Tourism: economic, physical and social impacts*, Longman, London.

Matrix (1984) *Making Space: women and the man-made environment*, Pluto Press, London.

Matthews, F. (1977) *Quest for an American Sociology: Robert Park and the Chicago School*, McGill-Queens University Press, Montreal.

Matthews, M.H. (1981) Children's perception of urban distance, *Area* **13**, 333–43.

Matthews, M.H. (1984a) Cognitive maps: a comparison of graphic and iconic techniques, *Area* **16**, 33–40.

Matthews, M.H. (1984b) Environmental cognition in young children: images of journey-to-school and home area, *Transactions of the Institute of British Geographers* **9**, 89–105.

Matthews, M.H. (1987) Gender, home range and environmental cognition, *Transactions of the Institute of British Geographers* **12**, 43–56.

Matthews, M.H. (1992) *Making Sense of Place: children's understanding of large-scale environments*, Harvester Wheatsheaf, Hemel Hempstead.

Maurer, R., Baxter, J.C. (1972) Images of the neighbourhood and city among Black–, Anglo–, and Mexican–American children, *Environment and Behaviour* **4**, 351–88.

Maw, R., Cosgrove, D. (1972) *Assessment of Demand for Recreation – Modelling Approach*, Polytechnic of Central London, Leisure Model Unit Working Paper No. 2.

Mawby, R.I. (1977) Defensible space: a theoretical and empirical appraisal, *Urban Studies* **14**, 169–79.

May, K.O. (1954) Transitivity, utility and aggregation in preference patterns, *Econometrica* **22**, 1–13.

Meegan, R. (1988) A crisis of mass production? in Allen, J., Massey, D. (eds) *Restructuring Britain: the economy in question*, Sage, London, pp. 135–83.

Meinig, D.W. (ed.) (1979a) *The Interpretation of Ordinary Landscapes*, Oxford University Press, New York.

Meinig, D.W. (1979b) Introduction. In: Meinig, D.W. (ed.) *The Interpretation of Ordinary Landscapes*, Oxford University Press, New York, pp. 1–10.

Meinig, D.W. (1979c) The beholding eye: ten versions of the same scene. In: Meinig, D.W. (ed.) *The Interpretation of Ordinary Landscapes*, Oxford University Press, New York, pp. 33–48.

Menchik, M. (1972) Residential environment preference and choice: empirically validating preference measures, *Environment and Planning A* **4**, 455–8.

Mercer, C. (1975) *Living in Cities: psychology and the urban environment*, Penguin, Harmondsworth.

Mercer, D. (1971) Discretionary travel behaviour and the urban mental map, *Australian Geographical Studies* **9**, 133–43.

Mercer, D. (1972) Behavioural geography and the sociology of social action, *Area* **4**, 48–52.
Mercer, D. (1976) Motivational and social aspects of recreation behaviour. In: Altman, I., Wohlwill, J.F. (eds) *Human Behaviour and Environment*, Plenum Press, New York, pp. 123–61.
Mercer, D. (ed.) (1977) *Leisure and Recreation in Australia*, Sorrett, Melbourne.
Mercer, D. (1980) *In Pursuit of Leisure*, Sorrett, Melbourne.
Mercer, D. (ed.) (1981) *Outdoor Recreation – Australian Perspectives*, Sorrett, Melbourne.
Merry, S.E. (1989) Crowding, conflict and neighbourhood regulation. In: Altman, I., Wandserman, A. (eds) *Neighbourhood and Community Environments*, Plenum Press, New York, pp. 36–68.
Meyer, G. (1977) Distance perception of consumers in shopping streets, *Tijdschrift voor Economische en Social Geografie* **68**, 355–61.
Meyer, R. (1980) A descriptive model of constrained residential search, *Geographical Analysis* **12**, 21–32.
Meyer, R., Speare, A. (1985) Distinctively elderly migration: types and determinants, *Economic Geography* **61**, 79–88.
Michelson, W. (1977) *Environmental Choice, Human Behaviour, and Residential Satisfaction*, Oxford University Press, New York.
Milgram, S. (1973) Introduction to Chapter II. In: Ittelson, W.H. (ed.) *Environment and Cognition*, Seminar Press, New York, pp. 21–8.
Miller, G.A. (1956) The magical number seven, plus or minus two: some limits on our capacity for processing information, *Psychological Review* **63**, 81–97.
Miller, G.A., Galanter, E., Pribam, K.H. (1960) *Plans and the Structure of Behaviour*, Holt, New York.
Miller, V.P., Quigley, J.M. (1990) Segregation by racial and demographic group: evidence from the San Francisco Bay Area, *Urban Studies* **27**, 3–21.
Mills, C.W. (1970) *The Sociological Imagination*, Penguin, Harmondsworth.
Mitchell, A. (1983) *The Nine American Life-Styles*, Macmillan, New York.
Mitchell, J.C. (ed.) (1969) *Social Networks in Urban Situations*, Manchester University Press, Manchester.
Mitchell, J.C. (1983) Case and situation analysis, *Sociological Review* **31**, 187–211.
Mitchell, J.K. (1984) Hazard perception studies: convergent concerns and divergent approaches during the past decade. In: Saarinen, T.F., Seamon, D., Sell, J.L. (eds) *Environmental Perception and Behaviour: an inventory and prospect*, University of Chicago, Department of Geography Research Paper No. 209, Chicago, pp. 33–59.
Moeser, S.D. (1988) Cognitive mapping in a complex building, *Environment and Behaviour* **20**, 21–49.
Montague, M.F.A. (ed.) (1968) *Man and Aggression*, Oxford University Press, London.
Mooney, E.A. (1991) Access to housing in rural Scotland. In: Lewis, G.J., Sherwood, K.B. (eds) *Rural Mobility and Housing*, Occasional Paper, Department of Geography, University of Leicester.
Moore, G.T. (1975) Spatial relations ability and developmental levels of urban cognitive mapping: a research note, *Man–Environment Systems* **5**, 247–8.
Moore, G.T. (1976) Theory and research on the development of environmental knowing. In: Moore, G.T., Golledge, R.G. (eds) *Environmental Knowing*, Dowden, Hutchinson & Ross, Stroudsburg, pp. 138–64.
Moore, G.T. (1979) Knowing about environmental knowing: the current state of theory and research on environmental cognition, *Environment and Behaviour* **11**, 33–70.
Moore, G.T., Golledge, R.G. (1976a) Preface. In: Moore, G.T., Golledge, R.G. (eds) *Environmental Knowing*, Dowden, Hutchinson & Ross, Stroudsburg, pp. xi–xvi.
Moore, G.T., Golledge, R.G. (1976b) Environmental knowing: concepts and theories. In: Moore, G.T., Golledge, R.G. (eds) *Environmental Knowing*, Dowden, Hutchinson & Ross, Stroudsburg, pp. 3–24.
Moore, P.G., Thomas, H. (1976) *The Anatomy of Decisions*, Penguin, Harmondsworth.
Moore, P.G., Thomas, H., Bunn, D.W., Hampton, J.M. (1976) *Case Studies in Decision Analysis*, Penguin, Harmondsworth.
Moore, R.C. (1986) *Childhood's Domain: play and place in child development*, Croom Helm, London.
Moos, A.I., Dear, M.J. (1986) Structuration theory in urban analysis. 1: theoretical exegesis, *Environment and Planning A* **18**, 231–52.
Morgan, B.S. (1976) The basis of family status segregation: a case study of Exeter, *Transactions of the Institute of British Geographers* **1**, 83–107.
Morgan, M.A. (1975) Values and political geography. In: Peel, R., Chisholm, M., Haggett, P. (eds) *Processes in Physical and Human Geography*, Heinemann, London, pp. 286–97.
Morrill, R.L. (1965a) *Migration and the Spread and Growth of Urban Settlement*, Lund Studies in Geography No. 26, Lund.
Morrill, R.L. (1965b) The negro ghetto: problems and alternatives, *Geographical Review* **55**, 339–61.
Morrill, R.L. (1965c) Expansion of the urban fringe: a simulation experiment, *Papers of the Regional Science Association* **15**, 185–202.
Morrill, R.L. (1968) Waves of spatial diffusion, *Journal of Regional Science* **8**, 1–19.
Morrill, R.L., Manninen, D. (1975) Critical parameters of spatial diffusion processes, *Economic Geography* **51**, 269–77.
Morrill, R., Downing, J., Leon, W. (1986) Attribute preferences and the non–metropolitan migration decision (USA), *Annals of Regional Science* **20**, 33–53.
Morris, D. (1968) *The Naked Ape*, Corgi, London.

Morris, J. (1988a) *Industrial Restructuring and Region: multinationals and the branch plant economy*, Wheatsheaf, Brighton.

Morris, J. (1988b) Research in industrial geography: a theoretical critique, *Transactions of the Institute of British Geographers* **13**, 337–44.

Moses, L.N. (1958) Location and the theory of production, *Quarterly Journal of Economics* **72**, 259–72.

Moulaert, F., Swyndedonw, E., Wilson, P. (1988) Spatial responses to Fordist and post–Fordist accumulation and regulation, *Papers and Proceedings of the Regional Science Association* **64**, 11–23.

Moulden, M., Bradford, M.G. (1984) Influences on educational attainment: the importance of the local residential environment, *Environment and Planning A* **16**, 49–66.

Mounfield, P.R., Guy, K., Unwin, D.J. (1982) *Processes of Change in the Footwear Industry of the East Midlands*, Leicester, University of Leicester, Department of Geography.

Muller, P.O. (1976) *The Outer City*, Association of American Geographers, Commission on College Geography Resource Paper No. 22, Washington.

Munroe, R.H., Munroe, R.L., Brasher, A. (1984) Precursors of spatial ability: a longitudinal study among the Logoli of Kenya, *Journal of Social Psychology* **125**, 23–33.

Murdie, R.A. (1969) *The Factorial Ecology of Metropolitan Toronto 1951 and 1961*, University of Chicago, Geography Research Paper 116.

Murie, A. (1983) *Housing Inequality and Deprivation*, Heinemann, London.

Murie, A. (1986) Social differentiation in urban areas: housing or occupational class at work? *Tidjschrift voor Economische en Sociale Geografie* **77**, 345–57.

Murphy, P.E. (1985) *Tourism: a community approach*, Methuen, London.

Murphy, P.E., Rosenblood, L. (1974) Tourism: an exercise in spatial search, *Canadian Geographer* **18**, 201–10.

Murray, D., Spencer, C.P. (1979) Individual differences in the drawing of cognitive maps: the effects of geographical mobility, strength of mental imagery and basic graphic ability, *Transactions of the Institute of British Geographers* **4**, 385–91.

Musterd, S. (1991) Neighbourhood change in Amsterdam, *Tijdschrift voor Economische en Sociale Geografie* **82**, 30–9.

Newman, 0. (1972) *Defensible Space*, Macmillan, New York.

New South Wales State Pollution Control Commission (1978) *Recreational Use in Botany Bay*, Government Printer, Sydney.

Niner, P. (1975) *Local Authority Housing Policy and Practice*, Centre for Urban and Regional Studies, Occasional Paper **31**, University of Birmingham.

Nisbet, R.A. (1970) *The Sociological Tradition*, Heinemann, London.

Norberg-Schultz, C. (1971) *Existence, Space and Architecture*, Studio Vista, London.

Norberg-Schultz, C. (1980) *Genius Loci: towards a phenomenology of architecture*, Rizzoli, New York.

Norcliffe, G.B. (1974) Territorial influences in urban political space: a study of perception in Kitchener-Waterloo, *Canadian Geographer* **18**, 311–29.

North, D.J. (1974) The process of locational change in different manufacturing organizations. In: Hamilton, F.E.I. (ed.) *Spatial Perspectives on Industrial Organization and Decision-making*, Wiley, London, pp. 213–44.

O'Keefe, J., Nadel, L. (1978) *The Hippocampus as a Cognitive Map*, Oxford University Press, London.

Olsson, G. (1965) *Distance and Human Interaction*, Regional Science Research Institute, Philadelphia.

Olsson, G. (1969) Inference problems in location analysis. In: Cox, K.R., Golledge, R.G. (eds) *Behavioural Problems in Geography: a symposium*, Northwestern Studies in Geography **17**, Evanston, pp. 14–34.

Olsson, G. (1970) Logics and social engineering, *Geographical Analysis* **2**, 361–95.

Onibokun, A.G. (1976) Social system correlates of residential satisfaction, *Environment and Behaviour* **8**, 323–44.

Ontario Research Council on Leisure (1977) *Analysis Methods and Techniques for Recreation Research and Leisure Studies*, Ontario Research Council on Leisure, Ottawa.

Opacic, S., Potter, R.B. (1986) Grocery store cognitions of disadvantaged consumer groups: a Reading case study, *Tijdschrift voor Economische en Sociale Geografie* **77**, 288–98.

Openshaw, S. (1973) Insoluble problems in shopping model calibration when trip pattern is not known, *Regional Studies* **7**, 367–71.

Openshaw, S. (1984) Ecological fallacies and the analysis of areal census data, *Environment and Planning A* **16**, 17–31.

Orleans, P. (1973) Differential cognition of urban residents: effects of social scale on mapping. In: Downs, R., Stea, D. (eds) *Image and Environment*, Aldine, Chicago, pp. 115–30.

Ormrod, R.K. (1990) Local content and innovation diffusion in a well-connected world, *Economic Geography* **66**, 109–22.

Ornstein, R.E. (1972) *The Psychology of Consciousness*, Freeman, New York.

Osgood, C.E. (1957) A behaviouralistic analysis of perception and language as cognitive phenomena. In: *Contemporary Approaches to Cognition: a symposium at the University of Colorado*, Harvard University Press, Cambridge, Mass., pp. 75–118.

Ossowicz, T., Slawski, J. (1989) The allocation model ORION: its development and application, *Papers of the Regional Science Association* **66**, 31–46.

Outdoor Recreation Resources Review Commission (1962) *Outdoor Recreation for America*, Government Printing Office, Washington.

Owens, P.L. (1984) Rural leisure and recreation research: a retrospective evaluation, *Progress in Human Geography* **8**, 157–88.

Oxley, D., Haggard, L.M., Werner, C.M., Altman, I. (1986) Transactional qualities of neighbourhood social networks: a case study of 'Christmas Street', *Environment and Behaviour* **18**, 640–77.

Pacione, M. (1975) Preference and perception: an analysis of consumer behaviour, *Tijdschrift voor Economische en Sociale Geografie* **66**, 84–92.

Pacione, M. (1976) Shape and structure in cognitive maps of Great Britain, *Regional Studies* **10**, 275–83.

Pacione, M. (1978) Information and morphology in cognitive maps, *Transactions of the Institute of British Geographers* **3**, 548–68.

Pacione, M. (1982a) Space preferences, locational decisions, and the dispersal of civil servants from London, *Environment and Planning A* **14**, 323–33.

Pacione, M. (1982b) The use of objective and subjective measures of life quality in human geography, *Progress in Human Geography* **6**, 495–514.

Pacione, M. (1990a) The ecclesiastical community of interest as a response to urban poverty and deprivation, *Transactions of the Institute of British Geographers* **15**, 193–204.

Pacione, M. (1990b) A tale of two cities: the migration of the urban crisis in Glasgow, *Cities*, November, 304–14.

Pahl, R.E. (1966) The rural–urban continuum, *Sociologia Ruralis* **6**, 314–29.

Pahl, R.E. (1968) Is mobile society a myth? *New Society*, 11 January, 46–8.

Pahl, R.L. (1977) Managers technical experts and the state: forms of mediation, manipulation and dominance in urban and regional development. In: Harloe, M. (ed.) *Captive Cities: studies in the political economy of cities and regions*, Wiley, New York, pp. 73–86.

Pahl, R.E. (1984) *Divisions of Labour*, Basil Blackwell, Oxford.

Palm, R. (1976a) The role of real estate agents as information mediators in two American cities, *Geografiska Annaler* **58B**, 28–41.

Palm, R. (1976b) Real estate agents and geographical information, *Geographical Review* **66**, 266–280.

Palm, R. (1985) Ethnic segmentation of real estate agent practice in urban housing markets, *Annals of the Association of American Geographers* **75**, 58–68.

Park, R.E. (1916) The city: suggestions for the investigation of human behaviour in the urban environment, *American Journal of Sociology* **20**, 577–612.

Park, R.E. (1929) The city as a laboratory. In: Smith, T., White, L. (eds) *Chicago: an experiment in social science research*, University of Chicago Press, Chicago, pp. 24–37.

Park, R.E. (1936) Human ecology, *American Journal of Sociology* **20**, 577–612.

Park, R.E., Burgess, E.W., McKenzie, R.D. (1925) *The City*, University of Chicago Press, Chicago.

Parker, S. (1976) *The Sociology of Leisure*, International Publishing Service, New York.

Parker, S. (1983) *Leisure and Work*, Allen & Unwin, London.

Parkes, D. and Thrift, N. (1978) Putting time in its place. In: Carlstein, T., Parker, D., Thrift, N.J. (eds) *Timing Space and Spacing Time*, Vol. 1 *Making Sense of Time*, Arnold, London, pp. 119–29.

Parkes, D., Thrift, N. (1980) *Times, Spaces and Places: a chronogeographic perspective*, Wiley, New York.

Parkes, D., Wallis, W.D. (1978) Graph theory and the study of activity structure. In: Carlstein, T., Parker, D., Thrift, N.J. (eds) *Timing Space and Spacing Time*, Vo1. 2 *Human Activity and Time Geography*, Arnold, London, pp. 75–99.

Parkin, F. (1979) *Marxism and Class Theory: a bourgeois critique*, Tavistock, London.

Passini, R., Proulx, G. (1988) Wayfinding without vision: an experiment with congenitally totally blind people, *Environment and Behaviour* **20**, 227–52.

Passini, R., Proux, G., Rainville, C. (1990) The spatio-cognitive abilities of the visually impaired population, *Environment and Behaviour* **22**, 91–118.

Patmore, J.A. (1970) *Land and Leisure*, David & Charles, Newton Abbot.

Patmore, J.A., Collins, M.F. (1980) Recreation and leisure, *Progress in Human Geography* **4**, 91–7.

Patricios, N.N. (1978) An agentive model of person–environment relations, *International Journal of Environmental Studies* **13**, 43–52.

Pearce, D. (1987) *Tourism Today: a geographical analysis*, Longman, London.

Pearce, P.L. (1977) Mental souvenirs: a study of tourists and their city maps, *Australian Journal of Psychology* **29**, 203–10.

Pearce, P.L. (1981) Route maps: a study of travellers' perceptions of a section of countryside, *Journal of Environmental Psychology* **1**, 141–55.

Pearce, P.L. (1982a) *The Social Psychology of Tourist Behaviour*, Pergamon, Oxford.

Pearce, P.L. (1982b) Perceived changes in holiday destinations, *Annals of Tourism Research* **9**, 145–64.

Pearce, P.L. (1988) *The Ulysses Factor: evaluating visitors in tourist settings*, Springer-Verlag, New York.

Pearce, P.L., Moscardo, G.M. (1986) The concept of authenticity in tourist experiences, *Australia and New Zealand Journal of Sociology* **22**, 121–32.

Peck, J.A. (1989) Reconceptualizing the local labour market: space, segmentation and the state, *Progress in Human Geography* **13**, 42–61.

Peet, R. (1975) Inequality and poverty: a Marxist–geographic theory, *Annals of the Association of American Geographers* **65**, 564–71.

Peet, R. (1977) The development of radical human geography in the United States, *Progress in Human Geography* **1**, 240–63.

Peet, R.J., Lyons, J.V. (1981) Marxism: dialectical materialism, social formation and the geographic relations. In: Harvey,

M.E., Holly, B.P. (eds) *Themes in Geographic Thought*, Croom Helm, London, pp. 187–205.

Peet, R., Thrift, N. (eds) (1989) *New Models in Geography: the political-economy perspective*, Unwin Hyman, London.

Perry, C. (1929) The neighbourhood unit formula. In: Whenton, W.C.L., Milgram, G., Henson, M.E. (eds), *Urban Housing*, Free Press, New York, pp. 36–43.

Phelps, A. (1986) Holiday destination image – the problem of assessment: an example developed in Menorca, *Tourism Management* **7**, 168–79.

Philo, C. (1989) Thoughts, words, and 'creative locational acts'. In: Boal, F.W., Livingstone, D.N. (eds) *The Behavioural Environment*, Routledge, London, pp. 205–34.

Phipps, A.G. (1979) Scaling problems in the cognition of urban distances, *Transactions of the Institute of British Geographers* **4**, 94–102.

Phipps, A.G. (1988) Rational versus heuristic decision-making during residential search, *Geographical Analysis* **20**, 231–48.

Phipps, A.G., Meyer, R.J. (1985) Narrative versus heuristic models of residential search behaviour: an empirical comparison, *Environment and Planning A* **17**, 761–76.

Piaget, J. (1970) *Structuralism* (trans. C. Maschler), Basic Books, New York.

Piaget, J. Inhelder, B. (1956) *The Child's Conception of Space*, Routledge & Kegan Paul, London.

Piche, D. (1981) The spontaneous geography of the urban world. In: Herbert, D.T., Johnston, R.J. (eds) *Geography and the Urban Environment*, Vol. 4, Wiley, Chichester, pp. 229–56.

Pick, H.L. (1976) Transactional–constructivist approach to environmental knowing: a commentary. In: Moore, G.T., Golledge, R.G. (eds) *Environmental Knowing*, Dowden, Hutchinson & Ross, Stroudsburg, pp. 185–6.

Pickles, J. (1985) *Phenomenology, Science and Geography*, Cambridge University Press, Cambridge.

Pickles, J. (1988) From fact-world to life-world: the phenomenological method and social science research. In: Eyles, J., Smith, D.M. (eds) *Qualitative Methods in Human Geography*, Polity Press, Cambridge, pp. 233–54.

Pierce, R.C. (1980) Dimensions of leisure, *Journal of Leisure Research* **12**, 5–19, 132–41, 273–84.

Pigram, J.J. (1983) *Outdoor Recreation and Resource Management*, Croom Helm, London.

Pile, S. (1990) Depth hermeneutics and critical human geography, *Environment and Planning D: Society and Space* **8**, 211–32.

Pinch, S.P. (1989) The restructuring thesis and the study of public services, *Environment and Planning A* **21**, 905–26.

Pinch, S.P., Mason, C.M., Witt., S.J.G. (1989) Labour flexibility and industrial restructuring in the UK 'Sunbelt': the case of Southampton, *Transactions of the Institute of British Geographers* **14**, 418–34.

Piore, M., Sabel, C.F. (1984) *The Second Industrial Divide: possibilities for prosperity*, Basic Books, New York.

Pipkin, J.S. (1981) Cognitive behavioural geography and repetitive travel. In: Cox, K.R., Golledge, R.G. (eds) *Behavioural Problems in Geography Revisited*, Methuen, London, pp. 145–81.

Pirie, G.H. (1976) Thoughts on revealed preferences and spatial behaviour, *Environment and Planning A* **8**, 947–55.

Pirie, G.H. (1988) Spatial choice behaviour reconsidered. In: Golledge, R.G., Timmermans, H. (eds) *Behavioural Modelling in Geography and Planning*, Croom Helm, London, pp. 91–5.

Pizam, A. (1978) Tourism's impacts: the social costs to the destination community as perceived by its residents, *Journal of Travel Research* **16**, 8–12.

Plog, S. (1973) Why destinations rise and fall in popularity, *Cornell Hotel and Restaurant Administration Quarterly* nos 13–16, 101–17.

Plog, S. (1990) A carpenter's tools: an answer to Smith's review of psychocentrism/allocentrism, *Journal of Travel Research* **28**, 43–5.

Pocock, D.C.D. (1973) Environmental perception: process and product, *Tijdschrift voor Economische en Sociale Geografie* **69**, 251–7.

Pocock, D.C.D. (1975) *Durham: image of a cathedral city*, University of Durham, Department of Geography Occasional Paper No. 6.

Pocock, D.C.D. (1976a) A comment on images derived from invitation-to-map exercises, *Professional Geographer* **28**, 161–6.

Pocock, D.C.D. (1976b) Some characteristics of mental maps: an empirical study, *Transactions of the Institute of British Geographers* **1**, 493–512.

Pocock, D.C.D. (1978) The cognition of intra-urban distance: a summary, *Scottish Geographical Magazine* **94**, 31–5.

Pocock, D.C.D. (1979) The novelist's image of the North, *Transactions of the Institute of British Geographers* **4**, 62–76.

Pocock, D.C.D. (1981a) Sight and knowledge, *Transactions of the Institute of British Geographers* **6**, 385–93.

Pocock, D.C.D. (1981b) Place and the novelist, *Transactions of the Institute of the British Geographers* **6**, 337–47.

Pocock, D. (ed.) (1981c) *Humanistic Geography and Literature*, Croom Helm, London.

Pocock, D.C.D. (1988) Geography and literature, *Progress in Human Geography* **12**, 87–102.

Pocock, D., Hudson, R. (1978) *Images of the Urban Environment*, Macmillan, London.

Popenoe, D. (1981) Women in the suburban environment: a US-Sweden comparison. In: Wekerle, G.R., *et al.* (eds) *New Space for Women*, Westview Press, Boulder, pp. 165–74.

Popp, H. (1976) The residential location decision process: some theoretical and empirical considerations, *Tijdschrift voor Economische en Sociale Geografie* **67**, 300–5.

Porteous, J.D. (1973) The Burnside gang: territoriality, social space and community planning, *Western Geographical Series* **5**, 130–48.

Porteous, J.D. (1976) Home: the territorial core, *Geographical Review* **66**, 383–90.

Porteous, J.D. (1977) *Environment and Behaviour: planning and everyday urban life*, Addison-Wesley, Reading, Mass.

Porteous, J.D. (1985) Smellscape, *Progress in Human Geography* **9**, 356–78.

Porteous, J.D. (1988) Topocide: the annihilation of place. In: Eyles, J., Smith, D.M. (eds) *Qualitative Methods in Human Geography*, Polity Press, Cambridge, pp. 75–93.

Potter, R.B. (1976) Directional bias within the usage and perceptual fields of urban consumers, *Psychological Reports* **38**, 988–90.

Potter, R.B. (1977a) The nature of consumer usage fields in an urban environment: theoretical and empirical perspectives, *Tijdschrift voor Economische en Sociale Geografie* **68**, 168–76.

Potter, R.B. (1977b) Spatial patterns of consumer behaviour and perception in relation to the social class variable, *Area* **9**, 153–6.

Potter, R.B. (1979) Perception of urban retailing facilities: an analysis of consumer information fields, *Geografiska Annaler* **61B**, 19–27.

Potter, R.B. (1982) *The Urban Retailing System*, Gower, Aldershot.

Potts, L. (1990) *The World Labour Market: a history of migration*, Zed Books, London.

Powell, J.M. (1978) *Mirrors of the New World: images and image- makers in the settlement process*, Australian National University Press, Canberra.

Pratt, G. (1989) Quantitative techniques and humanistic–historical materialist perspectives. In: Kobyashi, A., MacKenzie, S. (eds) *Remaking Human Geography*, Unwin Hyman, London, pp. 101–15.

Pred, A. (1967) *Behaviour and Location: foundations for a geographic dynamic location theory: Part 1*, Lund Studies in Geography, Series B, No. 27.

Pred, A.R. (1973a) The growth and development of systems of cities in advanced economies. In: Pred, A.R., Törnqvist, G.E. *Systems of Cities and Information Flows: two essays*, Lund Studies in Geography, Series B, No. 38, pp. 9–82.

Pred, A.R. (1973b) Urbanization, domestic planning problems and Swedish geographic research, *Progress in Geography* **5**, 1–76.

Pred, A. (1977a) *City-systems in Advanced Economies: past growth, present processes and future development options*, Hutchinson, London.

Pred, A. (1977b) The choreography of existence: comments on Hägerstrand's time-geography and its usefulness, *Economic Geography* **53**, 207–21.

Pred, A. (1981) Of paths and projects: individual behaviour and its societal context. In: Cox, K.R., Golledge, R.G. (eds) *Behavioural Problems in Geography Revisited*, Methuen, London, pp. 231–55.

Pred, A. (1984) Place as historically contingent process: structuration and the time-geography of becoming places, *Annals of the Association of American Geographers* **74**, 279–97.

Pred, A. (1986) *Place, Practice, and Structure*, Polity Press, Cambridge.

Pred, A. (1990) *Lost Words and Lost Worlds: modernity and the language of everyday life in late nineteenth century Stockholm*, Cambridge University Press, Cambridge.

Pred, A. (1991) Introduction to the Special Issue 'Meaning and modernity – cultural geographies of the invisible and the concrete', *Geografiska Annaler* **73B**, 3–5.

Pred, A., Kibel, B.M. (1970) An application of gaming simulation to a general model of economic locational processes, *Economic Geography* **46**, 136–56.

Pred, A., Palm, R. (1978) The status of American women: a time-geographic view. In: Lanegran, D.A., Palm, R. (eds) *An Invitation to Geography*, McGraw-Hill, New York, pp. 99–109.

Preston, V. (1987) Spatial choice behaviour in urban environments, *Urban Geography* **8**, 374–9.

Preston, V., Taylor, S.M. (1981) Personal construct theory and residential choice, *Annals of the Association of American Geographers* **71**, 437–51.

Prince, H.C. (1971) Real, imagined and abstract worlds of the past, *Progress in Geography* **3**, 1–86.

Proshansky, H.M. (1978) The city and self-identity, *Environment and Behaviour* **10**, 147–69.

Proshansky, H.M., Ittelson, W.H., Rivlin, L.G. (1970a) Basic psychological processes and the environment. In: Proshansky, H.M., Ittelson, W.H., Rivlin, L.G. (eds) *Environmental Psychology: man and his physical setting*, Holt, Rinehart & Winston, New York, pp. 101–4.

Proshansky, H.M., Ittelson, W.H., Rivlin, L.G. (1970b) Introduction. In: Proshansky, H.M. *et al.* (eds) *Environmental Psychology: man and his physical setting*, Holt, Rinehart & Winston, New York, pp. 1–6.

Proshansky, H.M., Fabian, A.K., Kaminoff, R. (1983) Place-identity: physical world socialization of the self, *Journal of Environmental Psychology* **3**, 57–83.

Pryor, R.J. (ed.) (1979) *Residence History Analysis, Studies in Migration and Ubanization*, No. 3, Department of Demography, Australian National University.

Punter, J.V. (1982) Landscape aesthetics: a synthesis and critique. In: Gold, J., Burgess, J. (eds) *Valued Environments*, Allen & Unwin, London, pp. 100–23.

Pyle, G.F. (1974) *The Spatial Dynamics of Crime*, University of Chicago, Department of Geography Research Paper No. 159.

Pyle, G.F. (1976) Geographic perspectives on crime and the impact of anticrime legislation. In: Adams, J. (ed.) *Urban Policymaking and Metropolitan Dynamics*, Ballinger, Cambridge, pp. 257–92.

Rapoport, A. (1977) *Human Aspects of Urban Form: towards a man–environment approach to urban design*, Pergamon, Oxford.

Rapoport, A. (1982) *The Meaning of the Built Environment: a nonverbal communication approach*, Sage, Beverly Hills.

Rapoport, A., Hawkes, R. (1970) The perception of urban complexity, *Journal of the American Institute of Planners* **36**, 106–11.

Rapoport, R., Rapoport, R. (1975) *Leisure and the Family Life Cycle*, Routledge & Kegan Paul, Henley.

Ravenstein, E.G. (1885) The laws of migration, *Journal of the Royal Statistical Society* **48**, 167–235.

Ravenstein, E.G. (1889) The laws of migration, *Journal of the Royal Statistical Society* **52**, 241–305.

Ray, D.M. (1967) Cultural differences in consumer travel behaviour in East Ontario, *Canadian Geographer* **11**, 143–56.

Reckless, W.C. (1934) *Vice in Chicago*, University of Chicago Press, Chicago.

Rees, G., Rees, T. (1981) *Migration and Industrial Restructuring and Class relations: the case of South Wales*, Papers in Planning Research, UWIST, Cardiff.

Rees, J. (1974) Decision-making, the growth of the firm and the business environment. In: Hamilton, F.E.I. (ed.) *Spatial Perspectives on Industrial Organization and Decision-making*, Wiley, London, pp. 189–211.

Rees, P.H. (1970) Concepts in social space. In: Berry, B.J., Horton, F.E. (eds) *Geographic Perspectives on Urban Systems*, Prentice-Hall, Englewood Cliffs, New Jersey.

Reilly, W.J. (1931) *The Law of Retail Gravitation*, Putman & Sons, New York.

Reitzes, D.C. (1983) Urban images: a social psychological approach, *Sociological Inquiry* **53**, 314–32.

Relph, E. (1976) *Place and Placelessness*, Pion, London.

Relph, E. (1981a) Phenomenology. In: Harvey, M.E., Holly, B.P. (eds) *Themes in Geographic Thought*, Croom Helm, London, pp. 99–114.

Relph, E. (1981b) *Rational Landscapes and Humanistic Geography*, Croom Helm, London.

Relph, E. (1987) *The Modern Urban Landscape*, Croom Helm, London.

Rex, J.A., Moore, R. (1967) *Race, Community and Conflict*, Oxford University Press, Oxford.

Richmond, A.H. (1973) *Migration and Race Relations in an English City*, Oxford University Press, London.

Riley, S., Palmer, J. (1976) Of attitudes and latitudes: a repertory grid study of perceptions of seaside resorts. In: Slater, P. (ed.) *The Measurement of Intrapersonal Space by Grid Technique*, Vol. 1 *Explorations of Intrapersonal Space*, Wiley, London, pp. 153–65.

Rivlin, L.G. (1989) The neighbourhood, personal identity, and group affiliation. In: Altman, I., Wandersman, A. (eds) *Neighbourhood and Community Environments*, Plenum Press, New York, pp. 184–204.

Roberts, K. (1978) *Contemporary Society and the Growth of Leisure*, Longman, London.

Robertson, I.M.L. (1984) Single parent life style and peripheral estate residence, *Town Planning Review* **55**, 197–213.

Robinson, M.E. (1982) Absolute and relative strategies in urban distance cognition, *Area* **14**, 283–6.

Robinson, M.E., Dicken, P. (1979) Cloze procedure and cognitive mapping, *Environment and Behaviour* **11**, 351–74.

Robson, B.T. (1969) *Urban Analysis*, Cambridge University Press, Cambridge.

Robson, B.T. (1988) *Those Inner Cities*, Clarendon Press, Oxford.

Rochford, E.B., Blocher, T.J. (1991) Coping with natural hazards as stressors: the predictors of activism in a flood disaster, *Environment and Behaviour* **23**, 27–46.

Rodgers, H.B. (1969) Leisure and recreation, *Urban Studies* **6**, 31–42.

Root, J.D. (1975) Intransitivity of preferences: a neglected concept, *Proceedings of the Association of American Geographers* **7**, 185–9.

Rose, G. (1988) Locality, politics and culture: Poplar in the 1920s, *Environment and Planning D: Society and Space* **6**, 151–68.

Rose, H. (1976) *Black Suburbanization*, Ballinger, Cambridge, Mass.

Roseman, C.C. (1971) Migration as a spatial and temporal process, *Annals of the Association of American Geographers* **61**, 589–98.

Roseman, C.C., Oldakowski, R.K. (1984) Place ties and migration expectations of a central city population, *Urban Geography* **5**, 573–98.

Rossi, I. (1981) Transformational structuralism: Lévi-Strauss's definition of social structure. In: Blau, P.M., Merton, R.K. (eds) *Continuities in Structural Inquiry*, Sage, Beverly Hills, pp. 51–80.

Rossi, P.H. (1955) *Why Families Move*, Free Press, Glencoe, Ill.

Rowe, S., Wolch, J. (1990) Social networks in time and space: homeless women in Skid Row, Los Angeles, *Annals of the Association of American Geographers* **80**, 184–204.

Rowland, D.T. (1979) *Internal Migration in Australia*, Australian Government Publishing Service, Canberra.

Rowles, G.D. (1978) *Prisoners of Space? Exploring the Geographical Experience of Older People*, Westview Press, Boulder.

Rowntree, L. (1986) Cultural/humanistic geography, *Progress in Human Geography* **10**, 580–6.

Rowntree, L. (1988) Orthodoxy and new directions: cultural/humanistic geography, *Progress in Human Geography* **12**, 576–86.

Royal Commission on Local Government in England (1969) *Community Attitudes Survey*, HMSO, London.

Rumley, D. (1979) The study of structural effects in human geography, *Tijdschrift voor Economische en Sociale Geografie* **70**, 350–60.

Rushton, G. (1969a) Analysis of behaviour by revealed space

preference, *Annals of the Association of American Geographers* **59**, 391–400.
Rushton, G. (1969b) The scaling of locational preferences. In: Cox, K.R., Golledge, R.G. (eds) *Behavioural Problems in Geography*, Northwestern University Studies in Geography No. **17**, Evanston, pp. 197–227.
Rushton, G. (1976) Decomposition of space-preference functions. In: Golledge, R.G., Rushton, G. (eds) *Spatial Choice and Spatial Behaviour*, Ohio State University Press, Columbus, pp. 119–34.
Rushton, G. (1979) Commentary on behavioural and perception geography, *Annals of the Association of American Geographers* **69**, 463–4.
Russell, J.A., Ward, L.M. (1982) Environmental psychology, *Annual Review of Psychology* **33**, 651–88.

Saarinen, T.F. (1966) *Perception of Drought Hazard on the Great Plains*, University of Chicago, Department of Geography Research Paper No. 106.
Saarinen, T.F. (1973) The use of projective techniques in geographic research. In: Ittelson, W.H. (ed.) *Environment and Cognition*, Seminar Press, New York, pp. 29–52.
Saarinen, T.F. (1976) *Environmental Planning: perception and behaviour*, Houghton Mifflin, Boston.
Saarinen, T.F. (1979) Commentary-critique of Bunting–Guelke paper, *Annals of the Association of American Geographers* **69**, 464–8.
Saarinen, T.F. (1988) Centering of mental maps of the world, *National Geographic Research* **4**, 112–27.
Saarinen, T.F. *et al.* (1984) *Environmental Perception and Behaviour: an inventory and prospect*, University of Chicago Department of Geography Research Paper No. 209, Chicago.
Sack, R.D. (1983) Human territoriality: a theory, *Annals of the Association of American Geographers* **73**, 55–74.
Sack, R.D. (1986) *Human Territoriality: its theory and history*, Cambridge University Press, Cambridge.
Sadalla, E.K., Magel, S.G. (1980) The perception of traversed distance, *Environment and Behaviour* **12**, 65–80.
Sadalla, E.K., Montello, D.R. (1989) Remembering changes in direction, *Environment and Behaviour* **21**, 346–63.
Sadalla, E.K., Staplin, L.J. (1980a) The perception of traversed distance: intersections, *Environment and Behaviour* **12**, 167–82.
Sadalla, E.K., Staplin, L.J. (1980b) An information storage model for distance cognition, *Environment and Behaviour* **12**, 183–93.
Saegert, S., Hart, R.A. (1978) The development of environmental competence in girls and boys. In: Salter, M. (ed.) *Play: anthropological perspectives*, Leisure Press, New York, pp. 43–51.
Sagan, C. (1977) *The Dragons of Eden*, Ballantine Books, New York.
Salins, P.D. (1971) Household location patterns in American metropolitan areas, *Economic Geography* **47**, 234–48.
Salling, M., Harvey, M.E. (1981) Poverty, personality, and sensitivity to residential stressors, *Environment and Behaviour* **13**, 131–64.
Salter, C.L., Lloyd, W.J. (1976) *Landscape in Literature*, Association of American Geographers Commission on College Geography Resource Paper 76–3, Washington.
Samuels, M.S. (1978) Existentialism and human geography. In: Ley, D., Samuels, M.S. (eds) *Humanistic Geography: prospects and problems*, Maaroufa Press, Chicago, pp. 22–40.
Samuels, M.S. (1981) An existential geography. In: Harvey, M.E., Holly, B.P. (eds) *Themes in Geographic Thought*, Croom Helm, London, pp. 115–32.
Sarkissian, W., Doherty, T. (1984) *Living in Public Housing*, New South Wales Housing Commission, Sydney.
Saunders, P. (1981) *Social Theory and the Urban Question*, Hutchinson, London.
Saunders, P. (1984) Beyond housing classes: the sociological significance of private property rights in the means of production, *International Journal of Urban and Regional Research* **8**, 202–27.
Savage, L.J. (1951) The theory of statistical decision, *Journal of the American Statistical Association* **46**, 55–67.
Sayer, A. (1984) *New Developments in Manufacturing and their Spatial Implications*, University of Sussex Urban and Regional Studies Working Paper No. 49, Brighton.
Sayer, A. (1989) The 'new' regional geography and problems of narrative, *Environment and Planning D: Society and Space* **7**, 253–76.
Sayer, A. (1990) Post-Fordism in question, *International Journal of Urban and Regional Research* **13**, 666–95.
Sayer, A. (1991) Behind the locality debate: deconstructing geography's dualisms, *Environment and Planning A* **23**, 283–308.
Schiefloe, P.M. (1990) Networks in urban areas: lost, saved or liberated communities, *Scandinavian Housing and Planning Research* **7**, 93–103.
Schiller, R. (1986) Retail decentralization – the coming third wave, *The Planner* **72**(7), 13–15.
Schmid, C.F. (1960) Urban crime areas, *American Sociological Review* **25**, 527–42, 655–78.
Schmitt, R.C. (1966) Density, health, and social disorganization, *Journal of the American Institute of Planners* **32**, 38–40.
Schneider, M. (1959) Gravity models and trip distribution theory, *Papers and Proceedings of the Regional Science Association* **5**, 51–6.
Schnore, L.F. (1965) *The Urban Scene*, Free Press, New York.
Schoenberger, E. (1989) Thinking about flexibility: a response to Gertler, *Transactions of the Institute of British Geographers* **14**, 98–108.
Schorr, A.L. (1963) *Slums and Social Insecurity*, Department of Health, Education and Welfare, Washington.
Schorr, A.L. (1970) Housing and its effects. In: Proshansky, H.M., Ittelson, W.H., Rivlin, L.G. (eds) *Environmental Psychology*, Holt, Rinehart & Winston, New York, pp. 319–33.

Schreyer, R., Beaulieu, J.T. (1986) Attribute preferences for wildland recreation settings, *Journal of Leisure Research* **18**, 231–47.

Schuler, H.J. (1979) A disaggregate store-choice model of spatial decision-making, *Professional Geographer* **31**, 146–56.

Schutz, A. (1960) The social world and the theory of social action, *Social Research* **27**, 205–21.

Schutz, A. (1967) *Phenomenology and the Social World* (trans. G. Walsh and F. Lehnert), Northwestern University Press, Evanston.

Schwirian, K.P. (1983) Models of neighbourhood change, *American Review of Sociology* **9**, 83–102.

Schwirian, K.P., Hankins, F.N., Ventresca, C.A. (1990) The residential decentralization of social status groups in American metropolitan communities 1950–80, *Social Forces* **68**, 1143–63.

Scott, A.J. (1982) Locational patterns and dynamics of industrial activity in the modern metropolis, *Urban Studies* **19**, 111–42.

Scott, A.J. (1983) Industrial organization and the logic of intra-metropolitan locations: 1. Theoretical considerations, *Economic Geography* **59**, 233–50.

Scott, A.J. (1986) Industrialization and urbanization: a geographical agenda, *Annals of the Association of American Geographers* **76**, 25–37.

Scott, A.J. (1988) *New Industrial Spaces*, Pion, London.

Scott, A.J., Angel, D.P. (1987) The US semiconductor industry: a locational analysis, *Environment and Planning A* **19**, 875–912.

Seabrook, J. (1988) *The Leisure Society*, Basil Blackwell, Oxford.

Seamon, D. (1979) *A Geography of the Lifeworld*, St Martin's Press, New York.

Seamon, D. (1984) Philosophical directions in behavioural geography with an emphasis on the phenomenological contribution. In: Saarinen, T.F., Seamon, D., Sell, J.L. (eds) *Environmental Perception and Behaviour: an inventory and prospect*, Research Paper No. 209, Department of Geography, University of Chicago, pp. 167–78.

Seamon, D. (1987) Phenomenology and environmental research. In: Zube, E.H., Moore, G.T. (eds) *Advances in Environment, Behaviour, and Design*, Vol. 1, Plenum, New York, pp. 3–27.

Sell, R.R. (1983) Analysing migration decisions: the first step – who decides? *Demography* **20**, 299–311.

Seltzer, J., Wilson, S.A. (1980) Leisure patterns among four day workers, *Journal of Leisure Research* **12**, 116–27.

Shannon, C.E., Weaver, W. (1949) *The Mathematical Theory of Communication*, University of Illinois Press, Urbana.

Sharpe, B. (1988) Informal work and development in the west, *Progress in Human Geography* **12**, 315–36.

Shaw, C.R., McKay, H.D. (1929) *Delinquency Areas*, University of Chicago Press, Chicago.

Shaw, C.R., McKay, H.D. (1942) *Juvenile Delinquency and Urban Areas*, University of Chicago Press, Chicago.

Shaw, R.P. (1975) *Migration Theory and Fact: a review and bibliography of current literature*, Regional Science Research Institute, Philadelphia.

Shepherd, I.D.H., Thomas, C.J. (1980) Urban consumer behaviour. In: Dawson, J.A. (ed.) *Retail Geography*, Croom Helm, London, pp. 18–94.

Sherman, R.C., Croxton, J., Giovanatto, J. (1979) Investigating cognitive representations of spatial relationships, *Environment and Behaviour* **11**, 209–26.

Shevky, E., Bell, W. (1955) *Social Area Analysis*, Stanford University Press, Stanford.

Shevky, E., Williams, M. (1949) *The Social Areas of Los Angeles*, University of California Press, Los Angeles.

Shippee, G., Burrough, J., Wakefield, S. (1980) Dissonance theory revisited: perception of environmental hazards in residential areas, *Environment and Behaviour* **12**, 33–52.

Shoard, M. (1980) *The Theft of the Countryside*, Temple Smith, London.

Short, J.R. (1978) Residential mobility in the private housing market of Bristol, *Transactions of the Institute of British Geographers* **3**, 533–47.

Shortridge, J.R. (1984) The emergence of 'Middle West' as an American regional label, *Annals of the Association of American Geographers* **74**, 209–20.

Siegel, A.W., Herman, J.F., Allen, G.L., Kibasic, H.C. (1979) The development of cognitive maps of large- and small-scale spaces, *Child Development* **50**, 582–85.

Siegel, A.W., White, S.H. (1975) The development of spatial representations of large-scale environments. In: Reese, H.W. (ed.) *Advances in Child Development and Behaviour*, Academic Press, New York, pp. 56–65.

Sillitoe, K.K. (1969) *Planning for Leisure*, HMSO, London.

Simmons, I.G. (1975) *Rural Recreation in the Industrial World*, Arnold, London.

Simon, H.A. (1952) A behavioural model of rational choice, *Quarterly Journal of Economics* **69**, 99–118.

Simon, H.A. (1957) *Models of Man: social and rational*, Wiley, New York.

Simon, H.A. (1959) Theories of decision-making in economics and behavioural science, *American Economic Review* **49**, 253–83.

Simon, H.A. (1976) *Administrative Behaviour: a study of decision making processes in administrative organization* (3rd edn), Free Press, Glencoe.

Skinner, B.F. (1971) *Beyond Freedom and Dignity*, Knopf, New York.

Skinner, B.F. (1974) *About Behaviourism*, Jonathan Cape, London.

Smith, C.A., Smith, C.J. (1978) Locating natural neighbours in the urban community, *Area* **10**, 102–10.

Smith, C.D. (1984) The relationship between the pleasingness of landmarks and the judgement of distance in cognitive maps, *Journal of Environmental Psychology* **4**, 229–34.

Smith, C.J. (1978) Self-help and social networks in the urban community, *Ekistics* **46**, 106–15.

Smith, D. (1989) *The Chicago School: a liberal critique of capitalism*, St Martin's Press, New York.

Smith, D.J. (1986) *Black and White*, Policy Studies Institute, London.

Smith, D.M. (1971) *Industrial Location: an economic geographical analysis*, Wiley, New York.

Smith, D.M. (1977) *Human Geography: a welfare approach*, Arnold, London.

Smith, D.M. (1979a) Modelling industrial location: towards a broader view of the space economy. In: Hamilton, F.E.I., Linge, G.J.R. (eds) *Spatial Analysis, Industry and the Industrial Environment: progress in research and applications*, Vol. 1 *Industrial Systems*, Wiley, Chichester, pp. 37–55.

Smith, D.M. (1979b) *Where the Grass is Greener: living in an unequal world*, Penguin, Harmondsworth.

Smith, D.M. (1979c) The identification of problems in cities: applications of social indicators. In: Herbert, D.T., Smith, D.M. (eds) *Social Problems and the City*, Oxford University Press, London, pp. 13–32.

Smith, G.C. (1976) The spatial information fields of urban consumers, *Transactions of the Institute of British Geographers* **1**, 175–89.

Smith, G.C. (1989) Elderly consumer cognitions of urban shopping centres, *Canadian Geographer* **33**, 353–9.

Smith, G.C. (1991) Grocery shopping patterns of the ambulatory urban elderly, *Environment and Behaviour* **23**, 86–114.

Smith, G.C., Shaw, D.J.B., Hinckle, P.R. (1979) Children's perception of a downtown shopping centre, *Professional Geographer* **31**, 157–64.

Smith, M.A., Parker, S., Smith, C.S. (eds) (1973) *Leisure and Society in Britain*, Allen Lane, London.

Smith, N. (1979) Geography, science and post-positivist modes of explanation, *Progress in Human Geography* **3**, 356–83.

Smith, N. (1987) Of yuppies and housing: gentrification, social restructuring, and the urban dream, *Environment and Planning D: Society and Space* **5**, 139–55.

Smith, N., Wilson, D. (1979) *Modern Linguistics: the results of Chomsky's revolution*, Penguin, London.

Smith, P.F. (1974) *The Dynamics of Urbanism*, Hutchinson, London.

Smith, P.F. (1976) *The Syntax of Cities*, Hutchinson, London.

Smith, S.J. (1984) Practizing humanistic geography, *Annals of the Association of American Geographers* **74**, 353–74.

Smith, S.J. (1986) *Crime, Space and Society*, Cambridge University Press, Cambridge.

Smith, S.J. (1987) Fear of crime: beyond a geography of deviance, *Progress in Human Geography* **11**, 1–23.

Smith, S.J. (1988) Constructing local knowledge: the analysis of self in everyday life. In: Eyles, J., Smith, D.M. (eds) *Qualitative Methods in Human Geography*, Polity Press, Cambridge, pp. 17–38.

Smith, S.L.J. (1980) Intervening opportunities to urban recreation centres, *Journal of Leisure Research* **12**, 64–72.

Smith, S.L.J. (1983) *Recreation Geography*, Longman, London.

Smith, S.L.J. (1990) A test of Plog's allocentric/psychocentric model: evidence from seven nations, *Journal of Travel Research* **28**, 40–3.

Smith, T.R., Clark, W.A.V., Huff, J.O., Shapiro, P. (1979) A decision making and search model for intra-urban migration, *Geographical Analysis* **11**, 1–22.

Smith, V.K., Kopp, R.J. (1980) The spatial limits of the travel cost recreational demand model, *Land Economics* **56**, 64–72.

Smith, V.K., Wilde, P. (1977) The multiplier effect of tourism in Western Tasmania. In: Mercer, D. (ed.) *Leisure and Recreation in Australia*, Sorrett, Melbourne, pp. 163–72.

Soja, E.W. (1987) The postmodernization of geography: a review, *Annals of the Association of American Geographers* **77**, 289–94.

Soja, E.W. (1989) *Postmodern Geographies: the reassertion of space in critical social theory*, Verso, London.

Sommer, R. (1969) *Personal Space: the behavioural basis of design*, Prentice-Hall, Englewood Cliffs.

Sonnenfeld, J. (1982) Egocentric perspectives on geographic orientation, *Annals of the Association of American Geographers* **72**, 68–76.

Sonnenfeld, J. (1985) Tests of spatial skill: a validation problem, *Man–Environment Systems* **15**(3–4), 107–20.

Speare, A. Jr (1974) Residential satisfaction as an intervening variable in residential mobility, *Demography* **11**, 173–88.

Spector, A.N., Brown, L.A., Malecki, E.J. (1976) Acquaintance circles and communication: an exploration of hypotheses relating to innovation adoption, *Professional Geographer* **28**, 267–76.

Spencer, C. and Blades, M. (1986) Patterns and process: a review essay on the relationship between behavioural geography and environmental psychology, *Progress in Human Geography* **10**, 230–48.

Spencer, C., Darvizeh, Z. (1981) The case for developing a cognitive environmental psychology that does not underestimate the abilities of young children, *Journal of Environmental Psychology* **1**, 21–31.

Spencer, C., Weetman, M. (1981) The microgenesis of cognitive maps: a longitudinal study of new residents of an urban area, *Transactions of the Institute of British Geographers* **6**, 375–84.

Spencer, C., Blades, M., Morsley, K. (1989) *The Child in the Physical Environment: the development of spatial knowledge and cognition*, Wiley, Chichester.

Spencer, D. (1973) *An Evaluation of Cognitive Mapping in*

Neighbourhood Perception, University of Birmingham, Centre for Urban and Regional Studies Research Memorandum No. 23.

Sprout, H., Sprout, M. (1957) Environmental factors in the study of international politics, *Journal of Conflict Resolution* **1**, 309–28.

Stabler, M. (1990) The concept of opportunity sets as a methodological framework for the analysis of selling tourism places: the industry view. In: Ashworth, G., Goodall, B. (eds) *Marketing Tourism Places*, Routledge, London, pp. 23–41.

Stacey, M. (1969) The myth of community studies, *British Journal of Sociology* **20**, 134–47.

Stafford, H.A. (1972) The geography of manufacturers, *Progress in Geography* **4**, 181–215.

Stankey, G.H., Cole, D.N., Lucas, R.C., Petersen, M.E., Frissel, S.S. (1985) *The Limits of Acceptable Change (LAC) System for Wilderness Planning*, USDA Forest Service, Ogden, Utah.

Stapleton, C.M. (1980) Reformulation of the family life-cycle concept: complications for residential mobility, *Environment and Planning A* **12**, 1103–18.

Stapleton, C.M. (1984) Intra-urban residential mobility of the aged, *Geografiska Annaler* **66B**, 99–109.

Starr, C., Whipple, C. (1980) Risks of risk decisions, *Science* **208**, 1114–19.

Stea, D. (1969) The measurement of mental maps: an experimental model for studying conceptual places. In: Cox, K.R., Golledge, R.G. (eds) *Behavioural Problems in Geography*, Northwestern University Studies in Geography No. **17**, Evanston, pp. 228–53.

Stea, D. (1973) Rats, men and spatial behaviour, all revisited or what animal geographers have to say to human geographers, *Professional Geographer* **25**, 106–12.

Stea, D. (1976) Program notes on a spatial fugue. In: Moore, G.T., Golledge, R.G. (eds) *Environmental Knowing*, Dowden, Hutchinson & Ross, Stroudsburg, pp. 106–20.

Stea, D., Blaut, J.M. (1973) Some preliminary observations on spatial learning in school children. In: Downs, R.M. Stea, D. (eds) *Image and Environment*, Aldine, Chicago, pp. 226–34.

Stea, D., Downs, R.M. (1970) From the outside looking in at the inside looking out, *Environment and Behaviour* **2**, 3–12.

Stein, M. (1960) *The Eclipse of Community*, Harper & Row, New York.

Steinitz, C. (1968) Meaning and the congruence of urban form and activity, *Journal of the American Institute of Planners* **34**, 233–48.

Stern, E., Leiser, D. (1988) Levels of spatial knowledge and urban travel modeling, *Geographical Analysis* **20**, 140–55.

Stokols, D. (ed.) (1977) *Perspectives on Environment and Behaviour*, Plenum, New York.

Stokols, D. (1978) Environmental psychology, *Annual Review of Psychology* **29**, 253–95.

Stokols, D., Altman, I. (eds) (1987) *Handbook of Environmental Psychology*, Wiley, New York.

Stone, G.P. (1954) City shoppers and urban identification, *American Journal of Sociology* **60**, 36–45.

Storper, M. (1988) Big structures, small events, and large processes in economic geography, *Environment and Planning A* **20**, 165–85.

Storper, M., Christopherson, S. (1987) Flexible specialization and regional industrial agglomeration: the case of the US motion picture industry, *Annals of the Association of American Geographers* **77**, 104–17.

Storper, M., Scott, A.J. (1989) The geographical foundations and social regulation of flexible production complexes. In: Wolch, J., Dear, M. (eds) *The Geography of Power: how territory shapes social life*, Unwin Hyman, Winchester, Mass., pp. 21–40.

Stouffer, S.A. (1940) Intervening opportunities: a theory relating mobility and distance, *American Sociological Review* **5**, 845–67.

Stouffer, S.A. (1960) Intervening opportunities and competing migrants, *Journal of Regional Science* **2**, 1–26.

Strasser, S. (1963) *Phenomemology and the Human Sciences: a contribution to a new scientific ideal*, Duquesne University Press, Pittsburgh.

Suttles, G.D. (1972) *The Social Construction of Communities*, University of Chicago Press, Chicago.

Svart, L. (1974) On the priority of behaviour in behavioural research: a dissenting view, *Area* **6**, 301–5.

Swyngedouw, E.A. (1989) The heart of the place: the resurrection of locality in an age of hyperspace, *Geografiska Annaler* **71B**, 31–42.

Symanski, R. (1979) Hobos, freight trains, and me, *Canadian Geographer* **23**, 103–18.

Szalai, A. (ed.) (1972) *The Use of Time: daily activities of urban and suburban populations in twelve countries*, Mouton, The Hague.

Taeuber, K.E. (1965) Residential segregation, *Scientific American* **213**, 12–19.

Talarchek, G.M. (1982) Sequential aspects of residential search and selection, *Urban Geography* **3**, 34–57.

Taylor, A. (1984) The planning implications of new technology in retailing and distribution, *Town Planning Review* **55**, 161–76.

Taylor, C.C., Townsend, A.R. (1976) The local 'sense of place' as evidenced in North-East England, *Urban Studies* **13**, 133–46.

Taylor, M.J. (1973) Local linkage, external economies and the iron-foundry industry of the West Midlands and East Lancashire conurbations, *Regional Studies* **7**, 387–400.

Taylor, M.J. (1977) Corporate space preferences: a New Zealand example, *Environment and Planning A* **9**, 1157–67.

Taylor, M.J. (1978a) Spatial competition and the sales linkages of Auckland manufacturers. In: Hamilton, F.E.I. (ed.) *Contemporary Industrialization: spatial analysis and regional development*, Longman, London, pp. 144–57.

Taylor, M.J. (1978b) Perceived distance and spatial interaction, *Environment and Planning A* **10**, 1171–7.

Taylor, M.J. (1984) Industrial geography, *Progress in Human Geography* **8**, 263–74.

Taylor, M.J., McDermott, P.J. (1977) Perception of location and decision-making: the example of New Zealand manufacturing, *New Zealand Geographer* **33**, 26–33.

Taylor, M.J., Neville, R.J.W. (1980) The malleability of managerial attitudes: the case of Singapore's plastics and electronics manufacturers, *Singapore Journal of Tropical Geography* **1**, 55–68.

Taylor, M.J., Thrift, N.J. (1983a) Industrial geography in the 1980s: entering a decade of differences, *Environment and Planning A* **15**, 1287–91.

Taylor, M.J., Thrift, N. (1983b) Business organization, segmentation and location, *Regional Studies* **17**, 445–65.

Taylor, P.J. (1982) A materialist framework for political geography, *Transactions of the Institute of British Geographers* **7**, 15–34.

Taylor, P.J. (1985) *Political Geography: world-economy, nation-state, and locality*, Longman, London.

Taylor, R.B. (1988) *Human Territorial Functioning*, Cambridge University Press, Cambridge.

Taylor, R.B., Gottfredson, S.O., Blower, S. (1984) Neighbourhood naming as an index of attachment to place, *Population and Environment* **7**, 103–25.

Taylor, S. (1981) The interface of cognitive and social psychology. In: Harvey, J.H. (ed.) *Cognition, Social Behaviour and the Environment*, Laurence Erlbaum, Hillsdale, NJ.

Teulings, A.W.M. (1984) The internationalization squeeze: double capital movement and job transfer within Philips worldwide, *Environment and Planning A* **16**, 597–614.

Thomas, C.J. (1976) Sociospatial differentiation and the use of services. In: Herbert, D.T., Johnston, R.J. (eds) *Social Areas in Cities*, Vol. II *Spatial Perspectives on Problems and Policies*, Wiley, London, pp. 17–64.

Thomas, C.J., Bromley, R.D.F. (1987) The growth and functioning of an unplanned retail park: the Swansea enterprise zone, *Regional Studies* **21**, 287–300.

Thomas, M. (1980) Explanatory frameworks for growth and change in multiregional forms, *Economic Geography* **56**, 1–17.

Thomas, R.W. (1990) Some spatial representation problems in disease modelling, *Geographical Analysis* **22**, 209–23.

Thompson, C. (1989) The geography of venture capital, *Progress in Human Geography* **13**, 62–98.

Thompson, D.L. (1963) New concept: subjective distance, *Journal of Retailing* **39**, 1–6.

Thompson, E.P. (1978) *The Poverty of Theory and Other Essays*, Merlin Press, London.

Thorndyke, P.W., Hayes-Roth, B. (1982) Differences in spatial knowledge acquired from maps and navigation, *Cognitive Psychology* **14**, 560–89.

Thorngren, B. (1970) How do contact systems affect regional development? *Environment and Planning* **2**, 409–27.

Thorngren, B. (1973) Swedish office dispersal. In: Bannon, M. (ed.) *Office Location and Regional Development*, An Foras Forbartha, Dublin, pp. 9–20.

Thorns, D.C. (1980) Constraints versus choices in the analysis of housing allocation and residential mobility. In: Ungerson, C., Karn, V. (eds) *The Consumer Experience of Housing*, Gower, Farnborough, pp. 50–68.

Thorns, D.C. (1985) Age time and calendar time: two facets of the residential mobility process, *Environment and Planning A* **17**, 829–44.

Thorns, D.C. (1988) New solutions to old problems: housing affordability and access within Australia and New Zealand, *Environment and Planning A* **20**, 71–82.

Thorns, D.C. (1989) The production of homelessness: from individual failure to system inadequacies, *Housing Studies* **4**, 253–66.

Thorpe, D. (1983) Changes in the retail sector. In: Davies, R.L., Champion, A.G. (eds) *The Future for the City Centre*, Academic Press, London, pp. 32–47.

Thorpe, D. (1991) The development of British superstore retailing – further comments on Davies and Sparks, *Transactions of the Institute of British Geographers* **16**, 354–67.

Thrift, N. (1979) Unemployment in the inner city: urban problem or structural imperative? A review of the British experience. In: Herbert, D.T., Johnston, R.J. (eds) *Geography and the Urban Environment*, Vol. 3, Wiley, Chichester, pp. 125–226.

Thrift, N.J. (1981) Behavioural geography. In: Wrigley, N., Bennett, R.J. (eds) *Quantitative Geography: a British view*, Routledge & Kegan Paul, London, pp. 352–65.

Thrift, N. (1983) On the determination of social action in space and time, *Environment and Planning D: Society and Space* **1**, 23–57.

Thrift, N. (1985) Bear and mouse or bear and tree? Anthony Giddens's reconstitution of social theory, *Sociology* **19**, 609–23.

Thrift, N. (1986) The geography of international economic disorder. In: Johnston, R.J. (ed) *A World in Crisis*, Basil Blackwell, Oxford, pp. 12–67.

Thrift, N. (1989) Images of social change. In: Hamnett, C., McDowell, L., Sarre, P. (eds) *Restructuring Britain: the changing social structure*, Sage, London, pp. 12–42.

Timmermans, H. (1979) A spatial preference model of regional shopping behaviour, *Tijdschrift voor Economische en Sociale Geografie* **70**, 45–8.

Timmermans, H. (1980) Unidimensional conjoint measurement models and consumer decision-making, *Area* **12**, 291–300.

Timmermans, H. (1981) Spatial choice behaviour in different environmental settings: an application of the revealed preference approach, *Geografiska Annaler* **63B**, 57–67.

Timmermans, H. (1983) Noncompensatory decision rules and consumer spatial choice behaviour: a test of predictive ability, *Professional Geographer* **35**, 449–55.

Timmermans, H. (1984a) Decompositional multiattribute preference models in spatial choice analysis: a review of some recent developments, *Progress in Human Geography* **8**, 189–221.

Timmermans, H. (1984b) Decision models for predicting preferences among multi attribute choice alternatives. In: Bahrenberg, G., Fischer, M.M., Nijkamp, P. (eds) *Recent Developments in Spatial Data Analysis*, Gower, Aldershot, pp. 337–54.

Timmermans, H., Golledge, R.G. (1990) Applications of behavioural research on spatial problems II: preference and choice, *Progress in Human Geography* **14**, 311–54.

Timmermans, H., Van der Heisden, R., Westervald, H. (1982) Cognition of urban retailing structures: a Dutch case study, *Tijdschrift voor Economische en Sociale Geografie* **73**, 2–12.

Timmermans, H., Van der Heisden, R., Westervald, H. (1984) Decision-making experiments and real–world choice behaviour, *Geografiska Annaler* **66B**, 39–48.

Timms, D.W.G. (1971) *The Urban Mosaic: towards a theory of residential differentiation*, Cambridge University Press, Cambridge.

Tinsley, H.E., Kass, R.A. (1979) The latent structure of the need satisfying properties of leisure activities, *Journal of Leisure Research* **11**, 278–91.

Tivers, J. (1985) *Women Attached: the daily lives of women with young children*, Croom Helm, London.

Tolman, E.C. (1932) *Purposive Behaviour in Animals and Men*, Century Crofts, New York.

Tolman, E.C. (1948) Cognitive maps in rats and men, *Psychological Review* **55**, 189–208.

Tönnies, F. (1887) *Community and Society* (trans. C.P. Loomis), Harper, New York (1963 edn).

Torkildsen, G. (1986) *Leisure and Recreation Management* (2nd edn), E. & F.N. Spon, London.

Törnqvist, G.E. (1970) *Contact Systems and Regional Development*, Lund Studies in Geography, Series B, No.35.

Törnqvist, G.E. (1973) Contact requirements and travel facilities in contact models of Sweden and regional development alternatives in the future. In: Pred, A.R., Törnqvist, G.E. *Systems of Cities and Information Flows: two essays*, Lund Studies in Geography, Series B, No. **38**, pp. 85–121.

Törnqvist, G.E. (1977) The geography of economic activities: some criticial viewpoints on theory and application, *Economic Geography* **53**, 153–62.

Törnqvist, G.E. (1978) Swedish industry as a spatial system. In: Hamilton, F.E.I. (ed.) *Contemporary Industrialization: spatial analysis and regional development*, Longman, London, pp. 86–109.

Townroe, P.M. (1974) Post-move stability and the location decision. In: Hamilton, F.E.I. (ed.) *Spatial Perspectives on Industrial Organization and Decision-making*, Wiley, London, pp. 287–307.

Townroe, P.M. (1975) Approaches to the study of industrial location. In: Massey, D., Morrison, W.I. (eds) *Industrial Location: alternative frameworks*, Centre for Environmental Studies Conference Paper No. **15**, London, pp. 32–40.

Townroe, P.M. (1991) Rationality in industrial location decisions, *Urban Studies* **28**, 383–92.

Townsend, A.R., Taylor, C.D. (1975) Regional culture and identity in industrialized societies: the case of north-east England, *Regional Studies* **9**, 379–94.

Townsend, P. (1979) *Poverty in the United Kingdom*, Penguin, Harmondsworth.

Trowbridge, C.D. (1913) On fundamental methods of orientation and 'imaginary maps', *Science* **88**, 888–96.

Tuan, Y-F. (1971) Geography, phenomenology, and the study of human nature, *Canadian Geographer* **15**, 181–92.

Tuan, Y-F. (1972) Structuralism, existentialism, and environmental perception, *Environment and Behaviour* **4**, 319–31.

Tuan, Y-F. (1973) Ambiguity in attitudes toward environment, *Annals of the Association of American Geographers* **63**, 411–23.

Tuan, Y-F. (1974a) *Topophilia: a study of environmental perception, attitudes, and values*, Prentice-Hall, Englewood Cliffs.

Tuan, Y-F. (1974b) Space and place: humanistic perspective, *Progress in Geography* **6**, 211–52.

Tuan, Y-F. (1974c) Commentary on 'Values in geography'. In: Buttimer, A. *Values in Geography*, Association of American Geographers Resource Paper No. 24, Washington, pp. 54–8.

Tuan, Y-F. (1975a) Images and mental maps, *Annals of the Association of American Geographers* **65**, 205–13.

Tuan, Y-F. (1975b) Place: an experiential perspective, *Geographical Review* **65**, 151–65.

Tuan, Y-F. (1976a) Literature, experience, and environmental knowing. In: Moore, G.T., Golledge, R.G. (eds) *Environmental Knowing*, Dowden, Hutchinson & Ross, Stroudsburg, pp. 260–71.

Tuan, Y-F. (1976b) Humanistic geography, *Annals of the Association of American Geographers* **66**, 266–76.

Tuan, Y-F. (1977) *Space and Place: the perspective of experience*, Arnold, London.

Tuan, Y-F. (1978a) Literature and geography: implications for geographical research. In: Ley, D., Samuels, M.S. (eds) *Humanistic Geography: prospects and problems*, Maaroufa, Chicago, pp. 194–206.

Tuan, Y-F. (1978b) Sign and metaphor, *Annals of the Association of American Geographers* **68**, 363–72.

Tuan, Y-F. (1979) Thought and landscape: the eye and the mind's eye. In: Meinig, D.W. (ed) *The Interpretation of Ordinary Landscapes*, Oxford University Press, New York, pp. 89–102.

Tuan, Y-F. (1980) Rootedness versus sense of place, *Landscape* **24**, 3–7.

Tuan, Y-F. (1982) *Segmented Worlds and Self: group life and*

individual consciousness, University of Minnesota Press, Minneapolis.
Tuan, Y-F. (1989) Environment, behaviour, and thought. In: Boal, F.W., Livingstone, D.N. (eds) *The Behavioural Environment*, Routledge, London, pp. 77–81.
Tuan, Y-F. (1990) Realism and fantasy in art, history and geography, *Annals of the Association of American Geographers* **80**, 435–46.
Turner, A. (1981) National parks and pressure groups in New South Wales. In: Mercer, D. (ed.) *Outdoor Recreation: Australian perspectives*, Sorrett, Melbourne, pp. 156–69.
Turner, B.S. (ed.) (1990) *Theories of Modernity and Postmodernity*, Sage, New York.
Turner, C., Manning, P. (1988) Placing authenticity – on being a tourist: a reply to Pearce and Moscardo, *Australia and New Zealand Journal of Sociology* **24**, 136–9.
Tversky, A., Kahneman, D. (1974) Judgement under uncertainty: heuristics and biases, *Science* **185**, 1124–31.
Tversky, B. (1981) Distortions in memory for maps, *Cognitive Psychology* **13**, 407–33.

Ulrich, R.S., Addoms, D.L. (1981) Psychological and recreational benefits of a residential park, *Journal of Leisure Research* **13**, 43–65.
Unger, D.G., Wandersman, A. (1985) The importance of neighbours: the social, cognitive and affective components of neighbouring, *American Journal of Community Psychology* **13**, 139–60.
Unkel, M.B. (1981) Physical recreation participation of females and males during life cycle, *Leisure Sciences* **4**, 1–27.
Urry, J. (1981) Localities, regions and social class, *International Journal of Urban and Regional Research* **5**, 455–74.
Urry, J. (1986) Locality research: the case of Lancaster, *Regional Studies* **20**, 233–42.
Urry, J. (1988) Cultural change and contemporary holiday-making, *Theory, Culture and Society* **5**, 35–55.
Urry, J. (1990a) *The Tourist Gaze: leisure and travel in contemporary societies*, Sage, London.
Urry, J. (1990b) The 'consumtion' of tourism, *Sociology* **24**, 23–35.

Van der Knaap, G.A., Linge, G.J.R. (1989) Labour, environment and industrial change. In: Linge, G.J.R., Van der Knaap, G.A. (eds) *Labour, Environment and Industrial Change*, Routledge, London, pp. 1–19.
Van Lierop, W., Nijkamp, P. (1980) Spatial choice and interaction models: criteria and aggregation, *Urban Studies* **17**, 299–311.
Van Valey, T.L., Roof, W.C., Wilcox, J.E. (1977) Trends in residential segregation 1960–1970, *American Journal of Sociology* **82**, 826–44.
Varady, D.P. (1986) Neighbourhood confidence: a critical factor in neighbourhood revitalization, *Environment and Behaviour* **18**, 480–501.
Vickerman, R.W. (1975) *The Economics of Leisure and Behaviour*, Macmillan, London.

Walker, R. (1978) The transformation of urban structure in the nineteenth century and the beginnings of suburbanization. In: Cox, K. (ed.) *Urbanization and Conflict in Market Societies*, Maaroufa Press, Chicago, pp. 165–212.
Walker, R. (1989) A requiem for corporate geography: new directions in industrial organization, the production of place and the uneven development, *Geografiska Annaler* **71B**, 43–68.
Walmsley, D.J. (1972a) *Systems Theory*, Australian National University, Canberra, Department of Human Geography Publication HG7.
Walmsley, D.J. (1972b) The influence of spatial opportunities on the journey-to-consume: a Sydney case study, *Royal Australian Planning Institute Journal* **10**, 144–8.
Walmsley, D.J. (1973) The simple behaviour system: an appraisal and an elaboration, *Geografiska Annaler* **55B**, 49–56.
Walmsley, D.J. (1974a) Emotional involvement and subjective distance: a modification of the inverse square root law, *Journal of Psychology* **87**, 9–19.
Walmsley, D.J. (1974b) Positivism and phenomenology in human geography, *Canadian Geographer* **18**, 95–107.
Walmsley, D.J. (1974c) Retail spatial structure in suburban Sydney, *Australian Geographer* **9**, 401–18.
Walmsley, D.J. (1975) Normalized distance: an illustration of the interdependence of mobility and opportunity, *Australian Journal of Marketing Research* **8**, 66–8.
Walmsley, D.J. (1976) Territory and neighbourhood in suburban Sydney, *Royal Australian Planning Institute Journal* **14**, 68–9.
Walmsley, D.J. (1977) Congruence of overt behaviour with preference structures, *Psychological Reports* **41**, 1082.
Walmsley, D.J. (1978) Stimulus complexity in distance distortion, *Professional Geographer* **30**, 14–19.
Walmsley, D.J. (1979) Time and human geography, *Australian Geographical Studies* **17**, 223–9.
Walmsley, D.J. (1980) Spatial bias in Australian news reporting, *Australian Geographer* **14**, 342–9.
Walmsley, D.J. (1982a) Personality and regional preference structures: a study of introversion–extraversion, *Professional Geographer* **34**, 279–88.
Walmsley, D.J. (1982b) Mass media and spatial awareness, *Tijdschrift voor Economische en Sociale Geografie* **73**, 32–42.
Walmsley, D.J. (1983) Public information flows in rural Australia, *Environment and Planning A* **15**, 255–63.
Walmsley, D.J. (1988a) *Urban Living: the individual in the city*, Longman, London.
Walmsley, D.J. (1988b) Review of R.G. Golledge and R.J. Stimson Analytical Behavioural Geography, *Australian Geographer* **19**, 310–11.
Walmsley, D.J. (1989) Country town newspapers and regional

consciousness: a New England case study, *Urban Policy and Research* **6**, 60–7.

Walmsley, D.J., Epps, W.R. (1988a) Direction-finding in humans: ability of individuals to orient towards their place of residence, *Perceptual and Motor Skills* **64**, 744–6.

Walmsley, D.J., Epps, W.R. (1988b) Do humans have an innate sense of direction? *Geography* **73**, 31–40.

Walmsley, D.J., Jenkins, J.M. (1991) Mental maps, locus of control, and activity: a study of business tourists in Coffs Harbour, *Journal of Tourism Studies* **2**, 36–42.

Walmsley, D.J., Jenkins, J.M. (1992a) Cognitive mapping of unfamiliar environments: the tourist experience, *Annals of Tourism Research* **19**, 268–86.

Walmsley, D.J., Jenkins, J.M. (1992b) Cognitive distance: a neglected issue in travel behaviour, *Journal of Travel Research*, in press.

Walmsley, D.J., Lewis, G.J. (1984) *Human Geography: behavioural approaches*, Longman, London.

Walmsley, D.J., Lewis, G.J. (1989) The pace of pedestrian flows in cities, *Environment and Behaviour* **21**, 123–50.

Walmsley, D.J., Weinand, H.C. (1991) Changing retail structure in southern Sydney, *Australian Geographer* **22**, 57–66.

Walmsley, D.J., Boskovic, R.M., Pigram, J.J. (1981) *Tourism and Crime*, University of New England Department of Geography, Armidale.

Walmsley, D.J., Saarinen, T.R., MacCabe, C.L. (1990) Down under or centre stage? The world images of Australian students, *Australian Geographer* **21**, 164–73.

Walsh, J.A., Webber, M.J. (1977) Information theory: some concepts and measures, *Environment and Planning A* **9**, 395–417.

Wapner, S., Kaplan, B., Cohen, S.B. (1973) An organismic–developmental perspective for understanding transactions of men in environments, *Environment and Behaviour* **5**, 255–89.

Wapner, S., Kaplan, B., Cohen, S.B. (1980) An organismic–developmental perspective for understanding transactions of men and environments. In: Broadbent, G., Bunt, R., Llorens, T. (eds) *Meaning and Behaviour in the Built Environment*, Wiley, New York, pp. 223–52.

Warburton, P.B. (1981) Cognitive images of water resources with specific references to water-based recreation in Leicester, unpublished MA Thesis University of Leicester.

Ward, S.L., Newcombe, N., Overton, W.F. (1986) Turn left at the church, or three miles north: a study of direction giving and sex differences, *Environment and Behaviour* **18**, 192–213.

Wardwell, J.M. (1977) Equilibrium and change in nonmetropolitan growth, *Rural Sociology* **42**, 156–79.

Warf, B. (1990) The reconstruction of social ecology and neighbourhood change in Brooklyn, *Environment and Planning D: Society and Space* **8**, 73–96.

Warnes, A.M. (1986) The residential mobility histories of parents and children, and relationships to present proximity and social integration, *Environment and Planning A* **18**, 1581–94.

Warnes, A.M. (1987) Geographical location and social relationships among the elderly. In: Pacione, M. (ed.) *Social Geography: progress and prospects*, Croom Helm, London, pp. 252–74.

Warnes, A.M. (1989) Social problems of the elderly in cities. In: Herbert, D.T., Smith, D.M. (eds) *Social Problems and the City*, Oxford University Press, Oxford, pp. 197–212.

Warren, D.I. (1978) Exploration in neighbourhood differentiation, *Sociological Quarterly* **19**, 310–31.

Warren, R. (1978) *The Community in America*, Rand McNally, Chicago.

Watson, J.B. (1924) *Behaviourism*, Norton, New York.

Watson, J.W. (1969) The role of illusion in North American geography: a note on the geography of North American settlement, *Canadian Geographer* **13**, 10–27.

Watson, J.W. (1976a) The image of nature in America. In: Watson, J.W., O'Riordan, T. (eds) *The American Environment: perceptions and policies*, Wiley, New York, pp. 63–75.

Watson, J.W. (1976b) Image regions. In: Watson, J.W., O'Riordan, T. (eds) *The American Environment: perceptions and policies*, Wiley, New York, pp. 15–28.

Watson, J.W., O'Riordan, T. (eds) (1976) *The American Environment: perceptions and policies*, Wiley, New York.

Watson, M.K. (1978) The scale problem in human geography, *Geografiska Annaler* **60B**, 36–47.

Watts, H.D. (1981) *The Branch Plant Economy: a study of external control*, Longman, London.

Watts, H.D. (1987) *Industrial Geography*, Longman, London.

Watts, H.D., Stafford, H.A. (1986) Plant closures and the multiplant firm: some conceptual issues, *Progress in Human Geography* **10**, 206–27.

Webber, M.J. (1969) Sub-optimal behaviour and the concept of maximum profits in location theory, *Australian Geographical Studies* **7**, 1–8.

Webber, M.J. (1971) Empirical verifiability of classical central place theory, *Geographical Analysis* **3**, 15–28.

Webber, M.J. (1972) *Impact of Uncertainty on Location*, Australian National University Press, Canberra.

Webber, M.J. (1978) *Information Theory and Urban Spatial Structure*, Croom Helm, London.

Webber, M.M. (1963) Order in diversity: community without propinquity. In: Wingo, L. (ed.) *Cities and Space*, Johns Hopkins Press, Baltimore, pp. 23–56.

Webber, M.M. (1964a) The urban place and the non-place urban realm. In: Webber, M.M. *et al.* (eds) *Explorations in Urban Structure*, University of Pennsylvania Press, Philadelphia, pp. 79–153.

Webber, M.M. (1964b) Culture, territoriality and the elastic mile, *Papers and Proceedings of the Regional Science Association* **13**, 59–70.

Weber, A. (1929) *Theory of the Location of Industries* (trans C.J. Friedrich), University of Chicago Press, Chicago.

Wekerle, G.R. (1980) Women in the urban environment, *Signs* **5**(3), Supplement S188–214.

Wellman, B. (1979) The community question, *American Journal of Sociology* **84**, 1201–31.

Wellman, B., Crump, B. (1978) *Networks, neighbourhoods and communities: approaches to the study of the community question*, University of Toronto, Centre for Urban and Community Studies Research Paper No. 97.

Wellman, B., Leighton, B. (1979) Networks, neighbours and communities: approaches to the study of the community question, *Urban Affairs Quarterly* **14**, 363–90.

West, P.C., Merriam, L.C. (1970) Outdoor recreation and family cohesiveness, *Journal of Leisure Research* **2**, 251–9.

Westin, A.F. (1967) *Privacy and Freedom*, Atheneum, New York.

Westover, T.N., Collins, J.R. (1987) Perceived crowding in recreation settings: an urban case study, *Leisure Sciences* **9**, 87–99.

Wheatley, P. (1976) Levels of space awareness in the traditional Islamic city, *Ekistics* **42**, 354–66.

White, A.L., Ratick, S.J. (1989) Risk, compensation, and regional equity in locating hazardous facilities, *Papers of the Regional Science Association* **67**, 29–42.

White, G.F. (1945) *Human Adjustment to Floods: a geographical approach to the flood problem in the United States*, University of Chicago Department of Geography Research Paper No. 29, Chicago.

White, G.F. (1964) *Choice of Adjustments to Floods*, University of Chicago Department of Geography Research Paper No. 93, Chicago.

White, G. (1985) Geography in a perilously changing world, *Annals of the Association of American Geographers* **75**, 10–18.

White, S.E. (1980a) A philosophical dichotomy in migration research, *Professional Geographer* **32**, 6–13.

White, S.E. (1980b) Awareness, preference and interurban migration, *Regional Science Perspectives* **10**, 71–86.

Whitelaw, J.S., Gregson, J.S. (1972) *Search Procedures in the Intra-urban Migration Process*, Monash University Publication in Geography No. 2.

Whorf, B.L. (1956) *Language, Thought, and Reality* (ed. J.B. Carroll), MIT Press, Cambridge, Mass.

Williams, C.H. (1977) Ethnic perceptions of Arcadia, *Cahiers de Géographie de Québec* **21**, 243–68.

Williams, J.D., McMillan, D.B. (1980) Migration decision making among non-metropolitan-bound migrants. In: Brown, D.L., Wardwell, J.M. (eds) *New Directions in Urban–Rural Migration*, Academic Press, New York, pp. 189–211.

Williams, J.D., McMillen, D.B. (1983) Location-specific capital and destination selection among migrants to non–metropolitan areas, *Rural Sociology* **48**, 448–57.

Williams, N.J. (1979) The definition of shopper types as an aid in the analysis of spatial consumer behaviour, *Tijdschrift voor Economische en Sociale Geografie* **70**, 157–63.

Williams, N.J. (1981) Attitudes and consumer spatial behaviour, *Tijdschrift voor Economische en Sociale Geografie* **72**, 145–54.

Williams, P.R. (1978) Building societies and the inner city, *Transactions of the Institute of British Geographers* **3**, 23–34.

Willis, K.G. (1974) *Problems in Migration Analysis*, Saxon House, Farnborough.

Willmott, P., Murie, A. (1988) *Polarization and Social Housing*, Policy Studies Institute, London.

Willmott, P., Young, M. (1960) *Family and Class in a London Suburb*, Routledge & Kegan Paul, London.

Wilman, E.A. (1980) The value of time in recreation benefit studies, *Journal of Environmental Economics and Management* **7**, 272–86.

Wilner, D.M., Walkley, R.P. (1963) Effect of housing on health and performance. In: Duhl, L. (ed.) *The Urban Condition*, Simon & Schuster, New York, pp. 215–28.

Wilson, A.G. (1970) *Entropy in Urban and Regional Modelling*, Centre for Environmental Studies Working Paper No. 26, London.

Wilson, A.G. (1971) A family of spatial interaction models and associated developments, *Environment and Planning* **3**, 1–32.

Wilson, A.G., Kirkby, M.J. (1975) *Mathematics for Geographers and Planners*, Clarendon Press, Oxford.

Wilson, C., Alexis, M. (1962) Basic frameworks for decisions. In: Gove, W.J., Dryson, J.W. (eds) *The Making of Decisions*, Free Press, New York, pp. 193–5.

Wilson, W.J. (1985) The urban underclass in advanced industrial society. In: Peterson, P.E. (ed.) *The New Urban Reality*, Brookings Institute, Washington, pp. 129–60.

Wingo, L. (1964) Recreation and urban development: a policy perspective, *Annals of the American Academy of Political and Social Science* 352, 129–40.

Wirth, L. (1938) Urbanism as a way of life, *American Journal of Sociology* **44**, 1–24.

Witt, P.A., Bishop, D.W. (1970) Situational antecedents to leisure behaviour, *Journal of Leisure Research* **2**, 64–72.

Wohlwill, J.F. (1970) The emerging discipline of environmental psychology, *American Psychologist* **25**, 303–12.

Wolpert, J. (1964) The decision process in spatial context, *Annals of the Association of American Geographers* **54**, 537–58.

Wolpert, J. (1965) Behavioural aspects of the decision to migrate, *Papers and Proceedings of the Regional Science Association* **15**, 159–72.

Wolpert, J. (1966) Migration as an adjustment to environmental stress, *Journal of Social Issues* **22**, 92–102.

Wolpert, J. (1970) Departures from the usual environment in locational analysis, *Annals of the Association of American Geographers* **60**, 220–9.

Womble, P., Studebaker, S. (1981) Crowding in a national park campground: Katmai movement in Alaska, *Environment and Behaviour* **13**, 557–73.

Wong, K.Y. (1979) Maps in mind: an empirical study, *Environment and Planning A* **11**, 1289–1304.

Wood, L.J. (1970) Perception studies in geography, *Transactions of the Institute of British Geographers* **50**, 129–41.

Wood, P.A. (1975) Are behavioural approaches to industrial location theory doomed to be descriptive?. In: Massey, D., Morrison, W.I. (eds) *Industrial Location: alternative frameworks*, Centre for Environmental Studies Conference Paper No. **15**, London, pp. 41–8.

Wood, P.A. (1991) Flexible accumulation and the rise of business services, *Transactions of the Institute of British Geographers* **16**, 160–72.

Wright, D.S., Taylor, A., Davies, D.R., Sluckin, W.G., Lee, S.G.M. (1970) *Introductory Psychology: an experimental approach*, Penguin, Harmondsworth.

Wright, J.K. (1947) Terrae incognitae: the place of the imagination in geography, *Annals of the Association of American Geographers* **37**, 1–15.

Wrigley, N. (1980) An approach to the modelling of shop-choice patterns: an exploratory analysis of purchasing patterns in a British city. In: Herbert, D.T., Johnston, R.J. (eds) *Geography and the Urban Environment*, Vol. III, Wiley, London, pp. 45–86.

Wrigley, N. (1982) Quantitative methods: developments in discrete choice modelling, *Progress in Human Geography* **6**, 547–62.

Wrigley, N. (ed.) (1988) *Store Choice, Store Location and Market Analyses*, Routledge & Kegan Paul, Andover.

Wrigley, N. (1989) The lure of the USA: further reflections on the internationalization of British grocery retailing, *Environment and Planning A* **21**, 283–8.

Young, G. (1973) *Tourism: blessing or blight*? Penguin, Harmondsworth.

Young, M., Willmott, P. (1962) *Family and Kinship in East London*, Penguin, Harmondsworth.

Young, M., Willmott, P. (1973) *The Symmetrical Family*, Penguin, Harmondsworth.

Zehner, R.B. (1972) Neighbourhood and community satisfaction. In: Wohlwill, J.F., Carson, D.H. (eds) *Environment and the Social Sciences*, American Psychological Association, Washington, pp. 169–83.

Zelinsky, W. (1971) The hypothesis of the mobility transaction, *Geographical Review* **61**, 219–49.

Zelinsky, W. (1983) The impasse in migration theory: a sketch map for potential escapees. In: Morrison, P. (ed.) *Population Movements*, IUSSP Ordina, Liege, pp. 19–48.

Zolberg, A.R. (1989) The next wave: migration theory for a changing world, *International Migration Review* **23**, 403–30.

Zube, E.H., Sell, J. (1986) Human dimensions of environmental change, *Journal of Planning Literature* **1**, 162–76.

Zukin, S. (1990) Socio-spatial prototypes of a new organization of consumption: the role of real cultural capital, *Sociology* **24**, 37–56.

Index